Animal Waste
and the
Land-Water
Interface

Edited by
Kenneth Steele

LEWIS PUBLISHERS

Boca Raton　New York　London　Tokyo

Library of Congress Cataloging-in-Publication Data

Catalog record is available from the Library of Congress.

Patti Snodgrass, editorial coordinator
Nancy Wyatt, technical editor/publication designer

ACKNOWLEDGMENTS

This book is an outgrowth of a conference by the same title held in Fayetteville, Arkansas, July 16-19, 1995. Many people were involved in planning and hosting the conference and are, therefore, indirectly responsible for this book. The following people served as committee members:

Steering Committee	Local Arrangements Committee
Ronnie Murphy	Kenneth Steele
Tom Wehri	Tommy Daniel
John Kovar	Richard Meyer
Karen Rylant	Don Scott
Tommy Daniel	David Parker
Kenneth Steele	Philip Moore
	Dwayne Edwards
	Duane Wolf
	Paul Vendrell

The conference was sponsored by:

Arkansas Water Resources Center
USDA Natural Resources Service
Tennessee Valley Authority
US Environmental Protection Agency
USDA Agricultural Research Service

and co-sponsored by:

Soil Science Society of America
University of Arkansas
American Water Resources Association
American Society of Agricultural Engineers
North American Benthological Society
American Society of Civil Engineers
Fayetteville Chamber of Commerce

Because of the breadth of this book, thirteen associate editors graciously volunteered their services to solicit and review chapters as follows:

David Correll	Stream Effects
Eldridge Collins and Theo Dillaha	Best Management Practices
J.W. Gilliam	Edge-of-Field Losses
Amy Purvis	Socio-Economic Considerations
K.R. Reddy and Robert Wetzel	Wetlands and Aquatic Systems
Andrew Sharpley	Watershed Management I
Lynn Shuyler	Regulatory versus Voluntary Programs
Tom Sims	Characteristics of Wastes and Waste-Amended Soils
Richard Strickland	Alternative Uses
Wes Wood	Nutrient Management
Robert Young	Watershed Management II

The editors enlisted the assistance of almost 100 reviewers to critique the individual chapters; unfortunately, space limitations do not allow us to list them individually.

Finally, the contributions of the authors must be acknowledged. Without their diligent work and timely response to editor and reviewer comments, this book would not have been possible.

The work by the Arkansas Water Resources Center staff, Tammy Berkey and Melpha Speak, with all of the clerical aspects associated with the production of this book is gratefully acknowledged.

The work of the editorial coordinator, Patti Snodgrass, and the technical editor/publication designer, Nancy Wyatt, made the individual manuscripts into a book.

PREFACE

In agricultural areas, the use or disposal of animal wastes is complexly intertwined with the quality of the land and both surface and ground water. Current research within a variety of disciplines is directed at learning more about this relationship and finding ways of controlling the quality of the environment.

Scientists and managers recognize the need to consider the entire watershed when delineating and solving today's environmental problems. The total watershed approach to management includes data not only on the land and surface water, but also on ground water, public perception, and political and economic factors.

The purpose of this book is to provide an interdisciplinary, holistic discussion of animal waste and its interactions with soil and water within the framework of the total watershed. This holistic approach requires input from agronomists, engineers, hydrologists, geographers, sociologists, soil scientists, stream ecologists, limnologists, geologists, and economists. The variety of topics covered in the chapters of this book demonstrate the diversity of information and expertise critical for total watershed management. Researchers, managers, regulators, farmers, and other stakeholders are learning that cooperation and interaction among many groups is essential to have truly comprehensive management of animal wastes in a watershed. Our goals for the wise use and protection of our natural resources depend on continued and expanded communication, cooperation, and coordination among all of the stakeholders.

Give us all a reverence for the earth as your own creation, that we may use its resources rightly in the service of others...
The Book of Common Prayer, The Church Hymnal
Corporation and The Seabury Press, 1979.

ACRONYMS AND ABBREVIATIONS

AFDM - ash-free dry mass
AGNPS - Agricultural Non-point Source Pollution Model
AML - ARC Macro Language
ASCS - Agricultural Stabilization and Conservation Service
B-IBI - benthic index of biotic integrity
BMP - best management practice
BOD - biological oxygen demand
BOD_5 - five-day biological oxygen demand
BOM - benthic organic matter
CAFO - Confined Animal Feeding Operation
CARE - Cost and Return Estimator
CEC - cation exchange capacity
COD - chemical oxygen demand
CREAMS - Chemicals, Runoff and Erosion from Agricultural Management
Systems
CSTR - continuous stirred tank reactors
CZARA - Coastal Zone Act Reauthorization Amendment
CZM - Coastal Zone Management
DER - Department of Environmental Regulation
DLLRMS - Dairy Loafing Lot Rotational Management System
EAA - Everglades Agricultural Area
EC - electrical conductivity
EPIC - Environmental Policy with Integrated Climate
EU - European Union
FDA - Food and Drug Administration
FDER - Florida Department of Environmental Regulation
FHANTM - Field Hydrology and Nutrient Transport Model
GAMS - General Algebraic Modeling System
GATT - General Tariff and Trade Agreements
GIS - geographical information systems
GLEAMS - Groundwater Loading Effects of Agricultural Management Systems
GRASS - Geographic Resources Analysis Support System
GUI - graphical user interface
HU/WQ - Hydrologic Unit Water Quality Tool
HUA - hydrologic unit area
HUMUS - Hydrologic Unit Model for the United States

IBI - index of biotic integrity
ICM - Integrated Crop Management
IDOC - Illinois Department of Conservation
IEPA - Illinois Environmental Protection Agency
LCM - leaf-chlorophyll meter
LOTAC - Lake Okeechobee Technical Advisory Committee
MDA - multiple discriminant analysis
MOP - multi-objective programming
NADB - North American Database
NEPA - National Environmental Policy Act
NLEAP - Nitrogen Leaching and Economic Analysis Package
NOAA - National Oceanic and Atmospheric Agency
NPDES - National Pollutant Discharge Elimination System
NPS - non-point source
NRCS - Natural Resources Conservation Service
NUE - nitrogen use efficiency
PAN - plant-available nitrogen
PCA - principal components analyses
PRA - Probablistic Risk Assessment
PSNT - pre-sidedress soil nitrate test
S-S-C-P - Situation-Structure-Conduct-Performance
SALT - Special Area Land Treatment
SAR - sodium adsorption ration
SFWMD - South Florida Water Management District
SWIM - Surface Water Improvement and Management
SWRRB - Simulation of Water Resources in Rural Basins
TCLP - toxicity characteristic leaching procedure
TKN - total Kjehldahl nitrogen
TMDLs - total maximum daily loads
TNRCC - Texas Natural Resources Conservation Commission
TOC - total soluble organic carbon
TOT - time of travel
TP - total phosphorus
TSS - total suspended solids
US EPA - United States Environmental Protection Agency
USDA - United States Department of Agriculture
USLE - Universal Soil Loss Equation
VFS - vegetated filter strip
WAMADSS - Watershed Management Decision Support System
WIPP - watershed index of pollution potential
WMD - water management districts
WSP - water-soluble phosphorus

CHARACTERISTICS OF WASTES AND WASTE-AMENDED SOILS

EDGE OF FIELD LOSSES

STREAM EFFECTS

WETLANDS AND AQUATIC SYSTEMS

WATERSHED MANAGEMENT I

NUTRIENT MANAGEMENT

REGULATORY VS. VOLUNTARY PROGRAMS

ALTERNATIVE USES

SOCIOECONOMIC CONSIDERATIONS

CHARACTERISTICS OF ANIMAL WASTES AND WASTE-AMENDED SOILS: AN OVERVIEW OF THE AGRICULTURAL AND ENVIRONMENTAL ISSUES

J. Thomas Sims

INTRODUCTION

Animal wastes have been successfully used as agricultural soil amendments since the dawn of civilization. Indeed, some of the earliest Greek and Roman agriculturists (Theophrastus, Varro, Columella) recognized not only that animal manures varied naturally in fertilizer value (ranking them as poultry > swine > goat > cow > horse) but that management practices (e.g., the use of bedding, type of animal feed) could improve manure quality. For almost 2,000 years, until the advent of chemical fertilizers in the 1940's, animal wastes were one of the primary sources of plant nutrients for the world's agriculture. Today, however, we view animal wastes as a mixed blessing. As we strive to develop a more "sustainable" agriculture that relies less on off-farm inputs, we recognize the potential of animal wastes to recycle nutrients, build soil quality, and maintain crop productivity. At the same time, however, the nature of modern animal agriculture, with its highly concentrated production facilities and reliance upon feed supplements to maintain animal health and productivity, has raised serious questions about the effects of animal wastes on the quality of our soil, water, atmosphere, and food supply. Agricultural research and technology transfer programs are now focusing intense efforts on the development of economically feasible and environmentally sound animal waste management practices. Because land application is the only practical alternative for much of animal-based agriculture, the cornerstone of most waste management programs will be a solid understanding of how the characteristics of animal wastes and waste-amended soils affect their agricultural value and potential deleterious effects on our environment.

PRESENT APPROACHES TO ANIMAL WASTE MANAGEMENT

Animal wastes are almost exclusively used as soil amendments for agricultural crop production. Manures, litters, composts, waste waters from processing plants, and lagoon effluents represent the most common type of animal wastes that are now applied to soils through a variety of tillage and irrigation practices.

1

Although animal wastes are known to improve the physical properties of soils by adding organic matter that helps to build soil structure and increase soil water holding capacity, application rates are usually based on a waste's nutrient value. Because most animal wastes are bulky, heterogeneous, relatively low-analysis fertilizer materials, the amount of a waste required can easily be 10- to 100-fold the rate of commercial inorganic fertilizer needed by the same crop. The logistical problems associated with the storage, handling, transport, and application of literally millions of tons of animal wastes each year create formidable economic problems for US farmers, even if environmental constraints are not considered. Nevertheless, the lack of widespread adoption of alternative end-uses for animal wastes, such as incineration as a fuel, composting for horticultural markets, enrichment and pelletizing with mineral fertilizers to facilitate wider agricultural use, and re-feeding to other animals, has meant that land application remains the only viable option for most farmers. Unfortunately, the long-term application of animal wastes to soils has been shown to create a number of environmental problems.

CURRENT ENVIRONMENTAL ISSUES

Most agricultural best management practices for animal wastes are now based on providing sufficient nitrogen (N), in a timely manner, to meet crop N requirements at a realistic yield. Documented problems with nitrate-N (NO_3-N) contamination of ground waters in areas with high animal densities have been the driving force behind this approach. However, a number of emerging environmental issues, such as the eutrophication of surface waters by phosphorus (P) in runoff; the fate of trace elements, antibiotics, pesticides, and growth hormones in wastes; and the effects of pathogens in wastes on human and animal health have forced us to re-evaluate the N-based management of animal wastes (Sims and Wolf, 1994). This quickly becomes an exercise in risk assessment, whereby the economic value of a waste constituent is balanced against its environmental impact, within the limits economics and technology impose on our ability to control its environmental fate.

Since most animal waste plans are based on N, a review of the rationale and practical difficulties of N-based waste management is useful. As stated above the fundamental premise is that the rate of animal waste applied should not provide more plant-available N than that required by the crop in order to avoid the contamination of ground waters by NO_3-N. This approach, therefore, assumes that we have the ability not only to analyze animal wastes for total N (all organic and inorganic forms, a relatively easy task) but also to predict the rate at which the various organic forms of N will mineralize, i.e., be converted to ammonium-N (NH_4-N) and NO_3-N. Because mineralization of organic N is microbially mediated, it also assumes that we can estimate, with reasonable accuracy, the effects that soil and environmental factors will have on the rate and extent of N mineralization. Knowledge of the rate of mineralization is critical because the timing of waste N availability must be closely synchronized with plant N uptake for maximum efficiency of N recovery. This approach further assumes that we can account for the slow mineralization of more resistant forms of waste organic N applied in previous years to arrive at a prediction of the size (and timing of availability) of the total pool of waste N. Considerable laboratory and field research has been directed toward the development of "decay series" for organic wastes that are used to approximate the amount of available N in wastes and waste-amended soils. As an example, a typical decay series reported for poultry litter was 60-20-10, indicating that 60% of the organic N in the litter would mineralize in first year, 20% in year two, and 10% in year three (Bitzer and Sims, 1988). Because poultry litter has relatively high concentrations of ammonia-N (NH_3-N), the total pool of plant-available N in year one must reflect both mineralizable organic N (N_o) and the effect management practices (e.g., tillage) have on NH_3-N losses by volatilization:

$$PAN = (e_f x\ NH_3\text{-}N) + (k_m\ x\ N_o)$$

where: PAN: Potentially available N
       e_f: Efficiency of recovery of NH_3-N, normally a function of time between application and incorporation. Delaying incorporation increases NH_3-N loss and thus decreases e_f.
      k_m: Mineralization constant, an estimate of the amount of organic N that will be converted to mineral N during year one.

Characterizing the N value of an animal waste requires that we make at least three rather broad assumptions:

- First, we have the capability to obtain a representative sample of an animal waste from a storage area and rapidly analyze the waste for the appropriate forms of organic and inorganic N.
- Second, the N content of the waste measured in a laboratory will be reasonably similar to that applied in the field (i.e., laboratory handling and field storage and handling will not cause large differences between the N content of the sample and the field-applied waste).
- Third, a research base is available that can provide quantitative estimates of the effects of soil and environmental conditions and agricultural management practices on N availability for the dominant cropping systems in an area.

Our willingness to use a N-based approach to animal waste management, given the complexity and high degree of uncertainty underlying each of these assumptions, clearly illustrates the importance we have placed on minimizing NO_3-N contamination of ground waters in many areas.

As our environmental monitoring efforts have expanded and our analytical capabilities have grown, we have begun to identify other criteria that should perhaps be considered when determining the most appropriate end-use for animal wastes. Two examples illustrate the nature of the problems that arise when we move away from N-based waste management to other criteria.

First, consider a P-based approach, one based on the assumption that the long-term use of animal wastes increases the delivery of environmentally significant levels of particulate and soluble P to surface waters in erosion, runoff, and drainage. Many studies have shown that, due to the unfavorable N:P ratio in animal wastes relative to crop N:P ratios, constant manuring of soils creates agronomically excessive levels of soil P. If a transport process exists to redistribute the P that originated in the animal wastes to a surface water body sensitive to eutrophication, a potential environmental problem also exists. Recent summaries of soil test P data in the US confirm that many soils in areas dominated by animal-based agriculture have high or excessive levels of P (Sims, 1994). In these situations, characterizing the P status of animal wastes and waste-amended soils relative to crop P requirements (a simple analytical task) results in a somewhat obvious initial interpretation--none is required, none should be applied. Two problems immediately arise with this interpretation. First, the concentration of P in a manure or soil sample may indicate that additional fertilizer or manure P is not required, but this measurement cannot reliably predict the loss of P from a field via runoff, erosion, or drainage. It is entirely possible that a soil rated as agronomically excessive in P could continue to receive animal waste applications for years with no unfavorable environmental effect whatsoever, simply because there is no transport process to carry sufficient amounts of biologically available P to the nearest sensitive surface water. Conversely, soils with moderately high P levels that are also highly susceptible to runoff could be significant non-point sources of P to lakes, ponds, and bays. More detailed information on soil properties, topography, and hydrology, as proposed in the Phosphorus Index System (Lemunyon and Gilbert, 1993), is required to characterize the potential of animal wastes and waste-amended soils to negatively impact surface waters. This is critical due to the second problem associated with P-based management: from an agronomic perspective, heavily manured soils rarely require P fertilization; therefore, there will be an inadequate land base available for animal wastes on many farms. Therefore, if the potential environmental impacts of high-P soils become the criteria for land application of animal wastes, farmers would be forced to find other end-uses for these wastes. This problem could persist for years as normal crop removal has been shown to deplete "excessive" soil P back to optimum values at very slow rates (McCollum, 1991). The crux of this issue is the lack of an infrastructure to re-distribute animal wastes to areas in which soil P levels are low enough to limit crop yields. Unfortunately, in areas in which animal agriculture is highly concentrated, it is often difficult to find large hectareages of low-P soils; further, the economics and logistics of transporting animal wastes even short distances are now unfavorable for individual farmers.

Next, consider basing land application rates of animal wastes on similar criteria as are used for other solid wastes such as municipal sewage sludges or composts. While N is commonly used to determine the most suitable sludge rate for agricultural crops, most state and federal agencies also require characterization of municipal wastes for trace elements, organics, and pathogens. Analytical methods include measuring both the total concentration of potentially hazardous elements or compounds in a solid waste and use of the toxicity characteristic leaching procedure (TCLP) to determine if the solubility of a waste constituent is great enough to be of concern. Rigorous analytical protocols must be followed when conducting these tests. In the recent National Sewage Sludge Rule (Part 503 of the Clean Water Act) adopted by the US EPA in 1993, "ceiling concentrations" and "cumulative loading rates" were established for pollutants in sewage sludges that are to be applied to agricultural land, and pathogen density limits were defined. State regulations can (and often do) specify more restrictive limits. For instance, the cumulative (lifetime) loading rate for copper, a common additive to poultry feed, set for sludge-amended soils in the 503 rule is 1500 kg/ha, while current regulations in Delaware use a value of 280 kg/ha for most soils in the state. Animal wastes have often been reported to have similar total concentrations of many trace elements identified as pollutants in the 503 rule as are found in some municipal wastes. Data on TCLP values for trace elements or organics in animal wastes are virtually non-existent. However, if the criteria in the 503 rule were now applied to animal wastes, it is very unlikely that major problems would occur. Reported trace element concentrations in manures and waste waters are usually well below the maximum concentrations (in mg/kg) of the pollutants defined in this rule: arsenic (As = 41), cadmium (Cd = 25), chromium (Cr = 1200), copper (Cu = 1500), lead (Pb = 300), mercury (Hg = 17), molybdenum (Mo = 18), nickel (Ni = 420), selenium (Se = 36), and zinc (Zn = 2800). Also, agronomic application rates of animal wastes are usually less than field application rates of sewage sludges; hence cumulative trace element loadings are normally lower for animal manures and waste waters.

Nevertheless, questions are now arising more frequently about the potential environmental effects not only of trace elements, but of antibiotics, pesticides, and hormones in animal wastes. Limited research is available that characterizes the form of these elements and compounds in animal wastes and even less on any subsequent transformations that might occur in waste-amended soils that could affect their bioavailability and transport. Because of this, basing land application of animal wastes on criteria similar to that used for other solid wastes is premature. A prudent course of action at this time would be to identify and characterize the concentrations of any potential "pollutants" present in animal wastes and, based on existing scientific literature with other solid wastes, prioritize any research or monitoring efforts on the environmental fate of these waste constituents.

ADVANCES NEEDED IN CHARACTERIZATION
OF ANIMAL WASTES

The agronomic importance of animal wastes and the increasing concerns about their environmental impacts make it apparent that we need to re-think the approaches used to "characterize" wastes. Practically speaking, at this point in time the only readily available tests for animal wastes are for total nutrient content. Other waste characteristics that can be obtained through more intensive testing (or literature reviews) include pH, moisture content, organic carbon, soluble salts, biological and chemical oxygen demand, essential and non-essential elements, some organics (antibiotics, pesticides, hormones), and some pathogens known to be associated with human health problems (e.g., *E. coli,* Salmonella). Some of the most needed advances in the characterization of animal wastes include the following:

Standardization of methods for sample collection, storage, and handling. The heterogeneous nature of animal wastes makes collection of a "representative" sub-sample from stockpiles, pits, and lagoons containing hundreds of tons or thousands of gallons of solid or liquid waste, an extremely difficult task. The cost of analysis, even for the simplest nutrient tests, usually precludes collecting large numbers of samples to characterize the variability that can be expected with differing waste management practices. Since this variability can be extreme, there is a need for industry-wide standards for the proper methods to collect animal waste samples. Similarly, the storage and handling of wastes can alter the content of nutrients, especially ammonical nitrogen, and other waste constituents sensitive to temperature, moisture or oxygen (e.g., BOD). Hence, the proper techniques to store a waste sample and to handle it in the laboratory also need standardization. Cooperative extension fact sheets are available describing "recommended" methods to collect and transfer samples from farm to laboratory, and most analytical laboratories have developed reasonable, but different, protocols for sample handling, and preparation. A uniform approach for the collection, storage, handling and preparation of samples prior to analysis is badly needed for the most common types of animal wastes.

Standardization of Analytical Methods for Animal Wastes. Standard reference methods for the testing of soils, plants, waters and waste waters have been published by various organizations. However, no widely accepted reference document is available describing the analytical methods appropriate for animal wastes. Most laboratories simply use methods developed for sewage sludges, plants, or soils with little verification conducted on the suitability of these methods for situations where management decisions on animal waste use are being made. The issue is complex. Most laboratories are quite capable of drying, grinding, and digesting (or extracting) animal waste samples and obtaining reproducible analytical results for nutrients, trace elements, and organics through colorimetric, spectroscopic, or chromatographic methods, once provided with a standardized method. Analytical questions do exist, such as the extent of "within sample" variability that can be expected for different waste constituents and the acceptable degree of reproducibility of analytical results. What is needed at this time is a publication describing the standard methods that should be used to measure the key parameters for the most common types of animal wastes, such as pH, BOD, total and soluble elements, and organics. This would greatly facilitate interlaboratory data comparisons and the reliability of interpretations based on waste analyses.

Innovative Characterization Methods for Animal Wastes. More important than standardized methods for animal waste analyses, however, is the fact that total concentrations of elements or organic compounds are of much less use than some estimate of "bioavailability" that can be related to a waste's value as a nutrient source or its potential environmental effects. Methods to provide such estimates are simply not available and are also badly needed. There are several examples of this. Chemical oxidants have been evaluated, with limited success, as predictors of "mineralizable" organic N in soils. Tests such as these might be better predictors of the amount of "potentially available" organic N in an animal than total organic N content because they often measure more readily hydrolyzable forms of organic N. Distinguishing between total and "bioavailable" P is also important when assessing the potential for an animal waste to be an important factor in the degradation of surface water quality. Chemical extractants for "algal-available" P have been used with soils and sediments and may be applicable to animal wastes as well. A final example would be the use of chelates (e.g., DTPA, EDTA) to separate the more labile, organically bound forms of trace elements from the insoluble inorganic species. Research relating chemical tests such as these to "real-world" data from crop production and runoff studies and surface or ground water quality is also needed.

Summarizing and Interpreting Existing Animal Waste Characterization Data. Animal wastes have been "characterized" for decades both in research projects and by private or state-funded analytical laboratories. Because of this a rather voluminous amount of information is available on many waste properties. Some of these data have been roughly summarized in various review articles (Barrington, 1991; Overcash et al., 1983; Sims and Wolf, 1994; Stephenson et al., 1990). Unfortunately, the data also reflect widely differing preparatory and analytical procedures, many of which are often poorly documented. Historical data from private testing laboratories may not be available at all. Despite these limitations, a vast and potentially useful data base sits virtually untouched by those wrestling with the selection of methods for characterizing wastes and with proper interpretation of the results. A comprehensive summary of historical data on waste properties, carefully documenting the methods of analysis used, in combination with current interpretive strategies for waste analyses would be an excellent resource for those involved in the development of modern animal waste management programs.

ADVANCES NEEDED IN THE CHARACTERIZATION OF ANIMAL WASTE-AMENDED SOILS

Relative to animal waste characterization methods, the procedures used to test soils and make agronomic recommendations are well-known and widely accepted. Environmental interpretation of soil testing results, however, is still in its infancy. The main reason for this is that soil testing methods were developed not for environmental purposes but to predict if a soil needs fertilization or liming to optimize crop yields. Characterizing the fertility status of animal waste-amended soils is no different than for agricultural soils receiving only commercial fertilizers. However for both types of soils, but more for those frequently receiving animal wastes, there is a need for innovative testing methods to assess potential environmental impacts of nutrients, trace elements, and organics. Some advances have occurred recently; however, more research and development is needed.

Recent Advances in the Characterization of Animal Waste-Amended Soils. Some important advances in soil testing have improved our ability to manage animal waste-amended soils in a more environmentally sound manner. It is not surprising, given the emphasis on N-based management by agricultural research for the last 20 years, that most of these improvements are in soil and plant N testing. Perhaps the most promising, widely adopted new test has been the pre-sidedress soil nitrate test (PSNT), developed by Magdoff et al. (1984). While the PSNT was not specifically designed for waste-amended soils, a key assumption of this test was that soils frequently amended with organic sources of N (animal manures, sludges, legume residues) will have a large pool of plant-available organic N, often large enough to meet all (or most) of the N requirement of corn. Because of this, the NO_3-N concentration in a soil sample taken to a depth of 0-30 cm in mid to late June should be well-correlated with the "N supplying capacity" of the soil (i.e., the amount of "potentially mineralizable organic N"). Soils with PSNT values greater than 20-25 mg NO_3-N/kg are usually found to be non-responsive to sidedressed N fertilizer. The PSNT has been successfully used throughout the US, as summarized by Bock and Kelley (1992). Most researchers (Fox et al., 1992; Sims et al., 1995) have noted that the PSNT is particularly well-suited for animal-based agriculture because of the difficulty in predicting plant-available N in manured soils in advance of planting from a simple manure analysis for total and NH_4-N. Some new plant N tests, the leaf chlorophyll meter (LCM) and late-season stalk NO_3-N test, also offer promising means to characterize the N status of soils. The LCM, a hand-held meter that actually measures leaf "greenness," not chlorophyll content, has been shown to be a good predictor of the N supplying capacity of soils in the Mid-Atlantic region of the US (Piekielek and Fox, 1992; Sims et al., 1994). Because of its ease of use (i.e., soil sampling and laboratory analysis are not required), the LCM has more potential than the PSNT to rapidly characterize the need for N fertilization of soils in a large geographic area. The late-season stalk nitrate test, developed by Binford et al. (1990), is used to conduct a "post-mortem" evaluation of a N management program. Field calibration of this test has shown that if the NO_3-N concentration in the lower portion of the corn stalk at maturity exceeds 2.0 g/kg, excessive N was present in the soil during the growing season, i.e., overfertilization or manuring occurred. Tests such as these, in conjunction with improvements in animal waste N testing, should help us better characterize the N status of soils where animal wastes are commonly used as soil amendments.

Environmental soil testing procedures for P are also being developed and evaluated, as summarized by Sims (1994) and Sharpley et al. (1994). The major goals of these tests are either to measure the potential "bioavailability" of particulate and dissolved P in soils and runoff or to characterize the capacity of soil profiles to retain further P against leaching. Some recent advances in estimating the amount of bioavailable P in soils include simple extraction with 0.1N NaOH (originally referred to as "algal-available" P) and the use of anion exchange resins and Fe-oxide filter strips. Predicting the leachability of P in soils is important in areas with shallow water tables that discharge into sensitive surface waters (Mozaffari and Sims, 1993). For instance, in the Netherlands, land application of animal wastes is now regulated based on the percentage saturation of the P adsorption capacity of soils and subsoils (Breeuswma and Silva, 1992), as estimated by an oxalate extraction procedure. Some states are now establishing upper, environmentally based "critical limits" for soil test P because of concerns about non-point source pollution of surface waters (Gartley and Sims, 1994). Typically, these upper soil test limits are set at five- to ten-fold the agronomic critical level that identifies a soil as non-responsive to P fertilization. Once a soil exceeds this value, no further additions of P (manures or fertilizer) are recommended, often forcing farmers, as mentioned above, to find alternative end-uses for animal wastes. Because non-point source pollution of surface waters by P is as dependent on transport processes (runoff, erosion, drainage) as on soil P values, establishing these upper limits often meets with harsh criticism from the agricultural community as an overly simplistic approach to a complex problem.

Other than N and P tests, there have been few important advances in the characterization of animal waste-amended soils. Given the other environmental issues stated earlier, there is clearly a need for further soil characterization research.

Improvements Needed in the Characterization of Animal Waste-Amended Soils. The main focus of soil characterization research for animal waste-amended soils should probably remain with N and P because of the well-documented environmental impacts and agronomic importance of these nutrients. Other key areas in need of investigation include characterization of the short- and long-term fate of trace elements, organics, and pathogens in animal wastes. Trace element research should focus on those elements defined by US

EPA as being of the greatest environmental concern (As, Cd, Cr, Cu, Hg, Mo, Ni, Pb, Se, and Zn), particularly those elements that are added to animal feed as growth stimulants or biocides (e.g., As, Cu, Se, and Zn). The limited long-term research available with some trace elements (As, Cu) has generally shown that changes in plant concentrations or total element concentrations in soils are minor when wastes are applied at agronomic rates (Mullins et al., 1982; Sharpley et al., 1991). Hence future research should focus more on the long-term changes in metal fractionation in soils routinely amended with animal wastes than on short-term studies of metal uptake by plants. One key question to answer would be whether long-term manuring has significantly shifted trace element distribution to more or less bioavailable soil fractions. Other questions include the loss of trace elements in runoff and/or leaching and the value of current "routine" soil tests as predictors of trace element availability or mobility.

Research is also needed on the fate of antibiotics, pesticides, and hormones added to animal feed. What percentage is excreted in soluble or organic forms? Does the biological availability of these organics change during digestion or during manure storage and handling? How long, and in what form, do they persist in soils? How susceptible are they to losses in runoff, leaching, or even as volatile gases? What types of soil tests can best characterize their form and potential fate in soils?

Summarizing and Interpreting Data from Animal Waste-Amended Soils. Given the limited resources available, and the length of time required to conduct multi-site, multi-year, multi-crop studies with animal waste-amended soils, serious consideration should be given to the value of preparing careful summaries of existing data. Accurate, regional summaries of soil test P, when combined with topographic information on soils and surface water quality data, can quickly pinpoint the geographic areas with the greatest potential for nonpoint source pollution by P. Such data are readily available in most states and can help to prioritize the allocation of resources. Similarly, reviewing the scientific literature for studies on the fate of trace elements and organics in waste-amended soils can help identify the most critical gaps in our knowledge about these issues.

An equally important "characterization" step is the need to address the interpretation of existing data on soil and waste properties. One of the most important issues is the need for environmental limits on soil P. What criteria should be used to limit animal waste applications based on soil test P or "readily desorbable bioavailable P" in surface soils or P sorption capacity in subsoils? Can we set these limits now or is more research needed? What are the environmental implications of delaying the implementation of this type of soil test interpretation until the research has been conducted? In a similar vein, should the interpretation philosophy used for trace elements and organics in soils amended with municipal or industrial wastes be extended to heavily manured soils? Do we need to conduct TCLP tests for manures and measure total metal levels in

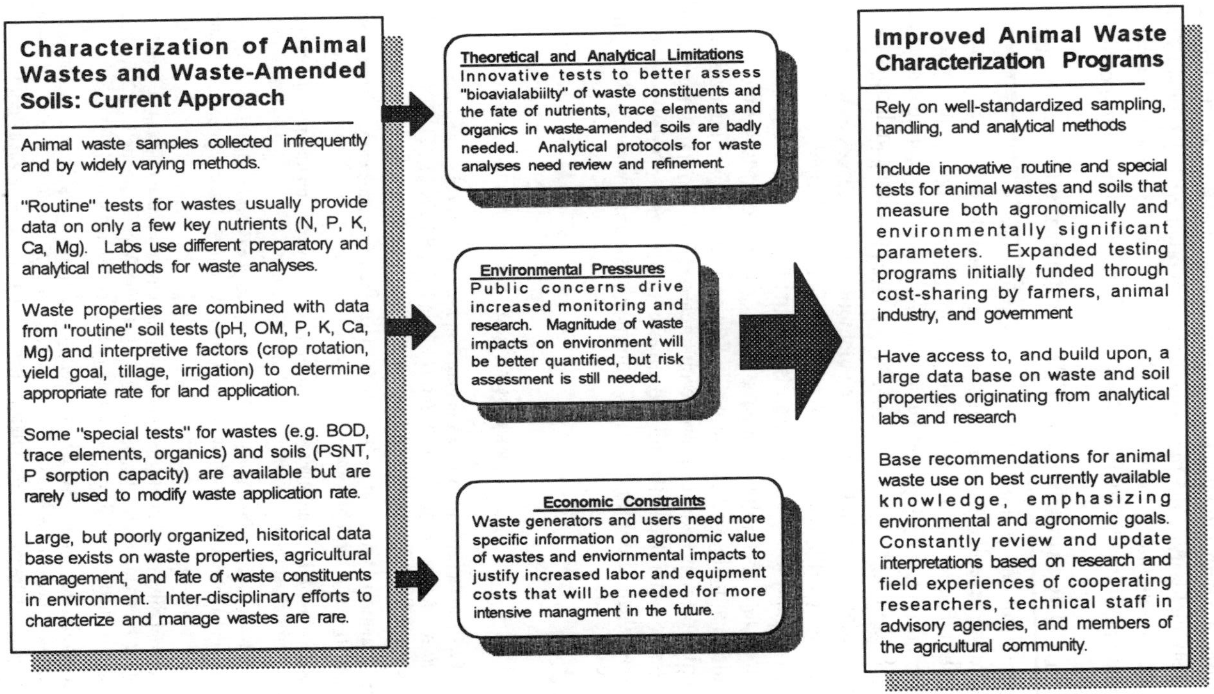

Characterization of Animal Wastes and Waste-Amended Soils: Current Approach

Animal waste samples collected infrequently and by widely varying methods.

"Routine" tests for wastes usually provide data on only a few key nutrients (N, P, K, Ca, Mg). Labs use different preparatory and analytical methods for waste analyses.

Waste properties are combined with data from "routine" soil tests (pH, OM, P, K, Ca, Mg) and interpretive factors (crop rotation, yield goal, tillage, irrigation) to determine appropriate rate for land application.

Some "special tests" for wastes (e.g. BOD, trace elements, organics) and soils (PSNT, P sorption capacity) are available but are rarely used to modify waste application rate.

Large, but poorly organized, historical data base exists on waste properties, agricultural management, and fate of waste constituents in environment. Inter-disciplinary efforts to characterize and manage wastes are rare.

Theoretical and Analytical Limitations

Innovative tests to better assess "bioavailability" of waste constituents and the fate of nutrients, trace elements and organics in waste-amended soils are badly needed. Analytical protocols for waste analyses need review and refinement

Environmental Pressures

Public concerns drive increased monitoring and research. Magnitude of waste impacts on environment will be better quantified, but risk assessment is still needed.

Economic Constraints

Waste generators and users need more specific information on agronomic value of wastes and environmental impacts to justify increased labor and equipment costs that will be needed for more intensive managment in the future.

Improved Animal Waste Characterization Programs

Rely on well-standardized sampling, handling, and analytical methods

Include innovative routine and special tests for animal wastes and soils that measure both agronomically and environmentally significant parameters. Expanded testing programs initially funded through cost-sharing by farmers, animal industry, and government

Have access to, and build upon, a large data base on waste and soil properties originating from analytical labs and research

Base recommendations for animal waste use on best currently available knowledge, emphasizing environmental and agronomic goals. Constantly review and update interpretations based on research and field experiences of cooperating researchers, technical staff in advisory agencies, and members of the agricultural community.

Figure 1. A summary of the current approach used to characterize animal wastes, factors influencing changes in this approach, and an overview of an improved animal waste characterization program.

manured soils? As with animal waste analyses, a wealth of characterization data are already available on soils that have been amended with manures, litters, composts and waste waters. These data should be used to guide current interpretations and prioritize future research.

CONCLUSIONS AND FUTURE DIRECTIONS

"Characterization" of animal wastes and animal waste-amended soils should be approached at several different scales, primarily emphasizing the evaluation of potential environmental impacts arising from the agricultural use of animal wastes. First, current sampling, testing, and interpretation procedures should be standardized. Next, advances are needed in analytical procedures to better assess the bioavailability of nutrients, trace elements, and organics in wastes and soils. Finally, more effective summarization of historical data on the properties of animal wastes and waste-amended soils should be done and used to prioritize future research and technology transfer efforts. These efforts should not be viewed as independent steps. As shown in Figure 1, the steps modern animal-based agriculture must take to improve characterization programs for animal wastes and waste-amended soils must be closely interrelated to balance the benefits of animal waste use with the many complex environmental problems that now exist.

REFERENCES

Barrington, S.F., Characteristics of livestock manures, in *Proc. Nat. Workshop on Land Application of Animal Manure*, Leger, D.A., Patni, N.K., and Ho, S.K. (ed.), 11-12 June 1991, Ottawa, Canada, Canadian Agric. Res. Council., 1991.

Binford, G.D., Blackmer, A.M., and El-Hout, N.M., Tissue test for excess nitrogen during corn production, *Agron. J.,* 82, 124, 1990.

Bitzer, C.C., and Sims, J.T., Estimating the availability of nitrogen in poultry manure through laboratory and field studies, *J. Environ. Qual.*, 17, 47, 1988.

Bock, B.R., and Kelley, K.R., *Predicting N Fertilizer Needs for Corn in Humid Regions*, National Fert. and Environ. Res. Center, Tenn. Valley Authority, Muscle Shoals, Alabama, 1992.

Breeuwsma, A., and Silva, S., Phosphorus fertilization and environmental effects in the Netherlands and the Po region (Italy), Report No. 57. DLO, The Winand Staring Centre for Integrated Land, Soil and Water Research, Wageningen (The Netherlands), 1992.

Fox, R.H., Meisinger, J.J., and Sims, J.T., Predicting N fertilizer needs for corn in humid regions: Advances in the Mid-Atlantic States, in *Predicting N Fertilizer Needs for Corn in Humid Regions,* National Fert. and Environ. Res. Center, Tenn. Valley Authority, Muscle Shoals, Alabama, 1992.

Gartley, K.L., and Sims, J.T., Phosphorus soil testing: Environmental uses and implications, *Commun. Soil Sci. Plant Anal.*, 25, 1565, 1994.

Lemunyon, J.L., and Gilbert, R.G., Concept and need for a phosphorus assessment tool, *J. Prod. Agr.,* 6, 483 , 1993.

McCollum, R.E., Buildup and decline in soil phosphorus: 30-year trends on a Typic umbraquult, *Agron. J.,* 83, 77, 1991.

Magdoff, F.R., Ross, D., and Amadon, J., A soil test for nitrogen availability to corn, *Soil Sci. Soc. of Am. J.,* 48, 1301, 1984.

Mullins, G.L., Martens, D.C., Miller, W.P., Kornegay, E.T., and Hallock, D.L., Copper availability, form, and mobility in soils from three annual copper-enriched hog manure applications, *J. Environ. Qual.,* 11, 316, 1982.

Mozaffari, P.M., and Sims, J.T., Phosphorus availability and sorption in an Atlantic Coastal Plain watershed dominated by animal based agriculture., *Soil Sci.,* 157, 97, 1993.

Overcash, M.R., Humenik, F.J., and Miner, J.R., Introduction to Livestock Waste Management, in *CRC Livestock Waste Management,* Vol. 1, M.R. Overcash et al. (eds.), CRC Press, New York, New York, 1983.

Piekielek, W.P., and Fox, R.H., Use of a chlorophyll meter to predict sidedress nitrogen requirements for maize, *Agron. J.,* 84, 59, 1992.

Sharpley, A.N., Carter, B.J., Wagner, B.J., Smith, S.J., Cole, E.L., and Sample, G.A., *Impact of Long-term Swine and Poultry Manure Application on Soil and Water Resources in Eastern Oklahoma,* Oklahoma State Univ. Agric. Exp. Stn. Tech. Bull. T-169, 1991.

Sharpley, A.N., Sims, J.T., and Pierzynski, G.M., Innovative soil phosphorus availability indices: Assessing inorganic phosphorus, in *Soil Testing: Prospects for Improving Nutrient Recommendations,* American Society of Agronomy, Madison, Wisconsin, 1994.

Sims, J.T., Environmental soil testing for phosphorus, *J. Prod. Agr.,* 6, 501, 1994.

Sims, J.T., Vasilas, B.L., Gartley, K.L., Milliken, B., and Green, V., Evaluation of soil and plant nitrogen tests for manured soils of the Atlantic Coastal Plain, *Agron. J.,* In Press, 1995.

Sims, J.T., and Wolf, D.C., Poultry waste management: Agricultural and environmental issues, *Adv. Agron.,* 52, 1, 1994.

Stephenson, A.H., McCaskey, T.A., and Ruffin, B.G., A survey of broiler litter composition and potential value as a nutrient resource., *Biol. Waste,* 34, 1, 1990.

WASTE-AMENDED SOILS: METHODS OF ANALYSIS AND CONSIDERATIONS IN INTERPRETATION OF ANALYTICAL RESULTS

S.M. Combs and L.G. Bundy

INTRODUCTION

There is considerable concern about the effect that agricultural practices have on the decline of surface water quality. Testing agricultural soils to predict this contribution may provide a means of assessing environmental impact of agricultural practices. Because of the long-term acceptance of soil testing and subsequent large historical database, routine soil fertility tests are being considered for environmental assessments. The focus of this chapter will be to explore the basis of routine and alternate soil tests for applicability to environmental use and to identify some of the lake and landscape parameters that need to be considered when interpreting results of any off-site field test with respect to impact on lake quality.

OVERVIEW OF SOIL TESTING

Approximately 3 million soil samples are analyzed by soil testing labs each year in the United States to determine nutrient availability to plants (Donohue, 1987). The extraction of a bulk sample with a simple mix of chemicals is an attempt to mirror the complex physiology of a root and its relationship to the nutrients in soil solution over the entire growing season.

Most soil tests are empirical--that is, a relative portion of the plant-available supply is measured--not the absolute supply. For this reason, chemical analysis must be related by correlation and calibration research to determine the potential for a crop to respond to additional nutrient.

Correlation is the process of determining if an extracted soil nutrient and the crop response to added nutrient are related in a predictable manner. Growth trials are necessary to evaluate the effectiveness of a method because plants absorb nutrients from soil considerably differently than synthetic chemicals extract nutrients from soils in the laboratory.

Calibration includes considerations of the portion of maximum yield achieved at a given soil test value and the amount of additional nutrient needed to achieve optimum yield. The laboratory results supply an index--a starting point--for making valid interpretations of the potential seriousness of a nutrient deficiency for a particular crop.

Soil scientists have conducted research for more than a century (Anderson, 1960) to predict the complex relationships among soil, nutrients, and plant growth. Bray (1937) introduced the concept of modern routine soil testing by recognizing that soil tests should be developed by considering what chemical forms of nutrients are present in soils, what the relative availability of each form is, and how the available form can best be extracted. Subsequent chemical tests especially for P, but also for K, were designed within geographic regions to extract a relative portion of the nutrient form of greatest importance to plant uptake and included Bray P1 (Bray and Kurtz, 1945), "double acid" or Mehlich variations (Nelson et al., 1953; Mehlich, 1984), and bicarbonate or Olsen P (Olsen et al., 1954). These methods differ in the mechanisms by which P is extracted and reflect those unique soil conditions inherent to different geographical regions.

The P determined by any of these methods measures a static pool of "available" P with the pool operationally defined by the extraction procedure (Corey, 1987). In many instances, methods using resins as sinks to simulate uptake by plant roots (Amer et al., 1955), or integrating diffusion kinetics (Bouldin, 1961; Barber, 1962; Nye and Tinker, 1977) have been shown to be a better indicator of "available" P, or the total amount of P removed by crops over the growing season. Unfortunately, the methods that relate both diffusion rate and concentration of nutrient in solution are not easily adapted to routine use. The methods of Bray and Kurtz (1945), Olsen et al. (1954) and Mehlich (1984) still remain as the best approach for routine analysis to explain the variations of yield and nutrient uptake by plants growing on different soils. This long-term development process defines the complexity of the modern soil testing system and expresses the development process a new test with environmental objectives may experience.

ADAPTING ROUTINE SOIL TESTING TO ENVIRONMENTAL NEEDS FOR PHOSPHORUS EVALUATION

Developing a new test or adapting a routine soil test to determine the potential P availability in lake systems and the processes that control availability in these systems are of high priority. In addition, environmental concerns for other nutrients, trace metal contaminants, pesticides, and organic residues may also be subjected to similar scrutiny with respect to water quality in the near future. Designing appropriate soil sampling, testing, and interpretative methods for routine use to determine agriculture's impact on water quality are imperative.

Dissolved and particulate P in runoff are considered major contributors to accelerated lake eutrophication, and minimizing P entry into lakes is the current focus of efforts to improve lake quality. Knowing total P has been the basis for many earlier strategies to control P input (Vollenweider, 1976); however, just as with soil testing, lake response to total P is not predictable. A substantial portion of the total P is not in a biologically available form (Sonzogni et al., 1982). Studies have shown that, even when total P inputs are decreased, little change in lake biological activity may result (Young and DePinto, 1982; Gray and Kirkland, 1986). Therefore, effective management strategies to limit P inputs must be based on determining the portion of total P actually available for biological growth.

Do routine soil tests have the potential to predict algal response in lakes? They do only if the routine soil test method and interpretation have meaning in a lake setting. Soil tests provide a relative index of the potential supply of a nutrient, such as P, to a plant. However, the extent to which a routine soil test could be used to predict lake response to P depends on whether lake chemistry and watershed characteristics important to algal response and transport of P can be inferred from the soil test. Routine soil tests by themselves probably do not have the research base to provide an estimate of P that may be potentially delivered and available to a lake system.

Routine soil tests may, however, provide one factor in a multiple-factor index to initially evaluate the relative potential of a site to contribute P to lakes (Lemunyon and Gilbert, 1993). The ultimate usefulness of a P-index will depend on its ability to integrate all factors important in the transport and delivery of P to a lake, the availability of the delivered P, and the lake sensitivity to P input. Stevens et al. (1993) reported that field use of the P-index showed it had the ability to rank soils on a relative basis for potential P loss but it was not sensitive enough to indicate lake benefits to implementing erosion and runoff control practices. Truman et al., (1993) reported that the P-index was dominated by the soil test P category and provided little information on which site characteristics were influencing P transport.

Bundy (1994) concluded from considering major losses and inputs of P from Wisconsin cropland that annual P losses represented 1 to 2% of the total fertilizer and manure P applied. For example, in 1991, P losses in Wisconsin were 2.83 million lb of P compared to the 2.07 million lb of P added to cropland. These results confirm that some natural waters can be exceptionally sensitive to P additions that represent only a small portion of the total added to cropland. A high percentage of the acreage in watersheds contributes little if any P to natural waters, and most of the P loss usually occurs in one or two high-runoff events each year. Controlling these annual losses will be difficult, and progress to reduce P losses to lakes will require site-specific management practices on sites with a high risk of P loss. Bundy (1994) concluded that general, state-wide management practices will not effectively reduce P inputs or achieve desired water quality improvement but will impose restrictions on crop acreage not contributing to P enrichment. Use of a field ranking system such as developed by Lemunyon and Gilbert (1993) may provide a way to identify specific sites most vulnerable to P loss. Long-term validation studies are needed to confirm that the use of an estimate of soil test P within the P-index approach can identify not only potential P loss from individual fields near lakes, but also transport of P to the lake and adverse effects on lake quality. The potential use of routine soil test P levels in regulatory programs makes these studies essential.

DEVELOPING A NEW SOIL TEST FOR ENVIRONMENTAL NEEDS

Using routine soil tests seems to provide, at best, one factor in a limited ranking of potential P loss from individual fields. An actual estimate of possible P delivery and interpretation of the potential for biological response may be a better basis for site-specific control practices to effectively improve water quality. Developing, correlating and calibrating a test will need to follow an evolution similar to that of routine soil testing. This evolution is necessary in order for a test to adequately predict cause and effect over many different watershed and lake conditions.

The first step in developing a test is to identify the forms of P in the lake system and which of these are important in biological response. Sonzogni et al. (1982) summarized the various soluble and particulate P forms and the chemical speciation found in fresh water lakes. Of the small reservoir of soluble P fractions, dissolved inorganic phosphate ($H_2PO_4^-$, HPO_4^{2-}) is directly bioavailable whereas the other dissolved P forms must be converted by hydrolysis or biological processes to the inorganic form. Of the particulate P forms, the P associated with Fe/Al hydrous oxides, minerals, and non-apatite Ca-P minerals represents a large reservoir of slowly available P.

Various types of chemical and biological tests have been proposed to determine the amount of P potentially available from sediments and suspended sediments deposited in lakes. Direct incubation studies to determine the amount of P available for algal uptake was initially introduced by Golterman et al. (1969) and showed partial availability of sediment P to algae. These tests are time consuming and confounded by several factors, including levels of other essential nutrients, algal species and abundance, pH, temperature, and light (Sonzogni et al., 1982).

Chemical extraction procedures of sediments are based on selective dissolution or desorption of inorganic particulate P. Success in extracting a discrete "pool" of inorganic P varies depending on extraction method (Sonzogni et al., 1982). Similar to soil testing, the usefulness of sediment extraction is determined by its correlation to a water quality parameter, i.e., algal response. Subsequent research has used algal response in a correlative way and found the uptake of P by algae incubated with soil, lake sediment, and suspended sediment as the sole P source to be closely correlated to the fraction of P extracted by 0.1 N NaOH (Dorich et al., 1985; Sagher, 1976; Williams et al., 1980). The extraction of P is closely correlated to algal uptake because the rate of P uptake by an abundance of P-deficient algae is determined by the rate of desorption of P from particles (Sonzogni et al., 1982). Bioavailable P can be defined as the amount of inorganic P a P-deficient algal population can utilize in a period of 48 hours or longer and corresponds to the dissolved reactive and a fraction of non-apatite inorganic particulate P (Sonzogni et al., 1982). A chemical test such as extraction of sediments with 0.1 N NaOH that reflects algal uptake can be more easily standardized than direct algal incubation tests. Anion resins (Armstrong et al., 1979) or Al-hydroxy resins (Huettl et al., 1979) were also evaluated as "sinks" that estimated bioavailable P by promoting desorption of P from sediment surfaces similar to algae. Anion resins were generally less efficient than the 0.1 N NaOH extraction and removed about 50% of the inorganic bioavailable P.

Using P-starved algae as the reference may, however, overestimate potential uptake in natural systems. Mild nutrient stress in plants has been suggested to cause increased nutrient carrier synthesis, thereby promoting higher uptake than when grown under "normal" nutrient levels (Lee, 1982; Drew and Slaker, 1984; Wild et al., 1979). Decreased uptake occurred several days after plants were subjected to high nutrient concentrations following nutrient stress conditions (Clarkson and Scattergood, 1982). These data indicate that mechanisms are activated or deactivated depending on what is needed to maintain required nutrient inflows in response to external concentration changes. Higher affinity or greater capacity would allow plants to be more efficient at acquiring nutrients at low solute concentrations than at higher. Therefore, using P-starved algae as the calibration base may overestimate actual uptake in lake systems.

Sharpley et al. (1991) modified the 0.1 N NaOH extraction of suspended sediments to allow estimation of bioavailable particulate P in agricultural runoff. The fraction of total available P is determined by extracting runoff with 0.1 N NaOH for 17 hours, soluble inorganic P is analyzed from a 0.45-um filtered runoff sample, and bioavailable particulate P is calculated as the difference between the two. Both physically and chemisorbed P are reported to be extracted from runoff samples. The extraction results can be influenced by ionic strength, cation species, extractant pH, and solution/soil ratio (Sharpley et al., 1981). Another modification (Robinson et al. 1994; Sharpley, 1993a,b) to avoid these problems introduced the use of an Fe-oxide impregnated strip as a "sink" to adsorb P from runoff solutions. The P removed from runoff samples by adsorption with the Fe oxide papers compared very well to P extracted by 0.1 N NaOH. This method is reported to be an alternative when high solution/runoff ratios cause analytical problems.

Adaptation by soil testing laboratories to accommodate the long extraction time, large solution volume, and sample dilution ratios will be necessary, and it is tempting to correlate the 0.1 N NaOH test to a routine soil test to eliminate the need for adaption. The correlatoin of the runoff extraction procedure to an established routine soil test would also allow continued use of relatively straightforward field soil sampling protocols instead of difficult field runoff sampling methods. However, correlating the 0.1 N NaOH bioavailability test to routine soil tests has shown limited success (Wolf et al., 1985). Correlating one method to another and assuming that the expected relationship to field response remains strong is risky, however. The correlation of test A to field response, and subsequent correlation of test A to test B, does not imply that conditions forming the basis of a significant statistical relationship are present between test B and field response. Munter and Schulte (1993) applied matrix algebra to address the question of what could be inferred about possible correlations between test B and field response when the correlations between test A and field response and test A and test B were known. They determined that if test A (0.1 N NaOH) and field response (algal bloom) have a simple correlation of 0.500 (like most soil tests and yield response), and test A (0.1N NaOH) and test B (routine soil test) have a simple correlation of 0.980, the range in correlation that could be expected between test B (routine soil test) and field response (algal bloom) would be 0.318 to 0.662. If the simple correlation between test A (0.1 N NaOH) and test B (routine soil test) was only 0.900, the expected range between test B (routine soil test) and field response would be even wider, 0.072 to 0.828. Therefore, expecting a strong relationship between test B (routine soil test) and field response (algal bloom) can indeed be risky and certainly would demand field verification.

INTERPRETING A BIOAVAILABLE PHOSPHORUS SOIL TEST

The 0.1 N NaOH extraction of agricultural runoff may have potential to be used as an assessment of a field soil's potential contribution to lake bioavailable P loading. The test may indicate the maximum amount of bioavailable P a lake may be receiving but would not account for factors affecting both the P transport and individual lake sensitivity.

More important is the need for interpretations to be developed that place the test result from either an accepted routine soil test or 0.1 N NaOH extraction of runoff into a context that accounts for dynamic lake and watershed characteristics unique to each site. It is simplistic to expect that a single test and set of interpretations will be applicable over the range of soil types, field management, watershed conditions, and lake chemistry within a broad region. Soil tests for agronomic purposes are regional in nature with interpretations for potential plant response to fertilizer application generally specific for local cropping and soil characteristics. Expecting a complex system such as a watershed to be any less dependent on local conditions is unrealistic.

LAKE CONSIDERATIONS

For most inland waters, P is the limiting nutrient to biomass production, but because of complex interactions in lake ecosystems, Stumm and Morgan (1981) report that the limiting factor concept needs to be applied with caution. Distinction between rate-determining factors (i.e., individual nutrient, temperature, light, etc.) that determine the rate of biomass production and a limiting factor (i.e., nutrient) that determines the maximum possible biomass standing crop should be made. Individual lake chemical, biological, and physical processes are dynamic, and not all lakes will respond similarly to the same P input or to management practices that reduce P inputs. Keizer and Sinke (1992) concluded that if lakes have downward seepage through the sediment layer, reducing external P loading will reduce total P relatively quickly. However, in comparable lakes without downward seepage, restoration of quality water conditions will be more difficult.

Lake sedimentation rate also affects the time available for changes in bioavailable P levels through sorption/desorption reactions in the water column and depends on particle size (shape) and density (Armstrong et al., 1979; Williams et al., 1980). These colloid chemical reactions can influence temporal and spatial distribution of both dissolved and suspended constituents (Stumm and Morgan, 1981). Sonzogni et al. (1982) reported that dilution of lakes receiving tributary water high in dissolved inorganic P would favor desorption of P from suspended sediments--thereby potentially increasing bioavailable P. In contrast, high levels of suspended solids occurring during peak runoff periods could inhibit P desorption from sediments and decrease the photic zone and, therefore, algal uptake of P. Increased biomass or soluble organic inputs could also alter the importance of P sources within the lake after delivery (Jones et al., 1993). They suggested that a decrease in directly available soluble P would occur due to formation of $FePO_4^-$ humic acid complexes. These complexes, however, could maintain a pool of soluble organic P dependent on lake pH and ionic strength.

Sonzogni et al. (1982) suggested considering the effect of positional constraints defined by lake depth and dissolved reactive P in management strategies in addition to the potential bioavailable particulate P. Deep lakes have little ability to release sediment P into the photic zone for algal utilization; therefore, management efforts directed at minimizing soluble P inputs may be more effective. Schindler (1974) demonstrated this lack of resuspension even when anoxic conditions persisted for several months. Shallow lakes frequently have sediment resuspension causing sediment P to become positionally available for algal use. Management for shallow lakes to reduce sediment loading would be especially useful (Sonzogni et al., 1982).

Grouping lakes having similar characteristics such as depth, trophic state, water inputs, bedrock geology, morphology, etc. should enable a more successful development of test result interpretations. Interpretations need to consider what the optimum concentration of P is in a lake, if the optimum is different than the critical P concentration (no algal bloom), and if the optimum concentration varies among lakes.

WATERSHED CONSIDERATIONS

A watershed responds to precipitation, weathering, agricultural runoff, and waste additions, and the mass flux of constituents reflects not only the extent of inputs but many other processes acting in the drainage area (Stumm and Morgan, 1981). By using watershed information and by considering the dependence of concentration on flow, the various processes contributing to the mass P flux to a lake may be identified.

Factors such as tillage, manure application, field slope, and crop residue have been reported to influence P loss and provide focus for many P management strategies to reduce field losses of P. These management issues are complex, interrelated, and sometimes contradicting. Developing a system that can incorporate this complexity is challenging. Lathrop (1990) reported annual P loadings to an alkaline Wisconsin lake to be highly variable, depending on runoff volume, storm intensity, and timing of runoff relative to watershed conditions. Andraski et al., (1985) reported reduced sediment loss and runoff from corn grown with reduced tillage under simulated rainfall. Other research has shown that ridge till actually increased runoff in spring when compared to conventional tillage; however, during heavy summer storms, excessive runoff occurred from conventionally tilled areas (Ginting et al., 1994).

Complicating the influence of reduced tillage on potential P transport is the effect of crop residue remaining on the surface to protect soils from erosion. Research has shown increased leaching of P from these residues (Wendt and Corey, 1980; Havis and Albert, 1993; Miller et al., 1994). In fact, Miller et al. (1994) found the P concentration in runoff to range from 1 to 16 mg/L from fields having different cover crop residues while dissolved P in associated studies by Spires and Miller, 1978 (reported by Miller et al., 1994) ranged from 0.16 mg/L on unmanured to 0.70 mg/L on manured fields. Miller et al. (1994) concluded that the presence of a cover crop increased the potential for dissolved P in runoff to a greater extent than surface manure applications. When the effect of the cover crop on reducing soil erosion is considered, the overall amount of P in runoff is probably as great from cover crop residue as from surface manure application. Sharpley et al. (1992) found that as vegetative cover improved, bioavailable P comprised a larger portion of the total P contained in runoff from both a peanut-sorghum rotation (29%) and from mature grasses (88%). Most of the increase could be attributed to soluble P.

Long-term manure application may alter soil characteristics influencing expected runoff. Percent organic C doubled (0.77 to 1.35%) and percent total N doubled (0.077% to 0.146%) when beef manure was applied annually at 12 T/acre since 1942 on continuous corn with furrow irrigation (Binford and Nielsen, 1992). Soils of southeastern South Dakota receiving long-term manure applications showed increased soil organic matter, soil nitrate-N, available P, available K, and electrical conductivity with effects persisting for many years (Vikekanandon and Fixen, 1990). Boyle et al. (1989) suggest that since organic amendments such as manure increase soil organic matter (fulvic acids, humic acids, and polysarcharides) that binds soil particles into aggregates, the resulting structure and pore size distribution enhances water infiltration. Therefore, less runoff would be expected with manure application, as shown by Smith et al., 1937 (reported by Boyle et al., 1989). An exception may be when manure is surface applied to frozen soil, which does not allow for enhancement of soil aggregation unless incorporated later by tillage.

CONCLUSIONS

Griffith (1973) states that it is impossible to conclude that the concentration of P has changed in fresh water lakes since no reliable data exist with which to compare current measurements. This does not mean that individual lakes have not experienced changes or that the total quantity of P entering lakes has remained constant. It does mean that extreme care is needed in applying general conclusions to the P status of specific lakes and in assessing the impact of P input on these lakes relative to a "reference" state of the water. It is not possible to describe the importance of P as a limiting factor for algal growth in general terms and have it hold for specific situations.

Predicting the loss of P from a field and the potential of that P to be transported to a lake and stimulate algal response does need to be evaluated in order to implement management strategies that will minimize this loss. However, extrapolating current routine soil tests developed for agronomic purposes to off-site losses of P to a lake is tenuous. Development of a new soil test requires extensive field calibration with interpretations that consider individual lake sensitivity to P input and complex landscape parameters affecting P transport. We need to exercise caution in assuming that a soil test can supply a defensible position from which to impose major changes in agricultural practices. An equitable environmental policy needs to consider the limitations of a test as well as its predictive capabilities when interpreting the potential P loss from individual fields to large, complex off-site drainage and lake systems.

REFERENCES

Amer, F., Bouldin, D.R., Black, C.A., and Duke, F.R., Characterization of soil phosphorus by anion exchange resin adsorption and ^{32}P equilibration, *Plant Soil*, 6(4), 391, 1955.

Anderson, M.S., History and development of soil testing, *J. Agric. Food Chem.*, 8, 84, 1960.

Andraski, B.J., Mueller, D.H., and Daniel, T.C., Phosphorus losses in runoff as affected by tillage, *Soil Sci. Soc. Am. J.*, 49, 1523, 1985.

Armstrong, D.E., Perry, J.R., and Flatness, D., Availability of pollutants associated with suspended or settled river sediments which gain access to the Great Lakes PLUARG Tech. Rep. In. Joint Commission, Great Lakes Regional Office, 1979.

Barber, S.A., A diffusion and mass flow concept of soil nutrient availability, *Soil Sci.*, 93, 39, 1962.

Binford, G.D., and Nielsen, R.A., *Long-term Applications of Manure and Fertilizer in Irrigated Corn*, Soil Sci. Research Dept., Dept. of Agron. University of Nebraska, Lincoln, Nebraska, 1992.

Bouldin, D.R., Mathematical desorption of diffusion processes in the soil-plant system, *Soil Sci. Soc. Am. J.*, 25, 476, 1961.

Boyle, M., Frankenberger, W.T., Jr., and Stolzy, L.H., The influence of organic matter on soil aggregation and water infiltration, *J. Prod. Agric.*, 2(4), 290, 1989.

Bray, R.H., New concepts in the chemistry of soil fertility, *Soil Sci. Soc. Am. Proc.*, 2, 175, 1937.

Bray, R.H., and Kurtz, L.T., Determination of total, organic and available forms of phosphorus in soils, *Soil Sci.*, 59, 39, 1945.

Bundy, L.G., *A Phoshorus Budget for Wisconsin Cropland*, Rept. for Wisconsin Dept. Ag, Trade and Consumer Protection, 1994.

Clarkson, D.T., and Scattergood, C.B., Growth and phosphate transport in barley and tomato plants during the development of, and recovery from, phosphate-stress, *J. Exp. Bot.*, 33, 865, 1982.

Corey, R.B., Soil test procedures: Correlation, in *Soil Testing: Sampling, Correlation, Calibration and Interpretations*, Brown, J.R., ed., SSSA Spec. Publ. 21, SSSA Madison, Wisconsin, 1987.

Donohue, S.J., The value and use of soil test summaries, in *Soil Testing: Sampling, Correlation, Calibration and Interpretations*, Brown, J.R., ed., SSSA Spec. Publ. 21, SSSA Madison, Wisconsin, 1987.

Dorich, R.A., Nelson, D.W., and Sommers, L.E., Estimating algal available phosphorus in suspended sediments by chemical extraction, *J. Environ. Qual.*, 14(3), 400, 1985.

Drew, M.C., and Saker, R.L., Uptake and long-distance transport of phosphate, potassium and chloride in relation to internal ion concentrations in barley: Evidence of nonallosteric regulation, *Planta*, 160, 500, 1984.

Ginting, D., Moncrief, J.F., Gupta, S.C., Evans, S.D., Nelson, G.A., Johnson, B.J., and Ranaivoson, A., Assessment of the effects of tillage and manure applicaton on sediment and P loss due to runoff field research, in *Soil Sci. 1994* Soils Series 140, Misc. Publ. 83, University of Minnesota, St. Paul, Minnesota, 1994.

Golterman, H.L., Bakels, C.C., and Jakobs-Mogelin, J.J., Availability of mud phosphates for growth of algae, *Vert. Int. Verein Limonol.*, 17, 467, 1969.

Gray, C.B.J., and Kirkland, R.A., Suspended sediment phosphorus composition in tributaries of the Okanagan Lakes, *B.C. Water Res.*, 20, 1193, 1986.

Griffith, E.J., Environmental phosphorus--an editorial, in *Environmental Phosphorus Handbook*, E.J. Griffith, Spencer, J.M., and Mitchell, D.T., eds., John Wiley and Sons, 1973.

Havis, R.N., and Albert, E.E., Nutrient leaching from field-decomposed corn and soybean residue under simulated rainfall, *Soil Sci. Soc. Am. J.*, 56, 211, 1993.

Huettl, P.J., Wendt, R.C., and Corey, R.B., Prediction of algal-available phosphorus in runoff suspensions, *J. Environ. Qual.*, 8, 130, 1979.

Jones, R.I., Shaw, P.J., and Gettann, H., Effects of dissolved humic substances on the speciation of iron and phosphate at different pH and ionic strength, *Environ. Sci. Tesh.*, 27(6), 1052, 1993.

Keizer, P., and Sinke, A.J.C., Phosphorus in the sediment of the Loosdrecht lakes and its implications for lake restoration perspectives, *Hydrobiologia*, 233, 39, 1992.

Lathrop, R., Response of Lake Mendota (Wisconsin, USA) to decreased phosphorus loadings and the effect on downstream lakes, *Verh. Internat. Verein. Limnol.*, 24, 457, 1990.

Lee, R.B., Selectivity and kinetics of ion uptake by barley plants following nutrient deficiency, *Ann. Bot.*, 50, 429, 1982.

Lemunyon, J.L., and Gilbert, R.G., The coneept and need for a phosphorus assessment tool, *J. Prod. Agric.*, 6(4), 483, 1993.

Mehlich, A., Mehlich 3 soil test extractant: A modification of Mehlich 2 extractant, *Commun Soil Sci. Plant Analy.*, 15, 1409, 1984.

Miller, M.H., Beauchamp, E.G., and Lauzon, J.D., Leaching of nitrogen and phosphorus from biomass of three cover crop species, *J. Environ. Qual.*, 23, 267, 1994.

Munter, R.C., and Schulte, E.E., An NCR-13 Summary of Mehlich III research, in *Proc. NCR-13 13th Soil Plant Analysts Workshop*, St. Louis, Missouri, Act. 26-27, 1993.

Nelson, W.L., Mehlich, A., and Winters, E., The development, evaluation and use of soil tests for phosphorus availability, in *Soil and Fertilizer Phosphorus*, W.H. Pierre and A.G. Norman, eds., Vol. IV, Academic Press, New York, 1953.

Nye, P.H., and Tinker, P.B., *Solute Movement in the Soil-root System*, Univ. Calif. Press, Berkeley, California, 1977.

Olsen, S.R., Cole, C.V., Watanabe, F.S., and Dean, L.A., Estimation of available phosphorus in soils by extraction with sodium bicarbonate, USDA Circ. 939, 1954.

Robinson, J.S., Sharpley, A.N., and Smith, S.J., Development of a method to determine bioavailable phosphorus loss in agricultural runoff, *Agriculture, Ecosystems and Environment*, 47, 287, 1994.

Sagher, A., *Availability of Soil Runoff Phosphorus to Algae*, Ph.D. Thesis, University of Wisconsin, Madison, Diss. Abstr. 37, 4877, 1976.

Schindler, D.W., Eutrophication and recovery in experimental lakes: Implications for lake management, *Science*, 184, 897, 1974.

Sharpley, A.N., An innovative approach to estimate bioavailable phosphorus in agricultural runoff using iron oxide-impregnated paper, *J. Environ. Qual.*, 22, 597, 1993a.

Sharpley, A.N., Estimating phosphorus in agricultural runoff available to several algae using ion oxide paper strips, *J. Environ Qual.*, 22, 678, 1993b.

Sharpley, A.N., Ahuja, L.R., Yamamoto, M., and Menzel, R.G., The kinetics of phosphorus desorption from soil, *Soil Sci. Soc. Am. J.*, 45, 493, 1981.

Sharpley, A.N., Smith, S.J., Jones, D.R., Berg, W.A., and Coleman, G.A., The transport of bioavailable phosphorus in agricultural runoff, *J. Environ. Qual.*, 21, 30, 1992.

Sharpley, A.N., Troeger, W.W., and Smith, S.J., The measurement of bioavailable phosphorus in agricultural runoff, *J. Environ. Qual.*, 20, 235, 1991.

Sonzogni, W.D., Shapra, S.C., Armstrong, D.E., and Logan,T.J., Bioavailability of phosphorus inputs to lakes, *J. Environ. Qual.*, 11(4), 555, 1982.

Stevens, R.G., Sobecki, T.M., and Spofford, T.L., Using the phosphorus assessment tool in the field, *J. Prod. Agric.*, 6(4), 487, 1993.

Stumm, W., and Morgan, J.J., *Aquatic Chemistry*, John Wiley and Sons, New York, 1981.

Truman, C.C., Gascho, G.J., Davis, J.G., and Wachope, R.D., Seasonal phosphorus losses in runoff from a Central Plain soil, *J. Prod. Ag*, 6, 507, 1993.

Vikekanandon, M., and Fixen, P.E., Effect of large manure applications on soil P intensity Commun, *Soil Sci. Plant Anal.*, 21(3,4), 287, 1990.

Vollenweider, R.A., Advances in defining critical loading levels for phosphorus in lake eutrophication, *Mem. 1st Hal. Idrobiol*, 33, 53, 1976.

Wendt, R.C., and Corey, R.B., Phosphorus variations in surface runoff from agricultural lands as a function of land use, *J. Environ. Qual.*, 9, 130, 1980.

Wild, A., Woodhouse, P.J., and Hopper, M.J., A comparison between the uptake of potassium by plants from solutions of constant potassium concentration and during depletion, *J. Exp. Bot.,* 30, 697, 1979.

Williams, J.D., Shear, H.H., and Thomas, R.L., Availability to *Scenedesmus quadricauda* of different forms of phosphorus in sedimentary materials from the Great Lakes, *Limnol. Oceanogr.,* 251, 1, 1980.

Wolf, A.M., Baker, D.E., Pionke, H.B., and Kunishi, H.M., Soil tests for estimating labile, soluble, and algae-available phosphorus in agricultural soils, *J. Environ. Qual.,* 14, 341. 1985.

Young, T.C., and DePinto, J.V., Algal-availability of particulate phosphorus from diffuse and print sources in the lower Great Lakes basin, *Hydrogiologia,* 91, 111, 1982.

NITROGEN TRANSFORMATIONS IN SOIL AMENDED WITH POULTRY LITTER UNDER AEROBIC CONDITIONS FOLLOWED BY ANAEROBIC PERIODS

W.F. Johnson, Jr. and D.C. Wolf

INTRODUCTION

Arkansas leads the nation in the production of broilers (*Gallus gallus domesticus*) with over 1 billion produced in 1993 (Klugh and Abbe, 1994). These broilers produced over 1 x 10^9 kg of poultry manure on a dry weight basis. The majority of the manure is applied to pastures in the form of poultry litter, which is a mixture of the manure and bedding material. The litter supplies the soil with nutrients that increase forage production. Excessive N application is of environmental concern because high NO_3-N levels may reach ground and surface waters and can result in health problems in young livestock and infants. Consuming drinking water that exceeds 10 mg NO_3-N/L increases the potential for health problems (US Public Health Service, 1962).

Nitrogen transformations in litter-amended soils are important since the litter contains little or no NO_3-N. The total N in the litter can be classified into three groups: inorganic N, labile organic N, and complex organic N (Sims and Wolf, 1994). The inorganic N form is mainly NH_4-N with minimal amounts of NO_3-N. The labile organic N fraction consists of uric acid and urea, which are mineralized to inorganic N forms. The complex organic N fraction requires months for mineralization to supply the inorganic N forms that plants can use.

Once organic N is mineralized, several N transformations are possible. Nitrification can occur wherein chemoautotrophic bacteria oxidize NH_4-N to NO_3-N under aerobic conditions. If the NO_3-N is subjected to anaerobic conditions and available C is present, denitrification is possible. In denitrification, microorganisms utilize the NO_3-N as a final electron acceptor during respiration (Stanford et al., 1975). This reaction results in NO_3-N being converted to gaseous forms, N_2O and N_2, that are evolved into the atmosphere.

When poultry litter is surface-applied to pastures, aerobic conditions typically occur that would allow for rapid mineralization of labile organic N and nitrification of NH_4-N. However, subsequent precipitation may saturate the soil and result in short periods of anaerobic conditions that would be favorable for denitrification. Thus, the objectives of this study were to investigate the conversion of organic N to inorganic N and to determine the NO_3-N levels after both aerobic and anaerobic incubations in soils amended with poultry litter.

MATERIAL AND METHODS

SOILS AND POULTRY LITTER CHARACTERIZATION

The soils used in the laboratory incubations were a Captina silt loam (fine-silty, siliceous, mesic Typic Fragiudult) and a Nixa silt loam (loamy-skeletal, siliceous, mesic Glossic Fragiudult). The two soils represent a major portion of the litter-amended soils in northwestern Arkansas and had different cropping histories. The Captina soil was collected from a bermudagrass (*Cynodon dactylon* L.) pasture, and the Nixa soil was collected from a tall fescue (*Festuca arundinacea* Schreb.) pasture. The field-moist soil was collected at a depth of 0 to 15 cm, passed through a 2-mm sieve, and stored in sealed polyethylene bags at 2 to 4°C for no longer than 7 days. The Captina soil contained 1.4 g/kg organic N and <1 g/kg inorganic N. The soil contained 14% sand, 63% silt, and 23% clay. The organic matter content was 2.6%, pH was 6.1, and the cation exchange capacity (CEC) was 12.1 cmol/kg (meq/100 g). The Nixa soil contained 1.6 g/kg organic N and <1 g/kg inorganic N. The soil contained 12% sand, 73% silt, and 15% clay and had a pH of 5.5. The organic matter content was 2.3%, and the CEC was 9.5 cmol/kg.

The poultry litter used was a manure-pine (*Pinus* spp.) shavings mixture that had accumulated in a broiler house over five flock cycles or for approximately one year. The poultry litter contained total N, organic N, NH_4-N, and NO_3-N levels of 48.2, 43.7, 4.5, and <1 g/kg, respectively. The pH of a 1:5 litter-to-water mix was 8.1, and the total C was 403 g/kg. The soil and litter were analyzed using standard methods for total-N (Bremner and Mulvaney, 1982), inorganic-N (Keeney and Nelson, 1982), total C (Nelson and Sommers, 1982), and mechanical analysis (Gee and Bauder, 1986). The organic matter content, pH, and CEC were analyzed by the protocols of the University of Arkansas Soil Testing and Research Laboratory.

INCUBATION PROCEDURES

The incubations were accomplished by placing in 50-mL centrifuge tubes 10-g (dry weight equivalent) portions of soil amended with poultry litter at a rate of 4 g/kg. Paired tubes were established with unamended soil. The moisture content of the soil was adjusted to a moisture potential of -30 kPa.

Aerobic incubation periods ranged from 0 to 336 hours to facilitate C mineralization and N mineralization and nitrification. By varying the incubation periods, different levels of inorganic-N were produced, and varying levels of available C remained in the soils. By evaluating various levels of both parameters, it was anticipated that a greater understanding of the potential for denitrification activity would result. Preliminary studies demonstrated that a lack of available C adversely impacted denitrification activity, and, therefore, short incubation periods were used.

The mouths of the centrifuge tubes were covered with Saran Wrap[R] during the incubation to reduce evaporation and allow air exchange. Anaerobic conditions were established after each aerobic period by adding 5 mL of distilled water to each tube and sealing by means of a rubber stopper. Anaerobic conditions were verified by measuring redox potentials in selected tubes. Flooded soil conditions were maintained from 0 to 72 hours to facilitate denitrification. All incubations were conducted in the dark at 21°C. Net NH_4-N, NO_3-N, and total soluble organic carbon (TOC) levels are defined as the differences between the levels in amended and control soils.

CARBON ANALYSIS

Samples from the different aerobic and anaerobic combinations were analyzed for TOC using a PHOTOchem Organic C Analyzer. A 1:1 soil to solution ratio was used, and this solution was shaken for 30 min at 232 oscillations/min. Supernatant was removed and placed in a 5-mL centrifuge tube and centrifuged at 2000x g for 10 min. The aliquot was filtered through a 0.45-μm Millipore[R] membrane filter before analysis.

STATISTICAL EVALUATION

The randomized complete block design was used in this experiment. Three replications of each treatment were established. Means were separated by LSD using the GLM procedure in SAS (SAS Institute, 1985).

RESULTS AND DISCUSSION

NITROGEN TRANSFORMATIONS

Aerobic incubation of the litter-amended soils resulted in N mineralization and nitrification (Table 1). Rapid hydrolysis of labile organic N forms resulted in significant increases in the net NH_4-N levels during the initial anaerobic incubations, and the net NH_4-N production was greater in the Captina than in the Nixa soil. All values include the 18 mg NH_4-N/kg soil that was present in and added to the soil when the litter was applied. The difference between the initial 28 mg NH_4-N/kg and the 18 mg NH_4-N/kg added with the litter was most likely due to hydrolysis of urea during the extraction of the time zero samples. Net NH_4-N levels represent the balance between additions due to mineralization of organic N and losses due to nitrification of NH_4-N.

During the aerobic incubation, the net NO_3-N levels increased with time due to nitrification of NH_4-N to NO_3-N (Table 1). The longer incubation time for the Nixa soil resulted in higher NO_3-N concentrations. Anaerobic conditions and the resulting denitrification resulted in significant NO_3-N losses in the Captina soil with the exception of the anomaly that occurred in the 240-hour aerobic incubation followed by the 66-hour anaerobic conditions. The greatest loss occurred in the 48-hour anaerobic incubation of the litter-amended soil that

Table 1. Net NH$_4$-N, NO$_3$-N, and total soluble organic carbon (TOC) in Captina and Nixa soils amended with poultry litter and incubated under aerobic and anaerobic conditions.

Aerobic	Anaerobic	NH$_4$-N	NO$_3$-N	TOC
----------------h----------------		-----------------------mg/kg-----------------------		
Captina silt loam				
0	0	28bc†	1d	82a
0	48	75a	0d	81a
0	66	73a	0d	64a
120	0	45bc	14c	20b
120	48	45bc	0d	20b
120	66	51ab	0d	7b
240	0	19c	34a	10b
240	48	22c	24b	8b
240	66	35bc	35a	8b
Nixa silt loam				
0	0	28c	2c	83b
0	72	53a	0c	98a
168	0	21d	22b	15c
168	72	41b	17b	21bc
336	0	20d	46a	11c
336	72	33c	41a	12c

†Means in a column for a soil followed by the same letter are not significantly different at the 5% level.

had been previously incubated for 120 hours aerobically. Aerobic incubation for 240 hours followed by 48 hours of anaerobic incubation also showed a significant reduction in NO$_3$-N levels, but 24 mg NO$_3$-N/kg soil remained when the incubation was terminated. It appeared that denitrification was limited due to a lack of available C to support the denitrifying bacteria. The net nitrate losses for the120- and 240-hour aerobic incubations followed by anaerobic conditions were 14 and 10 mg/kg, respectively. Nitrate levels in the Nixa soil were not significantly decreased by anaerobic incubation and resulted in a NO$_3$-N level of 41 mg/kg following 336 hours of aerobic and 72 hours of anaerobic conditions. The most likely explanation for the higher NO$_3$-N levels is that the longer incubation times resulted in reduced available C to support denitrification activity. Longer aerobic incubation would result in greater microbial respiration activity or decomposition of the organic components of the poultry litter and thus decrease the amount of C available for use by the denitrifying bacteria. The data would suggest that for denitrification to occur, the soil must become anaerobic after nitrification has occurred but before the available C has been depleted.

To evaluate the hypothesis that available C was limiting denitrification activity, the Nixa soil was amended with 50 mg NO$_3$-N/kg and glucose at 100 or 1,000 mg C/kg soil and incubated under anaerobic conditions for 72 h. The results showed that NO_3-N levels decreased to 49 and 3 mg/kg at glucose rates of 100 and 1,000 mg C/kg, respectively. Yoshinari et al. (1977) demonstrated that glucose addition stimulated denitrification. Cabrera and Chiang (1994) suggested that 100% of the initial NO_3-N present in poultry litter was lost due to denitrification when the samples were incubated at 2400 g H_2O/kg litter.

Net N mineralization for the Captina was 41 and 35 mg N/kg for the 120- and 240-hour aerobic incubations, respectively. The decreased N mineralization after 240 hours could be due to microbial assimilation of the inorganic N. Gale and Gilmour (1986) reported that N mineralization kinetics in soil amended with poultry litter had a rapid phase of 0 to 168 h. The subsequent intermediate phase was a period of static or reduced inorganic N levels in litter-amended soil. The Nixa soil had values of 25 and 48 mg N/kg for the 168- and 336-hour aerobic incubations, respectively.

The percentage net N mineralization was defined as:

$$\% \text{ Net Mineralization} = \frac{[(NH_4^+\text{-N} + NO_3^-\text{-N})_{trt} - (NH_4^+\text{-N} + NO_3^-\text{-N})_{ck} - (NH_4^+\text{-N} + NO_3^-\text{-N})_{litter}]}{[(\text{Total-N})_{litter} - (NH_4^+\text{-N} + NO_3^-\text{-N})_{litter}]} \times 100$$

where $(NH_4\text{-N})_{trt}$ and $(NO_3\text{-N})_{trt}$ were the NH_4-N and NO_3-N concentrations in mg/kg in the soil amended with poultry litter; $(NH_4\text{-N})_{ck}$ and $(NO_3\text{-N})_{ck}$ were the NH_4-N and NO_3-N levels in the soils that did not receive poultry litter; and $(\text{Total-N})_{litter}$, $(NH_4\text{-N})_{litter}$, and $(NO_3\text{-N})_{litter}$ were the total N, NH_4-N, and NO_3-N in the poultry litter and added at time zero.

The percentage organic N mineralized was 23 and 20% for 120- and 240-hour aerobic Captina incubations, respectively, and 14 and 27% for the 168- and 336-hour aerobic Nixa incubations, respectively. Bitzer and Sims (1988) evaluated 20 poultry manures during a 140-day incubation study and reported a mean value for organic N mineralization of 67%. Castellanos and Pratt (1981) developed regression equations to describe relationships between available N and N mineralized. Their data showed that approximately 22% of the organic N in chicken manure added to a San Emigdio fine sand was mineralized in a 168-hour laboratory incubation. This value is very similar to the reported values in our study for the 120- and 240-hour incubations. Hadas et al. (1983) determined that 34 to 44% of the total N in poultry manure was recovered as inorganic N following a 168-hour incubation. In the Captina soil, 31 and 27% of the total N was recovered in the inorganic form after incubation periods of 120 and 240 h, respectively. The higher recovery of inorganic N reported by Hadas et al. (1983) was most likely due to their use of poultry manure whereas our study used poultry litter that contained pine shavings and would most likely result in greater N immobilization and thus a lower net inorganic N recovery.

CARBON TRANSFORMATIONS

The C and N cycles are intimately linked in soil, and thus the level of microbially available C will have a large influence on N transformations in soils amended with litter. Addition of poultry litter to the two soils resulted in high TOC levels that significantly decreased with time (Table 1). Since denitrification occurred in the litter-amended soil incubated for 120 hours aerobically followed by a 48-hour anaerobic period, denitrification was likely limited by C availability in the 240-hour aerobic plus 48-hour anaerobic incubation. It is also likely that the availability of the total soluble organic C to soil microorganisms could have changed with time of incubation. Stanford et al. (1975) stated that denitrification activity was dependent upon the amounts of decomposable organic materials in soils. Reddy et al. (1980) determined that, as incubation periods increased, the denitrification potential of waste-amended soil decreased due to the loss of C during decomposition.

CONCLUSIONS

Aerobic incubations of Captina and Nixa silt loams amended with poultry litter resulted in net mineralization of organic N in the 240- and 336-hour incubations of 20 and 27%, respectively. The net NO_3-N levels increased with time of aerobic incubation. Denitrification in the Captina subjected to the 120-hour aerobic incubation and 48 hours of anaerobic conditions resulted in complete loss of NO_3-N. The 240-hour aerobic plus 48-hour anaerobic conditions resulted in less denitrification and was likely due to a lack of available C.

Results from this study suggest that denitrification can be rapid in soils amended with poultry litter if environmental conditions are suitable. The magnitude of NO_3-N loss via denitrification in soil amended with poultry litter is determined by the presence of NO_3-N, absence of O_2, and level of available C. The soil must be aerobic for sufficient time to allow nitrification to occur, but not long enough for the heterotrophic microbial population to deplete the pool of available C. If NO_3-N is present, the soil becomes anaerobic, and if available C is present, denitrification will be rapid. Because of the complexity of the denitrification process, it is difficult to make generalizations or predictions regarding the magnitude of N loss via denitrification in soils amended with surface applications of poultry litter.

REFERENCES

Bitzer, C.C., and Sims, J.T., Estimating the availability of nitrogen in poultry manure through laboratory and field studies., *J. Environ. Qual.,* 17, 47, 1988.

Bremner, J.M., and Mulvaney, C.S., Nitrogen-total, in *Methods of Soil Analysis,* Part 2, 2nd ed., A.L. Page, ed., *Agronomy,* 9, 595, 1982.

Cabrera, M.L., and Chiang, S.C., Water content effect on denitrification and ammonia volatilization in poultry litter, *Soil Sci. Soc. Am. J.,* 58, 811, 1994.

CHARACTERISTICS OF WASTES AND WASTE-AMENDED SOILS 33

Castellanos, J.Z., and Pratt, P.F., Mineralization of manure nitrogen--correlation with laboratory indexes, *Soil Sci. Soc. Am. J.*, 45, 354, 1981.

Gale, P.M., and Gilmour, J.T., Carbon and nitrogen mineralization kinetics for poultry litter, *J. Environ. Qual.*, 15, 423, 1986.

Gee, G.W., and Bauder, J.W., Particle-size analysis, in *Methods of Soil Analysis*, Part I, 2nd ed, A. Klute, ed., *Agronomy*, 9, 383, 1986.

Hadas, A., Bar-Yosef, B., Davidov, S., and Sofer, M., Effect of pelleting, temperature, and soil type on mineral nitrogen release from poultry and dairy manures, *Soil Sci. Soc. Am. J.*, 47, 1129, 1983.

Keeney, D.R., and Nelson, D.W., Nitrogen-inorganic forms, in *Methods of Soil Analysis*, Part 2, 2nd ed., A.L. Page, ed., *Agronomy*, 9, 643, 1982.

Klugh, Jr., B.F., and Abbe, D.S., *1993 Arkansas Agricultural Statistics*, Arkansas Agric. Exp. Stn. Report Series 327, Univ. of Arkansas, Fayetteville, Arkansas, 1994.

Nelson, D.W., and Sommers, L.E., Total carbon, organic carbon, and organic matter, in *Methods of Soil Analysis*, Part 2, 2nd ed., A.L. Page, ed., *Agronomy*, 9, 539, 1982.

Reddy, K.R., Khaleel, R., and Overcash, M.R., Nitrogen, phosphorus, and carbon transformations in a costal plain soil treated with animal manures, *Agric. Wastes*, 2, 225, 1980.

Sims, J.T., and Wolf, D.C., Poultry waste management: Agricultural and environmental issues, *Adv. Agron.*, 52, 1, 1994.

SAS Institute, Inc., *SAS User's Guide: Statistics*, SAS Institute, Inc., Cary, N.C., 1985.

Stanford, G., Vander Pol, R.A., and Dzienia, S., Denitrification rates in relation to total and extractable soil carbon, *Soil Sci. Soc. Am. Proc.*, 39, 284, 1975.

US Public Health Service, *Drinking Water Standards*, Public Health Service Publ. 956, 48, 1962.

Yoshinari, T., Hynes, R., and Knowles, R., Acetylene inhibition of nitrous oxide reduction and measurement of denitrification and nitrogen fixation in soil, *Soil Biol. Biochem.*, 9, 177, 1977.

PHOSPHORUS RETENTION IN SELECTED INDIANA SOILS USING SHORT-TERM SORPTION ISOTHERMS AND LONG-TERM AEROBIC INCUBATIONS

T.L. Provin, B.C. Joern, D.P. Franzmeier, and A.L. Sutton

INTRODUCTION

The application of livestock and poultry manures on agricultural land has received considerable attention in the past decade as concerns about non-point source pollution prevention have increased. Repeated applications of phosphorus (P)-rich manures in excess of crop removal can increase extractable soil P (Sutton et al., 1978; Barber, 1979; Sutton et al., 1986). As extractable soil P concentrations increase, the potential for P leaching can also increase (Kudeyarov et al., 1981). Understanding the potential downward movement of P is very important because more livestock producers are injecting manure below the soil surface to control odor, runoff, and nitrogen losses.

Bushnell (1952) found available soil P at 90 cm to be 2.5 times higher in a manured pasture than in an untreated control. Schwab and Kulyingyong (1989) reported Bray P_1 concentrations of 350 mg/kg in plots fertilized with 40 kg P/ha/year compared to 13 mg P/kg in unfertilized control plots in the surface 10 cm of soil after 40 years. Although available P concentrations decreased rapidly with depth, they also found that fertilized treatments had significantly higher Bray P_1 levels at 45 cm than did unfertilized treatments (10 mg P/kg compared to 2.5 mg P/kg). Hergert et al. (1981) found increased P concentrations in field drain tile effluent from manured plots compared to unfertilized control plots. Effluent P concentrations were actually highest during the peak flow period and dropped off when flow rates decreased. The authors attributed the increased P concentrations during peak flows to increased dissolution of P compounds during these drainage events.

Kovar and Barber (1988) studied the P supplying characteristics of 33 North American soils. The authors found good relationships between inorganic P added and anion exchangeable P, but poor agreement was found between P added and P found in the solution. Mendoza and Barrow (1987) studied the recovery of added P by soil extractants and found that recovery of P decreased as the soil P sorption capacity increased.

The relationship between short-term laboratory isotherm studies and longer-term incubations that more closely simulate field conditions has not been widely studied. The objectives of this study were 1) to determine the amount of P retained by five Indiana soils prior to increases in solution P using 24-hour sorption isotherms and one-year aerobic incubations and 2) to determine the usefulness of the Bray P_1 soil test in predicting water-soluble P (WSP) from the long-term incubation study.

MATERIALS AND METHODS

SOILS

Five soils were sampled to approximately 1.0 m in July 1992. Each soil was selected based on its chemical and physical properties and its abundance in the state. The Elston series was the only soil with any known history of manure or other biosolids applications. Soils were separated into their respective surface and sub-surface horizons (25 horizons in total), sieved to < 2 mm, and stored in their sampled moisture state. A sub-sample of each soil was taken and dried for chemical and physical analyses. Each soil was analyzed for particle size (Kilmer and Alexander, 1949), specific surface area (Carter et al., 1986), pH (soil:water, 1:1), Bray P_1 (Bray and Kurtz, 1945), and WSP (soil:0.01M $CaCl_2 \cdot 2H_2O$, 1:25, shaken 24 hours at 120 epm). These data are presented in Table 1. With the exception of the Elston Ap horizon, none of the soil horizons had initial Bray P_1 levels that would be classified as very high or excessive for plant growth in most states.

SHORT-TERM PHOSPHORUS SORPTION ISOTHERMS

To determine P retention using a rapid method, inorganic P sorption isotherms were conducted using a procedure similar to Nair et al. (1984). Approximately 0.75 g of soil was placed in a 50-ml Nalgene centrifuge tube with 20 ml of 10 mM $CaCl_2 \cdot 2H_2O$ solution (not adjusted for pH) spiked with 0, 50, 100, 175, 250, or 500 umol P as KH_2PO_4. After adding 0.1 ml chloroform to inhibit microbial activity, the centrifuge tubes were capped and shaken for 24 hours on an inline shaker at 120 epm. Following centrifugation at 3000 x g for 10 min, the solutions were decanted and P concentrations analyzed colorimetrically using the ascorbic acid method described by Murphy and Riley (1962).

LONG-TERM AEROBIC INCUBATIONS

In order to study the P retention of these soils in a more natural setting, each soil horizon was also incubated with $NH_4H_2PO_4$ (25, 50, 100, and 200 mg P/kg) at 25°C for one year. Sufficient water was added periodically to maintain moisture near field capacity (-33 kPa, as determined experimentally). Following the one-year incubation period, soils were air-dried, sieved to < 2 mm, and stored. Each soil was then analyzed for Bray P_1 and WSP. WSP was determined as above, except two additional extractions were performed following one-week incubations at 30°C between extractions.

Table 1. Soil chemical and physical properties evaluated during study.

Soil	Horizon	depth	Text.	pH	SA	Initial ppm Bray P₁	Phosphorus availability indexes after one year incubations with											
							25 mg P/kg			50 mg P/kg			100 mg P/kg			200 mg P/kg		
							A⁵	B⁶	C⁷	A	B	C	A	B	C	A	B	C
		cm			m²/g		mg P (/kg or /L)											
Clermont Typic Glossaqualf	Ap	0-15	SiL	5.02	96	172 4.51	12.5	0.09	0.00	17.2	0.44	0.26	35.3	1.33	0.17	73.0	3.31	0.77
	E	15-30	SiL	4.91	91	150 8.03	14.5	0.17	0.35	17.9	0.51	0.09	37.9	1.14	0.26	78.6	5.66	1.55
	Bg	30-56	SiL	5.11	114	278 0.65	3.08	0.00	0.09	10.4	0.25	0.00	20.6	0.00	0.00	54.2	0.68	0.17
	Btg1	56-79	SiL	5.25	155	340 0.17	0.98	0.09	0.00	6.80	0.08	0.00	13.5	0.25	0.00	44.5	0.85	0.52
	Btg2	79-97	SiL	4.40	180	345 0.11	0.48	0.08	0.00	1.35	0.17	0.17	6.60	0.09	0.00	56.7	0.34	0.09
	2Btg3	97-109	L	4.44	175	405 0.04	–⁸	–	–	–	–	–	–	–	–	–	–	–
Elston Typic Argiudoll	Ap	0-33	L	6.32	75	144 56.5	16.8	1.73	1.04	83.2	3.55	1.29	113	6.82	2.15	125	18.7	2.84
	A	33-56	L	4.92	71	222 10.5	14.0	0.00	0.00	27.3	0.17	0.00	54.1	0.62	0.62	89.5	1.77	0.34
	Bt1	56-74	SiL	5.17	48	159 2.90	9.77	0.00	0.00	18.5	0.18	0.00	41.4	1.32	0.17	64.2	4.09	0.78
	Bt2	74-97	SL	5.24	45	144 2.01	–	–	–	–	–	–	–	–	–	–	–	–
	BC	97-108	SL	5.41	82	165 2.15	–	–	–	–	–	–	–	–	–	–	–	–
Sebawa Ultic Hapludalf	Ap	0-18	L	6.90	133	156 15.7	7.98	0.34	0.17	6.29	0.63	0.00	19.1	1.47	0.60	70.4	7.88	0.94
	A	18-33	L	6.68	171	183 6.05	7.34	0.00	0.00	13.5	0.08	0.00	31.2	0.66	0.09	66.0	7.24	1.02
	Btg	33-53	SCL	5.80	115	162 5.29	5.42	0.00	0.00	16.7	0.09	0.00	34.0	0.66	0.00	76.3	4.85	0.69
	2Btg2	53-79	SL	6.53	57	79 6.65	13.0	0.00	0.00	26.7	0.69	0.09	44.0	2.76	0.60	89.8	18.3	2.43
	2Cg	79-91	SL	6.90	52	94 6.55	–	–	–	–	–	–	–	–	–	–	–	–
Toronto Udollic Achraqualf	Ap	0-15	SiL	6.43	105	150 13.9	20.3	0.62	0.43	33.2	0.18	0.00	51.2	3.48	1.82	94.1	9.90	2.25
	A	15-30	SiL	6.38	127	159 6.31	9.80	0.00	0.00	18.2	0.17	0.00	32.0	0.18	0.34	63.8	6.35	1.21
	Bt1	30-53	SiCL	6.04	197	250 0.37	1.31	0.00	0.09	4.04	0.18	0.00	16.4	0.00	0.00	26.9	0.98	0.00
	2Bt2	53-71	SiCL	6.05	265	310 0.08	0.21	0.00	0.00	1.10	0.00	0.00	5.04	0.00	0.00	12.9	0.45	0.00
	2Bt3	71-102	SiCL	6.51	212	310 0.06	0.81	0.00	0.00	0.08	0.00	0.00	5.99	0.17	0.00	22.1	0.25	0.00
Zipp Psammentic Hapludalf	Ap	0-18	SCL	6.57	137	172 1.07	13.7	0.09	0.00	17.9	0.60	0.17	35.1	2.36	0.09	51.4	4.31	1.55
	Bg1	18-35	SCL	6.70	155	190 0.93	2.96	0.00	0.00	8.60	0.27	0.00	18.3	0.42	0.00	37.1	3.29	0.43
	Bg2	35-69	SCL	6.88	151	228 0.25	0.87	0.00	0.00	3.51	0.00	0.00	12.0	0.67	0.17	22.9	0.81	0.00
	Cg	69-104	CL	7.06	200	278 0.08	0.50	0.00	0.00	2.15	0.16	0.00	4.05	0.00	0.00	18.9	0.16	0.00

1, USDA horizon designation; 2, USDA soil textural class; 3, surface area; 4, added P retained by soil (mg P/kg) at an equilibrium solution concentration of 1 mg P/L in short-term P sorption isotherm experiment; 5, Bray P₁ (mg P/kg) following 1 year incubation; 6, water-soluble P (WSP) in 1st extraction (mg P/L); 7, WSP in 3rd extraction (mg P/L); 8, insufficient soil.

RESULTS AND DISCUSSION

SHORT-TERM PHOSPHORUS SORPTION ISOTHERMS

The amount of added P retained at an equilibrium solution concentration of 1 mg P/L ranged from 79 to 405 mg P/kg for the 25 soil horizons evaluated in this study. Due to space limitations, only the Sebawa and Clermont soils will be discussed in detail. Complete isotherms for these two soils are presented in Figure 1. Isotherm data were used to determine the amount of P retained at an equilibrium solution concentration of 1 mg P/L. This solution P level matches sewage treatment plant effluent discharge requirements in Indiana.

The Sebawa soil series showed substantial differences in P retention between soil horizons (Figure 1). Phosphorus retention in the top three horizons of this soil was much higher than in the two lower horizons at a 1-mg P/L equilibrium solution concentration (156, 183, and 162 mg P/kg for horizons 1, 2, and 3; and 79 and 94 mg P/kg for horizons 4 and 5). In contrast, P retention in the Clermont soil increased with depth (Figure 1). The amount of added P retained at a 1-mg P/L equilibrium solution concentration was 172 and 150 mg P/kg for horizons 1 and 2 compared to 278, 340, 345, and 405 mg P/kg for horizons 3, 4, 5, and 6.

The reduced P retention in the lower two horizons of the Sebawa soil compared to the three upper horizons was likely due to their coarser soil texture and reduced surface area (Table 1). Phosphorus retention in the Clermont soil increased with depth as the clay content and surface area increased (Table 1). We

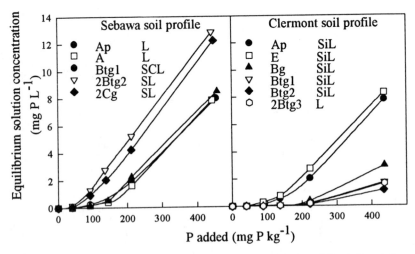

Figure 1. Relationship between phosphorus (P) added and final P solution concentrations for soil horizons in Sebawa and Clermont soil profiles.

found that the amount of added P retained at a 1-mg P/L equilibrium solution concentration was related to surface area for all soil horizons evaluated in this study. This relationship between P retention and surface area was not observed in other soils (data not presented) when initial Bray P_1 concentrations were > 50 mg/kg.

LONG-TERM AEROBIC INCUBATIONS

Bray P_1 levels

Increases in Bray P_1 levels after one-year incubations varied greatly between soil horizons (Table 1). The surface and fourth horizons of the Sebawa soil had initial Bray P_1 levels of 15.7 and 6.55 mg P/kg, respectively. Following the addition of 100 mg P/kg, the Bray P_1 level increased only 3.4 mg P/kg for the surface horizon compared to over 37 mg P/kg for the fourth horizon. The surface and fifth horizons of the Clermont soil had initial Bray P_1 levels of 4.51 and 0.11 mg P/kg, respectively. Following the addition of 100 mg P/kg, Bray P_1 levels for the surface and fifth horizons increased 30.8 and 6.49 mg P/kg, respectively.

Increases in Bray P_1 levels with P additions were greatest for soil horizons with initial Bray P_1 levels > 1 mg P/kg. Most of these soil horizons also had lower P retention capacities as determined in the short-term P isotherm experiment. In general, soil horizons with initial Bray P1 levels < 1 mg P/kg had higher Bray P_1 buffering capacities (resistance to change in Bray P_1 level per unit P addition). This increased buffering capacity may be due to P sorption by "high energy" fixation sites that are not extracted with a Bray P_1 test.

Water-soluble P levels

WSP levels in soils incubated for one year decreased greatly between the first and third extractions (Table 1 and Figure 2). WSP in the surface horizon of the Sebawa soil amended with 200 mg P/kg (Bray P_1 = 70.4 mg P/kg) decreased from 7.88 mg P/L in the first extraction to 0.94 mg P/L in the third extraction. Similar reductions in WSP between the first and third extractions were observed in the other soil horizons with Bray P_1 levels > 50 mg P/kg. All soil horizons with Bray P_1 levels < 50 mg P/kg had WSP levels at or below 0.6 mg P/L by the third extraction (Table 1).

The critical value, or Bray P_1 level at which WSP reaches 1 mg P/L, more than doubled between the first and third extractions. Averaged over all soil horizons and P additions evaluated in this study, the critical Bray P_1 value was 26 mg P/kg for the first WSP extraction and 53 mg P/kg for the third extraction (Figure 2).

The 25:1 solution to soil dilution factor used to determine WSP complicates the interpretation of these data. While a 1:1 solution to soil ratio would best mimic the soil environment during conditions of saturated flow, a 25:1 ratio allowed us to lower WSP levels with fewer extractions. We believe the

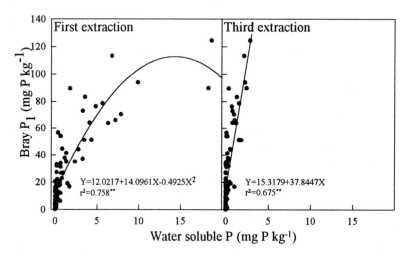

Figure 2. Relationship between Bray P_1 levels and water-soluble P in two sequential extractions from soils incubated with $NH_4H_2PO_4$ for one year.

increase in critical Bray P_1 values between the first and third extractions may be related to the ability of soils to maintain high solution P levels over an extended number of natural leaching events.

CONCLUSIONS

Our short-term P sorption isotherm study has demonstrated that the amount of added P retained prior to exceeding some threshold solution concentration can vary greatly between soils and among horizons within the same soil series. The amount of added P retained at an equilibrium solution concentration of 1 mg P/L ranged from 79 to 405 mg P/kg for the 25 soil horizons evaluated in this study. Differences in P retention using short-term P isotherms were primarily associated with differences in soil surface area.

Increases in Bray P_1 levels after one-year incubations also varied greatly between soils and among horizons within the same soil series. Increases in Bray P_1 and WSP levels following this incubation period were greatest in soils with initial Bray P_1 levels > 1 mg P/kg and soils exhibiting low retention capacities in the short-term P isotherm study. Soil horizons that initially had Bray P_1 levels < 1 mg P/kg showed the highest Bray P_1 buffering capacities. We believe this increased buffering capacity is due to P sorption by "high energy" fixation sites that are not extracted with a Bray P_1 soil test. We are currently developing a soil extraction procedure to quantify these "high energy" P fixation sites in soils.

In the incubated soils, WSP levels determined after one and three sequential extractions were related to Bray P_1 levels measured at the end of the one-year incubation period. The Bray P_1 level at which WSP exceeded 1 mg P/L increased from 26 mg P/kg after the first extraction to 53 mg P/kg after the third extraction.

Based on this limited data set, we found surface area to be an important property to consider when assessing P sorption capacities in soils. We also found the Bray P_1 soil test to be a useful tool for predicting WSP levels in the soils evaluated in this study. WSP levels measured after a certain number of 25:1 solution:soil ratio extractions may also provide valuable information about the mineral phases or other P fractions that control WSP in soils. The ability of a soil to maintain high solution P levels over an extended number of extractions may be far more useful in predicting potential P inputs to surface waters than a single extraction showing a relatively high WSP level.

At field scale, even when sorption capacities in the surface horizon are exceeded and WSP concentrations become elevated, the soil horizons below it may be able to sorb the leaching P and minimize the potential for P movement to surface waters via field drainage tile effluent. From a management standpoint, controlling P runoff to surface waters is extremely important. If P runoff can be minimized, it may be possible to utilize that portion of the soil profile above the drainage tile or seasonal high water table as a P sink in fields receiving repeated applications of manures and other P-rich materials.

REFERENCES

Barber, S.A., Soil phosphorus after 25 years of cropping with five rates of phosphorus application, *Commun. in Soil Sci. and Plant Anal.,* 10, 1459, 1979.

Bray, R.H., and Kurtz, L.T., Determination of total, organic and available forms of phosphorus in soils, *Soil Sci.,* 59, 39, 1945.

Bushnell, T.M., Soil conditions after 60 years in a Purdue pasture lot, *Proc. Indiana Acad. of Sci.,* 61, 180, 1952.

Carter, D.L., Mortland, M.M., and Kemper, W.D., Specific surface, in *Methods of Soil Analysis,* Part 1, A. Klute, ed., American Society of Agronomy, Madison, Wisconsin, 1986.

Hergert, G.W., Bouldin, D.R., Klausner, S.D., and Zwerman, P.J., Phosphorus concentration-water flow interactions in tile effluent from manured land, *J. Environ. Qual.,* 10, 338, 1981.

Kovar, J.L., and Barber, S.A., Phosphorus supply characteristics of 33 soils as influenced by seven rates of phosphorus addition, *Soil Sci. Soc. Am. J.,* 52, 160, 1988.

Kilmer, V.J., and L.T. Alexander, L.T., Method of making mechanical analysis of soil, *Soil Sci.,* 68, 18, 1949.

Kudeyarov, V.N., Dashkin, V.N., and Kudeyarova, A.YU, Losses of nitrogen, phosphorus and potassium from agricultural watersheds of minor rivers in the Oka Valley, *Air and Soil Poll.,* 16, 267, 1981.

Mendoza, R.E., and Barrow, N.J., Ability of three soil extractants to reflect the factors that determine the availability of soil phosphate, *Soil Sci.,* 144, 319, 1987.

Murphy, L.J., and Riley, J.P., 1962. A modified single solution method for determination of phosphate in natural waters, *Anal. Chim. Acta.,* 27, 31, 1962.

Nair, P.S., Logan, T.J., Sharpely, A.N., Sommers, L.E., Tabatabai, M.A., and Yuan, T.L., Interlaboratory comparison of a standard adsorption procedure, *J. Environ. Qual.,* 13, 591, 1984.

Schwab, A.P., and Kulyingyong, S., Changes in phosphate activities and availability indexes with depth after 40 years of fertilization, *Soil Sci.,* 147, 179, 1989.

Sutton, A.L., Nelson, D.W., Mayrose, V.B., and Nye, J.C., 1978. Effects of liquid swine waste applications on corn yield and soil chemical composition., *J. Environ. Qual.,* 7, 325, 1978.

Sutton, A.L., Nelson, D.W., Kelly, D.T., and Hill, D.L., Comparison of solid vs. liquid manure applications on corn yield and soil composition, *J. Environ. Qual.,* 15, 370, 1986.

THE EFFECT OF ANIMAL MANURE APPLICATIONS ON THE FORMS OF SOIL PHOSPHORUS

J.S. Robinson, Andrew N. Sharpley, and S.J. Smith

INTRODUCTION

Since the late 1980's, beef feedlot, poultry, and swine production in Oklahoma and Texas has almost doubled (Oklahoma Agricultural Statistics Service, 1990; Texas Agricultural Statistics Service, 1989). Manure from these operations is a low-cost alternative to mineral fertilizers for many farmers in the Southern Plains, providing a valuable source of nutrients for crop and forage production (Huhnke, 1982). However, the large amounts of manure produced in localized areas can exceed crop phosphorus (P) requirements. This can increase the potential for P loss in surface runoff (Sharpley et al., 1993). Thus, efficient utilization of manure to avoid potential water quality degradation is one of the main problems facing many farmers.

In order to make reliable recommendations for land application of animal manure and management alternatives, information is needed on the fate of manure P in soil. In recent studies, we found that continual long-term application (eight to 35 years) of beef feedlot, poultry litter, and swine manure increased the total P (TP) content of several soils (0 - 5 cm depth) in Oklahoma and Texas up to eight-fold (Sharpley et al., 1984, 1991, 1993). These studies did not evaluate the effect of manure on inorganic and organic forms of soil P in terms of potential sources of the nutrients for plant uptake or loss in runoff. As no other information is available, we determined the distribution and relative availabilities of P forms in soils receiving beef feedlot, poultry litter, and swine manure for up to 35 years in the Southern Plains area of Oklahoma and Texas.

MATERIALS AND METHODS

SITE CHARACTERISTICS

Soils that received beef feedlot manure in the Texas Panhandle (Potter County), poultry litter in southeastern Oklahoma (LeFlore and McCurtain Counties), and swine manure in northeastern Oklahoma (Delaware County) were selected. Management history is given in Table 1. The sites reflect typical area soils receiving manure and were all on ground with gentle slopes (< 2% slope) so prior changes in soil properties due to erosion or deposition would be minimal. Soils treated with poultry litter and swine manure were under 'Coastal' or 'Midland' bermudagrass (*Cynodon dactylon* L. Pers.) cut for hay approximately twice a year. Beef feedlot manure was applied to irrigated grain sorghum (*Sorghum bicolor* L. Moench). No mineral fertilizer was applied during the period of manure application. Information on the rate and duration of manure application at each site was obtained from the land owner. For poultry litter, pine shavings were used as bedding material in the broiler houses.

The same soil types on adjacent areas that had not received manure (untreated) were also sampled for background information. Untreated soils similar to those receiving poultry litter and swine manure were all under unfertilized native grass. Soils adjacent to those receiving beef feedlot manure were under irrigated, unfertilized grain sorghum.

Treated and untreated soils were sampled on the same day by taking six 2.5-cm-diam. cores at each site in 5-cm increments to a 150-cm depth. Soil cores were air-dried, composited, and sieved (2 mm). The samples were stored in air-tight containers for later analysis.

CHEMICAL ANALYSES

Soil pH was determined using a glass electrode (soil/water ratio, 1:2; weight/volume). Soil inorganic (IP) and organic P (OP) was fractionated according to

Table 1. Site characteristics.

Soil	Classification	Estimated manure applied[†]	Duration	Estimated P applied
		Mg/ha/year	year	kg/ha/year
Beef Feedlot Manure				
Pullman cl	Torrertic Paleustoll	22	8	90
		67	8	270
Poultry Litter				
Cahaba vfsl[‡]	Typic Hapludult	9.0	12	130
Carnasaw fsl	Typic Hapludult	5.6	20	85
Kullit fsl	Aquic Paleudult	9.0	12	130
Muskogee l	Aquic Paleudalf	9.0	12	130
Neff sil	Aquultic Hapludalf	4.5[#]	35	35
Rexor l	Ultic Hapludalf	6.7	12	100
Ruston fsl	Ultic Hapludalf	6.7	12	100
Shermore	Typic Fragiudalf	5.6	20	85
Swine Slurry				
Captina sl	Typic Fragiudalf	22.1	9	100
Sallisaw sl	Typic Paleudalf	47.8	15	80
Stigler sl	Aquic Paleudalf	61.1	9	40

†Poultry litter and beef feedlot manure applied on a dry weight basis and swine manure as a slurry with units of m³/ha/year.
‡vfsl, fsl, sl, sil, l, and cl represent very fine sandy loam, fine sandy loam, sandy loam, silt loam, loam, and clay loam, respectively.
#Litter applied every other year.

the procedure described by Hedley et al. (1982). This involved sequential extraction of 0.5 g soil with 30 ml of 0.5 M $NaHCO_3$ (pH 8.5), 0.1 M NaOH, and 1.0 M HCl each for 16 h. The residual soil was finally digested with conc. H_2SO_4 and H_2O_2. The IP content of each filtered and neutralized extract and total P content of the digest was determined by the molybdenum-blue method of Murphy and Riley (1962). The total P content of the bicarbonate and hydroxide extracts was also determined following perchloric acid digestion (Olsen and Sommers, 1982). The organic P content of the bicarbonate and hydroxide extracts was calculated as the difference between total P and IP contents. The fractions are subsequently referred to as bicarbonate IP, hydroxide IP, acid IP, bicarbonate OP, hydroxide OP, and residual P. All analyses were conducted in duplicate and the means presented.

The sequential extraction procedure removes IP and OP of increasing chemical stability in terms of soil P fertility. Hedley et al. (1982), and more recently Tiessen et al., (1984), reported that bicarbonate IP is the most biologically available IP form, while hydroxide IP is associated with amorphous and some crystalline Al and Fe phosphates. Acid IP is mainly relatively stable Ca-bound P. Bicarbonate OP is easily mineralized and may contribute to plant-available P. Hydroxide OP constitutes chemically and physically protected organic forms that are involved in long-term soil P transformations. Residual P includes a resistant mixture of occluded IP covered with sesquioxides, calcium-bound IP included in other minerals, and non-extracted stable OP.

RESULTS AND DISCUSSION

The TP content of manure collected from the land owners prior to application averaged 4 g/kg for beef feedlot manure, 15 g/kg for poultry litter, and 1.7 g/L for swine slurry (dry weight basis). These values are similar to those reported by Gilbertson et al. (1979) for beef feedlot manure (8 g/kg), poultry litter (14 g/kg), and swine slurry (1.6 g/L). The major change in soil TP content following manure application to the soils of the present study occurred in the surface 5 cm of soil (Sharpley et al., 1984, 1993). Thus, P fractionation of only surface soil (0 to 5 cm) was carried out.

Total P content increased almost eight-fold (Carnasaw soil and poultry litter) with manure application (Table 2). The sequential fractionation of soil P into forms of differing availability allows further evaluation of the disposition of P in soils treated with manure. Manure application resulted in a significant ($P < 0.05$) increase in all fractions (Table 2). However, the increase was greater for inorganic than for organic fractions. For example, the bicarbonate IP content of manure-treated soils was on average 11-fold, hydroxide IP five-fold, and acid IP 12-fold greater than for untreated soils, while bicarbonate OP in treated soils was only three-fold and hydroxide OP two-fold greater (Table 2).

Thus, manure application resulted in a shift in dominance of OP forms in untreated soils to IP in treated soils. The proportion of TP as IP increased from

Table 2. The distribution of P forms in the surface 5 cm of soil untreated and treated with beef feedlot manure, poultry litter, or swine slurry.[†]

| | Total P | | Inorganic | | | | | | Organic P | | | | Residual P | |
| | | | Bicarb. | | Hydrox. | | Acid | | Bicarb. | | Hydrox. | | | |
Soil	Unt.	Tr.	Unt.	Tr.	Unt.	Tr.	Unt.	Tr.	Unt.	Tr.	Unt.	Tr.	Unt.	Tr.
							——— mg/kg ———							
Beef Feedlot Manure														
Pullman														
22/8	353	538	28	63	92	121	42	110	39	58	88	103	64	83
67/8	353	996	28	149	92	219	42	289	39	100	88	169	64	70
Poultry Litter														
Cahaba	162	625	11	114	23	89	6	236	14	34	55	72	53	80
Camasaw	324	2439	8	257	55	1028	50	321	32	142	74	228	105	463
Kullit	212	1410	12	237	49	227	28	346	23	185	42	167	58	249
Muskogee	552	1298	21	203	140	296	58	223	26	124	149	214	149	238
Neff	530	812	13	66	66	129	18	74	12	32	213	248	208	245
Rexor	395	696	27	136	69	118	39	94	28	60	124	148	109	140
Ruston	245	1091	22	289	44	177	21	303	20	60	28	63	111	199
Shermore	250	1349	8	191	32	432	4	144	63	124	89	194	53	265
Average	334a	1215b	15a	187b	60a	312b	28a	218a	27a	95b	97a	167b	106a	235b
Swine Slurry														
Captina	273	566	13	97	65	129	9	119	15	21	39	39	132	162
Sallisaw	226	486	12	96	31	80	12	89	19	26	56	60	96	135
Stigler	302	536	23	90	70	127	23	71	16	20	66	93	104	135
Average	267a	529b	16a	94b	57a	112b	15a	93b	17a	22b	54a	64b	111a	144b

†Averages followed by a different letter are significantly different ($P < 0.05$) as determined by analysis of variance for paired data.

46 to 61% in soils treated with beef feedlot manure and from 31 to 58% for soils treated with either poultry litter or swine slurry (Table 3). Overall, most of the applied manure P accumulated as weakly bound, plant-available IP and IP associated with hydrous Al and Fe oxides and Ca precipitates (Table 3).

For all manure types and most treated soils, the most dramatic increase occurred in acid IP (associated with Ca phosphates), which was 12-fold greater than untreated soils compared to two- to 11-fold increases for the other IP and OP fractions (Tables 2 and 3). This was the case even though the pH of treated soils ranged from 5.9 to 7.2. While beef feedlot manure and poultry litter increased soil pH only 0.4 and 0.2 units, respectively, swine slurry application increased pH an average 0.9 units (Table 3). It is likely that the increase in acid IP with manure application results from the large amounts of Ca added in manure. The average Ca contents of beef feedlot manure, poultry litter, and swine manure slurry were 20, 60, and 2 g/kg, respectively (Gilbertson et al., 1979).

CONCLUSIONS

Manure application increased the amounts of P in several fractions with a change in dominance of OP in untreated (64%) to IP in treated soils (60%). This influences both soil P fertility status and vulnerability of manure-treated soil to enrich runoff with P. A shift from OP to IP forms, particularly the more labile or weakly bound fractions, not only provides more P for plant uptake but can increase the potential for P transport in runoff.

For each manure type and all soils, manure P increased IP to a greater extent than OP. Although amounts of P in all fractions increased, there was a shift in dominant IP form from Al- and Fe-bound to Ca-bound IP with application of each manure type. Thus, it is possible that current soil test methods, such as acid Bray and Mehlich extractants, may overestimate plant-available P in soils treated with manure by dissolution of Ca - P complexes. This may also lead to an overestimation of the potential for soil to release P to runoff.

Table 3. Phosphorus fractions in untreated surface soil (0 - 5 cm) and treated with manure, averaged for each manure type.[†]

Manure type	Soil pH		Inorganic			Organic		
			Bicarb	NaOH	Acid	Bicarb.	NaOH	Residual
			------------------------- % -------------------------					
Beef feedlot	6.8a	Untreated	8a	26a	12a	11a	25a	18a
	7.2b	Treated	13b	22a	26b	10a	18b	11b
Poultry litter	5.9a	Untreated	5a	18a	8a	10a	28a	31a
	6.1a	Treated	16b	22a	20b	8a	15b	19b
Swine slurry	5.0a	Untreated	6a	20a	5a	7a	20a	42a
	5.9b	Treated	18b	21a	19b	4a	12b	27b

†For each manure type, fractions of untreated and treated soils followed by the same letter are not significantly different ($P < 0.05$) as determined by analysis of variance.

REFERENCES

Gilbertson, C.B., Norstadt, F.A., Mathers, A.C., Holt, R.F., Barnett, A.P., McCalla, T.M., Onstad, C.A., and Young, R.A., *Animal Waste Utilization on Cropland and Pastureland*, USDA Utilization Research Report No. 6, 1979.

Hedley, M.J., White, R.E., and Nye, P.H., Plant-induced changes in the rhizospere of rape (*Brassica napus* Var. Emerald) seedlings: III, Changes in L value, soil phosphate fractions and phosphate activity, *New Phytol,* 91, 45, 1982.

Huhnke, R.L., Land application of livestock manure, Cooperative Extension Service, Oklahoma State Univ., Extension Facts 1710, 1982.

Murphy, J., and, Riley, J.P., A modified single solution method for the determination of phosphate in natural waters, *Anal. Chim. Acta,* 27, 31, 1962.

Olsen, S.R., and, Sommers, L.E., Phosphorus, in *Methods of Soil Analysis*, Part 2, 2nd edition A.L. Page et al., eds., *Agronomy,* 9, 403, 1982.

Oklahoma Agricultural Statistics Service, *Oklahoma Agricultural Statistics*, Oklahoma Dept. Agric., Oklahoma City, Oklahoma, 1990.

Sharpley, A.N., Carter, B.J., Wagner, B.J., Smith, S.J., Cole, E.L., and Sample, G.A., *Impact of Long-term Swine and Poultry Manure Applications on Soil and Water Resources in Eastern Oklahoma*, Okla. State Univ., Tech. Bull. T169, 1991.

Sharpley, A.N., Smith, S.J., and Bain, W.R., Nitrogen and phosphorus fate from long-term poultry litter applications to Oklahoma soils, *Soil Sci. Soc. Am. J.,* 57, 1131, 1993.

Sharpley, A.N., Smith, S.J., Stewart, B.A., and Mathers, A.C., Forms of phosphorus in soil receiving cattle feedlot waste, *J. Environ. Qual.,* 13, 211, 1984.

Texas Agricultural Statistics Service., *Texas Agricultural Statistics,* Texas Dept. of Agriculture, Austin, Texas, 1989.

Tiessen H., Stewart, J.W.B., and Cole, C.V., Pathways of phosphorus transformations in soils of differing pedogenesis, *Soil Sci. Soc. Am. J.,* 48, 853, 1984.

CHARACTERISTICS OF ANIMAL WASTES AND WASTE-AMENDED SOILS: NUTRIENT MANAGEMENT PLANNING BEYOND THE FARM BOUNDARY

Joseph P. Zublena

INTRODUCTION

The concepts and principles of on-farm nutrient management are similar to those involved when nutrient management planning expands to larger geographic areas. However, the legal, social, and scientific decisions, as they relate to the issue, become more complex as the number of parties involved increases. Large-scale nutrient management programs have been directed primarily at surface water quality because of the potential for eutrophication.

POINT SOURCE NUTRIENT MANAGEMENT

Traditionally, the focus of many state and regional programs has been on point sources, such as municipal and industrial waste water treatments plants, and on phosphorus (P) reductions (Reckhow and Stow, 1994). Emphasis has been given to the most visible and most manageable sources of pollution. For example, P detergent bans and point source P limitations implemented in many states resulted in about a one-third reduction in both influent and effluent P concentrations (DiFiore, 1988). Many of these early nutrient management policies and programs, such as the Clean Water Act[1], considered non-point sources only peripherally, and no consideration was given to nitrogen (N) controls (Rader, 1994). Early funding policies of US EPA actually discourage implementation of N removal or multiple nutrient removal (STAC, 1986). As the contributions of point sources decreased, the increase in the relative proportion of non-point source (NPS) pollution made impairments of surface water quality from NPS, such as sediment and erosion, more evident (Crowder and Young; 1988; US EPA, 1992a).

---

[1]Clean Water Act Amendments of 1990, P.L. 101-549 secs. 401-416, 104 Stat. 2584-631 (codified at 42 U.S.C. sec. 7651-7651o).

NON-POINT SOURCE NUTRIENT MANAGEMENT

Non-point source pollution is more difficult to monitor and to enforce mitigating controls than point sources because of the heterogeneity of natural systems and our limited knowledge of how NPS pollution impacts aquatic systems (Cabe and Herriges, 1991; Segerson, 1988; Xepapadeas, 1991; Rader, 1994). Lack of scientifically based, uniform water quality standards that can be quantitatively measured has also hampered nutrient management policy making across the nation (Rader, 1994; Lynch, 1994). One example is the measurement of water quality impaired by nutrients. New Jersey has water quality standards for total P (0.05 mg/L in lakes; 0.10 mg/L in streams, unless it can be shown P is not limiting) and for dissolved oxygen[2]. North Carolina has a standard for chlorophyll-A and dissolved oxygen (Rader, 1994). Florida has the following narrative standard: "In no case shall nutrient concentrations of a body of water be altered to cause an imbalance in natural populations of aquatic flora and fauna[3]." To add to the controversy, different water quality standards can be imposed based on the beneficial use assigned to each specific body of water. Oklahoma, for example, includes aesthetic values as part of their beneficial use criteria (Lynch, 1994). Subjective criteria are extremely difficult to enforce or to assess.

Since many water bodies and their accompanying nutrient management problems are not confined to state borders, interstate agreement on water quality standards and policy issues should be pursued. Until that time, political and scientific controversy may occur. Lynch (1994) discussed several controversies that arose between Oklahoma and Arkansas over nutrient enrichment of a common river system. These included, but were not limited to the relative contribution of nutrients from each state, the ability to prove nutrient-based impairment, the magnitude of point versus non-point source inputs involved in eutrophication, and the application of downstream standards to upstream polluters, particularly when the standards exist in different political and legal jurisdictions. He also noted from his experiences that the most appropriate method for solving interstate problems appears to be through interstate working groups with the US EPA as an arbitrator of disputes. In Oklahoma and Arkansas, both states agree that a strong effort is being made to base future water quality decisions on science rather than on politics.

The lack of common goals and objectives is cited as a problem with many nutrient management programs at geographic scales greater than the farm level. The desire to react quickly often leads one to focus on perceived and easily studied criteria rather than the development of carefully determined objectives and attributes that can be used to weigh management alternatives and probabili-

---

[2]N.J.A.C. 7:9-4 (c)(6)
[3]Rules and regulations of the State of Florida, Title 17, Chapter 17-3.091; also see id. 17-3.111 (16) and 17-3.121 (19).

ties of success (Reckhow and Stow, 1994). One tool to assist in complex decision making is decision analysis (Keeney and Raiffa, 1976; von Winterfeldt and Edwards, 1986). Reckhow and Stow (1994) explain the benefits of this tool for decisions with multiple issues or objectives:

> The problem is first dissected into single management objectives and single measurable attributes. Attributes should be meaningful to the issue, predictable, comprehensive and non-overlapping. Subsequent analysis should focus on estimating the effects of the different management actions on the levels of the attributes. Ideally, the analysis should quantify all uncertainties, which can be expressed as a probability. A measure of value, utility, or net benefit must be assigned to each outcome, and attributes must be weighed with respect to each other.

They further state that "in many instances, a final solution within the decision theoretical framework is not necessary, as the primary value is the insight and understanding provided by the analysis."

TARGETING PROGRAMS

After criteria affecting water quality are determined and standards are developed, efforts can be made to collect information within a watershed, or other defined area, to identify or target major problem areas or individuals (NPS Evaluation Panel, 1990). Targeting is a term often used in pollution-related literature that implies a focused effort. The focus is often quite specific with a high probability for desired results. Targeting has been used to increase cost effectiveness of federal soil erosion programs (Park and Sawyer, 1985). It also has been identified as a key management option for a watershed-based approach to control non-point sources of pollution (US EPA, 1993; Water Quality 2000, 1992). Zublena and Barker (1992) in North Carolina developed a state-wide nutrient assessment program that balanced manure nutrients available for crop use with crop nutrient need estimates. Results of this assessment were used to target counties with potential nutrient surpluses and to focus educational programs on animal waste management. On a more localized scale, Lemunyon and Gilbert (1993) developed an indexing assessment tool that can be used to evaluate specific sites for their vulnerability to contamination of surface waters with phosphorus. With this tool sites with the highest potential for NPS pollution can be identified and targeted for management alternatives to reduce the risk of phosphorus losses. In a similar study, Bosch et al. (1994) evaluated the effectiveness of information and targeting in the context of a performance-based standard that placed a limit on allowable losses of one or more pollutants. Their case study assessed the relative importance of distance from streams, slope within a field, and initial soil test P levels in determining costs of reducing sediment P. Results indicated that within-the-field slope was the most important determinant of the per-unit cost of reducing sediment P deliveries from sites distant from water bodies and initial soil test P levels was the next most important factor.

CONGRESSIONAL ISSUES

"Nationally, congressional discussions on water quality no longer revolve around whether there is a problem or what policy goals should be, but rather [focus on] how to effectively address the problem to achieve the goals" (Zinn, 1994). According to Zinn (1994), key issues in the nutrient management debate at the national level include the following:

1. Should agriculture be treated as another business?
2. Is the USDA part of the issue or solution?
3. How do you define success in water quality, and who will decide when it is enough ?
4. How will issues of water quality in agriculture, including nutrient management, be redefined by a broader definition of the way to frame environmental questions?

In seeking answers to these questions, Zinn mentions three solutions that are being debated nationally: (1) application of regulations where the polluter pays; (2) application of more holistic solutions to problems that deal on a larger scale such as watershed planning and management; and (3) a combination of a holistic approach with efforts to maintain the importance of the farm boundary through the implementation of on-farm total resource management programs. The following quote by Zinn emphasizes the current and future perspective as it relates to environmental quality and the farming community: "the preeminence of the farm boundary is being replaced by a boundary far away, and individual farmers must consider the effects of their actions far from their property."

NUTRIENT TRADING

One holistic approach to watershed nutrient management that is gaining the interest of regulators, dischargers, and environmental groups is nutrient trading (US EPA, 1992b; Bartfield, 1994). Nutrient trading, like pollution trading, is a market-based alternative to traditional control regulations (Hahn and Stavins, 1991; Stavins and Whitehead, 1992). Traditional regulations treat all dischargers alike, forcing identical control burdens regardless of their relative impacts on the problem. In contrast, pollution trading sets a limit on the overall level of pollution in a given area and issues permits to individual polluters for their share based on actual discharge data. Individuals can buy or trade allocations among themselves, letting market forces produce a cost-effective outcome (Montgomery, 1972). Pollution trading has been used on a limited basis in the United States (Jacobson et al., 1994). Examples include US EPA's Emissions Trading Program and the nation-wide lead fuel phase down during the 1980s (Dudek and Palmisano, 1988; Hahn and Hester, 1989). In theory, pollution trading can also achieve the goal of controlling NPS nutrient pollution. Nevertheless, in practice these programs have not performed well (Jacobson et al., 1994). Currently, four water pollution control trading programs have been implemented--

two in Colorado, one in Wisconsin, and one in North Carolina (US EPA, 1992b). To date, no trades have been completed in any of these programs (Jacobson et al., 1994). In trading systems that encompass point and NPS pollutants, it is important to establish an appropriate trading ratio to reflect impact. The trading ratio is the amount of NPS control that a point source discharger must undertake to create a credit for a given unit of point source discharge. For example, In North Carolina's Tar-Pamlico nutrient trading program, a ratio of 3:1 (NPS/PS) is used for cropland best management practices (BMPs) and 2:1 for confined animal operations. (Jacobson et al., 1994). Ratios greater than 1:1 were used as a safety factor because the predictability and reliability of NPS control is less than for point sources.

SUMMARY

In summary, interest in nutrient management will continue to grow at local, regional, and national levels. As the geographic area and number of individuals involved increase, management and policy become more complex. Policy implementation is hampered by the lack of scientific knowledge that can objectively assess water quality criteria and link the assessment to specific conditions and violators. This is especially true for NPS nutrients. Current water quality standards are not standardized between states or within regional watersheds. Methods to improve nutrient management decisions are available and include decision analysis and targeting. Innovative nutrient trading within a specified water body may offer a more market-driven compliance system that increases cost efficiency for treating NPS pollution. Lastly, environmental policy and concerns may weaken the farm boundary as more holistic environmental management objectives are accepted over watersheds and other large geographic areas.

REFERENCES

Bartfield, E. *Point/Nonpoint Source Trading: Looking Beyond Potential Costs Savings,* Env. Law 23, in press.

Bosch, D.J., Batie, S.S., and Carpentier, L.C., The value of information for targeting water quality protection programs within watersheds, in *Economic Issues Associated With Nutrient Management Policy,* Proc. SRIEG Group-10. No. 32. SRDC No. 180.93-107, 1994.

Cabe, R., and Herriges, J.A., The regulation of nonpoint source pollution under imperfect and asymmetric information, *J. Environ. Econom. Management,* 20, 113, 1991.

Crowder, B., and Young, C.E., *Managing Farm Nutrients: Tradeoffs for Surface- and Groundwater Quality,* USDA Agric. Econ. Rep. no. 583, U.S. Gov. Print. Office, Washington, D.C.

Difiore, R.S., *The Phosphate Detergent Ban and its Impacts on Wastewater Treatment Plants in North Carolina,* NC AWWA/WPCA conference, Nov. 1988.

Dudek, D. and Palmisano J., Emissions trading: Why is this thoroughbred hobbled? *Columbia J. Environmental Law,* 13, 217, 1988.

Hahn, R. and Hester, G., Marketable permits: Lessons for theory and practice, *Ecology Law Quarterly,* 16, 361, 1989.

Hahn R., and Stavins, R.N., Incentive-based environmental regulation: A new era from an old idea, *Ecology Law Quarterly,* 18, 1, 1991.

Jacobson, E.M., Hoag, D.L., and Danielson, L.E., The theory and practice of pollution credit trading in water quality, in *Economic Issues Associated With Nutrient Management Policy,* Proc. SRIEG Group-10. No. 32, SRDC No. 180.69-92, 1994.

Keeney, R.L., and Raiffa, H., *Decisions with Multiple Objectives: Preferences and Value Tradeoffs,* Wiley, New York, 1976.

Lemunyon, J.L., and Gilbert, R.G., The concept and need for a phosphorus assessment a tool, *J. Prod. Agric.,* 6(4), 483, 1993.

Lynch, R., Inter-state nutrient management, in *Economic Issues Associated with Nutrient Management Policy,* Proc. SRIEG Group-10, No. 32, SRDC No. 180, 43-52, 1994.

Montgomery, W.D., Markets in licenses and efficient pollution control programs, *Econom. Theory,* 5, 395, 1972.

Nonpoint Source Evaluation Panel, *Report and Recommendations of the Nonpoint Source Evaluation Panel,* CBP/TRS 56/91, Washington, D.C.: Chesapeake Bay Program U.S. Environmental Protection Agency, 1990.

Park, W.M., and Sawyer, D.G., *Targeting Soil Erosion Control Efforts in a Critical Watershed,* ERS Staff Report No. AGES850801, Washington, D.C., Natural Resource Economic Division, Economic research service, U.S. Department of Agriculture, December 1985.

Rader, D.N., Nutrient trading as a management option: The Tar-Pamlico experiment, in *Economic Issues Associated With Nutrient Management Policy,* Proc. SRIEG Group-10, No.32, SRDC No. 180, 53-67, 1994.

Reckhow, K.H., and Stow, C., Ecological impacts of excess nutrients in the environment: Issues, management, and decision making, in *Economic Issues Associated With Nutrient Management Policy,* Proc. SRIEG Group-10, No. 32, SRDC No. 180, 5-19, 1994.

Scientific and Technical Advisory Committee (STAC), *Nutrient Control in the Chesapeake Bay, 1992,* 1986.

Segerson, K., Uncertainty and incentives for nonpoint source pollution control, *J. Environ. Econom. Management,* 15, 87, 1988.

Stavins, R.N., and Whitehead, B.W., Dealing with pollution: Market-based incentives for environmental protection, *Environment,* 31, 7, 1992.

US Environmental Protection Agency, Office of Water, *National Water Quality Inventory, 1990 Report to Congress,* Washington, DC, EPA 503/9-92/006, 1992a.

US Environmental Protection Agency, *Incentive Analysis for Clean Water Act Reauthorization: Point Source/ Nonpoint Source Trading for Nutrient Discharge Reduction,* Office of Water and Office Policy, Planning and Evaluation, 1992b.

US Environmental Protection Agency, *A Watershed Approach,* Office of Water, U.S. Print. Office, Wash. D.C., 1993.

von Winterfieldt, D., and Edwards, W., *Decision Analysis and Behavioral Research,* Cambridge University Press, Cambridge UK, 1986.

Water Quality 2000, *A National Water Agenda for the 21st Century,* Water Quality 2000, Alexandria, Virginia, 1992.

Xepapadeas, A.P., Environmental policy under imperfect information: Incentives and moral hazard, *J. Environ. Econom. Management,* 10, 113, 1991.

Zinn, J.A., Nutrient management: The congressional setting, in *Economic Issues Associated With Nutrient Management Policy,* Proc. SRIEG Group-10, No. 32, SRDC No. 180, 27-35, 1994.

Zublena, J.P., and Barker, J.C., Livestock and poultry manure nutrient assessment and distribution in North Carolina, *Soil Sci. Soc. N.C. Proc.*, 35, 46, 1992.

ANIMAL WASTE MANAGEMENT AND EDGE OF FIELD LOSSES

Robert L. Mikkelsen and J. Wendell Gilliam

INTRODUCTION

L and application of animal wastes is an ancient technique used to recycle nutrients, improve the soil physical properties, and enhance the fertility of agricultural soils (Parr and Hornick, 1992). However, with the continuing increase in animal production, there are growing concerns about the safe and appropriate handling and disposal of the associated animal waste products. When animal wastes are applied in excess of the capacity of the land to receive them, components of the waste (such as N, P, C, and pathogens) may be detrimental for crop production and water quality. To protect agricultural productivity and surrounding water bodies, it is important that animal wastes be managed properly.

The management of animal wastes applied to soil is more complex than is the management of traditional inorganic fertilizers. With inorganic fertilizers, the chemistry and reaction are much better understood and thus more predictable, and the nutrient composition and content are guaranteed by law. In contrast, the nutrient content of wastes is extremely variable, and not all nutrients present in wastes are immediately available to the plant for uptake. Factors such as the method and rate of application, soil properties, cropping patterns, and season of the year are all important considerations in determining the most efficient and environmentally appropriate use of animal wastes. Similarly, the rate of biological decomposition and nutrient release from organic wastes is governed by complex interactions of environmental factors such as soil pH, temperature, and moisture status, which may fluctuate throughout the year.

When manures are applied to soil at agronomically acceptable rates, they serve as an excellent source of plant nutrients. However, due to the inherent variability of wastes and the uncertainty associated with the nutrient release rate, it is often difficult to apply exactly the amount of each nutrient required to meet plant demands. When over-application occurs, undesirable impacts may occur as the manure moves off the receiving site. Losses may occur via gas emissions (e.g., NH_3 volatilization and denitrification), leaching, or lateral surface and sub-surface flow from the site. Excessive losses of N and P have been linked to potential eutrophication of surface water and accompanying problems. Pathogenic organisms may move from the receiving site with surface runoff water. Leaching losses of nutrients may serve to degrade the quality of ground water.

One technique for reducing the pollution potential from surface water leaving manure-receiving sites is to establish buffer strips of vegetation between the potential source of pollution and the adjacent water body. These buffer zones generally consist of a permanently vegetated area (generally >5 m wide) that is managed separately from the remainder of the field and dedicated for the primary use as a nutrient and sediment filter. The vegetated filter, consisting of trees, grass, or wetland plants, can serve as a chemical and physical barrier between the agricultural field and the body of water.

This chapter will review the production and chemical composition of animal manures and the potential use of vegetated filter strips (VFSs) for minimizing off-site impacts due to land application of animal manures.

PRODUCTION AND NUTRIENT CONTENT OF ANIMAL WASTES

Animal production continues to increase in the US, with a particularly rapid increase in poultry and swine production recently. Different animals produce waste with greatly different chemical and physical properties. For example, production of broilers and turkeys may result in the generation of litter, a combination of manure and bedding material. Swine production commonly results in the generation of anaerobic lagoon waste that is disposed of as a dilute liquid. Dairy cow waste is commonly handled as a slurry or kept dry. There are regional concentrations of animal production in the US where specific types of animal waste are of primary concern (Figure 1).

Animal wastes typically contain nutrients that are valuable for plant growth, but the concentration of nutrients can be extremely variable (Table 1). Although the wastes contain many essential plant nutrients, they are typically applied to land at a rate intended to meet the agronomic N requirement of the crops growing on the receiving site. However, due to the ratio of N to P in most manures, application of sufficient N to meet current crop demand will generally result in an application of P that exceeds crop requirements. In soils with a long history of manure application, soil P concentrations are typically quite high (Sims and Wolf, 1994).

Table 1. Typical daily production and chemical characteristics of fresh manures (based on 1000 kg live animal mass per day; NCSU, 1990).

Category	Dairy	Beef	Swine	Layer	Broiler	Turkey
	----------------------------kg/day----------------------------					
Typical Mature Weight	640	360	61	1.8	0.9	6.8
Total Manure	86	58	84	64	85	47
Total Solids	12	8.5	11	16	22	12
Total N	0.45	0.34	0.52	0.84	1.1	0.62
Total P	0.09	0.18	0.34	0.61	2.80	0.51

Various approaches have been used to predict the fraction of N that will mineralize and become available for plant uptake during the first year after application; however, it is difficult to estimate this accurately. Since the plant availability of N from all organic nutrient sources is the sum of the initial inorganic N and mineralized N minus loss (e.g., NH_3 volatilization, immobilization, denitrification, leaching, and runoff), there is considerable year-to-year variation as well as variation between manure sources. Because of the inherent variability found in both manures and the environment, there is no consensus on what mathematical model or coefficients should be used to predict N release from various manures, but general regional guidelines are used.

The method of animal waste application also plays an important role in determining the fate of the manure-derived nutrients. It is commonly found that application methods that result in placement of the manure beneath the soil surface result in much lower NH_3 volatilization losses. Similarly, runoff

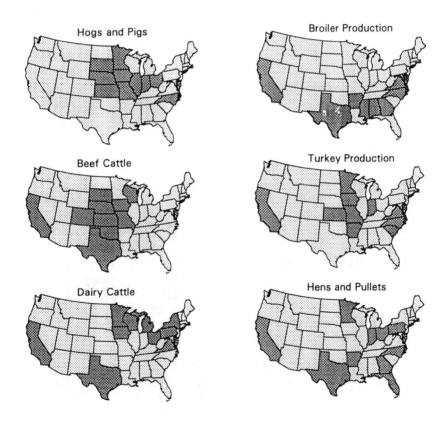

Figure. 1. Top ten animal producing states in the United States (US Census of Agriculture, 1993).

water from sites receiving surface applications of waste generally contain higher concentrations of N, P, and C compared with water from sites that received wastes that were incorporated into the soil.

Animal wastes contain significant amounts of P, since approximately 70-90% of P ingested by farm animals is excreted (Church, 1979). Phosphorus in manure is found in both organic and inorganic compounds. The organic fraction of manure P can account for as much as two-thirds of the total P found in animal waste (Figure 2). The availability of manure P for plant nutrition depends on the mineralization of organic P compounds and the specific adsorption characteristics of the soil.

Many experiments have been conducted to compare the use of manure P with inorganic P fertilizer for crop production; however, the results have not always been consistent. Several researchers have reported that manure is not as effective as inorganic fertilizer as a P source. For example, Gracey (1984) reported that neither swine, cattle, nor sheep manure was as effective as monoammonium phosphate at increasing grass yields or increasing soil-extractable P when applied at the same rate. Similarly, Gunary (1968) noted that P in sheep manure was only about 70% as available as the same amount of P from superphosphate fertilizer when they were incorporated into the soil. Additionally, when animal manure remains on the soil surface, its immediate value as a P source may be even more limited by a lack of contact and reaction with the soil below.

Numerous researchers have reported that animal manures are equivalent or superior to inorganic P sources for plant nutrition. Jensen (1991) reported

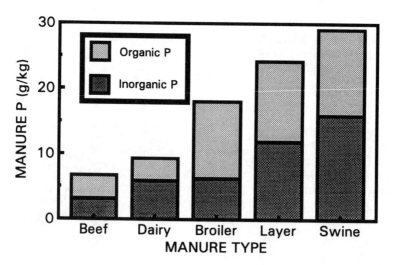

Figure 2. Ratio of organic and inorganic phosphorus (P) compounds present in animal manure (Barnett, 1994).

that in a field experiment, application of cattle slurry as the P source resulted in increased crop yields compared with monoammonium phosphate. Similar results have been reported by Sharma et al. (1980) who found higher crop yields and plant P recovery following manure application compared with an equivalent superphosphate application. Goss and Stewart (1979) reported that cattle manure served as an excellent P source for alfalfa and maintained adequate plant P concentrations for a longer period than an application of superphosphate.

It is difficult to explain the reasons for the differences in effectiveness between organic and inorganic P sources, but it must be remembered that application of animal manures may greatly change the physical and chemical condition of the soil. It has been demonstrated that the soil micro-environment surrounding the manure is greatly different than the surrounding bulk soil. Factors such as the soil atmosphere, pH, microbial population, texture, organic matter content, and water holding capacity may all be changed following manure application (e.g., Amberger and Amann, 1984).

Animal manures applied to soil may be a potential source of pathogens and parasites. Heavy loads of fecal pollution are quite common from outdoor feedlots where storm runoff may be equivalent to the discharge of raw sewage from a city of 10,000 inhabitants (Geldreich, 1990). The severity of some diseases may be increased following manure application (Osunlaja, 1990 a,b). Land application of manure is particularly associated with *Salmonella* and *Escherichia coli*, but other organisms such as *Bacillus anthracis*, *Mycobacterium tuberculosis*, *Clostridia* spp. and *Leptospira* spp., can survive and be spread in manure (Sainsbury, 1983). Baxter-Potter and Gilliland (1988) reported that runoff water from agricultural land regularly exceeds water quality standards regardless of management practices. Since bacteria are not transported appreciably through soils, overland flow is the primary mechanism for movement and dispersion. The population of fecal organisms generally decreases rapidly as manure is heated, dried, and exposed to sunlight on the soil surface (Crane et al., 1980).

VEGETATED FILTER STRIPS

Many studies have shown that VFSs can potentially remove N, P, and sediment from water running off of agricultural fields. This filtering mechanism has important implications for water quality and land use management. Nitrogen removal from runoff water reduces the potential for problems associated with eutrophication and exceeding the water quality standards. Phosphorus removal from runoff is receiving renewed attention because of its role in aquatic productivity and eutrophication. Sediment pollution of waterways is also a major water quality concern.

Vegetated filter strips are effective in improving water quality via two major mechanisms: 1) treatment of surface runoff and 2) treatment of sub-surface drainage.

SURFACE LOSSES FROM MANURE

Surface runoff from agricultural fields occurs when the rainfall intensity exceeds the natural infiltration rate of the soil or when the soil surface becomes saturated and the rainfall intensity exceeds the hydraulic conductivity of the soil. Surface runoff may also occur from compacted or disturbed soils without intense rainfall if the infiltration capacity has been reduced (e.g., aggregate destruction, animal paths, wheel traffic, etc.). Many studies have reported that transport of nutrients and sediment can be quite high during storm events, resulting in surface runoff. For example, Roberts (1987) reported high NO_3 losses in surface runoff in storms following fertilizer application. Phosphorus losses are also common in runoff water in association with sediment loss. Likewise, herbicide losses can be high in surface runoff in storms following application (Bengtson et al., 1990).

In many regions in which animal production occurs, waste products are commonly applied to agricultural fields for manure disposal in order to keep waste products from accumulating near the animals and to utilize the manure-derived nutrients for crop production. Much of this waste is applied directly to the soil surface, leaving it potentially vulnerable to surface losses. The potential loss of N, P, C, and microorganisms from the surface of soil receiving animal manures has been frequently demonstrated (Khaleel et al., 1980). However, the effects of storm intensity, antecedent soil properties, manure characteristics, and subsequent transformations on runoff losses have not received adequate attention.

Surface-applied manure is subject to chemical and physical changes following application, although the changes are generally slower than with soil-incorporated manures. Most manure contains considerable amounts of NH_4 or NH_4-forming compounds, such as urea or uric acid. Ammonia volatilization can result in large N losses when manure remains on the soil surface. Many P compounds present in manure are initially water soluble and have the potential to move with rain or irrigation water (Gerritse, 1977). As the manure ages, the organic P will hydrolyze to inorganic phosphate-containing compounds. In the inorganic ortho-phosphate form, P is strongly adsorbed by soil and is less susceptible to direct movement with water.

When manure is applied to soil with sub-surface injection or incorporation following addition, nutrient losses via gaseous emission and runoff losses are greatly reduced compared with surface applications. Water running from surface-application sites generally contains higher concentrations of N, P, C, and pathogens than water from sites where manure has been incorporated into the soil (Mueller et al., 1984).

SUB-SURFACE LOSSES FROM MANURE

Sub-surface leaching of nutrients following manure application can result in large losses and potential contamination of ground water (Simpson, 1991).

Nitrogen is the nutrient by which manure application rates are most commonly based and is generally the plant nutrient in the highest concentration in animal wastes. Following mineralization and nitrification of the manure N, it is susceptible to leaching losses in well-drained soils. Nitrate leaching losses may be accelerated in coarse-textured soils or in a highly aggregated soil. When manure-derived N is applied in excess of crop removal, there is further potential for N leaching losses. In soils prone to wetness, NO_3 may be lost via denitrification to the atmosphere.

Much of the manure P is present initially in organic compounds. These organic P compounds are generally more water soluble, and they are more subject to leaching than inorganic phosphate (Gerritse, 1977). Until the organic P compounds are adsorbed or enzymatically hydrolyzed, they may be subject to leaching and runoff as soluble P compounds.

VEGETATED FILTER STRIPS TO REDUCE MANURE LOSSES

Vegetated filter strips can be effective in reducing losses of nutrients and sediment leaving agricultural fields that have received either animal manures or inorganic fertilizers. Many of these VFSs consist of natural, unmanaged riparian vegetation or forest surrounding cleared agricultural land. Other VFSs are commonly established by planting grass or other vegetation (such as trees) on the down-hill slope from the field as a planned management practice to improve water quality. Many studies have demonstrated that VFSs can effectively remove N, P, and sediment from runoff water leaving cropland (Muscutt et al., 1993). Natural, pre-existing VFSs and riparian vegetation are already serving as an important line of protection for water quality (Lowrance et al., 1985). In this role, VFSs are effective in removing dissolved nutrients in runoff through mechanical, chemical and biological reactions.

NITROGEN

The use of VFS to remove N from runoff water has been quite successful; however, the mechanism of N removal is often quite complex. When animal wastes are surface-applied on soil, much of the N may be initially present in organic compounds. When intense storms follow shortly after surface applications of manure, the majority of N loss from treated soil is in the form of organic N compounds (Edwards and Daniel, 1993). Losses as organic N will decrease in importance as the manure ages and the N-containing organic compounds are mineralized to NH_4 and subsequently nitrified. Therefore, to be effective, VFSs must be able to trap nutrients in various forms and in various chemical states throughout the year.

The NO_3 concentration in water leaving the field edge is reduced through three major mechanisms: plant uptake, denitrification, and sediment trapping. The relative importance of these mechanisms will change depending on the source of N, the characteristics of the VFS, and the surrounding hydrogeology.

Researchers often report a reduction in N concentration of surface or sub-surface water as it passes through a VFS without reporting the primary mechanism responsible for the N removal. Although documentation of VFS effectiveness is important, the causal factors must also be understood in order to appropriately use these buffer areas for protecting water quality.

Many researchers have noted that plant uptake in the VFS is extremely effective in removing NO_3 leaving the field edge (e.g., Haycock and Pinay, 1993). Muscutt et al. (1993) summarized much of the recent work on N removal via VFS and reported a consistent reduction of NO_3 concentration to <1 mg NO_3-N/L in the water leaving the VFS. Lowrance (1992) concluded that N accumulation by vegetation is the most significant N removal process in many circumstances. However, plant uptake should be viewed as a temporary storage of acquired nutrients unless the plant is ultimately harvested and removed from the site. Although effective at nutrient removal, VFSs consisting of annual plants such as grasses will die and subsequently release the nutrients following decomposition. Similarly, a portion of the nutrients stored in woody perennial plants will be redeposited back to the soil.

Plant uptake alone cannot account for all of the N removed by many VFSs. For example, Peterjohn and Correll (1984) estimated that plant uptake could only account for 33% of the N removed from surface and sub-surface water in a VFS. In these cases, denitrification may also play an important role in N removal from water. Jacobs and Gilliam (1985) reported that NO_3 removal from a VFS continued during the winter, when plant uptake was minimal. The presence of a high water table and abundant organic matter resulted in conditions favorable for denitrification during much of the year.

Denitrification occurs primarily in soils that are anaerobic or contain anaerobic microsites. In a VFS, this generally results from decreased O_2 diffusion into soil as the water content increases. The shortage of O_2 causes denitrifying bacteria to utilized NO_3 as an alternative electron acceptor during respiration. Oxygen is commonly limited in VFSs since they are frequently saturated and contain considerable organic matter as a result of slow mineralization because of the wet conditions. Carbon is continually being added due to the presence of perennial vegetation (Groffman et al., 1992).

The relative importance of plant uptake and denitrification on NO_3 removal from VFS depends on the site characteristics and the season of the year. In soil environments with a high water table and poor internal drainage, denitrification may account for a considerable amount of the NO_3 removed from the runoff water, especially during the dormant season (Jordan et al., 1993). However, when plants are actively growing, the dominant pathway for NO_3 removal may shift, and plant uptake may become the major mechanism for N removal from runoff (Groffman et al., 1992). However, even when the vegetation is not actively growing, the presence of plant debris is still valuable by providing the organic C substrate required for denitrification.

The significance of these two removal mechanisms appears to be quite specific for a given site and season. For example, Clausen et al. (1993) reported that in a 35-m riparian zone planted to maize, neither plant uptake nor denitrification accounted for significant N removal from ground water or surface water. However, Haycock and Pinay (1993) reported significant amounts of NO_3 removal from sub-surface field runoff within the first few meters of their forested VFS.

A major pathway for N movement through a VFS may occur via sub-surface routes. The flow of ground water and the denitrification potential may thereby determine N movement and loss through this route (Jacobs and Gilliam, 1985). For example, the longer the transport time through the sub-surface zone, the higher the denitrification losses may be. Warwick and Hill (1988) found only relatively low denitrification potential in some riparian soils and suggested that a long retention time within this zone was critical for denitrification to be a major pathway for NO_3 removal. Similarly, microbial activity is generally greatest in the surface layer of the soil, so denitrification is most rapid when ground water inundates this zone (Groffman et al., 1992).

Additional research is clearly needed to determine the relative importance of these two pathways of N removal from surface runoff and the environmental factors that govern them. When the N transport and removal processes are better understood, it will be possible to manipulate the conditions to achieve the desired water quality. However, there is ample evidence to show that VFSs are generally effective in N removal from surface and sub-surface water.

SEDIMENT CONTROL

Surface runoff from agricultural land is commonly accompanied by sediment transport. Sedimentation of downstream rivers and lakes may pose a serious form of water pollution. Depending on the landscape characteristics, the storm intensity, and the soil properties, sediment loss from fields may range from minimal to extremely high. Runoff losses may be particularly high when the soil is bare and recently tilled.

The effectiveness of VFSs at sediment removal is well documented. For example, Chescheir et al. (1991) measured sediment removal from agricultural drainage and found that approximately 90% of the sediment was removed after passing through a forested wetland. Similarly, Cooper et al. (1987) used ^{137}Cs tracers to determine that 80% of the sediment leaving agricultural fields in North Carolina was trapped within a wooded VFS. However, Dillaha et al. (1989) warned that the ability of a VFS to remove sediment may decrease over time. In a short-term study, a monitored VFS was able to remove 90% of the sediment during the early stages of measurement. Later in the experiment, the VFS was only able to remove 5% of the sediment, due to sediment inundation of the VFS. They also warned that sediment removal is effective only if the runoff is spread out and not channelized. Dickey and Vanderholm (1981) also reported a sharp drop in sediment removal when shallow overland flow was compared with channelized flow through a VFS.

In addition to problems related to sediment losses directly, losses of organic and inorganic chemicals bound to the surface of the sediment may also be an important source of water contamination. Losses of sediment-bound P may account for as much as 70% of P losses (e.g., Sharpley and Smith, 1990). The relative proportions of sediment-bound P and water-soluble P will vary with soil properties, the specific storm events, and the season of the year (Ng et al.,1993), and their removal via VFS will also vary. Sediment-bound N can also be a significant source of N loss in some circumstances (Hubbard et al., 1982). Manure solids may also be lost directly via runoff, with or without sediment losses.

The use of VFSs for reduction of sediment losses is quite cost effective in many circumstances when compared with other watershed management practices to control sediment loss (Pritchard et al., 1993). Vegetative filter strips are an approved management practice in many parts of the world for protection of water quality.

Future work with removal of manure constituents via VFS must include expanded use of modelling. Selecting appropriate sites for manure application should by approached systematically using Geographical Information Systems (e.g., Xu et al., 1993). Similarly, design and construction of VFS can be optimized for specific purposes with computer modelling (Edwards et al., 1994; Yoon et al., 1994). The use of models will aid in understanding the function of VFSs in minimizing losses from manure-amended land.

REFERENCES

Amberger, A., and Amann, C., Wirkugnen organischer Substanzen auf Boden- und Dungerphospat. *Z. Pflanzenernaehr. Bodenk.*, 147, 60, 1984.

Barnett, G.M., Phosphorus forms in animal manure, *Bioresource Technol.,* (in press), 1994.

Baxter-Potter, W.R., and Gilliland, M.W., Bacterial pollution in runoff from agricultural lands, *J. Environ. Qual.,* 17, 27, 1988.

Bengtson, R.L., Carter, C.E., Morris, H.F., and Bartkiewicz, S.A., The influence of subsurface drainage practices on herbicide losses, *Trans. ASAE,* 33, 415, 1990.

Chescheir, G.M., Gilliam, J.W., Skaggs, R.W., and Broadhead, R.G., Nutrient and sediment removal in forest wetlands receiving pumped agricultural drainage, *Wetland,* 11, 87, 1991.

Church, D.C., Digestive physiology and nutrition of ruminants, *Nutrition,* 2, 62, 1979.

Clausen, J.C., Wayland, K.G., Saldi, K.A., and Guillard, K., Movement of nitrogen through an agricultural riparian zone: 1. Field studies, *Water Sci. Tech.,* 28, 605, 1993.

Cooper, J.R., Gilliam, J.W., Daniels, R.B., and Robarge, W.P., Riparian areas as filters for agricultural sediment, *Soil Sci. Soc. Amer. J.,* 51, 416, 1987.

Crane, S.R., Westerman, P.W., and Overcash, M.R., Die-off of fecal organisms following land application of poultry manure, *J. Environ Qual.,* 9, 531,1980.

Dickey, E.D., and Vanderholm, D.H., Vegetative filter treatment of livestock feedlot runoff, *J. Environ. Qual.,* 10, 279, 1981.

Dillaha, T.A., Reneau, R.B., Mostaghimi, S., and Lee, D., Vegetative filter strips for agricultural nonpoint source pollution control, *Trans ASAE,* 32, 513, 1989.

Edwards, D.R., Benson, V.W., Williams, J.R., Daniel, T.C., Lemunyon, J., and Gilbert, R.G., Use of the EPIC model to predict runoff transport of surface-applied inorganic fertilizer and poultry manure constituents, *Trans ASAE,* 37, 403, 1994.

Edwards, D.R., and Daniel, T.C., Environmental impacts of on-farm poultry waste disposal--A review, *Bioresource Technol.,* 41, 9, 1992.

Edwards, D.R., and Daniel, T.C., Effects of poultry litter application rate and rainfall intensity on quality of runoff from fescuegrass plots, *J. Environ. Qual.,* 22, 361, 1993.

Geldreich, E.E., Microbiological quality of source waters for water supply, in *Drinking Water Microbiology,* G.A. McFeters, ed., Springer Verlag, New York, 1990.

Gerritse, R.G., Phosphorus compounds in pig slurry and their retention in the soil, in *Utilization of Manure by Land Spreading,* Commission European Commun., London, 1977.

Goss, D.W., and Stewart B.A., Efficiency of phosphorus utilization by alfalfa from manure and superphosphate, *Soil Sci. Soc. Amer. J.,* 43, 523, 1979.

Gracey, H.I., Availability of phosphorus in organic manures compared with monoammonium phosphate, *Agric. Wastes,* 11, 133, 1984.

Groffman, P.M., Gold, A.J., and Simmons, R.C., Nitrate dynamics in riparian forests: Microbial studies, *J. Environ. Qual.,* 21, 666, 1992.

Gunary, D., The availability of phosphate in sheep dung, *J. Agric. Sci.,* 70, 33, 1968,

Haycock, N.E., and Pinay, G., Groundwater nitrate dynamics in grass and poplar vegetated riparian buffer strips during the winter, *J. Environ. Qual.,* 22, 273, 1993.

Hubbard, R.K., Erickson, A.E., Ellis, D.G. and Wolcott, A.R., Movement of diffuse source pollutants in small agricultural watersheds-Great Lake basin, *J. Envir. Qual.,* 11, 117, 1982.

Jacobs, T.C., and Gilliam, J.W., Riparian losses of nitrate from agricultural drainage waters, *J. Environ. Qual.,* 14, 472, 1985.

Jensen, I., The after-effect of P from cattle slurry and superphosphate on yield and nutrient uptake in sugar beets, *Acta Agric. Scand.,* 41, 259, 1991.

Jordan, T.E., Correll, D.L., and Weller, D.E., Nutrient interception by a riparian forest receiving inputs from adjacent cropland, *J. Environ. Qual.,* 22, 467, 1993.

Khaleel, R., Reddy, K.R., and Overcash, M.R., Transport of potential pollutants in runoff water from land areas receiving animal wastes: A review, *Water Res.,* 14, 421, 1980.

Lowrance, R., Leonard, R. and Sheridan, J., Managing riparian ecosystems to control nonpoint pollution., *Soil Water Conserv.,* 40, 87, 1985.

Lowrance, R., Groundwater nitrate and denitrification in a Coastal Plain riparian forest, *J. Environ Qual.,* 21, 401, 1992.

Mueller, D.H., Wendt, R.C. and Daniel, T.C., Soil and water losses as affected by tillage and manure application, *Soil Sci. Soc. Amer. J.,* 48, 896, 1984.

Muscutt, A.D., Harris, G.L., Bailey, S.W., and Davies, D.B., Buffer zones to improve water quality: A review of their potential use in UK agriculture, *Agric. Ecosystems and Enviro.,* 45, 59, 1993.

NCSU (North Carolina State University), *Livestock Waste Characteristics,* Biological and Agricultural Engineering Department, Raleigh, North Carolina, 1990.

Ng, H.Y.F., Mayer, T., and Marsalek, J., Phosphorus transport in runoff from a small agricultural watershed, *Water Sci. Tech.,* 28, 3, 1993.

Osunlaja, S.O., Effect of organic soil amendments on the incidence of stalkrot of maize, *Plant Soil*, 127, 237, 1990a.

Osunlaja, S.O., Effect of organic soil amendments on the incidence of stalk rot of maize caused by *Macrophomina phaseolina* and *Fusarium moniliforme*, *J. Basic Micro.*, 30, 753, 1990b.

Parr, J.F., and Hornick, S.B., 1992. Agricultural use of organic amendments: A historical perspective, *Amer. J. Altern. Agric.*, 7, 181, 1992.

Peterjohn, W.T., and Correll, D.L., Nutrient dynamics in an agricultural watershed: Observations on the role of a riparian forest, *Ecology*, 65, 1466, 1984.

Pritchard, T.W., Lee, J.G., and Engel, B.A., Reducing agricultural sediment: An economic analysis of filter strips versus micro-targeting, *Water Sci. Tech.*, 28, 561, 1993.

Roberts, G., Nitrogen inputs and outputs in a small agricultural catchment in the eastern part of the United Kingdom, *Soil Use Manage.*, 3, 148, 1987.

Sainsbury, D., *Animal Health: Disease and Welfare of Farm Livestock*, Collins, London, 1983.

Sims, J.T., and Wolf, D.C., Poultry waste management: Agricultural and environmental issues, *Adv. Agron*, 52, 1, 1994.

Simpson, T.W., Agronomic use of poultry industry waste, *Poultry Sci.*, 70, 1126, 1991.

Sharma, R.C., Grewal, J.S., and Singh, M., Effects of annual and biannual applications of phosphorus and potassium fertilizer and farmyard manure on yield of potato tubers, on nutrient uptake and on soil properties, *J. Agric. Sci. Camb.*, 94, 533, 1980.

Sharpley, A.N., and Smith, S.J., Phosphorus transport in agricultural runoff: The role of soil erosion, in *Soil Erosion on Agricultural Land*, J. Bordman et al., eds., London, 1990.

US Census of Agriculture, *Annual Agricultural Statistics*, US Dep. Agric. Washington, D.C., 1993.

Warwick, J., and Hill, A.R., Nitrate depletion in the riparian zone of a small woodland stream, *Hydrobiologia*, 157, 231, 1988.

Yoon, K.S., Yoo, K.H., Wood, C.W., and Hall, B.M., Application of GLEAMS to predict nutrient losses from land application of poultry litter, *Trans ASAE*, 37, 453, 1994.

Xu, F., Prato, T., and Fulcher, C., Broiler litter application to land in an agricultural watershed: A GIS approach, *Water Sci. Technol.*, 28, 111, 1993.

NITRATE LOSS VIA GROUND WATER FLOW, COASTAL SUSSEX COUNTY, DELAWARE

A. Scott Andres

INTRODUCTION

Coastal Sussex County, Delaware (Figure 1), is the focus of scientific interest and political conflict related to nitrate contamination of water. The political conflict relates to assessment of liability and paying for changing long-established agricultural and waste water disposal practices. Until about 1970, land use was almost entirely agricultural. The area is the birthplace of the modern poultry industry and a part of the number one poultry producing county in the country. Production has been steadily increasing over the past decade with over 220 million broilers produced in 1991 (Delmarva Poultry Ind., written commun.).

Over the past 20 years, significant suburban residential and resort development has occurred around the ocean and bay shores, largely over former agricultural lands. Recent evaluations estimate land use to be about 43% agricultural, 45% forest and wetlands, and 11% residential (Ritter, 1986). Much of the area still relies on on-site sewage disposal, although the most intensely developed areas now have modern sewage treatment and disposal facilities.

In the 1970s, researchers at the University of Delaware began to systematically study the occurrence of nitrate contamination of ground and surface waters in Delaware and its relationships to soils, land use, and agricultural practices (Robertson, 1977; Ritter, 1986; Ritter and Chirnside, 1982; Andres, 1991, 1992, 1993; Ullman et al., 1993). Over the past 10 years, the US Geological Survey also has conducted studies of ground water and surface water quality and the relationships between land use and water quality (Bachman, 1984; Denver, 1986, 1989; Hamilton et al., 1993). To date, over 3000 ground water samples have been collected by researchers in support of nitrate studies. In addition, the Delaware Department of Natural Resources and Environmental Control has been conducting a surface water sampling program since 1972 to comply with the Clean Water Act. This work has documented significant occurrences of nitrate contamination of ground and surface waters, including two coastal lagoons, Rehoboth and Indian River bays (Figure 1). Total nitrogen concentrations in the meso- to polyhaline portion of upper Indian River Bay

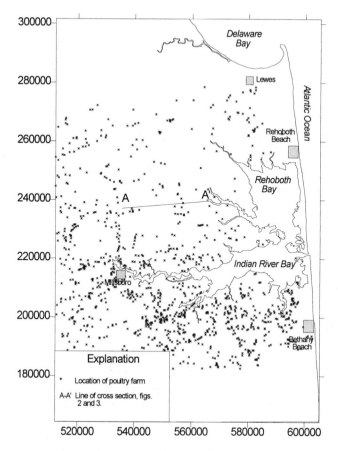

Figure 1. Map of coastal Sussex County, Delaware. Coordinates are Delaware state plane.

range from non-detectable to about 15 mg/L with a long-term mean of about 2 mg/L (Weston, Inc., 1993). Almost all of the nitrogen is in the form of nitrate. Ground and surface water problems in coastal Sussex County are similar to those found over much of the Delmarva Peninsula (Bachman, 1984; Hamilton et al., 1993; Reay et al., 1992).

GROUND WATER

HYDROLOGY

Climate in the area is warm temperate, with long-term average annual precipitation of about 1.14 m (Talley and Andres, 1987). In general, topography is relatively flat, and moderately to excessively well drained soils are common (Ireland and Matthews, 1974). Similar annual and seasonal water budgets have been calculated by both stream flow hydrograph separation and climatic water budget methods; 0.58 to 0.66 m are lost through evapotranspiration, 0.076 to 0.10 m through overland runoff, and 0.38 to 0.40 m through ground water runoff (Johnston, 1976).

Ground water is a very important factor in evaluating most water resources issues in Delaware. On the average, base flow (i.e., fair weather ground water discharge) makes up about 75-80% of fresh water stream flow (Johnston, 1976). As a result, the quality and quantity of ground water discharge have a large effect on the quality and flow of surface water. The primary water bearing unit underlying much of southern Delmarva is the Columbia aquifer (Andres, 1991; Hamilton et al., 1993). In addition to providing water for stream flow, this aquifer is the predominant source of fresh water for all uses. It is composed of quartz sand and gravel with regionally discontinuous clay and silt lenses. The Columbia typically behaves as an unconfined aquifer although semi-confined conditions exist in some areas. The aquifer ranges in thickness from 15-50 m. Measured transmissivities range from 460 to 1100 m^2/d (Talley and Andres, 1987).

Andres (1991) and Hamilton et al. (1993) have reported on the water chemistry in the Columbia aquifer. Water unaffected by man is low in dissolved solids (less than 50 mg/L) and contains less than 1 mg/L NO_3-N. There are two water chemistries in the more permeable parts of the aquifer: a Na-K, Cl-HCO_3 (type I) and a Ca-Mg-Fe, HCO_3 (type II). Type I water always has significant dissolved oxygen (> 5 mg/L) and commonly contains nitrate. Type II water occurs where the geologic units contain appreciable organic carbon, more clay and silt beds, and sulfide minerals. Type II water contains no dissolved oxygen and rarely contains nitrate, even where there are large sources of nitrogen. Both chemical and biological actions are thought to remove NO_3 from type II waters.

The effects of agricultural activities on ground water quality are presented by Denver (1986, 1989) and Andres (1991, 1993). Agricultural activities tend to add dissolved solids in the form of NO_3, SO_4, Cl, Ca, and Mg. Waters affected by agricultural activities tend to have more than 100 mg/L of dissolved solids. Waters contaminated by NO_3 are usually depleted in HCO_3 because the nitrification reaction removes HCO_3.

Ground water flow has been found to be a three-dimensional process on local and regional scales (Andres, 1991; Denver, 1993). Individual ground water flow paths determined from flow-net analyses range from several tens of feet to several miles (Figure 2). Ground-water flow paths extend to the base of the aquifer to depths of 30 m below the water table. Ground-water flow velocities range from a tenth of a meter per day in the upper portions of a watershed to about one meter per day near discharge areas. Ground water residence time, determined from flow-net, and/or tritium and chlorofluorocarbon analyses, can range from months to decades (Andres, 1991; Denver, 1993).

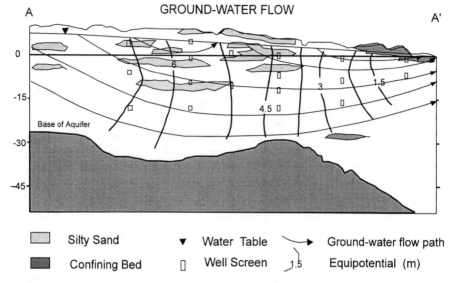

Figure 2. Flow net along cross section A-A' (shown in Figure 1). Horizontal distance approximately 9150 m.

NITRATE CONTAMINATION OF WATER

This discussion focuses on the effects of poultry production on ground water quality. In this discussion, watersheds and sub-watersheds that have had a poultry house density greater than 1/50 ha over the past 35 years are considered to be under intensive poultry production. Although domestic waste water disposal and fertilizer use are known to cause nitrate contamination of ground water, the amount of nitrogen discharged to ground water by these sources is relatively small compared to the amount produced by poultry production.

Analysis of aerial photography and maps has determined that there have been hundreds of poultry houses in the area for the past 40 years (see Figure 1). Over 350 are located within a 9325-ha area around Indian River Bay. Given recent production practices, about 3040 kg/day of manure-derived available (mineralized) nitrogen are generated in this area (Delaware Coop. Ext. Serv., 1989). Given available cropland, this corresponds to an average loading rate of about 270 kg N/ha, an amount that far exceeds the nitrogen needs of crops grown in the area. However, because of a lack of a better means of disposal, almost all of this material is applied to local fields. Even with the oversupply of manure, chemical-based nitrogen fertilizers are commonly used. Sims and Wolf (1993) estimated that between the poultry industry and fertilizer sales, there is a statewide annual excess of 48 kg/ha of nitrogen.

Because of land use, flat topography, well-drained, highly permeable soils, and aquifer characteristics, nitrate contamination is common throughout coastal

Sussex County (Table 1). Based on analyses of samples from 441 wells, Andres (1991) reports that nitrate concentrations in excess of the current US EPA drinking water standard (10 mg/L NO_3-N) occur over large areas and in wells up to 27 m deep. About 20% of all wells in the area have nitrate concentrations in excess of the standard. With time, ground water flow has transported nitrate considerable distances from the source to locations where the overlying land use may be different (Figure 3). Contamination is much more severe in areas with intensive poultry production than elsewhere (Table 1). About one-third of the wells in these areas have nitrate concentrations in excess of the drinking water standard (Andres, 1991).

Much of the ground water nitrate enters the fresh and salty surface waters, although it is possible that a significant portion of the nitrogen is lost or trapped as it passes through bay and stream bottom sediments because of biogeochemical processes. Andres (1992) estimated ranges of potential nitrate loads and

Table 1. Summary statistics of nitrate data (Andres, 1992)

Category	No. of Samples	Mini-mum	Maxi-mum	Mean	Median	Std. Dev.
			---mg/L NO_3-N---			
All Data	433	LD†	72	7.91	5.80	9.16
Non-intensive poultry area	202	LD	18	5.78	5.22	4.16
Intensive poultry area	231	LD	72	9.84	6.48	11.7

†LD = less than detection, 0.5 mg/L

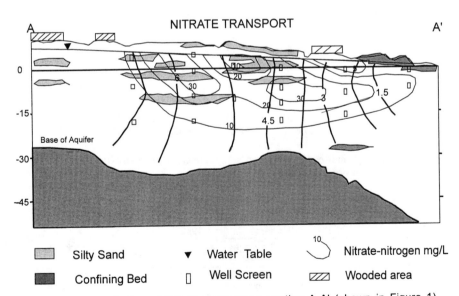

Figure 3. Distribution of NO_3-N along cross section A-A' (shown in Figure 1). Concentrations are in mg/L. Horizontal distance approximately 9150 m.

loading rates to Rehoboth and Indian River bays through the direct discharge of ground water (Table 2, Figure 4). These estimates were derived from a combination of flow-net and water-budget ground-water flux models and geostatistical models of areal nitrate concentrations.

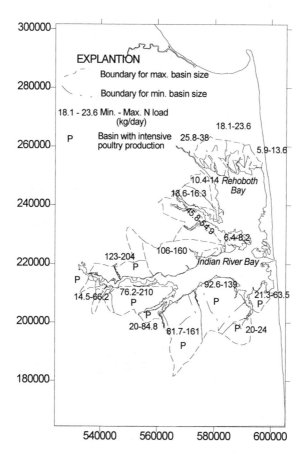

Figure 4. Direct ground water discharge drainage basins and ranges of potential daily N loads. Coordinates are Delaware State Plane in feet.

Table 2. Nitrate-nitrogen loading through direct ground-water discharge (Andres, 1992).

Category	Drainage area	NO$_3$-N load	Loading rate
	ha*1000	kg/d	kg/ha-yr
Total	10.6-15.4	617-1134	20.2-34.6
Non-intensive poultry area	5.26-5.98	245-308	15.0-21.3
Intensive poultry area	5.39-9.38	372-826	24.0-47.7

The ranges of nitrate load and loading rate values in Table 2 and Figure 4 are largely due to uncertainties in the sizes of the sub-watersheds contributing to direct ground-water discharge. Given the ratio of the areas of the total and direct ground-water discharge watersheds, fresh water streams may transport an additional 2900 kg/d of nitrogen into Indian River Bay. The combination of direct and indirect input of nitrate contaminated ground water has resulted in severe eutrophic conditions in portions of Indian River Bay and many fresh water ponds (Ullman et al., 1993). This has been manifested as algal blooms, fish kills, excessive turbidity, low dissolved oxygen concentrations, and other environmental ills (Weston, Inc., 1993). In fact, the upper reaches of Indian River Bay have higher nitrate concentrations than the worst segments of Chesapeake Bay (Weston, Inc., 1993).

CONCLUSIONS

Because of the slow rate of ground water flow and the large volume of contaminated water, significant nitrate contamination of ground water will likely persist for 20 to 40 years even if all nitrate input were to stop today. However, given the recent production history of the poultry industry, it is likely that increasing amounts of excess nitrogen will be generated. Innovative means of using this resource will need to developed if ground water contamination is not to become more severe than present.

Given the causes and characteristics of nitrate contamination of ground water, reductions of the incidence of well contamination and of nitrogen discharge to surface water will not occur in the foreseeable future. This has significant implications to perception of the level of success of land-use controls and agricultural waste-management practices in improving the local environment.

REFERENCES

Andres, A.S., *Results of the Coastal Sussex County, Delaware, Ground-water Quality Survey*, Delaware Geological Survey Rept. of Investigations No. 49, 1991.

Andres, A.S., *Estimate of Nitrate Flux to Rehoboth and Indian River Bays, Delaware*, Delaware Geological Survey Open-File Rpt. No. 35, 1992.

Andres, A.S., Nitrate contamination of ground and surface waters, coastal Sussex County, Delaware, *Hydrologic Science and Technology*, 9, 1993.

Bachman, L.J., *Nitrate in the Columbia Aquifer, Central Delmarva Peninsula, Maryland*, U. S. Geological Survey Water-Resources Investigations Report 84-4322, 1984.

Delaware Cooperative Extension Service, *Poultry Manure Management*, Cooperative Bulletin 24, 1989.

Denver, J.M., *Hydrogeology and Geochemistry of the Unconfined Aquifer, West-central and South-central Delaware*, Delaware Geological Survey Rept. of Investigations No. 41, 1986.

Denver, J.M., *Effects of Agricultural Practices and Septic System Effluent on the Quality of Water in the Unconfined Aquifer in Parts of Eastern Sussex County, Delaware*, Delaware Geological Survey Rept. of Investigations No. 45, 1989.

Denver, J.M., *Herbicides in Shallow Ground Water at Two Agricultural Sites in Delaware*, Delaware Geological Survey Rept. of Investigations No. 51, 1993.

Hamilton, P.A., Denver, J.M., Phillips, P.J., and Shedlock, R.J., *Water-quality Assessment of the Delmarva Peninsula, Delaware, Maryland, and Virginia--Effects of Agricultural Activities on, and Distribution of, Nitrate and Other Inorganic Constituents in the Surficial Aquifer*, US Geological Survey Open-File Rept. 93-40, 1993.

Ireland, W., Jr., and Matthews, E.D., *Soil Survey for Sussex County, Delaware*, US Dept. of Agriculture, 1974.

Johnston, R.H., *Relation of Ground Water to Surface Water in Four Small Basins of the Delaware Coastal Plain*, Delaware Geological Survey Rept. of Investigations No. 24, 1976.

Reay, W.G., Gallagher, D.L., and Simmons, G.M., Groundwater discharge and its impact on surface water quality in a Chesapeake Bay Inlet, *Water Resources Bulletin*, 28, 6, 1992.

Ritter, W.F., Nutrient budgets for the Inland Bays, Newark, Delaware, Univ. of Delaware Agricultural Engineering Dept., 1986.

Ritter, W.F., and Chirnside, A.E.M., *Ground-water quality in selected areas of Kent and Sussex Counties, Delaware*, Newark, Delaware, Univ. of Delaware Agricultural Engineering Dept., 1982.

Robertson, F.N., *The quality and potential problems of the ground water in coastal Sussex County, Delaware*, Newark, Delaware, Univ. of Delaware Water Resources Center, 1977.

Sims, J.T., and Wolf, D.C., Poultry waste management, *Advances in Agronomy*, 52, 1993.

Talley, J.H., and Andres, A.S., *Basic Hydrologic Data for Coastal Sussex County, Delaware*, Delaware Geological Survey, Special Publication 14, 1987.

Ullman, W.J., Geider, R.J., Welch, S.A., Graziano, L.M., and Overman, B., Nutrient fluxes and utilization in Rehoboth and Indian River bays, Univ. of Delaware College of Marine Studies, 1993.

Weston, Inc., *Characterization of the Inland Bays Estuary*, West Chester, Pennsylania, Roy F. Weston, Inc., 1993.

FECAL BACTERIA IN SURFACE RUNOFF FROM POULTRY-MANURED FIELDS

M.S. Coyne and R.L. Blevins

INTRODUCTION

Recent growth in Kentucky's poultry industry (from 1.5 million broilers in 1990 to over 43 million broilers in 1993) (Kentucky Agricultural Statistics, 1994) has created an equally large waste disposal problem. As in many states, most of this waste is land-applied without prior processing (Edwards and Daniel, 1992). Manure spreading, while it is a traditional and effective agricultural waste disposal practice, frequently exceeds the rate at which wastes can be processed in agricultural ecosystems. The subsequent runoff of nutrients and fecal bacteria contributes to agricultural non-point source pollution and helps to degrade drinking water supplies.

Vegetative or grass filter strips are typically established to minimize surface runoff from agricultural fields. They principally reduce sediment runoff (Dillaha et al., 1989). Use of grass filters and grassed waterways to control fecal waste runoff from point sources, such as feedlots, has also been studied (Dickey and Vanderholm, 1981; Schellinger and Clausen, 1992; Young et al., 1980). However, there is little information specifically related to the control of fecal bacteria runoff from agricultural fields treated with poultry manure.

Giddens and Barnett (1980) suggested that if runoff water from manured cropland were allowed to flow over grass, its pollution potential would be greatly reduced. However, modeling studies by Walker et al. (1990) implied that buffer strips alone would not be sufficient to bring bacterial concentrations within acceptable limits, for example, below the EPA standard of 200 fecal coliforms/100 mL for primary contact water (bathing and swimming water). Crane et al. (1983) suggested that even after passing through a vegetative filter, runoff contaminated by fecal wastes would have concentrations of fecal indicator bacteria inexcess of 10^3 to 10^5/100 mL.

Because of the growth of the poultry industry in Kentucky, and because information related to grass filter strip use in controlling surface runoff of bacteria from manured fields is limited, we decided to see whether grass filters were an effective management practice to reduce fecal contamination of surface and ground water via runoff from poultry-manured fields. We began a continuing series of rain simulation studies in 1992 to examine this question. This chapter summarizes some of our results.

METHODS AND MATERIALS

LOCATION AND PLOT DESIGN

The rain simulations were done at the University of Kentucky Agricultural Experiment Station in Lexington during the summers of 1992, 1993, and 1994. The soil is a Maury silt loam (fine, mixed, mesic Typic Paleudalf) with an average natural slope of 9%.

Each rain simulation plot consisted of an erosion strip abutted by a grass filter at its lower edge. In 1992 the erosion strips were 22.1 m long and paired with 9.0-m-long grass filters, and only the erosion strips were rained on. In 1993 and 1994, 18.3-m erosion strips were paired with 4.5-m filter strips. The design change was made to facilitate raining simultaneously on both the erosion strip and the grass filter (Figure 1).

All erosion strips were tilled, and poultry litter was incorporated by chisel plowing followed by disking. Grass filters consisted of a mixture of tall fescue (*Festuca arundinacea* L.) and Kentucky bluegrass (*Poa pratensis* L.) sod and were mowed to a 4.0-cm height in each plot before rain simulations.

PLOT TREATMENT

Poultry litter that was a mixture of manure, sawdust, and shavings from a laying house was used for these studies. In 1992 and 1993 the application rate (wet weight) was 16.5 Mg/ha, and in 1994 the application rate was 10 Mg/ha. Litter was uniformly applied over the surface of the erosion strips and shal-

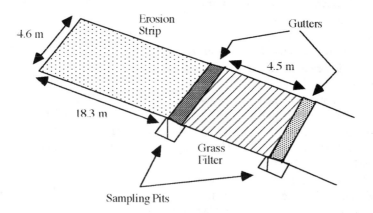

Figure 1. Diagram of a rain simulation plot. Erosion strip and grass filter length varied from year to year.

lowly incorporated (85% incorporation). In 1992 poultry litter was applied to all plots on a single date. In 1993 and 1994 the poultry litter was applied 48 h before each rain simulation.

Rain simulations were at a rate of 6.4 cm/h. This approximates the intensity of a one-in-ten-year storm in central Kentucky, but it was necessary to cause runoff within a reasonable period. In 1994, to improve the reproducibility of runoff, the erosion plots were pre-wet 24 h before rain simulations by raining on them for one hour at a rate of 3.3 cm/h. In addition, in 1994, a second rain simulation was performed on each plot five days after the first simulation.

SAMPLING PROTOCOL

Runoff from the erosion strips started 15 to 30 min after simulated rain began. Runoff from the grass filters began 30 to 45 min after that. Rain simulation continued until runoff from the grass filter was observed for 1 h. Runoff was periodically collected for 10 to 30 sec at 5-min intervals in 10-cm-wide gutters below the erosion strip and grass filter (Figure 1). The gutter below the erosion strip had a slide that directed runoff onto the grass filters or into the gutter for sampling (Fogle and Barfield, 1993).

Runoff samples from the erosion strip and grass filter were uniformly mixed before subsampling and then stored, in sterile plastic bags, on ice in the field and at 4°C in the laboratory to minimize bacterial growth and mortality. Fecal bacteria were enumerated within 24 h of collection. Sediment concentration was determined gravimetrically after representative samples were dried at 105°C.

MICROBIAL ANALYSES

Ten-fold serial dilutions in buffered saline solution (0.85% NaCl) were made of runoff samples to reduce bacterial concentrations to a measurable number. Fecal coliforms (i.e., *Escherichia coli*), fecal streptococci, and *Salmonella* concentrations were determined from manual counts of colony forming units on selective media after spiral plating or membrane filtration. Fecal coliforms were grown on Difco (Detroit, MI) mFC agar incubated at 44°C for 24 h. Fecal streptococci were grown on Difco KFS agar incubated at 37°C for 48 h. *Salmonella* were grown on a medium described by Cox (1993) and incubated at 37°C for 48 h.

CALCULATION OF TRAPPING EFFICIENCY

Trapping efficiency, T_r, of grass filter strips for surface runoff, sediment, and fecal bacteria was estimated using a variation of the trapezoidal rule used for hydrographs and sedigraphs (Barfield and Albrecht, 1982):

$$T_r = (M_i - M_o)/M_i$$

where M_i and M_o are the total liters of water, mass of sediment, or number of fecal bacteria in the inflow and outflow of the grass filter strip. The mass inflow was estimated from:

$$M_i = \sum_{j=1}^{n} C_{ij} \, q_{ij} \, \Delta t_j$$

where C_{ij}, q_{ij}, and Δt_j are the sediment or fecal bacteria concentrations, flow rate, and time interval of the jth measurement of inflow, respectively. M_o was estimated by:

$$M_o = \sum_{j=1}^{n} C_{oj} \, q_{oj} \, \Delta t_j$$

where C_{oj} and q_{oj} are the concentrations and flow rate of the jth measurement of outflow and Δt_j is the time interval of outflow. Concentration and flow were conservatively estimated by the average value of C_j and C_{j+1} or q_j and q_{j+1} for the period during which runoff occurred.

RESULTS AND DISCUSSION

1992 RAIN SIMULATIONS

Fecal bacteria concentrations were typically highest during the initial runoff from erosion strips, and ultimately bacteria concentrations in grass filter runoff exceeded concentrations in erosion strip runoff. Fecal coliform concentrations in grass filter runoff were well above water quality standards for primary contact (200 fecal coliforms/100 mL). Although the grass filters effectively trapped sediment and surface runoff (Table 1), they were less effective at trapping fecal bacteria in plots for which this could be determined.

1993 RAIN SIMULATIONS

Beginning in 1993, the grass filters were also rained on during simulations to better reflect an actual storm. Grass filters were again an effective sediment trap, even at lengths of only 4.5 m (Table 1). Surface runoff trapping

Table 1. Runoff, sediment, and bacterial trapping by grass filters during simulated rain.

Year	Filter Length	Runoff	Sediment	Fecal coliforms	Fecal streptococci
		---------------------- % Trapped ----------------------			
1992	9.0 m	89.3	99.1	74.0	29.4
		87.3	98.9	43.0	nd[†]
1993	4.5 m	86.3	97.2	95.0	91.9
		65.8	93.1	50.0	23.1

[†]nd = not determined

decreased in the shorter grass filters relative to the longer filters used in 1992. Fecal bacteria concentrations in runoff were greater in 1993 than 1992, but the runoff patterns in 1993 were similar to those we observed in 1992 (Figure 2).

Bacterial trapping efficiency appeared to be greater in 1993 than in 1992 even though the grass filters were shorter (Table 1). The erosion strip length was shorter in 1993 than in 1992, but dye tracing observations suggested that more of the filter surface area appeared to be used, presumably because moisture conditions were more uniform.

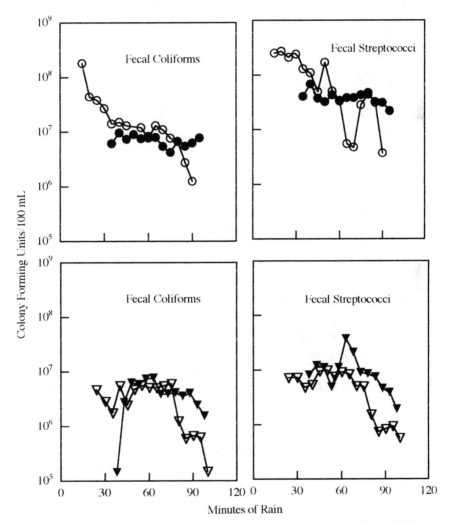

Figure 2. Fecal coliform and fecal streptococci concentrations in runoff from 18.3-m erosion strips (open symbols) and 4.5-m grass filters (closed symbols) in 1993. Each paired set of figures represents an individual plot.

Up to 95% of the fecal bacteria were trapped by the grass filters (Table 1), but their concentration in surface runoff was still high, well above the limit for primary contact water. The fecal bacteria concentrations in grass filter runoff again exceeded the concentrations in erosion strip runoff. This was not as noticeable as it was in 1992.

Crane et al. (1983) and Walker et al. (1990) suggested that grass filters alone were insufficient to reduce bacterial concentrations to meet water quality goals. Our research supports that conclusion. Most of the fecal bacteria in runoff can be held in relatively short grass filter strips by infiltration and sediment trapping. This is inconsequential since the standards for water quality (less than 1 fecal coliform/100 mL for drinking water and less than 200 fecal coliforms/100 mL for primary contact water) are based on bacterial concentration, not mass. It is unlikely that those standards will ever be met as long as the initial fecal bacteria concentration in runoff remains elevated.

At first glance it appears that grass filters have the opposite effect on water quality than intended, since we have consistently found that the concentration of fecal bacteria in grass filter runoff exceeds the concentration in erosion strip runoff. This implausible result can be explained two ways. First, the grass filters receive sediment from a much larger area than they represent themselves. The effect is to concentrated manure-laden sediment at the field/filter interface from which fecal bacteria are continually removed by the action of rain and flowing water. Second, the poultry litter is not itself a homogenous mixture of fecal bacteria; rather, it is a collection of large aggregates that are heterogeneously applied to the field. When these aggregates erode from the field and are trapped by the grass filters, they subsequently break up, releasing the bacteria held within.

1994 RAIN SIMULATIONS

Rain simulations were on-going in 1994 at the time of writing, so only preliminary information will be presented. In 1994 we began looking at *Salmonella* in runoff from poultry-manured fields as well as trying to relate sediment particle size to bacterial runoff. We also began looking at the effect of sequential rainfall on fecal bacteria concentrations in runoff. Runoff patterns in a representative plot are shown in Figure 3.

In 1994, one day prior to runoff studies, the erosion strips were rained on to create more uniform soil moisture conditions. This influenced fecal bacteria runoff in subsequent rain simulations because it apparently reduced the rate at which fecal bacteria concentrations declined as runoff continued. Litter may have been dispersed by the rain and caused fecal bacteria to be more uniformly distributed in soil. Some fecal bacteria growth may have also been stimulated, since the concentration of fecal bacteria in runoff in 1994 was only slightly less than it was the preceding year when 65% more poultry litter was applied (Figure 3).

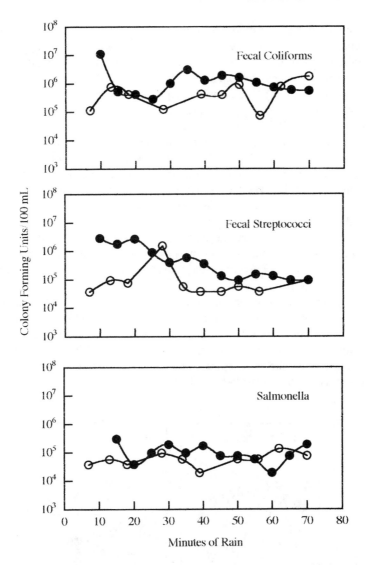

Figure 3. Fecal bacteria concentrations in runoff from 18.3-m erosion strips (open symbols) and 4.5-m grass filters (closed symbols) in 1994.

The patterns of fecal bacteria runoff we observed in previous years remained consistent. Fecal bacteria concentrations generally declined as runoff continued; the fecal bacteria concentration in grass filter runoff was typically greater than in erosion strip runoff; and fecal bacteria concentrations were in excess of $10^4/100$ mL for the duration of runoff.

The binding of bacteria to soil is a relatively rapid process, so we would expect to see most of the bacteria particle bound. We wanted to see if fecal bacteria concentrations in runoff were associated with a particular sediment size fraction. To do this we used a pipette method for particle size analysis and removed aliquots from each sample period at 0-, 5-, and 75-min intervals. With this method, after 5 min, coarse particles, greater than 20 μm in diameter, will settle below a 10-cm sampling depth, and by 75 min, only particles smaller than 5 μm will remain in suspension.

As runoff continued, the effect of settling time on bacterial concentrations, particularly fecal coliforms, became more apparent (Figure 4). The greater the settling time, the lower the bacterial concentration. Aggregates greater than 5 μm in size become progressively more important to the concentration of fecal bacteria in surface runoff the longer that runoff continues.

The implications of this result in terms of grass filters are apparent. First, it shows that by trapping sediment, the grass filters trap the bulk of fecal bacteria. However, these bacteria are subject to detachment by the impact of rainfall and flowing water. This accounts for our observation that fecal bacteria concentrations are greater in grass filter runoff than erosion strip runoff. The field/filter interface acts as a bacterial reservoir.

Settling time had less effect on bacterial concentrations in runoff from the grass filters than it did for erosion plot runoff. This suggests that the sediment particles to which bacteria were bound were smaller and more uniform (Figure 5). The implications of this for bacterial trapping by grass filters are ominous since it indicates that the particles that are mobile in grass filters (clay-sized or slightly larger) are those least likely to be trapped.

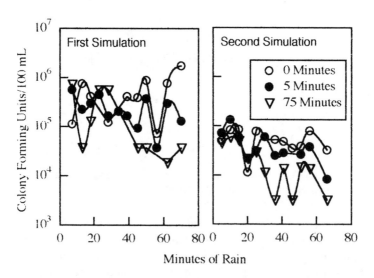

Figure 4. Effect of settling time on fecal coliform concentrations in surface runoff from poultry manured erosion strips collected from successive rain simulations in 1994. The second rain simulation occurred five days after the first simulation.

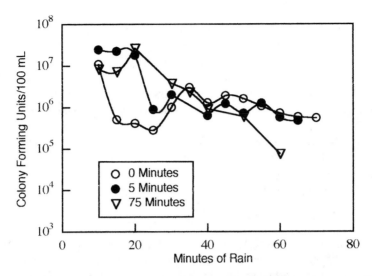

Figure 5. Effect of settling time on the fecal coliform concentration in surface runoff from 4.5-m grass filters following the first rain simulation.

Construction of grass filters with sufficient length to a) prevent any runoff from occurring and b) trap smaller than silt-sized particles would be impractical in agricultural settings. The fecal bacteria concentration in runoff will likely exceed water quality standards in those cases where surface application and even incorporation of poultry litter at rates of 10 Mg/ha or greater are followed shortly afterwards by rain.

The greatest runoff and constituent loss from surface-applied poultry litter occurs in the first runoff following application (Edwards and Daniel, 1993). When rain simulations were repeated on a plot, after a five-day interval, initial fecal bacteria concentrations in runoff were reduced approximately one order of magnitude. The effect of settling time was approximately the same (Figure 4). Clearly, the potential for fecal bacteria runoff at levels in excess of water quality standards was still present.

CONCLUSIONS

Grass filters have been used successfully to reduce soil erosion from agricultural fields and are a recommended practice for reducing concentrations of fecal constituents from point sources. Their use in minimizing the effects of fecal contamination by surface runoff from manured fields has been examined with mixed results. After three years of rainfall simulation studies, our results indicate that reducing surface runoff water to meet primary water contact standards of 200 fecal coliforms/100 mL using grass filters as the sole management practice will be an elusive goal. Even though fecal constituents are reduced up to 95%, the runoff still contains bacterial concentrations far in excess of existing primary water contact standards.

Our study used atypically intense rainfall to cause surface runoff. Depending on the length, slope, and management practices used on a manured field, grass filters as short as 4.5 m would probably trap runoff if it occurred. Consequently, grass filters should deter surface water contamination by fecal bacteria in runoff from manured fields on most occasions. It is unlikely that the grass filter length needed to ensure total compliance with primary contact water standards for surface runoff water will be practical. Runoff from poultry manured fields, if it occurs, will contribute to surface water contamination, particularly when manure application is followed by runoff-producing rainfall.

ACKNOWLEDGMENTS

We thank A. Villalba, R. Gilfillen, M. Murugesan, L. Dunn, C.E. Madison, and R. Rhodes. The work on which this chapter is based was supported, in part, by the US Department of the Interior, under the provisions of Public Law 101-397, and the Kentucky Water Resources Research Institute as grant no. 14-08-0001-G2021. Additional support was provided by Kentucky State Senate Bill 271. This chapter is a contribution of the Agronomy Department, University of Kentucky in connection with a project of the Kentucky Agricultural Experiment Station and is published with the approval of the director as Journal Article No. 94-3-140.

REFERENCES

Barfield, B.J., and Albrecht, S.C., Use of a vegetative filter zone to control fine-grained sediments from surface mines, in *Symposium on Surface Mining Hydrology, Sedimentology, and Reclamation*, p. 481, University of Kentucky, Lexington, Kentucky, 1982.

Cox, J.M., Lysine-mannitol-glycerol agar, a medium for the isolation of *Salmonella* spp., including *S. typhi* and atypical strains, *Applied and Environmental Microbiology*, 59(8), 2602, 1993.

Crane, S.R., Moore, J.A., Gismer, M.E., and Miller, J.R., Bacterial pollution from agricultural sources: A review, *Trans. ASAE*, 26, 858, 1983.

Dickey, E.C., and Vanderholm, D.H., Vegetative filter treatment of livestock feedlot runoff, *Journal of Environmental Quality*, 10, 279, 1981.

Dillaha, T.A., Reneau, R.B., Mostaghimi, S., and Lee, D.,, Vegetative filter strips for agricultural non point source pollution control, *Trans. ASAE*, 32, 513, 1989.

Edwards, D.R., and Daniel, T.C., Environmental impacts of on-farm poultry waste disposal--A review, *Bioresource Technology*, 41, 9, 1992.

Edwards, D.R., and Daniel, T.C., Effects of poultry litter application rate and rainfall intensity on quality of runoff from fescuegrass plots, *Journal of Environmental Quality*, 22, 361, 1993.

Fogle, A.W., and Barfield, B.J., A low head loss sampling device for monitoring inflow to natural vegetated filter strips, *Trans. ASAE*, 36, 791, 1993.

Giddens, J., and Barnett, A.P., Soil loss and microbiological quality of runoff from land treated with poultry litter, *Journal of Environmental Quality*, 9, 518, 1980.

Kentucky Agricultural Statistics Service, *Kentucky Agricultural Statistics, 1993-1994*, Kentucky Agricultural Statistics, Service, Louisville, Kentucky, 1994.

Schellinger, G.R., and Clausen, J.C., Vegetative filter treatment of dairy barnyard runoff in cold regions, *Journal of Environmental Quality*, 21, 40, 1992.

Walker, S.E., Mostaghimi, S., Dillaha, T.A., and Woeste, F.E., Modeling animal waste management practices: Impacts on bacteria levels in runoff from agricultural land, *Trans. ASAE*, 33, 807, 1990.

Young, R.A., Huntrods, T., and Anderson, W., Effectiveness of vegetated buffer strips in controlling pollution from feedlot runoff, *Journal of Environmental Quality*, 9, 483, 1980.

EDGE-OF-FIELD LOSSES
OF SURFACE-APPLIED ANIMAL MANURE

T.C. Daniel, D.R. Edwards, and D.J. Nichols

INTRODUCTION

Poultry (*Gallus gallus domesticus*) and swine (*Sus scrofa*) production are important aspects of Arkansas' agricultural economy. As these industries have grown, so have concerns regarding the proper utilization and disposal of the associated wastes produced. There is thus increasing emphasis by industry, individual producers, and governmental agencies on managing animal wastes to best protect the state's ground and surface water resources.

Our research has evaluated the potential impacts of pasture-applied poultry litter, poultry (caged-laying hen) manure, and swine manure constituents on runoff and sub-surface water. In addition, we investigated the effectiveness of such waste management practices as timing animal waste application to avoid storms occurring soon after application, substituting commercially-available inorganic fertilizer for poultry litter, and lightly incorporating litter and fertilizer. Some of our research efforts to date are described and summarized below.

RESEARCH RESULTS

EFFECTS OF APPLICATION RATE AND RAIN INTENSITY ON RUNOFF QUALITY[1]

Introduction

Broiler, egg, and swine production require periodic removal of the litter (manure plus bedding materials) or manure from production houses. The wastes are typically applied to pasture and can supply valuable nutrients to forage grasses.

The objectives of this study were to determine how runoff quality from fescue (*Festuca arundinacea* Schreb.) plots treated with poultry litter, poultry (caged-layer) manure, and swine manure slurry is impacted by application rate and rain intensity for storms occurring 1 day after application.

---

[1]Edwards and Daniel, 1992, 1993b,c.

Materials and Methods

Runoff plots with 5% slopes and dimensions of 1.5 m by 6.0 m (long axis oriented down-slope) were constructed on Captina silt loam (fine-silty, siliceous, mesic Typic Fragiudult) and established in fescue in the summer and fall of 1990. Steel borders surrounded each plot to isolate runoff, and an aluminum gutter at the down-slope end of the plot collected runoff.

A rain simulator (Edwards et al., 1992) was used to generate 0.5 h of continuous runoff from each plot. Discrete runoff samples were collected every 0.08 h during runoff. Collection times and sample volumes were recorded to compute runoff rates and volumes. One flow-weighted sample per plot was composited from six discrete samples for analysis.

Composite samples were analyzed by the University of Arkansas Water Quality Laboratory for total Kjeldahl nitrogen (TKN), ammonia-N (NH_3-N), nitrate-N (NO_3-N), total phosphorus (TP), ortho-P (PO_4-P), total suspended solids (TSS), chemical oxygen demand (COD), and electrical conductivity (EC) using standard methods (Clesceri et al., 1989). Aliquots of composites were passed through a 0.45-μm pore size filter for PO_4-P analysis.

Broiler litter (manure plus sawdust and rice hulls bedding material) was uniformly applied to the plots at 0 (control), 218, 435, and 870 kg N/ha. Poultry manure slurry from shallow scrape alleys of a laying hen facility was applied to the surface of the plots at application rates of 0 (control), 220, and 879 kg N/ha. Swine manure slurry from an unagitated pit beneath a production facility was applied at 0 (control), 217, and 435 kg N/ha. Simulated rain was applied 24 h after litter and manure application at intensities of 50 and 100 mm/hour. All treatments were replicated three times.

Results and Discussion

Mean runoff concentrations of poultry litter, poultry manure slurry, and swine manure slurry constituents are given in Table 1. The mass losses of poultry litter, poultry manure slurry, and swine manure constituents transported off the plots in runoff are shown in Table 2.

Poultry litter application rate significantly ($\alpha = 0.05$) affected runoff concentrations, with concentrations increasing approximately linearly with application rate. Rain intensity had a significant ($\alpha = 0.05$) influence only on runoff concentrations of TKN, TP, PO_4-P, and COD, with the concentrations of these constituents decreasing with increasing rain intensity.

Mass losses of all litter constituents were significantly ($\alpha = 0.05$) influenced by both litter application rate and rain intensity and increased with increases in either variable.

Proportions of applied litter constituents lost in runoff depended primarily on rain intensity. At the 50-mm/hour rain intensity, litter constituents losses were generally low (except NH_3-N). At the 100-mm/hour intensity, however, total N and TP losses were as high as 18.7 and 7.3%, respectively.

Table 1. Mean runoff concentrations as a function
of application rate and rain intensity.

Treatment	TKN	NH$_3$-N	NO$_3$-N	TP	PO$_4$-P	COD	TSS
				mg/L			
Poultry litter							
C 50[†]	5.7[‡]	1.3	0.1	1.1	0.8	78.3	6.6
C 100	4.7	2.3	0.1	1.0	0.9	82.7	7.8
L 50	159.7	66.0	0.2	22.0	21.0	786.0	64.6
L 100	181.0	43.7	0.1	14.2	12.7	608.7	83.2
M 50	363.3	114.3	0.3	50.1	43.2	1033.3	236.5
M 100	295.0	94.0	0.3	36.7	30.0	1057.0	148.9
H 50	564.0	210.3	1.1	76.3	62.9	1544.7	337.8
H 100	406.7	169.3	0.6	59.8	47.0	1092.3	302.4
Poultry manure slurry							
C 50	5.7	1.3	0.1	1.1	0.8	78.3	6.6
C 100	4.7	2.3	0.1	1.0	0.9	82.7	7.8
L 50	99.3	48.3	0.0	29.1	16.2	887.3	229.4
L 100	70.3	34.0	0.1	18.2	11.7	648.3	241.8
H 50	232.3	123.3	0.0	54.1	34.5	1120.7	768.0
H 100	190.0	93.7	0.0	45.3	21.7	1086.7	749.8
Swine manure slurry							
C 50	5.7	1.3	0.1	1.1	0.8	78.3	6.6
C 100	4.7	2.3	0.1	1.0	0.9	82.7	7.8
L 50	42.7	32.0	0.1	11.9	11.9	282.3	50.7
L 100	28.7	18.7	0.1	9.5	8.0	206.7	50.7
H 50	77.3	51.7	0.1	29.7	29.4	504.0	83.0
H 100	38.0	28.3	0.1	15.8	13.9	315.3	74.2

[†]C, L, M, and H represent control, 218, 435, and 870 kg N/ha application rates for poultry litter; C, L, and H represent control, 220, and 879 kg N/ha application rates for poultry manure; and C, L, and H represent control, 217, and 435 kg N/ha application rates for swine manure. 50 and 100 represent rain intensities of 50 and 100 mm/h.
[‡]All values represent the mean of three samples.

Poultry manure slurry application rate had a significant ($\alpha = 0.05$) effect on runoff concentrations of all manure constituents except NO$_3$-N, but simulated rain intensity affected only TKN, NH$_3$-N, and PO$_4$-P concentrations. Runoff concentration of manure constituents increased approximately linearly with increasing slurry application rate. For a given slurry application rate, runoff concentrations of manure constituents decreased with increasing rain intensity due to dilution.

Mass losses of all poultry manure constituents except NO$_3$-N were significantly ($\alpha = 0.05$) influenced by slurry application rate and rain intensity. Mass losses of NO$_3$-N were affected only by rain intensity. Even though the high simulated rain intensity led to lower runoff concentrations of manure constituents than the low intensity, the increased runoff resulting from the high intensity produced greater mass losses of all manure constituents investigated.

Table 2. Mean mass losses as a function of application rate and rain intensity

Treatment	TKN	NH$_3$-N	NO$_3$-N	TP	PO$_4$-P	COD	TSS
				kg/ha			
Poultry litter							
C 50[†]	0.2[‡]	0.0	0.0	0.0	0.0	2.7	0.3
C 100	0.9	0.5	0.0	0.2	0.2	18.6	1.6
L 50	9.0	3.7	0.0	1.2	1.2	44.2	3.6
L 100	40.1	10.2	0.0	3.3	2.9	143.3	19.2
M 50	20.4	6.5	0.0	2.7	2.4	52.0	13.6
M 100	63.8	20.9	0.1	7.8	6.3	231.8	30.7
H 50	41.4	16.2	0.1	5.8	4.7	113.0	22.8
H 100	98.0	42.1	0.1	14.5	11.1	255.8	68.8
Poultry manure							
C 50	0.2	0.0	0.0	0.1	0.0	2.6	0.3
C 100	0.9	0.5	0.0	0.2	0.2	18.6	1.6
L 50	6.9	3.4	0.0	2.0	1.1	60.1	13.5
L 100	19.0	9.5	0.0	5.0	3.3	172.0	60.4
H 50	28.7	15.1	0.0	6.6	4.3	137.7	90.5
H 100	62.8	30.9	0.0	14.9	7.2	358.4	248.6
Swine manure							
C 50	0.2	0.0	0.0	0.0	0.0	2.7	0.3
C 100	0.9	0.5	0.0	0.2	0.2	18.6	1.6
L 50	5.8	4.3	0.0	1.5	1.5	35.7	7.0
L 100	8.4	5.5	0.0	2.8	2.3	59.7	15.5
H 50	12.6	8.4	0.0	4.8	4.8	80.2	13.0
H 100	12.2	9.0	0.0	5.2	4.5	103.8	24.7

[†]C, L, M, and H represent control, 218, 435, and 870 kg N/ha application rates for poultry litter; C, L, and H represent control, 220, and 879 kg N/ha application rates for poultry manure; and C, L, and H represent control, 217, and 435 kg N/ha application rates for swine manure. 50 and 100 represent rain intensities of 50 and 100 mm/h.
[‡]All values represent the mean of three samples.

Proportions lost of applied poultry manure constituents were 3.3% or less at the 50-mm/hour intensity, but manure constituent losses approximately doubled at the 100-mm/hour intensity.

Swine manure slurry application rate significantly ($\alpha = 0.05$) affected run-off concentrations of TKN, NH$_3$-N, TP, PO$_4$-P, COD, TSS, and EC. In general, concentrations of these constituents increased linearly with slurry application rate. Rain intensity was a significant variable for the same constituent concentrations except for TSS. As rain intensity increased, slurry constituent concentrations decreased due to increased runoff and associated dilution. The interaction between the two variables significantly ($\alpha = 0.05$) affected runoff concentrations of NH$_3$-N, TP, PO$_4$-P, COD, and EC. For the slurry-treated plots, approximately 60% of TKN consisted of NH$_3$-N and PO$_4$-P accounted for more than 80% of TP.

Event mass losses of swine manure slurry constituents TKN, NH$_3$-N, TP, PO$_4$-P, COD, and TSS were significantly ($\alpha = 0.05$) affected by only slurry application rate. For these constituents, the higher runoff amounts from the 100-mm/hour rain intensity were offset by dilution. Mass losses of affected constituents increased approximately linearly with application rate.

Conclusions

Runoff occurring 1 day after poultry litter, poultry manure slurry, and swine manure slurry application to grassed areas can cause runoff concentrations and mass losses of litter and manure constituents to be significantly higher than from untreated areas. In general, both runoff concentrations and mass losses of constituents increased with application rate. Increased rain intensity decreased runoff concentrations, but increased runoff caused net increases in mass losses except in runoff from swine manure-treated plots.

EFFECT OF DRYING INTERVAL ON RUNOFF FROM PLOTS RECEIVING SWINE MANURE[2]

Introduction

In some states (e.g., Arkansas) that regulate swine manure application, operating permits prohibit application of swine manure when significant rain is anticipated within 24 h following manure application. Longer drying intervals may allow time-dependent processes to occur that influence the susceptibility of manure constituents to transport in runoff.

This study was conducted to assess the influences of swine manure treatment and drying interval between manure application and first runoff event on quality of runoff from fescue plots.

Materials and Methods

In this study, the runoff plots, rain simulators, and methods used to generate, collect, and analyze runoff samples were the same as described above.

Swine manure slurry was applied to the plots at 0 (control) and 220 kg N/ha. Simulated rain was applied at 50 mm/hour to manure-treated and control plots at intervals of 4, 7, or 14 days following manure application.

Results and Discussion

Runoff concentrations of TKN, NH_3-N, NO_3-N, TP, COD, and TSS were significantly ($\alpha = 0.05$) higher from manure-treated plots than from control plots for all drying intervals of 4 to 14 days following application. Runoff NO_3-N concentrations increased with drying interval due to nitrification, but concentrations of other manure constituents were independent of drying interval. Mass losses of all constituents were higher from manure-treated plots than from control plots and were also independent of drying interval.

Although data collected by Westerman and Overcash (1980) and Edwards and Daniel (1993a) provide evidence of decreases in runoff TKN, NH_3-N, and COD concentrations for drying intervals up to 3 days, this study suggests that

---

[2]Edwards and Daniel, 1993a

longer drying intervals (up to 14 days) are ineffective in further reducing run-off concentrations of these manure constituents. Also, when the results of this study are combined with the results from previous research, it seems likely that runoff TP is unaffected by drying intervals greater than 1 day.

Conclusions

In terms of the practical significance of the findings from this study, if N and/or organic matter are of greatest runoff quality concern, then there are advantages to timing swine manure application to avoid a runoff-producing storm occurring 3 to 4 days following manure application. Little runoff quality benefit seems to accompany drying intervals greater than 4 days. If P is of greatest concern, there seems to be no runoff quality advantage associated with avoiding manure application more than 1 day in advance of a storm.

COMPARISON OF POULTRY LITTER AND INORGANIC FERTIL-IZER AND RUNOFF QUALITY FROM SUBSEQUENT STORMS[3]

Introduction

One potential technique that has been suggested for reducing the environmental impacts of poultry litter application in regions with dense poultry production is to sell the litter for transport to other regions. The proceeds from the sale of the litter would then be used to purchase inorganic fertilizer to maintain forage production. The desirability of implementing this strategy, should it prove economically feasible, would depend on whether runoff quality from areas amended with inorganic fertilizer compares favorably with that from areas receiving poultry litter.

The objective of this study was to examine how quality of runoff varies between inorganic fertilizer and poultry litter applied as fertilizer in the short term (up to about 8 weeks) following application. Another objective was to determine the changes in runoff water quality from fertilizer- or litter-treated plots with subsequent storms.

Materials and Methods

In this study, the runoff plots, rain simulators, and methods used to generate, collect, and analyze runoff samples were the same as described above.

This experiment included three fertilizer treatments (no fertilizer [control], poultry litter, and inorganic fertilizer) and four simulated rains applied 7, 14, 36 and 68 days after amendment application. The amounts of N and P added to the plots in the litter were 218 and 87 kg/ha, respectively. Inorganic fertilizer (13-13-13 mixed with ammonium nitrate) was added at a comparable N and P rate. Simulated rain was applied at 50 mm/hour.

---

[3]Edwards and Daniel, 1994.

Results and Discussion

For the first rain simulation, runoff concentrations of NH_3-N, NO_3-N, TP, and PO_4-P were highest from plots receiving inorganic fertilizer, while the highest concentrations of COD and TSS occurred in runoff from plots treated with poultry litter. Runoff concentrations of TKN from plots receiving inorganic fertilizer and poultry litter were not significantly ($\alpha = 0.05$) different for the first simulated rain but were both significantly ($\alpha = 0.05$) greater than observed for the control plots. Mean concentrations of TKN, NH_3-N, NO_3-N, TP, and PO_4-P in runoff from the litter-treated plots during the second rain simulation were not significantly ($\alpha = 0.05$) different than background (control) levels. By the third simulation, runoff from both litter and fertilizer treatments did not significantly ($\alpha = 0.05$) differ from background (control).

Runoff losses of litter and fertilizer constituents corresponded to runoff concentrations. Losses of litter constituents occurred mainly during the first rain simulation and represented 66 and 54% of total losses of TKN and TP, respectively. The majority of losses of all inorganic fertilizer constituents also occurred during the first rain simulation except for NO_3-N and represented approximately 70% of applied TKN, TP, and PO_4-P, and 80% of total NH_3-N losses. Losses of N and P from all treated plots summed over all four rains represented only approximately 1.4 and 2.7% of the amounts of these nutrients applied to the plots.

Conclusions

Runoff quality from areas treated with poultry litter or inorganic fertilizer can be expected to be worst following the first storm. Runoff quality can approach background levels after relatively few (two to five) runoffs. Environmentally based management practices should thus focus primarily on reducing runoff losses during the first post-application runoff. Since substituting inorganic fertilizer for poultry litter does not appear to provide improvements in runoff quality, this management practice cannot be recommended based on the data obtained in this study.

EFFECT OF INCORPORATION OF POULTRY LITTER OR INORGANIC FERTILIZER ON RUNOFF QUALITY[4]

Introduction

Incorporation of fertilizers and other soil amendments is commonly practiced to aid retention of nutrients and minimize losses in runoff. To date, information on the efficacy of incorporating poultry litter or inorganic fertilizer on pastures to improve runoff quality is lacking.

This study was conducted to evaluate the hypothesis that incorporation of surface-applied poultry litter and inorganic fertilizer would improve runoff water quality from fescue pasture.

---

[4]Nichols et al., 1994.

Materials and Methods

In this study, the runoff plots, rain simulators, and methods used to generate, collect, and analyze runoff samples were the same as described above.

This experiment involved evaluation of two amendments (poultry litter or inorganic fertilizer) in conjunction with two application methods (surface-application or incorporation by rotary tillage).

The inorganic fertilizer treatment consisted of 13-13-13 and ammonium-nitrate fertilizers. Inorganic fertilizer was applied at 218 kg N/ha and 87 kg P/ha to match the N and P amounts applied to the litter-treated plots.

Litter and inorganic fertilizer were uniformly applied to the respective plot surfaces. Incorporation treatments were immediately tilled across the slope using a gasoline-powered rotary tiller with tines on approximately 150-mm centers to a soil depth of 2 to 3 cm.

Simulated rain was applied to all plots at 50 mm/hour 7 days after litter and fertilizer were either surface-applied or incorporated.

Results and Discussion

Mean runoff concentrations and mass losses from undisturbed surface-applied poultry litter or inorganic fertilizer treatments did not differ significantly ($\alpha = 0.05$) from incorporated litter or fertilizer mean runoff concentrations and mass losses. Incorporation apparently did not adequately turn under the surface-applied litter and fertilizer to decrease their contributions to runoff compared to undisturbed fescue plots.

Conclusions

Shallow (2 to 3 cm) incorporation of poultry litter or inorganic fertilizer by rotary tillage had no significant effect on runoff quality for a storm occurring 7 days after application. The inherent nutrient retention and infiltration capabilities of undisturbed fescue pasture may have been impaired by the rotary tillage, as the fescue thatch covering the soil surface was disrupted by the tillage. The ability of the thatch to retain applied litter and fertilizer nutrients may have been impaired or diminished as a result.

In addition, tillage brought soil and thatch materials higher in the fescue crop profile and may have increased the susceptibility of these nutrients to removal by runoff. A tillage practice that more thoroughly turns the litter or fertilizer under might reduce nutrients in runoff more effectively than the practice evaluated in this study. However, more aggressive tillage could be more injurious to pasture grasses, more expensive to implement, and more likely to increase soil erosion.

POULTRY LITTER AND MANURE CONTRIBUTIONS TO NITRATE LEACHING THROUGH THE VADOSE ZONE[5]

Introduction

In addition to impacting surface waters, poultry litter or manure application to pastures may increase NO_3-N levels in sub-surface water.

Our objective was to determine the effect of application rate of poultry litter or manure on NO_3-N concentration in the unsaturated (vadose) zone under fescue plots as a function of depth and time.

Materials and Methods

In August 1991, poultry litter was applied to the runoff plots (described above) at 0 (control), 10, and 20 Mg/ha and poultry manure at 17.7 Mg/ha. In June 1992, an additional 4.5 Mg/ha of litter was applied to the plots that received 10 Mg/ha litter in 1991, and 3.8 Mg/ha manure was applied to plots receiving 17.7 Mg/ha manure in 1991.

Six pits were installed between 12 runoff plots (one pit for each pair of plots) to permit sampling of the unsaturated zone. Water samples were obtained from the unsaturated zone by pan microlysimeters installed at a depth of 60 cm and cup suction lysimeters installed at depths of 60 and 120 cm. An 8- to 12-kPa vacuum was maintained on the lysimeters. Soil water was removed from the lysimeters weekly for NO_3-N analysis. The fescue was periodically clipped to a height of about 7.5 cm and removed.

Results and Discussion

Nitrate-N concentrations in vadose water increased with higher litter and manure application rates. The type of poultry waste had no significant ($\alpha = 0.05$) effect on NO_3-N levels when equal amounts of N were applied. Plots receiving litter at either 10 or 20 Mg/ha or manure at 17.7 Mg/ha produced NO_3-N concentrations as high as 13, 54, and 41 mg/L at 60 cm and 8, 24, and 37 mg/L at 120 cm, respectively. In 1992, the 4.5 Mg/ha litter and 3.8 Mg/ha manure treatments produced NO_3-N concentrations less than 1 mg/L at the 60- and 120-cm depths.

Conclusions

The recommended poultry litter application rate in Arkansas is not more than 11.2 Mg/ha/year, split in two 5.6 Mg/ha applications. The litter application of 10 Mg/ha in 1991 and the 4.5 Mg/ha in 1992 were similar to a maximum full-year application and a single split-application, respectively, and neither resulted in NO_3-N concentrations above the drinking water standard (10 mg/L) at the 120-cm depth. The litter application rate recommended by the state may be acceptable under hay production conditions similar to those of our study if the major concern is preventing ground water NO_3-N concentrations above the drinking water standard.

---

[5]Adams et al., 1994.

OVERALL CONCLUSIONS

Pasture-applied poultry litter, poultry manure, and swine manure can produce runoff concentrations and mass losses significantly higher than untreated areas and can elevate sub-surface water NO_3-N levels. Management practices thus need to be employed to minimize impacts on water resources.

Timing litter or manure application to avoid storms occurring within 4 days appears to significantly reduce runoff of TKN, TP, and COD. Substituting inorganic fertilizer for litter or lightly incorporating litter or fertilizer do not appear to improve runoff quality. In terms of sub-surface NO_3-N levels, the data support the state of Arkansas' recommendation of applying poultry litter at not more than 11.2 Mg/ha/year in two 5.6 Mg/ha applications.

REFERENCES

Adams, P.L., Daniel, T.C., Edwards, D.R., Nichols, D.J., Pote, D.H., and Scott, H.D., Poultry litter and manure contributions to nitrate leaching through the vadose zone, *Soil Sci. Soc. Am. J.*, 58(4), 1206, 1994.

Clesceri, L.S., Greenburg, A.E., and Trussell, R.R., *Standard Methods for the Examination of Water and Wastewater*, 17th ed. APHA, Washington, DC, 1989.

Edwards, D.R., and Daniel, T.C., Potential runoff quality effects of poultry manure slurry applied to fescue plots, *Trans. ASAE*, 35(6), 1827, 1992.

Edwards, D.R., and Daniel, T.C., Drying interval effects on runoff from fescue plots receiving swine manure, *Trans. ASAE*, 36(6), 1673, 1993a.

Edwards, D.R., and Daniel, T.C., Effects of poultry litter application rate and rainfall intensity on quality of runoff from fescuegrass plots, *J. Environ. Qual.*, 22(2), 361, 1993b.

Edwards, D.R., and Daniel, T.C., Runoff quality impacts of swine manure applied to fescue plots, *Trans. ASAE*, 36(1), 81, 1993c.

Edwards, D.R., and Daniel, T.C., Quality of runoff from fescuegrass plots treated with poultry litter and inorganic fertilizer, *J. Environ. Qual.*, 23(3), 579, 1994.

Edwards, D.R., Norton, L.D., Daniel, T.C., Walker, J.T., Ferguson, D.L., and Dwyer, G.A., Performance of a rainfall simulator for water quality research, *Ark. Farm Res.* 41(2), 13, 1992.

Nichols, D.J., Daniel, T.C., and Edwards, D.R., Nutrient runoff from fescue pasture after incorporation of poultry litter and inorganic fertilizer, *Soil Sci. Soc. Am. J.*, 58(4), 1224, 1994.

Westerman, P.W., and Overcash, M.R., Short-term attenuation of runoff pollution potential for land-applied swine and poultry manure, in *Livestock Waste: A Renewable Resource. Proc. 4th Int. Symp. on Livestock Wastes*, 289-292, ASAE, St. Joseph, Michigan, 1980.

DAIRY LAGOON EFFLUENT IRRIGATION: EFFECTS ON RUNOFF QUALITY, SOIL CHEMISTRY, AND FORAGE YIELD

J.M. Sweeten, M.L. Wolfe, E.S. Chasteen, M. Sanderson,
B.A. Auvermann, and G.D. Alston

INTRODUCTION

Dairy industry expansion using open lot designs has impacted surface water quality and ground water usage in parts of north-central Texas. A research project was begun in 1988 that involved monitoring water use, waste water production, and wastewater quality at selected dairies in Erath County (Sweeten and Wolfe, 1993 and 1994b). These studies provided improved design information on manure and waste water collection, storage, lagoon treatment, and land application systems. The project was also used to educate dairy farmers on improved management practices including reduced fresh water usage for manure removal while achieving good sanitation practices.

This current project was implemented to evaluate the benefits and treatment effectiveness of land application of lagoon effluent. Monitoring stations were located below micro-watersheds receiving effluent irrigation. The project was designed to determine the water quality of runoff from effluent-treated micro-watersheds before the runoff would leave the farm or enter surface waters and water quality of percolate below the root zone (i.e., vadose zone). This project was conducted through a US EPA Section 319 grant in parallel with ongoing projects conducted as part of the five-year Upper North Bosque River Hydrologic Unit Area (HUA) project funded by the USDA Water Quality Initiative.

OBJECTIVES

The principal objectives of the project were to determine the effects of land treatment of dairy lagoon effluent on runoff quality, percolate quality in the vadose zone, forage yield, and quality for typical forage systems and at application rates that are similar to soil test recommendations for nitrogen.

METHODS, EQUIPMENT AND PROCEDURES

MANURE AND WASTE WATER MANAGEMENT SYSTEM

The site used for the project was Dairy B which was a 950-cow open lot dairy on 65 ha (160 acres). The milking herd size varied from 650 to 850 cows. Liquid manure and waste water from the milking parlor were routed by gravity flow into a 0.67-m (2.2-ft) deep dual-chambered concrete settling basin with 10:1 access ramps on each end for cleanout. The settling basins were decanted through a slotted plank outlet and perforated corrugated plastic riser into the primary lagoon/holding pond with 22,800 m³ (806,000 ft³) capacity. Open lots occupied 6.5 ha (16 acres) including feeding and cattle alleys and shade structures. Runoff flowed through the settling basin into the primary lagoon/holding pond, and any overflow entered a second holding pond with 10,200 m³ (362,000 ft³) liquid capacity. A separate settling basin collected runoff from the two north side dry lots, with supernatant drained through a perforated riser pipe into the second holding pond. Runoff from 3 dry lots (south side) did not enter a settling basin but discharged directly into the primary lagoon/detention pond. Waste water and runoff collected in the two lagoons were irrigated onto 48 ha (120 acres) of coastal bermudagrass pasture with a sandy loam soil.

EFFLUENT APPLICATION ON MICRO-WATERSHEDS

Fifteen micro-watersheds 12.2 m x 12.2 m (40 ft x 60 ft) were located at Dairy B in a bermudagrass pasture that had not been fertilized previously with manure or irrigated with dairy lagoon effluent. According to the *Soil Survey of Erath County, Texas* (Wagner et al., 1973), Dairy B is located within the Nimrod-Selden soil association. The soils on the micro-watershed site were mapped in 1973 as Nimrod fine sand (0-5% slopes) on the western 80% of plots area and transitioned into a Windthorst soil, 1-8% slope, severely eroded on the eastern 20% of plot area. More recent investigation indicated that these soils were composed of the following: deep Selden series, fine sand (fine-loamy, siliceous, thermic Aquic Paleustalfs); Duffau series, fine sandy loam (fine-loamy, siliceous, thermic Udic Paleustalfs); and a Chaney series, loamy sand (fine, mixed, thermic Aquic Paleustalfs) soil. Based on a topographic survey, the site sloped east to west at 1.5%, and from south to north (main slope of plots) at 3 to 5%.

A randomized complete block design with three replicates was used for statistical analysis of data (Steele and Torrie, 1980). Two forage systems were used: coastal bermudagrass [*Cynodon dactylon* (L.) Pers.] and coastal bermudagrass overseeded with wheat [*Triticum aestivum* (L.)]. These forage treatments were denoted as C and CW, respectively. Three nitrogen levels were applied as lagoon effluent. Application rates of lagoon effluent were selected to apply multiples of soil test recommendations for nitrogen. Data from the previous monitoring study at Dairy B were used as the basis for planning nutrient application rates (Sweeten and Wolfe, 1993 and 1994b). Planned application rates were designated as 0X—0 kg/ha (0 lb N/acre/year), 1X—224 kg/ha (200 lb N/acre/year), and 2X—448 kg/ha (400 lb N/acre/year) times the equivalent of plant-available nitrogen (PAN) from soil testing recommendations. All micro-watersheds were to receive the same amount of total irrigation water at the same hydraulic loading rate. The five treatment combinations were denoted as 0XC, 1XC, 1XCW, 2XC, and 2XCW. Each treatment was replicated three times. There was no 0XCW treatment established due to space and equipment limitations and the unlikelihood that double-cropped (summer/wheat) forages without fertilization would be a practical alternative for producer adoption.

Forage systems were irrigated from four Rainbird 40 Whizhead sprinklers centered 6.1 m (20 ft) apart on portable PVC laterals connected to risers at each micro-watershed. Effluent and water application rates were measured with a Doppler ultrasonic flow meter. Primary lagoon effluent was sampled from bleed-off valves in the irrigation line and from catch-cans placed randomly on the sod surface below each sprinkler path.

Type H-flumes of 0.3-m (1-ft) depth were instrumented with 14 automatic ISCO Model 2700 waste water samplers and one ISCO Model 3700 sampler. Two perforated glass bricks were placed at a soil depth of 1.5 m (5 ft) on each plot to capture vadose zone samples. Samples were obtained for water quality analysis after each rainfall using a hand pump. Effluent, runoff, and vadose zone samples were analyzed for nutrients, salts, electrical conductivity, nitrate, solids, and chemical oxygen demand in accordance with the Quality Assurance/Quality Control (QA/QC) Plan for the Section 319 project. Analysis was provided at two laboratories on the campus of Texas A&M University: (a) Agricultural Engineering Department Water and Waste water Laboratory; and (b) Extension Soil/Water/Forage Testing Laboratory, Soil and Crop Sciences Department.

Soils were sampled from three locations in each of the micro-watersheds on three occasions: April 1991, October 1992, and November 1993. The sampling depths were 0-152, 152-305, 305-610, and 610-914 mm (0-6, 6-12, 12-24, and 24-36 in.), except that 1991 sampling also included a 914-1016 mm (36-40 in.) depth.

Forage was harvested an average of five times during the growing season and was analyzed for nitrogen. Three 2.8-m^2 (30-ft^2) areas were randomly clipped in each plot with a sickle mower, samples were collected, and yield was determined on a forced air dry basis at 55°C (131°F) for 48 hours.

RESULTS

EFFLUENT APPLICATION ON MICRO-WATERSHEDS

Eleven irrigation events were applied to the 15 micro-watersheds (Table 1). Hydraulic loading rates for each event varied with the minimum hydraulic intake of the soil and antecedent soil moisture to avoid irrigation tailwater. Actual total nitrogen application values were calculated using the N concentration of irrigated effluent times the hydraulic-loading rate. The plant-avail-

Table 1. Irrigation events at Dairy B microwatersheds, Erath County, Texas.

| | Loading of Micro-watersheds | | | | | | |
| | Fresh water Application | | Effluent | | Nitrogen | | |
Date	OX	1X	1X	2x	0x	1x	2x
1991	---------------------mm---------------------				---------kg N/ha----------		
9/5-9/6	--	--	10.2	20.3	--	8.8	17.8
1992							
4/14-4/15	--	--	12.7	25.4	--	35.6	71.2
4/23	22.1	22.1	--	--	2.5	2.5	--
6/17	--	--	12.7	25.4	--	32.4	64.7
7/22-7/24	19.1	5.1	12.7	24.4	1.7	42.2	80.1
8/31	--	--	5.8	10.7	--	20.4	37.0
12/1	8.1	3.6	5.6	10.9	0.6	11.9	22.7
Total	49.3	30.7	49.5	96.8	4.7	144.9	275.7
1993							
1/25	--	--	3.3	6.6	--	10.6	20.8
3/8	--	--	6.4	12..4	--	16.8	32.9
5/14	11.2	7.4	8.6	18.0	3.2	18.7	34.5
7/1	19.1	10.9	6.1	17.3	5.5	7.5	26.2
7/19	14.5	7.1	7.1	14.5	4.1	20.6	37.0
Total	44.7	25.4	31.5	68.8	12.9	76.2	151.4
Total	94.0	56.1	91.2	185.9	17.6	229.9	445.0

able nitrogen averaged 90-95% of the total nitrogen applied as lagoon effluent. The nitrogen contribution of fresh water stored in a nearby pond was also computed.

There were initial delays in equipment procurement and installation that resulted in a late start in effluent irrigation in 1991. Several succeeding months of above-average rainfall caused further delays in effluent irrigation. Another limiting factor was the unavailability of lagoon effluent or pond water in dry-weather periods. This dairy had implemented a water conservation program that also limited the amount of effluent available in dry weather. For these reasons, the actual amount of effluent nutrients were substantially less than planned.

For the 1991 and 1992 crop years, the 1X forage system received 142 kg/ha (127 lb N/acre/year) and the 2X forage system received 271 kg/ha (242 lb N/acre/year). Hydraulic loadings of 41.1, 81.3, and 106.2 mm (1.62, 3.20, and 4.18 in.) were applied in split applications to the 0X, 1X, and 2X treatments from September 1991 through August of 1992.

Nutrient loadings for the 1993 crop year were as follows: 0X—13 kg/ha (12 lb/acre); 1X—88 kg/ha (79 lb/acre); 2X—174 kg/ha (156 lb/acre). Hydraulic loading were 52.8, 66.0, and 79.8 mm (2.08, 2.60, and 3.14 in.), respectively.

Results of soil sample analysis in April 1991 before effluent treatments began indicated low N and P throughout the soil profile. Soil chemical analysis in the fall of 1993 following two complete years of treatment (Table 2) still showed low nitrate values at four sampling intervals in the top 914 mm (36 in.) regardless of lagoon effluent and forage treatment.

The chemical quality of lagoon effluent, fresh pond water, and rainfall received were determined along with chemical results of sampling micro-watershed runoff and vadose zone water quality (Sweeten and Wolfe, 1994a). Constituent concentrations of lagoon effluent applied to micro-watersheds was compared with the surface runoff quality from the micro-watersheds (Table 3). For both cropping systems and lagoon effluent application rates, the micro-watersheds achieved high levels of nutrient and solids reduction as a result of land treatment/utilization of dairy lagoon effluent. Average concentration reductions were 95% volatile solids, 96% COD, and 96% for both TKN and P (Table 4). Similarly, soil percolate samples contained much lower concentrations of indicator parameters than applied from effluent. Concentration reductions through the soil profile as reflected in vadose zone samples were 94% VS, 99% COD, 96% TKN and 99.6% P (Sweeten and Wolfe, 1994a).

FORAGE YIELDS AND NITROGEN REMOVAL

Lagoon effluent treatments were not applied until September 1991. After interseeding wheat during the fall of 1991, effluent treatments were applied

Table 2. Average values of soil chemical properties as a function of lagoon effluent irrigation treatment and soil depth, 4 November 1993.

Treatment	Soil Depth	pH	NO$_3$-N	P$_2$O$_5$	K$_2$O	Na
	mm		-----------------mg/L-----------------			
0XC	0-152	7.1	1	33	203	10
	152–305	7.0	1	3	178	19
	305-610	7.1	1	1	190	45
	610-914	7.1	1	1	220	86
1XC	0-152	7.3	1	66	237	17
	152-305	6.9	1	15	174	50
	305-610	6.9	1	1	166	61
	610-914	6.9	1	1	215	86
1XCW	0-152	7.4	2	69	246	13
	152-305	7.3	1	11	155	38
	305-610	7.0	1	1	197	50
	610-914	6.9	1	1	203	38
2XC	0-152	7.7	2	90	293	11
	152-305	7.9	1	19	190	40
	305-610	7.7	1	1	191	64
	610-914	7.5	1	0	212	46
2XCW	0-152	7.8	2	93	310	15
	152-305	8.0	1	6	197	55
	305-610	7.8	1	1	181	60
	610-914	7.3	1	1	223	54

Table 3. Runoff water quality in relation to applied effluent quality, 1991-93
(Sweeten and Wolfe, 1994a).

Parameter	Applied Effluent	Runoff Quality				
		OXC	1XC	1XCW	2XC	2XCW
		----------------------------------mg/L----------------------------------				
Volatile Solids, mg/L	1651.6	79.1	57.7	59.8	85.0	99.0
Volatile Susp. Solids, mg/L	963.3	39.8	21.8	14.8	27.1	39.8
Chem. Oxy. Demand, mg/L	2131.7	40.7	33.6	34.5	57.7	65.5
Total Kjeldahl Nitro., mg/L	237.9	7.6	9.5	6.0	8.9	11.2
Total Phosphorus, mg/L	53.90	0.5	1.13	1.06	2.39	3.47
Potassium, mg/L	503.0	5.1	7.8	8.7	19.0	49.1
Sodium, mg/L	156.1	1.9	2.8	2.4	3.7	3.6
Chloride, mg/L	163.4	1.0	0.9	0.9	4.2	1.7
Elec. Con., μmhos/cm	4101.5	80.9	69.4	79.8	126.4	230.4

Table 4. Reductions in water quality parameters through land treatment systems for
two forage systems, applied effluent vs. runoff from micro-watersheds, 1991-93.

Parameters	Percent reduction, effluent vs. runoff					
	OXC	1XC	1CXW	2XC	2XCW	Mean ± SD
Volatile Solids, mg/L	95.2	96.5	96.4	94.9	94.0	95.5 ± 1.21
Volatile Susp. Solids, mg/L	95.9	97.7	98.4	97.2	95.8	96.4 ± 1.50
Chem. Oxy. Demand, mg/L	98.1	98.4	98.4	97.2	96.9	96.8 ± 1.40
Total Kjeldahl Nitro., mg/L	96.8	96.0	97.5	96.2	95.3	96.6 ± 1.29
Total Phosphorus, mg/L	99.1	97.8	98.0	95.6	93.6	96.6 ± 1.43
Potassium, mg/L	99.0	98.4	98.3	96.2	90.2	96.5 ± 1.93
Sodium, mg/L	98.8	98.2	98.5	97.6	97.7	96.7 ± 1.86
Chloride, mg/L	99.4	99.4	99.4	97.4	98.9	96.9 ± 1.90
Elec. Cond., μmhos/cm	98.0	98.3	98.0	96.9	94.4	96.9 ± 1.86

again in April 1992 after prolonged wet weather. Dry matter yields of wheat
and cool-season weeds were similar among cropping systems. Annual yields of
the Coastal-only and Coastal-wheat systems during 1992 appeared to be similar; however, forage yield increased as effluent rates increased.

Wheat harvested during April 1993 responded to effluent rates (Table 5).
Bermudagrass yields in the Coastal-wheat system were reduced during May by
competition of interseeded wheat. However, annual yields of Coastal-wheat
were greater than Coastal-only. Forage yields among both cropping systems
increased as effluent rate increased. Compared to annual forage yields during
1992, the amount of dry matter harvested in all cropping systems was lower
during 1993 because of a dry summer.

The concentration of N in dry matter harvested (Sweeten and Wolfe, 1994a)
was used to estimate N uptake by crops. Mean N concentration of forages among
all harvests responded to increasing effluent rates in both 1992 and 1993. Annual amounts of N removed by forages during 1992 increased as effluent rates
increased, but similar amounts of N were removed by the Coastal-only and

Table 5. Mean dry matter yield of forage systems† that received dairy lagoon effluent on micro-watersheds (Erath County, 1993).

Crop	Effluent N Rate	Harvest date, 1993						Total
		4-20‡	5-27	6-24	7-28	9-22	10-25	
		----------------------------kg/ha----------------------------						
Coastal-only	0X		323	893	156	426	370	2,165
Coastal-only	1X		2,368	2,045	906	687	665	6,671
Coastal-only	2X		3,512	3,004	2,158	1,859	1,210	11,743
Coastal-wheat	1X	1,882	454	2,493	1,167	1,140	847	7,962
Coastal-wheat	2X	3,256	616	3,624	1,747	1,693	1,440	12,433

†Values are means of three replicates.
‡Winter weeds were treated with glyphosate herbicide.

Coastal-wheat systems. During 1993, the amount of N removed by forages was greater under higher effluent rates but similar between cropping systems. The amount of nitrogen removed from the micro-watershed soils by forages were 61 and 88% greater with the 2X application rate than the 1X rate for the coastal-wheat and the coastal-only cropping systems, respectively (Table 6). Nitrogen uptake from the control treatment was only 22 to 42% as high as nitrogen uptake for any of the lagoon effluent treatments.

SUMMARY AND CONCLUSIONS

At a typical open-lot dairy farm in north-central Texas, micro-watersheds were established and monitored to evaluate the effects of primary lagoon effluent on runoff quality, percolate quality in the vadose zone, and forage yield and quality in a summer-only (coastal bermudagrass) or a summer-winter (coastal/wheat rotation) forage system. Irrigation with lagoon effluent enhanced forage yield and did not impair quality of runoff or vadose zone percolate under the conditions tested for two complete cropping years. Land application of lagoon

Table 6. Amount of nitrogen removed by forage systems† that received dairy lagoon effluent treatments on micro-watersheds (Erath County, 1991, 1992, 1993).

Crop	Effluent N Rate	Year and Number of Harvest Dates			Mean ± SD	Total
		1991 n = 3	1992 n = 5	1993 n = 6		
		----------------------------kg N/ha----------------------------				
C‡	0X	64	51	31	48 ± 17	149
C	1X	65	157	127	116 ± 48	349
C	2X	97	299	261	219 ± 107	657
CW	1X	75	172	139	129 ± 49	386
CW	2X	99	276	247	207 ± 95	631

†Values are totals for indicated number of harvest dates and means of three replicates per treatment.
‡C = coastal bermudagrass, CW = coastal bermudagrass overseeded with wheat.

effluent at rates that were at or below soil test recommendations for total or available nitrogen resulted in runoff quality and vadose zone percolate quality that was 94-99% lower in volatile solids, COD, nitrogen and phosphorus than the applied lagoon effluent.

REFERENCES

Steel, R.D.G., and Torrie, J.H., *Principles and Procedures of Statistics*, New York: McGraw-Hill Book Company, 1980.

Sweeten, J.M., and Wolfe, M.L., *The Expanding Dairy Industry: Impact on Groundwater Quality and Quantity with Emphasis on Waste Management Systems Evaluation for Open Lot Dairies*, Technical Report 155, Texas Water Resources Institute, Texas A&M University, College Station, Texas, 1993.

Sweeten, J.M., and Wolfe, M.L., *Animal Waste Management System Evaluation—USEPA Section 319 Agricultural Nonpoint Source Pollution Project. Final Report.* Texas Agricultural Extension Service, Texas A&M University System, College Station, Texas., July, 1994a.

Sweeten, J.M., and Wolfe, M.L., Manure and Wastewater Management Systems for Open Lot Dairy Operations, *Trans of the ASAE* 37(4), 1145, 1994b.

Wagner, B.J., Thomas, J.R., Harris, E.R., DeLeon, E., Ford, C.G., and Kelley, J.D., *Soil Survey of Erath County, Texas,* USDA-Soil Conservation Service, Washington, D.C., 1973.

LIVESTOCK AND PASTURE LAND EFFECTS ON THE WATER QUALITY OF CHESAPEAKE BAY WATERSHED STREAMS

David L. Correll, Thomas E. Jordan, and Donald E. Weller

INTRODUCTION

There have been surprisingly few studies in which the effects of livestock or managed grasslands on stream water quality could be unequivocally determined. Most of the research on stream water quality examined either relatively pristine streams with forested watersheds or larger streams with complex watersheds. These larger systems usually included both point and diffuse sources from multiple land uses. When these investigators attempted to relate land-based measurements of nutrient sources to stream water quality, the analyses were usually indirect. Nutrient loadings per land area and nutrient removals as livestock harvests were used to infer that most of the nutrients were unaccounted for and thus might appear in land discharges to streams (e.g., Wanielista et al., 1977). Measurements of nutrient accumulation in soil profiles under livestock areas or in infiltrating soil water were also used to infer eventual leaching to streams (e.g., Ryden et al., 1984).

In the 1970's and early 1980's, some studies measured land discharges, both overland storm flow and ground water, from managed grasslands, pasture lands, and concentrated livestock holding areas (Table 1). It is hard to generalize from these studies, since they were usually for periods of a year or two and were in different locations. Comparisons to other land uses in the same setting and weather conditions were usually lacking, and often the water quality parameters measured were different or not well specified. An interesting summary of nitrogen losses in land drainage for pastures as a function of cattle density in Connecticut found a fairly good fit to a linear relationship, where N loss increased 100 kg/ha/year for each increase of one cow/ha (Frink, 1970). Jones et al. (1976) measured discharges from 34 watersheds in Iowa for three years and regressed the concentrations or the discharge per area of total reactive phosphate and total ammonium against the number of standard cow units per ha. These regressions were statistically significant. Correlation coefficients varied from 0.41 to 0.66, and phosphate concentrations increased by 0.67 mg P/L with each animal unit

Table 1. Early studies of pasture land nutrient discharges.

Location	Time (yrs)	Paired water-sheds	Regression on land use/stock	Measured Discharge[†]	Nutrient Parameters	Reference
Conn.	3	No	Yes	V,C OF,GW	TN	Frink 1970
S. Dakota	2	Yes	No	V,C OF	TP,NO3,TKN OrgC	Harms et al., 1974
Oklahoma	1	Yes	No	V,C OF	TP,DTP,DPi TKN,NH4,NO3	Olness et al., 1975
Iowa	3	No	Yes	C OF,GW	NO3,TNH4,TPi	Jones et al., 1976
N. Carolina	2	No	Yes	C OF,GW	TP,NO3,NH4 TKN	Duda & Finan, 1983

†V=Volume, C=Concentration, OF=Overland Storm Flow, GW=Ground Water Flow,
DPi=Dissolved Ortho-Phosphate, TPi= Total Ortho-Phosphate

per ha. Duda and Finan (1983) compared nutrient discharges from ten widely scattered watersheds in North Carolina with livestock populations and land use composition. Watersheds with concentrated livestock populations discharged 5 to 10 times more nutrients. However, the ten study watersheds had only 0 to 15% pasture lands.

We are aware of three on-going, long-term studies of pasture water-sheds in which comparative data were also taken simultaneously for other land uses. One near Coshoctin, Ohio, used a series of sites to examine the effects of variations in cattle populations and the rate of mineral nutrient fertilization and to compare pastures with other land uses (e.g., Chichester et al., 1979; Owens et al., 1989, 1992). A second study in New Zealand compared pastures with native forest and pine plantations and studied the effects of riparian buffers and reforestation of pastures (e.g., Cooke, 1988; Cooke and Cooper, 1988; Cooper and Thomsen, 1988; Smith, 1992). The third site is the Rhode River watershed in the Coastal Plain of Maryland (Correll et al., 1977; Correll and Dixon, 1980; Correll, 1983; Correll et al., 1984; Jordan et al., 1986; Correll et al., 1992), with recent extension to the Piedmont and Appalachian physiographic provinces on the Chesapeake Bay watershed (Correll et al., in press). One other recently published study in England deserves special mention. Nitrate losses in overland flow and ground water from 14 hectare-sized experimental pastures were measured for up to 11 years and related to fertilization rates (Scholefield et al., 1994).

We now report a long-term summary of the nutrient discharges from a Rhode River pasture watershed, compare those with discharges from nearby forested and cropland/riparian forest watersheds, and report preliminary data from 47 sub-watersheds of the Gunpowder River in the Maryland Piedmont, the Conestoga River in the Great Valley of Pennsylvania, and Buffalo/White Deer Creeks in the Ridge & Valley of Pennsylvania. These 47 sub-watersheds have been placed in three groups dominated by pasture and livestock, cropland, or forest, respectively.

METHODS

Volume-integrated discharges of organic C, total N and P and various fractions of N and P were measured for 16 complete years for a completely forested watershed (#110) and a cropland/riparian forest watershed (#109) and for 14 complete years for a pasture-dominated watershed (#111), ending in 1993. These three Rhode River sub-watersheds are underlain with a clay aquiclude that perches local ground water and forces it to percolate to the stream channel. The stream draining each watershed was monitored with a V-notch weir with its foundations bedded into this clay aquiclude. Thus, both overland flow and ground water discharges were measured. The weirs were equipped with both volume-integrating composite samplers and fraction collectors, which were activated to take separate discrete samples at known times and stage heights during storm events. Descriptions of these watersheds, the weirs, water discharge monitoring techniques, and analytical chemistry methods have already been published (Correll, 1977, 1981, 1983; Correll et al., 1977).

Livestock population data for the pasture watershed were obtained from the farm owner. Hereford beef cattle were rotated with another pasture so that a herd averaging 0.86 cows/ha was on the watershed area 50% of the time. From late March until early October, an equal number of calves were also present. Thus, mean livestock density was 0.43 cows and 0.22 calves/ha/year. Little or no fertilizer other than livestock waste was applied. At the end of the spring of 1989, the livestock were removed, and the field was planted in pine seedlings but remained unfertilized.

For the 47 other Chesapeake Bay sub-watersheds, only data on nitrate; dissolved organic C, N, and P; and dissolved ammonium and inorganic phosphate will be given. These streams were sampled eight times from July 1992 through June 1993, and samples were immediately filtered through Millipore HA, nominal 0.4-um pore size, filters that had been prewashed with distilled water. Samples were immediately placed on ice until analysis within two weeks. Only preliminary, approximate land use composition and livestock populations are known for these watersheds (Correll et al., in press).

RESULTS

RHODE RIVER WATERSHED

On average less total organic C and total N and P were discharged per ha/year from the pasture than from either the forest- or cropland-dominated watershed (Table 2). However, nitrate and total phosphate-P discharges of the pasture were intermediate between those of the forest- and the cropland-dominated systems (Table 2). The same land use relationship was true, on average, for total organic C each season but not always for total N and P (Table 2). Total N and P discharges were higher for pasture than for forest in the winter. Nitrate discharges from pasture were higher than from forest in all

Table 2. Long-term mean nutrient fluxes and flux ratios
from Rhode River watersheds.

Watershed Discharge Fluxes (kg/ha)				Discharge Ratios (% or atomic ratio)		
	Cropland	Pasture	Forest		Cropland	Pasture	Forest
A. Winter							
Total Organic-C	7.68	3.38	3.66	% N	6.38	7.19	4.81
				% P	1.27	1.64	0.596
Total Organic-N	0.490	0.243	0.176	OrgC/N/P	200/11/1	160/10/1	430/18/1
Nitrate-N	1.28	0.324	0.0281	InorgN/OrgN	2.84	1.55	0.321
Total Ammonium-N	0.112	0.0531	0.0284	InorgP/OrgP	1.51	1.16	1.05
Total Nitrogen	1.88	0.620	0.233	InorgN/InorgP	20.9	13.0	5.46
Total Organic-P	0.0976	0.0554	0.0218	TN/TP	17.1	30.9	4.32
Total Phosphate-P	0.147	0.0640	0.0229				
Total Phosphorus	0.244	0.1194	0.0447				
B. Spring							
Total Organic-C	15.6	4.40	12.5	% N	8.00	8.18	5.46
				% P	2.26	2.27	0.902
Total Organic-N	1.25	0.360	0.681	OrgC/N/P	110/7.8/1	110/8.0/1	290/13/1
Nitrate-N	1.98	0.310	0.0948	InorgN/OrgN	1.73	1.05	0.267
Total Ammonium-N	0.205	0.0675	0.0870	InorgP/OrgP	1.91	0.797	0.793
Total Nitrogen	3.44	0.738	0.863	InorgN/InorgP	7.16	10.4	4.53
Total Organic-P	0.353	0.100	0.113	TN/TP	7.39	9.03	9.51
Total Phosphate-P	0.674	0.0803	0.0889				
Total Phosphorus	1.03	0.181	0.201				
C. Summer							
Total Organic-C	10.8	1.72	5.47	% N	11.2	7.31	6.65
				% P	6.71	1.55	1.39
Total Organic-N	1.20	0.126	0.364	OrgC/N/P	38/3.7/1	170/10/1	190/10/1
Nitrate-N	0.569	0.0259	0.0133	InorgN/OrgN	0.606	0.430	0.118
Total Ammonium-N	0.163	0.0282	0.0298	InorgP/OrgP	0.825	1.40	0.783
Total Nitrogen	1.94	0.180	0.407	InorgN/InorgP	2.99	3.19	1.84
Total Organic-P	0.723	0.0267	0.0759	TN/TP	3.58	6.21	7.39
Total Phosphate-P	0.542	0.0375	0.0520				
Total Phosphorus	1.20	0.0642	0.122				
D. Fall							
Total Organic-C	1.46	0.933	3.53	% N	10.1	6.46	6.12
				% P	2.70	1.14	0.919
Total Organic-N	0.147	0.0603	0.216	OrgC/N/P	96/8.2/1	230/12/1	280/15/1
Nitrate-N	0.127	0.0111	0.00469	InorgN/OrgN	1.06	0.398	0.0671
Total Ammonium-N	0.0288	0.0129	0.00980	InorgP/OrgP	1.46	2.75	0.590
Total Nitrogen	0.304	0.0843	0.230	InorgN/InorgP	6.00	1.82	1.68
Total Organic-P	0.0395	0.0106	0.0324	TN/TP	7.32	4.68	10.5
Total Phosphate-P	0.0576	0.0292	0.0191				
Total Phosphorus	0.0920	0.0399	0.0487				
E. Complete Year							
Total Organic-C	36.3	10.1	25.7	% N	8.71	7.55	5.72
				% P	3.31	1.85	0.949
Total Organic-N	3.16	0.763	1.47	OrgC/N/P	78/5.8/1	140/9.0/1	270/13/1
Nitrate-N	3.90	0.649	0.138	InorgN/OrgN	1.80	1.05	0.201
Total Ammonium-N	0.524	0.154	0.157	InorgP/OrgP	1.21	1.08	0.750
Total Nitrogen	7.58	1.57	1.77	InorgN/InorgP	6.75	8.80	3.56
Total Organic-P	1.20	0.187	0.244	TN/TP	6.31	8.94	9.18
Total Phosphate-P	1.45	0.202	0.183				
Total Phosphorus	2.66	0.389	0.427				

seasons, but total phosphate-P discharges were less than for forest in the spring and summer (Table 2). On average for the year, total N discharged by the pasture was 49% organic N, 41% nitrate N, and only 10% ammonium N, while total phosphorus discharged was 48% organic P and 52% inorganic phosphate P (Table 2).

Organic matter discharged from the pasture watershed had a higher percentage of N and P than organic matter discharged from the cropland or forest in the winter and spring, while the N and P content was intermediate between cropland and forest in the summer and fall (Table 2). The atomic ratio of inorganic N to organic N for pasture was intermediate between cropland and forest for all seasons, while the atomic ratio of inorganic P to organic P for pasture was higher than for the other land uses in the summer and fall but intermediate in the winter and spring (Table 2). The atomic ratio of inorganic N to P for pasture was intermediate between cropland and forest in the winter and fall but higher in the spring, summer, and overall for the year (Table 2).

Concentrations of total organic C in both overland storm and base flows were lower for pasture than for either forest or cropland (Table 3). Pasture base flow discharges had substantially higher dissolved organic C concentrations than storm flow, while storm flow had much higher particulate organic C concentrations. However, both dissolved and particulate organic C concentrations during base flow and storm flow were lower than corresponding concentrations from either forest or cropland (Table 3). Nitrate and dissolved ammonium concentrations for pasture base flow and storm discharges were intermediate between forest and cropland, but particulate ammonium concentrations in both base flow and storms were higher than for the other land uses. Dissolved organic N concentrations were lowest in pasture, but particulate organic N concentrations from pasture were highest in storm flow and intermediate for base flow (Table 3).

Dissolved phosphate concentrations during storms were higher from pasture than from forest or cropland, but base flow concentrations were 50% higher in forest than in pasture discharges (Table 3). Particulate phosphate concentrations were intermediate for pasture both in base flow and storm flow. Dissolved organic P concentrations from pasture were highest during base flow and lowest during storm events (Table 3). Particulate organic P concentrations from pasture were intermediate for both base flow and storm events.

Time series plots of volume-weighted mean seasonal nutrient concentrations of N (Figure 1) and P fractions (Figure 2) show both the seasonality and the high interannual variability in nutrient discharges for the pasture land watershed. Changes in N and P concentrations were often unrelated. For example, high N concentrations in 1977-78 were coincident with rather low P concentrations; however, P concentrations were high in 1988, but N concentrations were rather low.

Table 3. Mean proportions of nutrients in dissolved and particulate phases during storm events (Storm) and baseflow (Base) conditions. Concentrations are given in mg C, N, or P per liter.

Nutrient Fraction	Cropland Base	Cropland Storm	Pasture Base	Pasture Storm	Forest Base	Forest Storm
Diss. Organic-C	10.2	9.8	9.94	5.80	23.6	27.3
Part. Organic-C	18.4	77.3	11.7	20.6	13.1	33.9
Total Organic-C	28.6	87.1	21.6	26.4	36.7	61.2
Nitrate	1.21	1.61	0.201	0.402	0.0549	0.140
Diss. Ammonium	0.173	0.146	0.143	0.097	0.0856	0.0734
Part. Ammonium	0.0206	0.0291	0.0289	0.0520	0.0095	0.0501
Diss. Organic-N	0.165	0.461	0.0583	0.258	0.261	0.495
Part. Organic-N	0.538	2.51	0.350	2.90	0.266	0.892
Diss. Total-N	1.55	2.22	0.402	0.757	0.402	0.708
Part. Total-N	0.559	2.54	0.379	2.95	0.276	0.942
Total-N	2.11	4.76	0.781	3.71	0.678	1.65
Diss. Phosphate-P	0.0210	0.0279	0.0212	0.0346	0.0331	0.0227
Part. Phosphate-P	0.230	0.821	0.161	0.150	0.0785	0.0628
Diss. Organic-P	0.0107	0.0204	0.0145	0.0146	0.0120	0.0416
Part. Organic-P	0.202	1.45	0.0793	0.606	0.0371	0.224
Diss. Phosphorus	0.0317	0.0483	0.0357	0.0492	0.0451	0.0643
Part. Phosphorus	0.432	2.27	0.240	0.756	0.116	0.287
Total Phosphorus	0.464	2.32	0.276	0.805	0.161	0.351
% Diss. Organic-C	35.7	11.3	46.0	22.0	64.3	44.6
% Diss. Ammonium	89.4	83.4	83.2	65.1	90.0	59.4
% Diss. Organic-N	23.5	15.5	14.3	16.8	49.5	35.7
% Diss. Phosphate	8.4	3.3	11.6	18.7	29.7	26.5
% Diss. Organic-P	5.0	1.4	15.5	2.4	24.4	15.7

PIEDMONT AND APPALACHIAN WATERSHEDS

Dissolved nutrient concentrations in base flow from pasture-, cropland-, and forest-dominated watersheds in the Piedmont and Appalachian (PAP) physiographic provinces of the Chesapeake Bay watershed (Table 4) can be compared with analogous data from Rhode River (RR) watersheds (Table 3). Dissolved organic C concentrations in pasture-dominated PAP drainages were 59% higher than for the RR pasture watershed, while dissolved organic C was about the same for cropland watersheds in both places and was much lower for forested PAP drainages. The dissolved organic matter from the pasture-dominated PAP streams also contained over three times as much N but only 68% as much P as the organic matter from the pasture RR watershed. Differences in composition of dissolved organic matter had different patterns for the other land uses. Thus forested PAP watersheds discharged organic matter containing only 39% more N, but almost four times as much P, while cropland PAP watersheds released organic matter with 44% more N, but only 5% more P than RR watersheds (Tables 3, 4).

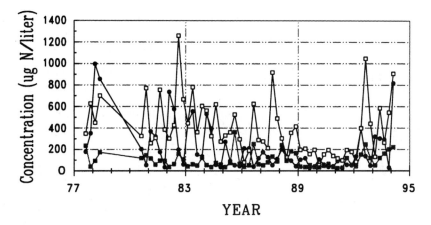

Figure 1. Seasonal volume-weighted concentrations of nitrate (solid points), total ammonium (shaded squares), and total organic N (open squares) in combined overland storm and perched ground water discharges from a Rhode River pasture land watershed (# 111). Beef cattle at an average density of 0.43 cows and 0.22 calves grazed the pasture until livestock were removed in spring of 1989. No mineral fertilizers were applied. No data were taken from the spring of 1978 until the summer of 1980.

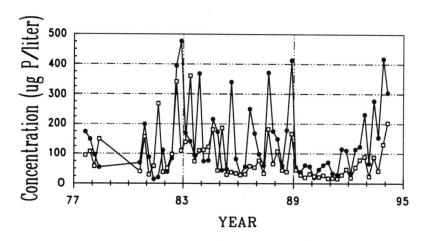

Figure 2. Seasonal volume-weighted concentration of total phosphate P (solid points) and total organic P (open squares) in combined overland storm flow and perched ground water discharges from a Rhode River pasture land watershed (# 111). Land management as in Figure 1.

Table 4. Dissolved nutrient concentrations in streams draining Piedmont
and Appalachian Regions of the Chesapeake Bay watershed.
Values are annual means ± 1 standard deviation.

Nutrient Parameter	Cropland-Dominated (17 streams)	Pasture-Dominated (13 streams)	Forest-Dominated (17 streams)
Organic C (mg C/L)	9.78±1.07	15.8±2.70	8.13±1.76
Organic C			
(% N)	2.33±0.52	1.86±0.51	1.54±1.51
(% P)	0.11±0.01	0.10±0.03	0.20±0.20
Nitrate (mg N/L)	5.58±2.57	7.99±4.17	0.54±0.51
Ammonium (ug N/L)	79.4±71.4	117±100	21.3±16.9
Organic N (ug N/L)	229±59.4	290±85.0	125±26.4
Phosphate (ug P/L)	25.3±21.7	55.1±43.2	6.32±9.27
Organic P (ug P/L)	10.6±1.64	15.4±4.77	16.1±16.6
Atomic Ratio (DIN/DIP)	942	653	314
Atomic Ratio (OrgC/OrgN/OrgP)	4800/83/1	6300/120/1	4000/55/1

Nitrate concentrations for pasture-dominated PAP watersheds were very high, comprising most of the dissolved nitrogen, and were 40 times higher than RR pasture lands (Tables 3, 4), but dissolved ammonium concentrations were somewhat lower for the pasture-dominated PAP watersheds than for RR. Dissolved organic N concentrations were about five times higher in pasture-dominated PAP streams than in the RR pasture stream. Dissolved phosphate concentration in PAP pasture watersheds was more than double that of the RR pasture, but dissolved organic P was about the same concentration in both. The atomic ratio of dissolved inorganic N to dissolved inorganic P was 650 for pasture PAP systems but only 36 for the pasture RR system. The atomic ratios of organic C to N to P were also much higher for pasture PAP systems (6300/120/1) than for pasture RR system (1800/8.9/1).

DISCUSSION

Our results seem to suggest several generalizations about water quality effects of livestock management in the mid-Atlantic region. First, the mere maintenance of managed grassland in this region where forest is the natural vegetative cover may result in some significant, but not very dramatic, differences in the nutrient composition of streams. For the Rhode River site, pasture lands, even when not grazed, have lower discharges per unit area of organic C, N, and P and higher discharges of nitrate and inorganic phosphate than undisturbed mature forest lands (Tables 2, 3). However, there was only one of each type of watershed available for comparison at the Rhode River site. Therefore, it is uncertain whether the differences arise from land use effects or other unknown factors. When grazing was discontinued on the Rhode River pasture in spring of 1989, there were no major changes in nutrient discharges (Figures 1, 2). The apparent decline in nutrient concentrations for the first few years after grazing was discontinued, and the subsequent increases in nutrient concentrations may or may not have been long-term effects of altered land management.

A second generalization seems to be that high nitrate discharges from intensively managed pasture lands may be the result of high rates of fertilization with mineral fertilizer and/or high inputs of nitrogen as winter feed supplements as reported by Owens et al. (1992) and Scholefield et al. (1994). Our nutrient concentration data for Piedmont and Appalachian watersheds (Table 4) dominated by livestock could be converted to rough estimates of flux by assuming that the combination of overland storm flow and ground water discharge was 35 cm/year, a value typical for this region (Correll, 1982). These pasture-dominated watersheds discharged approximately 28 kg nitrate N, 0.41 kg dissolved ammonium N, and 1.0 kg dissolved organic N/ha/year, respectively. This is a much higher rate for nitrate than the long-term average of 0.65 for the Rhode River pasture (Table 2) but lower than nitrate losses of 38.5 to 134 kg N/ha/year to ground water from pastures fertilized with 200 to 400 kg mineral N/ha/year (Scholefield et al., 1994). Ground water discharges from a New Zealand pasture (0.29 nitrate N, 0.05 ammonium N, and 0.81 organic N) were similar to those from the Rhode River pasture (Cooper and Thomsen, 1988). It would seem that the nitrogen discharges reported by Frink (1970), 60 to 230 kg N/ha/year, were unusually high.

It should be noted that in the Piedmont and Appalachian regions we sampled, it is common practice to allow livestock access to stream channels and sometimes even to fence them into stream channel/riparian areas during the day, while at the Rhode River site cattle were fenced out of the stream most of the time. Some of the Appalachian sites studied have among the highest livestock densities found in the United States (Correll et al., in press).

SUMMARY

Managing land for livestock production has significant effects on stream water quality. Even if a mid-Atlantic coastal watershed is managed only to maintain it as a grassland, the water quality of the streams will be altered in comparison with a watershed allowed to remain in the natural forest vegetation for this region. When livestock are managed at high densities using mineral fertilizers and imported food, the quality of both overland storm flow and ground water moving from these lands to local streams will be seriously affected.

There is a need for more information on the effects of livestock production on stream water quality. Studies need to be designed to accurately and quantitatively relate stream water quality to both overland storm discharges and ground water infiltration from livestock production areas. Good hydrologic data, as well as a complete suite of water quality parameters, should be measured on representative samples. Such studies need to be long-term in order to observe the effects of variations in weather. Livestock and land management practices need to be better documented.

ACKNOWLEDGMENTS

This research was supported by a series of grants from the Smithsonian Environmental Science Program and the National Science Foundation, including BSR92-06811 and DEB93-17968.

REFERENCES

Chichester, F.W., Van Keuren, R.W., and McGuiness, J.L., Hydrology and chemical quality of flow from small pastured watersheds: II. Chemical quality, *J. Environ. Qual.*, 8(2), 167, 1979.

Cooke, J.G., Sources and sinks of nutrients in a New Zealand hill pasture catchment: II. Phosphorus, *Hydrol. Proc.*, 2, 123, 1988.

Cooke, J.G., and Cooper, A.B., Sources and sinks of nutrients in a New Zealand hill pasture catchment: III. Nitrogen, *Hydrol. Proc.*, 2, 135, 1988.

Cooper, A.B., and Thomsen, C.E., Nitrogen and phosphorus in streamwaters from adjacent pasture, pine, and native forest catchments, *New Zealand J. Mar. Freshwater Res.*, 22, 279, 1988.

Correll, D.L., An overview of the Rhode River watershed program, chap. 7, in *Watershed Research in Eastern North America*, Vol. I, Correll, D.L., ed., Smithsonian Press, Wash. D.C., 1977.

Correll, D.L., Nutrient mass balances for the watershed, headwaters intertidal zone, and basin of the Rhode River estuary, *Limnol. Oceanogr.*, 26(6), 1142, 1981.

Correll, D.L., Seasonal and annual variation in Rhode River watershed hydrology, chap. 6, in *Environmental Data Summary for the Rhode River Ecosystem (1970-1978), Section A: Long-Term Physical/Chemical Data. Part I: Airshed and Watershed*, Correll, D.L. ed., Chesapeake Bay Center for Environmental Studies, Edgewater, Maryland, 1982.

Correll, D.L., N and P in soils and runoff of three coastal plain land uses, in *Nutrient Cycling in Agricultural Ecosystems*, Lowrance, R., Todd, R., Asmussen, L., and Leonard, R., eds., Univ. of Georgia. Agric. Exp. Sta. Spec. Pub. 23, 1983.

Correll, D.L., and Dixon, D., Relationship of nitrogen discharge to land use on Rhode River watersheds, *Agro-Ecosyt.*, 6, 147, 1980.

Correll, D.L., Goff, N.M., and Peterjohn, W.T., Ion balances between precipitation inputs and Rhode River watershed discharges, chap. 5, in *Geological Aspects of Acid Deposition*, Bricker, O.P., ed. Butterworth Press, Boston, 1984.

Correll, D.L., Jordan, T.E., and Weller, D.E., Nutrient flux in a landscape: Effects of coastal land use and terrestrial community mosaic on nutrient transport to coastal waters, *Estuaries* 15(4), 431, 1992.

Correll, D.L., Jordan, T.E., and Weller, D.E., The Chesapeake Bay watershed: Effects of land use and geology on dissolved nitrogen concentrations, in *Toward a Sustainable Watershed: The Chesapeake Experiment*, Mihurski, J., ed., Chesapeake Research Consortium, Solomons, Maryland, in press.

Correll, D.L., Wu, T.-L., Friebele, E.S., and Miklas, J., Nutrient discharge from Rhode River watersheds and their relationship to land use patterns, chap. 22, in *Watershed Research in Eastern North America*, Vol. I, Correll, D.L. ed., Smithsonian Press, Wash. D.C., 1977.

Duda, A.M., and Finan, D.S., Influence of livestock on nonpoint source nutrient levels of streams, *Trans. Amer. Soc. Agr. Eng.,* 26, 1710, 1983.

Frink, C.R., *Plant Nutrients and Animal Waste Disposal,* Circular 237, Conn. Agr. Exper. Sta., New Haven, Connecticut, 1970.

Harms, L.L., Dornbush, J.N., and Andersen, J.R., Physical and chemical quality of agricultural land runoff, *J. Water Poll. Contr. Fed.,* 46(11), 2460, 1974.

Jones, J.R., Borofka, B.P., and Bachmann, R.W., Factors affecting nutrient loads in some Iowa streams, *Water Res.,* 10, 117, 1976.

Jordan, T.E., Correll, D.L., Peterjohn, W.T., and Weller, D.E., Nutrient flux in a landscape: The Rhode River watershed and receiving waters, chap. 3 in *Watershed Research Perspectives,* Correll, D.L., ed. Smithsonian Press, Wash. D.C., 1986.

Olness, A., Smith, S.J., Rhoades, E.D., and Menzel, R.G., Nutrient and sediment discharge from agricultural watersheds in Oklahoma, *J. Envir. Qual.,* 4(3), 331, 1975.

Owens, L.B., Edwards, W.M., and Van Keuren, R.W., Sediment and nutrient losses from an unimproved, all-year grazed watershed, *J. Envir. Qual.,* 18, 232, 1989.

Owens, L.B., Edwards, W.M., and Van Keuren, R.W., Nitrate levels in shallow groundwater under pastures receiving ammonium nitrate or slow-release nitrogen fertilizer, *J. Envir. Qual.,* 21, 607, 1992.

Ryden, J.C., Ball, P.R., and Garwood, E.A., Nitrate leaching from grassland, *Nature,* 311, 50, 1984.

Scholefield, D., Tyson, K.C., and Garwood, E.A., Nitrate leaching from grazed grassland lysimeters: Effects of fertilizer input, field drainage, age of sward and patterns of weather, *J. Soil Sci.,* 44(4), 601, 1994.

Smith, C.M., Riparian afforestation effects on water yields and water quality in pasture catchments, *J. Envir. Qual.,* 21, 237, 1992.

Wanielista, M.P., Yousef, Y.A., and McLellon, W.M., Nonpoint source effects on water quality, *J. Water Poll. Contr. Fed.,* March, 441, 1977.

STREAM IMPACTS DUE TO FEEDLOT RUNOFF

E.O. Ackerman and A.G. Taylor

INTRODUCTION

L ivestock production is a principal component of Illinois' agricultural industry. The 1993 Illinois Agricultural Statistics Annual Summary (IDOA, 1993) reports that over half of the 80,000 farms in the state have livestock or poultry enterprises. There are 32,000 farms that have cattle, 13,500 have hogs, 4,000 have sheep, and 3,000 have milk cows. These livestock operations are distributed throughout the state outside the major metropolitan areas.

The Illinois Livestock Waste Regulations (IPCB, 1992) set forth the pollution control standards applicable to all animal feeding operations in the state. These rules, which are administered by the Illinois Environmental Protection Agency (IEPA), require the implementation of runoff control systems or practices such that discharges of livestock waste to navigable waters do not occur during a rainfall event of less intensity than a 25-year, 24-hour, storm. Approximately 40% of the facilities inspected by the IEPA's Agricultural Engineers do not comply with this provision (IEPA, 1992).

Most of the livestock facilities in Illinois are confined housing or concentrated feedlot operations. Those that are mismanaged or inadequately designed to control wastes have the potential to chronically impact surface water resources. Such negative effects were demonstrated in a study conducted during the Summer of 1991 by the IEPA and the United States Department of Agriculture Soil Conservation Service. Intensive surveys conducted in two rural Illinois watersheds showed depressed dissolved oxygen levels, elevated concentrations of nutrients, and exceedingly high fecal bacteria levels. Fish and macro invertebrate communities were degraded as a result of the dual effects of livestock waste runoff and limited stream flow (Hite et al., 1992).

More acute and observable effects, i.e., fish kills, tend to occur when concentrated amounts of manure are discharged to streams in the absence of precipitation or during limited rainfall events. IEPA records indicate that 133 fish kills, attributable to manure discharges, were investigated by IEPA personnel between 1979 and 1992. This chapter focuses upon the causative factors and water quality impacts of three fish kill incidents and discusses the potential mitigating measures that could prevent the recurrence of each of the events.

METHODS

The IEPA carries out field investigations of water pollution incidents related to livestock production. Citizen complaints are a frequent method by which the IEPA is informed of water pollution events. During the course of an investigation, it is common to collect samples for the purpose of enforcement. It is important that the samples are taken in a consistent manner and chain-of-custody is maintained. The techniques for sample collection and preservation are set forth in an IEPA procedures document (IEPA, 1991). Water samples collected for inorganic analysis are sent to an IEPA laboratory in Champaign, Illinois. Analysis for organic compounds are performed at a separate laboratory in Springfield, Illinois. Sample analysis is performed according to guidelines established by the U.S. Environmental Protection Agency (USEPA 1992).

The Illinois Department of Conservation (IDOC) supplements the IEPA inspection whenever a fish kill incident occurs. IDOC conducts fish kill investigations and determines fish losses according to established procedures (IDOC, 1992). Specific fish values used by IDOC are adopted from the American Fisheries Society (AFS 1992).

CASE HISTORY #1

In August of 1980, a swine operation in west-central Illinois was surveyed by IEPA field staff in response to an ongoing fish kill in an unnamed tributary of North Creek. The subject facility produced 1,000 market hogs annually. The swine were housed in an open-front structure situated on a 120 ft x 61-ft sloping concrete feedlot.

The facility lacked any type of waste collection structure. As illustrated in Figure 1, manure drained to an alley along one side of the feedlot then traveled down a hillside into a ditch that discharged into the tributary. Manure solids were observed to be as deep as 1 ft in the runoff path near the feedlot.

Although precipitation could readily flush the manure off the feedlot and into the stream, there was no precipitation occurring at the time of the field investigation. The apparent cause of the problem was a misting system used to continuously spray the animals to combat heat stress.

Due to the discharge from the feedlot, the tributary was turbid, covered in part with a brownish foam, and had a strong livestock waste odor. Several catfish were observed to be in distress at a location approximately 1/4 mile downstream from the feedlot.

Laboratory analyses showed a dissolved oxygen concentration of 0.0 milligrams per liter (mg/L) at two downstream sampling stations. As a comparison, the upstream dissolved oxygen measurement was 6.8 mg/L. To comply with the Illinois Water Quality Standards (IPCB, 1990), dissolved oxygen levels must not be less than 5.0 mg/L at any time.

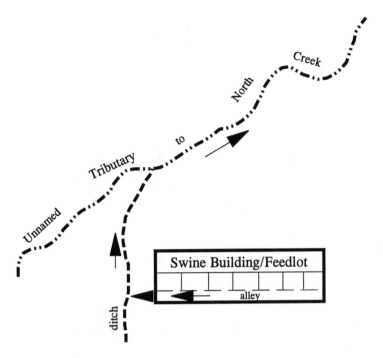

Figure 1. Plan view of swine feedlot and drainage pattern.

The instream sample concentrations of ammonia-N varied from 0.7 mg NH_4-N/L upstream to 56.0 mg NH_4-N/L 1/4 mile downstream of the discharge point. The water quality standard for ammonia-N varies depending upon the water pH and temperature; however, in no case does it exceed 15 mg/L. Figure 2 compares the upstream and downstream concentrations of ammonia-N to that in the discharge from the drainage ditch.

Other parameters that were measured were the biochemical oxygen demand (BOD) and phosphorus. A wastewater sample collected from the drainage ditch next to the feedlot had a BOD concentration of 822 mg/L. The phosphorus concentration upstream from the discharge point was 0.29 mg/L, while the concentration 1/4 mile downstream was 31.0 mg/L.

This incident illustrates two problem sources common among the small- to medium-sized swine operations in Illinois. First, many producers have constructed open-front facilities without including manure collection systems to contain feedlot runoff. Second, the misting and/or watering systems can generate conditions that cause runoff. In this situation the problem was corrected by installation of a concrete settling basin and earthen storage structure at a total cost of less than $4,000.

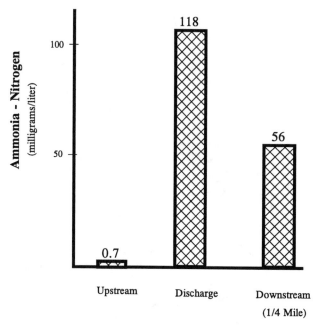

Figure 2. Ammonia-nitrogen concentrations in North Creek tributary resulting from swine waste discharge.

CASE HISTORY #2

The second incident involved a 500-head cattle feeding operation in west-central Illinois. The facility consisted of a large barn and an open concrete feedlot. The feedlot was sloped to one end to facilitate the drainage of manure into a waste collection pit.

Manure from the collection pit was periodically pumped into a 400,000-gal above-ground storage structure (approximately 60 ft in diameter x 20 ft tall). The pit was connected to the storage tank by a 12-in.-diameter buried pipe. In August of 1991, a valve on the transfer pipe was inadvertently left open. This allowed manure from the storage tank to backflow into the smaller pit, which ultimately overflowed. Approximately 88,000 gal of concentrated cattle manure was released during the incident.

The manure traversed overland through a field and discharged into Ellison Creek and, ultimately, the Mississippi River. More than 8 miles of the stream was adversely impacted by the undiluted manure. An extensive fish kill resulted that involved 19 species of fish. An IDOC Fisheries Division biologist estimated 34,900 fish killed at a total value of $6,700.

The ammonia-N concentration in the manure discharge was 1,423 mg NH_4-N/L. The instream ammonia-N concentrations ranged from 0.06 mg NH_4-N/L upstream to 60 mg NH_4-N/L 2 miles downstream from the discharge

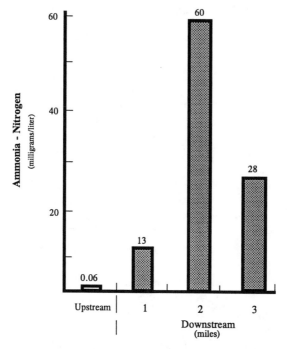

Figure 3. Ammonia-nitrogen concentrations in Ellison Creek resulting from cattle waste discharge (Units = mg NH$_4$-N/L).

point. Figure 3 graphically illustrates the instream ammonia-N concentrations. The BOD of the discharged waste was 28,740 mg/L. Dissolved oxygen concentrations were 0.2 mg/L at three downstream sampling stations 2 and 3 miles downstream of the discharge point. The upstream dissolved oxygen content measured 8.2 mg/L.

This incident points out the severe water quality impacts that may be caused by the release of large quantities of undiluted livestock waste. It also shows that even the better-designed waste handling systems are not failproof and require diligent management and operation. Preventive measures recommended for facilities having similar setups are the installation of high-water-level alarm systems, top-fill methods, and elimination of all bottom drain devices and/or multiple shutoff valves in the plumbing system. It is also advised that producers establish a routine procedure to inspect all system components and lock the valves shut once the waste transfer operation is complete.

CASE HISTORY #3

The third incident involves the land application of animal waste. Surface application of liquid livestock waste to cropland is a common agricultural practice in Illinois. Illinois livestock waste regulations do not prohibit surface application of livestock waste. However, these regulations do contain design criteria that recommend limiting surface application to areas where land slope is no greater than 5% or when the yearly average soil loss is less than 5 tons/acre. It is recognized that even on land with slopes less than 5%, soil losses and associated pollution may reach unacceptable levels.

Such was the case with a 500-head sow, farrow-to-finish swine facility located in west-central Illinois. During December, 1992 and January 1993, approximately 84,000 gal of liquid swine waste was surface-applied on frozen and ice-covered cropland in a 60-acre field. The application field was relatively flat with less than 5% slope. Approximately half of the 60-acre field was located within the watershed of a neighboring private pond. Thawing conditions and rainfall in early 1993 caused the manure to run off the field and enter the private pond. Severe water quality degradation occurred in the 1-acre pond and in Pigeon Hollow, the stream that received overflow from the pond. In March 1993, the pond was observed to be turbid and black colored and had a distinct swine waste odor.

A fish kill resulted in the private pond affecting several species, including channel catfish and largemouth bass. IDOC estimated that 2,300 fish were killed at a value of $3250. Laboratory results from five water samples taken from the pond showed ammonia concentrations ranging from 1.0 to 11 mg NH4-N/L. Ammonia concentrations were elevated for at least 2 miles downstream. The BOD concentration in the pond was 215 mg/L. The receiving stream contained 200 mg/L BOD about 1/8 mile downstream from the pond.

This incident occurred as a result of surface application of liquid livestock waste on frozen or snow-covered ground. In order to prevent such runoff, producers are encouraged to avoid surface application when frozen and snow-covered conditions prevail. Furthermore, waste application should be limited to those seasons when direct injection or incorporation is feasible. Knifing or injecting the waste into cropland has several advantages as it prevents surface runoff, increases fertilizer value to the crop, and prevents odor nuisance conditions. In order to maintain flexibility in the timing of application, producers may need to provide a minimum of 6 month's storage capacity, and they must ensure that an adequate land base exists upon which the waste can be beneficially used. To facilitate coordinating the timing of application with available land base, it is important for the producer to develop a written manure utilization plan.

SUMMARY

Studies and field investigations conducted by the IEPA demonstrate that confined animal feeding operation can adversely impact the surface water resources in Illinois. Long-term effects, which include water quality degradation and alteration of the biotic make-up in stream ecosystems, are associated with periodic precipitation/runoff events in watersheds densely populated with livestock. More visible and devastating effects result when large quantities of concentrated livestock waste are discharged into streams and surface impoundments.

Although Illinois Livestock Waste Regulations require all confined animal feeding operations to implement waste runoff controls, wastewater discharges do occur. The discharges are often the consequence of facilities not having the required manure containment structures, inattentive operation of equipment, insufficient design of equipment or waste handling systems, and poor waste utilization practices.

The proper management of livestock waste is critical to maintaining and improving the integrity of Illinois' lakes and streams. To ensure that this goal is realized, livestock producers in Illinois must diligently attend to both the containment and utilization aspects of managing the manure generated by their respective operations.

REFERENCES

American Fisheries Society (AFS), *Monetary Value of Freshwater Fish*, Special Publication # 24 1992

Hite. R., Bickers, C., King, M., and Brockamp, D., *Effects of Livestock Wastes on Small Illinois Streams: Lower Kaskaskia River Basin and Upper Little Wabash River Basin Summer 1991*, IEPA/WPC/92-114. Illinois Environmental Protection Agency, Springfied, Illinois., 1992.

Illinois Dept. of Agriculture (IDOA), *Illinois Agricultural Statistics Annual Summary* 1993, Bulletin 93-1, Springfield, Illinois, 1993.

Illinois Department of Conservation (IDOC), *Illinois Fisheries Division Manual*, Section 6200 Pollution-Caused Fish Kill Investigations, Springfield, Illinois, 1992.

Illinois Environmental Protection Agency (IEPA), *A Field Guide For Environmental Sampling*, 1991 Revised Edition, Springfield, Illinois, 1991.

Illinois Environmental Protection Agency (IEPA), *Illinois Water Quality Management Plan*, IEPA/WPC/92-220, Springfield, Illinois, 1992.

Illinois Pollution Control Board (IPCB), *35 Illinois Administrative Code Subtitle C: Water Pollution*, Chicago, Illinois, 1990.

Illinois Pollution Control Board (IPCB), *35 Illinois Administrative Code Subtitle E: Agriculture Related Pollution*, Chicago, Illinois, 1992.

US Environmental Protection Agency (US EPA), *Methods For Organic Chemical Analysis of Municipal and Industrial Wastewater*, Washington, D.C., 1992.

USING RIPARIAN BUFFERS
TO TREAT ANIMAL WASTE

R.K. Hubbard, G. Vellidis, R. Lowrance, J.G. Davis, and G.L. Newton

INTRODUCTION

A number of researchers have investigated use of filter strips for treatment of animal wastes. Doyle et al., (1977) applied 850 kg N/ha (90 t/ha of dairy manure) and found that 3.8 m of forest buffer or 4.0 m of grass buffer was useful in improving the water quality of manure-polluted runoff. Thompson (1977) tested a 24-m-long waste area that received approximately 600 kg N/ha (63 t/ha of dairy manure) and found significant reductions in concentrations with distance downslope for buffer strip lengths of 12.2 and 36.6 m. Dickey et al. (1977) and Swanson et al. (1975) found that vegetative filters and a serpentine waterway, respectively, permitted highly polluted runoff from barnlots and feedlots to be infiltrated into the soil and diluted by runoff from outside areas.

Riparian forests are known to be effective in reducing non-point pollution from agricultural fields. At least three separate studies at different sites in the Gulf-Atlantic Coastal Plain region have shown that concentrations and loads of nitrogen (N) in surface runoff and sub-surface flow are markedly reduced after passage through a riparian forest (Jacobs and Gilliam, 1985; Peterjohn and Correll, 1984).

Riparian zones reduce N concentrations in water arriving from uplands through vegetative uptake and denitrification. Several investigators (Vitousek and Reiners, 1975; Lowrance et al., 1984) have suggested that select harvest of "mature" trees in riparian forests is a method of perpetuating vigorous vegetative uptake of soil nutrients. Odum (1969) hypothesized that constant, pulsed, and annually increasing impacts of nutrients may keep the riparian forest in a "bloom" state, and the forest may respond by high growth and nutrient uptake rates for a considerable period of time.

The limited field data on using riparian forests to control agricultural non-point pollution have been integrated into draft national specifications for riparian buffer systems by the USDA-Soil Conservation Service and Forest Service. These draft specifications provide for a riparian buffer system of three zones (Welsch, 1991). Zone 1 is a narrow band of permanent trees (5-10 m wide) immediately adjacent to the stream channel that provides streambank stabilization, organic debris input to streams, and shading of streams. Zone 2 is a forest management zone where maximum biomass production is stressed, within limits placed by economic goals. Zone 3 is a grass buffer strip up to 10 m wide to provide control of coarse sediment and spreading of overland flow.

Ongoing research at the Coastal Plain Experiment Station, Tifton, Georgia, is focusing on determining the feasibility of using riparian zones for utilization and treatment of animal waste. In one study a riparian buffer zone downslope of a site receiving liquid dairy waste has been re-established by planting hardwoods adjacent to the stream and pines in the Zone 2 portion of the landscape. A second study is determining the feasibility of using Zones 3 and 2 of riparian buffer systems (Welsch, 1991) to utilize and treat swine lagoon waste. The objectives for both of these projects are to determine the overall effectiveness of riparian buffer systems in utilizing the wastes and to gain basic information on the nutrient and heavy metal filtering processes. These are two completely different projects with different design and sampling techniques, but they are related in that they both focus on the ability of riparian buffer systems to filter nutrients from animal wastes. This chapter describes both projects and presents preliminary results.

WETLAND RESTORATION AND FILTERING
OF DAIRY LAGOON WASTE

This project, initiated in 1991, is being conducted at a site where screened liquid dairy manure from a storage lagoon is applied on 5.6 ha by center pivot irrigation. The waste is applied to the east, west, north and south pivot quadrants at N application rates of 200, 400, 600, and 800 kg/ha/year, respectively (Figure 1). The cropping system consists of overseeding of abruzzi rye (*Secale cereale* L.) into 'Tifton 44' bermudagrass (*Cynodon dactylon* L.) sod in the fall, followed by minimum tillage planting of silage corn (*Zea mays* L.) into the bermudagrass and rye stubble in the spring, followed by summer crops of hay or silage from the residual bermudagrass.

The north quadrant of the pivot drains downslope into a natural wetland area (Figure 1). The wetland is being restored to a forested condition that allows determination of the effects of the wetland on water quality during the restoration process. The specific objectives for the restoration were (a) to measure nutrient (N, P) concentration changes in surface runoff and shallow ground water as they move through the wetland; (b) to determine nutrient uptake and removal in the wetland by soil microbial processes and vegetation; and (c) to evaluate the wetland as a potential bioremediation site.

The wetland was partially restored in February 1991 by reintroducing a combination of native trees. The trees are being grown for eventual harvest as pulpwood, timber wood, or both. Slash pine (*Pinus elliottii* Engelm.), swamp black gum (*Nyssa sylvatica* var. *biflora* Marsh.), and green ash (*Fraxinus pennsylvanica* (Borkh.) Sarg.) were selected to provide fast growth and year-

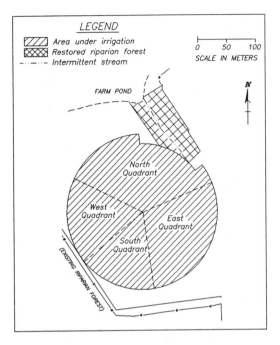

Figure 1. Dairy upland and restored riparian zone.

round nutrient uptake. Slash pine is commonly used in the coastal plain in the landscape position analogous to Zone 2 of the riparian forest buffer system, while swamp black gum and green ash commonly grow in the wetter areas near the streams.

Water quality measurements are made on water entering, moving through, and exiting the wetland. This is accomplished using a combination of monitoring wells, surface runoff collectors, and flumes. Ground water in and around the perimeter of the wetland is monitored by wells. Wells on the side adjacent to the animal waste application site are installed to a depth of 1 m and are fully slotted up to the soil surface. Plinthite and underlying materials of the Miocene age Hawthorn Formation form an effective aquitard at this site beginning at depths of approximately 0.7 m. A biweekly sampling schedule is used for measuring the depth to the water table and also for collecting samples for nutrient analyses.

Surface runoff is sampled at two locations entering the wetland and at two locations near the stream flow (Vellidis et al., 1993). At each location, the runoff is collected in a gutter, passed through a 200-mm Modified Tucson Flume, and redistributed through a slotted gutter. Composite water samples are collected from the flumes.

Water samples from the surface runoff collectors and shallow wells in the wetland are analyzed for NO_3-N, NH_4-N, total N, PO_4-P, and total P. These analyses are performed by standard methods on the Lachat Quikchem AE Flow Injection Analyzer[1] (APHA, 1989). Data from the study to date show that NO_3-N concentrations entering the upper end of the wetland in shallow ground water average 8.2 mg/L (SD = 5.6) while the mean NO_3-N concentration in ground water in the drainage way is 0.9 mg/L (SD = 3.0). Data from surface runoff have also shown reduction of both N and P concentrations as water passes through the wetland.

Evaluation of the wetland as a bioremediation site will be accomplished by maintaining a nutrient budget for the system over the life of the project (Hubbard et al., 1992). This budget will include the observations of surface and shallow ground water quality plus results from soil samples. Soil samples for denitrification and inorganic N measurements are being taken monthly at five depth increments to 0.3 m. Gaseous losses of N from the soil through denitrification are measured in intact core samples. It is anticipated that the effectiveness of the wetland ecosystem as a bioremediation system will increase as the trees mature.

OVERLAND FLOW-RIPARIAN ZONE TREATMENT OF SWINE WASTE

This project, initiated in January 1993, will (1) determine the ability of grass-riparian zone buffer strips to cleanse animal waste moving through the system via overland flow, and (2) compare the filtering effectiveness of naturally occurring riparian vegetation with that of recommended wetland species. Three different vegetative treatments are being used for the study: (1) 10-m grass buffer ('Tifton 78' bermudagrass (*Cynodon dactylon* L. Pers.) with an interseeding of 'Georgia 5' fescue, a heat-tolerant tall fescue (*Festuca arundinacea*) cultivar (for winter cover), draining into 20 m of natural riparian forest vegetation, (2) 20-m grass buffer draining into 10 m natural riparian forest vegetation, and (3) 10-m grass buffer draining into 20-m maidencane (*Panicum hematomon*), a species recommended for constructed wetlands (Figure 2).

The animal waste used for the study is swine lagoon waste, which is applied to replicated plots of each vegetative treatment at either high (2X) or low (1X) waste water application rates. Within the overall objectives the study will determine rates of sediment, N, P, Cu, and Zn removal. Movement of N (NO_3-N, NH_4-N, and total N), P (ortho, bioavailable, and total), sediment, Cu and Zn is measured in surface runoff-overland flow by collectors both within and at the bottom edge of the plots (Figure 2). Movement of N and ortho-P is

---

[1]Mention of trade names, commercial products, or companies in this publication is solely for the purpose of providing specific information and does not imply recommendation or endorsement by the U.S. Department of Agriculture over others not mentioned.

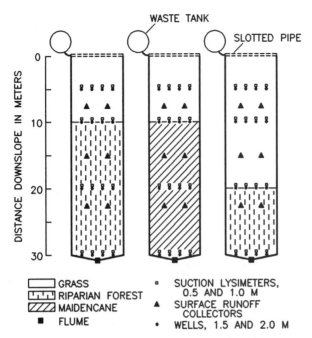

Figure 2. Vegetative treatments and instrumentation for swine waste project.

measured in the root zone and shallow ground water using suction lysimeters and shallow wells. Suction lysimeters are located at depths of 0.5 and 1.0 m, while shallow wells are located at depths of 1.5 and 2.0 m. This site also is underlain by plinthite and subsoil materials, which form an aquitard.

Waste has been applied to the plots on a schedule approximating twice per week to the 2X plots and once per week to the 1X plots. Analyses have shown that NO_3-N concentrations in the waste are quite low, all being less than 4 mg/L with a mean of 0.6 mg/L (SD = 0.6). Ammonium concentrations are much greater, ranging from 28 to 122 mg/L with a mean of 75 mg/L (SD = 23). Using average total N concentrations in the waste, the current 1X and 2X waste rates will supply approximately 600 and 1200 kg N/ha/year, respectively. The 600-kg/ha/year rate is the amount of N commonly applied to triple cropping systems in the Coastal Plain.

Evaluation of the data so far has primarily focused on N. Mean NH_4-N concentrations in surface runoff within the plots over all treatments at 7.5, 15, and 22.5 m downslope during fall 1993 and winter 1994 were 25, 14, and 5 mg/L (SD = 1.6, 2.3, and 2.7), respectively. Although high NH_4-N concentrations were occasionally observed 22.5 m downslope, most runoff events showed that the system was removing NH_4-N. Runoff samples collected at the ends of the plots in December 1993 through March 1994 showed little to no NH_4-N reaching the ends of the plots. Ammonium concentrations in these samples ranged from less than 0.1 to 3.6 mg/L with an overall mean of 0.3 mg/L (SD = 0.3).

Mean NO_3-N concentrations in surface runoff from all three within-plot sampling locations for events from October 1993 through March 1994 were all less than 20 mg/L. No distinct pattern relative to sampling position was apparent on any of the vegetative treatments. Nitrate concentrations in runoff samples collected at the ends of the plots from December 1993 through March 1994 ranged from less than 0.1 to 3.2 mg/L with a mean of 0.8 mg/L (SD = 0.5). Chloride concentrations in these same samples ranged from 0.1 to 16.8 mg/L with a mean of 3.7 mg/L (SD = 1.4). The Cl concentrations indicated that solutes from the waste were reaching the ends of the plots while the lower NO_3-N concentrations indicated that N species were being effectively filtered by the vegetation and by denitrification.

Mean NO_3-N concentrations in the samples collected from the 0.5-m suction lysimeters from all three vegetative treatments were highest 4.5 and 9.5 m downslope from the waste application pipe. Mean NO_3-N concentrations over all treatments were 30.9 and 22.7 mg/L (SD = 27.4, 24.2) at these two site positions, respectively. Mean NO_3-N concentrations at 19.5 and 29.5 m downslope from the waste application pipe were 7.4 and 4.4 mg/L (SD = 7.2, 5.5), respectively. Overall the suction lysimeter data showed that through April 1994 the overland flow-riparian zone systems effectively filtered NO_3-N such that concentrations in shallow ground water at 0.5-m depth at the ends of the plots did not exceed background levels. Mean NH_4-N concentrations in these lysimeters indicated some leaching at the 9.5-m site position but no leaching farther downslope.

Mean NO_3-N concentrations in the shallow ground water at 1.5-m depth were highest at the 5- and 10-m site positions and lower at the 20- and 30-m site positions. Mean NO_3-N concentrations at these four positions for the sampling period were 19.3, 17.4, 10.2, and 7.8 mg/L (SD = 18.3, 16.4, 5.7, 3.9), respectively. Some impact of the applied waste on the ground water at 20 m is already apparent, but the concentrations at 30 m under all treatments are still at background levels. Little or no leaching of NH_4-N to 1.5-m depth was observed on any of the treatments.

Denitrification measurements on cores collected from the plots showed that the denitrifiers are responding to the applied waste as a source of both N and carbon. Mean denitrification rates in the 0- to 12-cm soil depth over all treatments were 38 (SD = 83), 19 (SD = 39), 78 (SD = 148), and 440 (SD = 623) kg/ha/year for July 1993, November 1993, December 1993, and February 1994, respectively.

SUMMARY

Studies in a variety of physiographic areas have shown that riparian buffer systems are effective in filtering sediment and nutrients entering from upslope agricultural fields. The mechanisms involved are both physical and biological, including deposition, uptake by vegetation, and loss by microbiological processes such as denitrification. Our research is investigating the use of riparian buffer systems to filter nutrients from animal waste. In one study in which liquid dairy manure is applied by irrigation to an upland site, we are measuring the effects of a downslope restored riparian forest on filtering of nutrients from the animal wastes that enter the system via surface runoff or shallow lateral flow. We are measuring the effectiveness of three different vegetative treatments within riparian buffer systems for filtering nutrients from overland flows of swine lagoon waste. Waste is applied at the upper end of each plot at either a high or low rate and then allowed to flow downslope. Data collected since 1991 from the restored wetland, which is impacted by liquid dairy manure, show that the wetland buffer system is effectively removing nutrients. The swine waste overland flow grass riparian buffer plots first received waste in October 1993. Data collected to date indicate that the systems are utilizing N and P such that the high concentrations found in the waste are not observed in water at the bottom ends of the plots. Both of these research projects are designed to aid land managers in developing systems for utilizing and treating animal waste so that soils and waters are not degraded.

ACKNOWLEDGMENTS

The dairy upland research project was funded for three years by the United States Department of Agriculture-Cooperative State Research Service (USDA-CSRS) under its Low Input Sustainable Research and Education Program beginning in October 1991. The overland flow-riparian zone treatment of swine waste project was funded from 1992-96 through a USDA-CSRS National Research Initiative (NRI) competitive grant (92-37102-7399). The research reported herein also was supported by Hatch and State funds allocated to the Georgia Agricultural Experiment Stations and funds allocated to the USDA-ARS Southeast Watershed Research Laboratory.

REFERENCES

APHA, *Standard Methods for the Examination of Water and Wastewater,* 17th edition, American Public Health Association, Washington, D.C., 1989.

Dickey, E.C., Vanderholm, D.H., Jackobs, J.A., and Spahr, S.L., *Vegetative Filter Treatment of Feedlot Runoff,* ASAE Paper No. 77-4581, ASAE, St. Joseph, Michigan 49085, 1977.

Doyle, R.C., Stanton, G.D., and Wolf, D.C., *Effectiveness of Forest and Grass Buffer Strips in Improving the Water Quality of Manure Polluted Runoff,* ASAE Paper No. 77-2501, 1977.

Hubbard, R.K., Vellidis, G., and Lowrance, R., Wetland restoration for filtering nutrients from an animal waste application site, in *ASAE Symposium, Land Reclamation: Advances in Research and Technology,* 144, 1992.

Jacobs, T.C., and Gilliam, J.W., Riparian losses of nitrate from agricultural drainage waters, *J. Environ. Qual.,* 14, 472, 1985.

Lowrance, R.R., Todd, R.L., and Asmussen, L.E., Nutrient cycling in an agricultural watershed: I. Phreatic movement, *J. Environ. Qual.*, 13, 22, 1984.

Odum, E.P., The strategy of ecosystem development, *Science* 16, 262, 1969.

Peterjohn, W.T., and Correll, D.L., Nutrient dynamics in an agricultural watershed: Observations on the role of a riparian forest, *Ecology,* 65, 1466, 1984.

Swanson, N.P., Linderman, C.L., and Mielke, L.N., Direct land disposal of feedlot runoff, in *Managing Livestock Waste,* ASAE Pub. Proc-275, St.Joseph, Michigan, 255, 1975.

Thompson, Dale B., Nutrient movement during winter runoff from manure treated plots, M.S. Thesis, Michigan State University, East Lansing, Michigan, 1977.

Vellidis, G., Lowrance, R., Smith, M.C., and Hubbard, R.K., Methods to assess the water quality impact of a restored riparian wetland, *J. Soil and Water Conserv.*, 48(3), 223, 1993.

Vitousek, P.M., and Reiners, W.A., Ecosystem succession and nutrient retention: A hypothesis, *BioScience,* 25, 262, 1975.

Welsch, D.J., *Riparian forest buffers,* USDA-FS Publication No. NA-PR-07-91. USDA-FS, Radnor, Pennsylvania, 1991.

STREAM BIOASSESSMENT AND CONTRASTING LAND USES IN THE TENNESSEE VALLEY

Billie L. Kerans, Steven A. Ahlstedt, Thomas A. McDonough,
Frank J. Sagona, and Charles F. Saylor.

INTRODUCTION

Protection of our environment requires balanced assessments of both human and ecological problems associated with pollution (Karr, 1993). Natural streams are diverse, complex ecosystems. Catastrophic changes resulting from anthropogenic impacts are usually relatively easy to observe (e.g., massive fish kills). However, the effects of human society may often be less obvious and, therefore, much more difficult to observe and quantify. Subtle changes, for example a slight population decline in just one species, can indirectly affect ecosystems, causing changes that cascade through the biota and shift the structure and function of the aquatic assemblage. The complexity of biological systems and the diverse influences of human society require a broadly based approach to ecological assessment (Karr, 1991; Kerans and Karr, 1994).

Impacts to stream ecosystems from agriculture include riparian zone destruction, increased sediment loads and organic enrichment, and bacterial and pesticide contamination (Cooper, 1993). Near the site of the impact, stream biota can be strongly affected. Chronic non-point source (NPS) pollution may be a problem farther from the source of contamination, especially at the watershed or landscape level. Using the biota to assess ecological health is especially useful on the landscape level because the biota integrate the effects of all disturbances (Karr, 1991).

This study examines the effects of land use on macroinvertebrate assemblages of small sub-watersheds of the Middle Fork Holston River in the Tennessee Valley. Many sub-watersheds are disturbed by agriculture, especially livestock operations. Using aerial surveys, Sagona and Phillips (1993) determined the number of livestock operations, soil loss potential, and amount of riparian zone destruction in the sub-watersheds. These factors were then integrated into a watershed index of pollution potential (WIPP), which quantifies the relative potential for NPS pollution in the sub-watersheds. The purpose of this chapter is to relate the composition and functional structure of stream macroinvertebrate assemblages to differences in WIPP scores among sub-watersheds of the Middle Fork Holston. We answer two specific questions: Did sub-watersheds with similar WIPP scores have similar invertebrate assemblages? Do the invertebrate characteristics that define sub-watersheds match expectations concerning the responses of invertebrate assemblages to pollution?

BACKGROUND

The Middle Fork Holston (Table 1) has been identified as a watershed with significant potential for stream impacts attributed to agricultural NPS pollution (Sagona and Malone, 1989). Aerial color infrared photography at a scale of 2.54 cm to 610 m was used to characterize livestock operations, soil loss potential (from land use and cover data) and riparian zone destruction (Table 2). These data were combined into an watershed index of pollution potential (WIPP; Sagona and Phillips, 1993). Pollutant classes evaluated for the WIPP include: nutrients and pathogens, sediments and turbidity, and stream bank condition. WIPP scores range between 60 and 12 and are inversely related to potential NPS impacts.

Bear, Carlock, and Walker have the highest WIPP scores, indicating that they are potentially the least disturbed sub-watersheds (Table 2). All three sub-watersheds have low numbers of livestock operations, low soil loss potentials,

Table 1. Selected physical characteristics of Middle Fork Holston River sub-watersheds where macroinvertebrates were monitored.

Hydrologic Unit	Drainage Basin Area	Length of Streams	Bioassessment Years
	km²	km	
Bear	37.4	50.2	1990-1993
Walker	38.1	54.7	1992
Carlock	18.7	29.7	1992
Greenway	19.1	34.3	1991
Hutton	29.7	31.7	1990-1993
Hall	40.6	34.1	1990-1993
Upper Middle Fork	78.7	58.6	1991

From Sagona (1992) and Sagona and Phillips (1993)

Table 2. Land uses of Middle Fork Holston River sub-watersheds.

Hydrologic Unit	Livestock Operations		Soil Loss Potential	Disturbed Streambanks		WIPP
	Number	Density	Average Loss	Length	Percent	
		sites/km²	1000kg/h/yr	km		
Bear	11	0.3	8.2	7.1	14.1	47
Walker	13	0.3	7.1	10.3	18.8	44
Carlock	21	1.1	7.2	13.8	46.5	41
Greenway	31	1.6	39.1	10.1	29.6	34
Hutton	34	1.1	18.7	20.1	63.4	33
Hall	43	1.1	25.1	21.4	62.8	23
Upper Middle Fork	63	0.8	12.8	17.7	30.2	19

From Sagona (1992) and Sagona and Phillips (1993)

and few km of disturbed stream banks. However, the Carlock basin is small and may be more disturbed than the WIPP suggests as indicated by the high density of livestock operations and percentage of disturbed stream banks (Table 2). Greenway and Hutton sub-watersheds have intermediate WIPP scores resulting from intermediate numbers of livestock operations. Greenway sub-watershed has a higher soil loss potential and a lower percentage of disturbed stream banks than the Hutton sub-watershed. The Hall and Upper Middle Fork sub-watersheds have the lowest WIPP scores and the highest number of livestock operations (Table 2). The Hall sub-watershed has a higher soil loss potential, percentage of disturbed stream banks and density of livestock operations than the Middle Fork sub-watershed.

METHODS

From 1990 to 1993 TVA biologists sampled benthic invertebrates during April and May (Table 1). Three replicate samples were taken in pools and runs (Hess samples, 856 cm^2) and riffles (Surber samples, 929 cm^2). Substrates were disturbed to a depth of 9 cm, and organisms were captured in downstream nets (1000 μm). A D-frame net (900 μm) and handpicking were used to produce one qualitative sample of all habitats present at each site. Biologists limited qualitative sampling to a maximum of 2 h at each site. Invertebrate samples were preserved in the field and returned to the laboratory where all organisms were counted and identified through contract to TVA.

We investigated 14 characteristics of invertebrate assemblages whose usefulness in determination of the biological condition of streams has been previously explored (e.g., Karr and Kerans, 1992; Kerans and Karr, 1994). These included two suites of characteristics. The first suite described the structure and composition of the assemblage and included total, mayfly, stonefly, and caddisfly taxa richnesses, which generally decline with increasing human disturbance. The relative abundances (number of organisms in group/total abundance) of oligochaetes and chironomids and community dominance (number of organisms in the two most abundant taxa/total abundance), which we hypothesized would increase with increasing human disturbance, were also included in this group.

The second suite described the trophic and functional composition of the assemblage. These characteristics were related to ecological processes and the food base of the assemblage. Characteristics included relative abundances of shredders, grazers, and predators and total abundance, which we hypothesized would decline with increasing human disturbance, and relative abundances of omnivores, collector-gatherers, and collector-filterers, which we hypothesized would increase with increasing human disturbance.

In these exploratory analyses, we used principal components analyses (PCA) to separate sites based on invertebrate characteristics. PCA was used instead of the benthic index of biotic integrity (B-IBI; Kerans and Karr, 1994) because the B-IBI requires more data than we had available in this dataset and the B-IBI has not been calibrated for smaller streams such as those in this study. We did separate PCAs for riffles, pools, and the qualitative samples. PCAs used the correlation matrix and were based on means of the characteristics for each sub-watershed in each year. We used numerators of the proportional characteristics to avoid spurious correlations. Taxa counts were untransformed. All other characteristics were log transformed, $\log([x + 1])$. We confined our analyses to the first two principal axes, which explained the major variability among sub-watersheds. We determined the association between the PCA scores for each sub-watershed (means used when data included more than one year) and WIPP scores and NPS pollution measures (Table 2) using Spearman Rank correlations.

RESULTS

In riffles 71% of the variation among sub-watersheds was described by the first two PC axes. Sub-watersheds with similar WIPP scores clustered very clearly (Figure 1a). Sub-watersheds with the highest WIPP scores (Bear, Carlock, and Walker) were negatively associated with PC1 and positively associated with PC2. All invertebrate characteristics had positive associations with PC1 (Table 3), reflecting the positive correlations among total abundances and total taxa richness and abundances and richnesses of individual taxa. However, the strengths of the associations of invertebrate characteristics and PC1 varied. Thus, negative associations of the sub-watersheds with PC1 generally indicated low taxa richnesses and abundances of invertebrate groups positively

Table 3. Eigenvalues of principal components analyses of sites.

Grouping	Riffles		Pools		Qualitative	
	PC1	PC2	PC1	PC2	PC1	PC2
Assemblage Composition						
Total Taxa	0.267	0.270	0.319	0.239	0.560	0.070
Mayfly Taxa	0.125	0.367	0.175	0.366	0.371	0.845
Caddisfly Taxa	0.208	0.164	0.166	0.371	0.568	-0.215
Stonefly Tax	0.010	0.533	0.209	0.201	0.475	-0.485
% Oligochaetes	0.202	-0.408	0.114	-0.478	--†	--
% Chironomids	0.232	0.100	0.302	-0.153	--	--
Dominance	0.333	-0.229	0.253	-0.341	--	--
Functional Structure						
% Omnivores	0.360	-0.046	0.357	0.046	--	--
% Shredders	0.252	0.207	0.340	0.004	--	--
% Gatherers	0.356	-0.115	0.298	-0.292	--	--
% Filterers	0.311	-0.080	0.305	-0.030	--	--
% Grazers	0.336	0.086	0.297	0.228	--	--
% Predators	0.062	0.405	0.140	0.236	--	--
Total Abundance	0.362	-0.128	0.308	-0.257	--	--

†Abundance attributes were not calculated for qualitative samples.

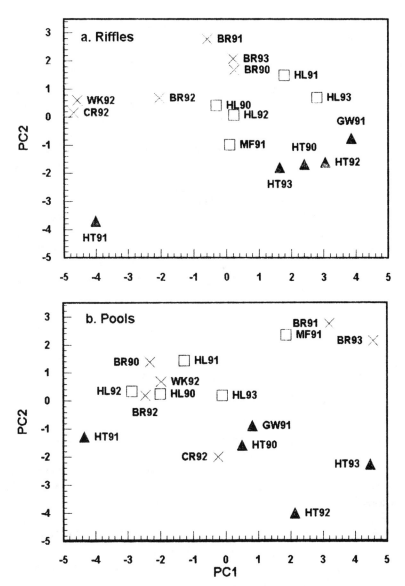

Figure 1. The sub-watersheds arrayed on the first two principal components based on the assemblage composition of (a) riffles and (b) pools. X indicates sub-watersheds with the lowest WIPP scores (CR92--samples taken in Carlock Creek in 1992; BR90, BR91, BR92, BR93--samples taken in Bear Creek from 1990 to 1993; WK92--samples taken in Walker Creek in 1992). ▲ indicates sub-watersheds with intermediate WIPP scores (HT90, HT91, HT92, HT93--samples taken in Hutton Creek from 1990 to 1993; GW91--samples taken in Greenway Creek in 1991). ☐ indicates sub-watersheds with the lowest WIPP scores (HL90, HL91, HL92, HL93--samples taken in Hall Creek from 1990 to 1993; MF91 samples taken in the Upper Middle Fork Holston in 1991).

associated with PC1 (Table 3). Positive scores on PC2 were associated with high predator abundance and stonefly and mayfly taxa richnesses and low oliogochaete abundance. Overall, sub-watersheds with the lowest NPS pollution potentials tended to have low total invertebrate abundance and low abundances of omnivores, gatherers, grazers, filterers, and dominance (negative scores on PC1) and high stonefly and mayfly taxa richnesses, high predator and low oligochaete abundances (positive scores on PC2). Sub-watersheds with intermediate WIPP scores (Hutton and Greenway) tended to have the opposite characteristics (except Hutton in 1991). Sub-watersheds with the lowest WIPP scores (Hall and Upper Middle Fork Holston) were postively associated with PC1 and tended to have little association with PC2 (Figure 1a). These sub-watersheds had high total abundance and high abundances of omnivores, gatherers, grazers, filterers and dominance (positive scores on PC2; Table 3).

Of the NPS pollution characteristics, only soil loss potential was positively associated with PC1 ($r = 0.954$, $P = 0.0005$, n = 7). Thus, the invertebrate characteristics that distinguished sub-watersheds with highest WIPP scores (Bear, Carlock, and Walker) in riffles tended to be associated with low soil loss potentials.

In pools 76% of the variation among invertebrate assemblages was explained by the first two PC axes. Invertebrate characteristics were arrayed on PC1 and PC2 in pools in a manner similar to the way the characteristics were arrayed on PC1 and PC2 in riffles (Table 3). The only major exception was that in pools high caddisfly taxa richness, rather than high stonefly taxa richness, was positively associated with PC2. Sub-watersheds with similar WIPP scores did not cluster as distinctly in pools as they did in riffles (Figure 1b). Sub-watersheds with intermediate WIPP scores (Hutton and Greenway) exhibited a positive association with PC1 (except for Hutton in 1991) and a negative association with PC2, which was similar to their associations with the PC axes in riffles. Thus, the Hutton and Greenway sub-watersheds exhibited invertebrate characteristics that were similar in riffles and in pools. Interestingly, the Carlock sub-watershed clustered with the Hutton and Greenway sub-watersheds. The other sub-watersheds tended to be positively associated with PC2 (except Carlock) and to have no clear association with PC1 (Figure 1b). Thus, sub-watersheds with the lowest and highest WIPP scores had high mayfly and caddisfly taxa richness and low dominance and abundance of oligochaetes.

Of the NPS pollution characteristics (Table 2), only livestock density was significantly correlated with any of the PC axes. Livestock density was negatively associated with PC2 ($r = -0.692$, $P = 0.085$, n = 7). Thus, sub-watersheds with intermediate WIPP scores (Hutton and Greenway) and the Carlock sub-watershed tended to have the highest livestock densities.

Only taxa richnesses were determined in the qualitative samples taken from all habitats. Eighty-five percent of the variation among sub-watersheds was explained by the first two PC axes. As in the analysis for pool samples, sub-watersheds with similar WIPP scores did not cluster as clearly as in riffles

(Figure 2). However, the Hutton, Greenway, and Carlock sub-watersheds appeared to separate from the other groups on PC1. Thus, these sites tended to have overall low taxa richnesses of groups thought to indicate good stream health (Table 3). Livestock density was negatively correlated with PC1 ($r = -0.748$, $P = 0.053$, n = 7).

DISCUSSION

Sub-watersheds with similar WIPP scores formed the most distinctive clusters using the invertebrate assemblages of riffles. Bear, Carlock, and Walker sub-watersheds, which the WIPP suggests have the least potential for NPS pollution, had high stonefly and mayfly taxa richnesses, high predator abundance, low total abundance, and low abundances of oligochaetes, omnivores, gatherers, filterers, grazers, and low dominance. These characteristics of invertebrate assemblages generally indicate streams with relatively high water quality (Lenat, 1984, 1988; Kerans and Karr, 1994). Hutton and Greenway sub-watersheds, which the WIPP suggests have intermediate potential for NPS pollution, had invertebrate assemblages with the opposite characteristics. Data from invertebrate assemblages were corroborated by the correlations with potential NPS pollution sources because the Hutton and Greenway assemblages were associated with high soil loss potentials, whereas the Bear, Carlock, and Walker assemblages were associated with low soil loss potentials.

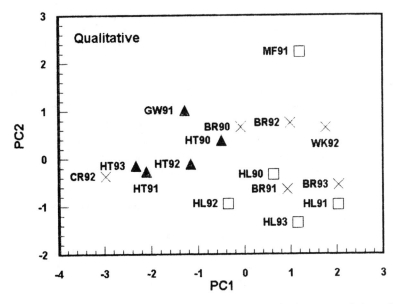

Figure 2. The sub-watersheds arrayed on the first two principal components based on assemblages of invertebrates in the qualitative samples. Symbols are as in Figure 1.

In contrast, only the Hutton and Greenway sub-watersheds formed a cluster that could be distinguished from the other sub-watersheds using invertebrate assemblages of pools and of the qualitative samples. The characteristics of invertebrate assemblages that separated these sub-watersheds from others were similar to those that separated sub-watersheds in riffles, indicating that quality of the invertebrate assemblages of the Hutton and Greenway sub-watersheds were similar in pools and riffles. However, in pools and in the qualitative samples, the Hutton and Greenway sub-watersheds (as well as the Carlock sub-watershed) were associated with high densities of livestock operations. These results suggest that although the invertebrate assemblages of pools, riffles, and the qualitative samples in the Hutton and Greenway sub-watersheds were similar, the NPS pollution sources that were associated with the assemblages differed. Invertebrate assemblages of riffles, pools, and in the qualitative samples may be responding to different NPS pollution sources. The invertebrate assemblages of pools and riffles of large rivers of the Tennessee Valley appear to exhibit different responses to disturbance (Kerans et al., 1992; Keran and Karr, 1994). In these small streams, the invertebrate assemblages of riffles, pools, and the qualitative samples provide slightly different information about the health of the sub-watersheds.

Interestingly, the Hall and Upper Middle Fork Holston sub-watersheds may actually be in better relative condition than indicated by the WIPP. Riffles of the Hall and Upper Middle Fork Holston sub-watersheds, which the WIPP suggests have the highest potential for NPS pollution, exhibited characteristics of invertebrate assemblages that were similar to both of the other WIPP groupings. Invertebrate assemblages were similar to the Bear, Carlock, and Walker sub-watersheds (high mayfly and stonefly taxa richnesses and predator abundance but low abundance of oligochaetes), indicating low levels of disturbance. Invertebrate assemblages were also similar to those of the Hutton and Greenway sub-watersheds (high abundances of filterers, gatherers, grazers, and omnivores and high dominance), indicating high levels of disturbance. Invertebrate assemblages of pools and in the qualitative samples of the Bear and Walker sub-watersheds also overlapped extensively with the Hall and Middle Fork Holston sub-watersheds. Using the WIPP, Sagona and Phillips (1993) showed that the Hall and Upper Middle Fork Holston sub-watersheds exhibited lower water quality than the level suggested by the IBI (index of biotic integrity; Karr et al., 1986; Saylor et al., 1988), which uses fish communities for bioassessment of stream quality. Thus, the WIPP suggests a higher potential for agricultural NPS impacts than that actually observed in the biota of these sub-watersheds.

The WIPP is based on occurrences of key land use and non-point source pollution features in the sub-watersheds (Sagona and Phillips, 1993). It is intended to indicate a potential for NPS impacts associated with observed land use features. It does not attempt to quantify transport or delivery of pollutants to streams. Spatial distribution and delivery processes may be such that the stream can transport or assimilate the watershed's NPS inputs. Consequently, large watersheds such as the Upper Middle Fork may have higher occurrences of non-point source pollution factors, but the effects on the biota may be smaller than in smaller watersheds with fewer sources of disturbance that are at higher densities (e.g., Greenway). Perhaps the types and actual numbers or densities of livestock (as opposed to livestock operations) may explain more variation in the invertebrate assemblages. Sagona and Phillips also suggest that Hall Creek has many springs that may improve its water quality through dilution. However, based on the invertebrate assemblages of riffles, it is clear that some level of disturbance is occurring in the Hall and Upper Middle Fork Holston watersheds.

Most invertebrate characteristics that were strongly associated with distinguishing sub-watersheds clustered matching our expectations concerning their response to pollution. Total abundance was the major exception because it was positively associated with many characteristics thought to indicate poor water quality. Although previous research in Tennessee Valley rivers suggests that high abundance often indicates good water quality (Kerans and Karr, 1994), the relationship between total abundance and water quality in these smaller sub-watersheds is complicated. Lenat (1984) has also shown that high abundances may be caused by nutrient enrichment by agriculture. Our results suggest a similar pattern. We suggest that total abundance must be used only cautiously in bioassessment.

Our work with these sub-watersheds of the Middle Fork Holston basin suggest that the actual measurements of NPS pollution features, the WIPP and the invertebrate assemblages of riffles, pools, and in the qualitative samples indicate the level of human disturbances to stream systems. The WIPP, densities of livestock operations and high soil loss potentials were associated with changes in invertebrate assemblages of sub-watersheds thought to indicate reduction in the quality of water resources. Using all of these pieces of information provides a powerful means of assessment of the quality of the water resources.

REFERENCES

Cooper, C.M., Biological effects of agriculturally derived surface water pollutants on aquatic systems --a review, *Journal of Environmental Quality*, 22, 402, 1993.

Karr, J.R., Biological integrity: A long-neglected aspect of water resource management, *Ecological Applications*, 1(1), 66, 1991.

Karr, J.R., Defining and assessing ecological integrity: Beyond water quality, *Environmental Toxicology and Chemistry*, 12, 1521, 1993.

Karr, J.R., Fausch, K.D., Angermeier, P.L., Yant, P.R., and Schlosser, I. J., *Assessment of Biological Integrity in Running Water: A Method and Its Rationale*, Illinois Natural History Survey Special Publication, Number 5, Champaign, Illinois, 1986.

Karr, J.R., and Kerans, B.L., Components of biological integrity: Their definition and use in development of an invertebrate IBI, in *Proceedings of the 1991 Midwest Pollution Control Biologists*

Meeting, Davis, W.S. and Simon, T.P., eds., U.S. Environmental Protection Agency, Chicago, Illinois, 1992, pages 1-16.

Kerans, B.L., and Karr, J.R., A benthic index of biotic integrity (B-IBI) for rivers of the Tennessee Valley, *Ecological Applications*, 4(4), 768, 1994.

Kerans, B.L., Karr, J.R. and Ahlstedt, S.A., Aquatic invertebrate assemblages: Spatial and temporal differences among sampling protocols, *Journal of the North American Benthological Society*, 11(4), 377, 1992.

Lenat, D.R., Agriculture and stream water quality: A biological evaluation of erosion control procedures, *Environmental Management*, 8, 33, 1984.

Lenat, D.R., Water quality assessment of streams using a qualitative collection method for benthic macroinvertebrates, *Journal of the North American Benthological Society*, 7, 222, 1988.

Sagona, F.J., *Watershed Management: The Connection Between Land Use and Water Quality*, paper presented at First Annual Southeastern Lakes Management Conference, Marietta, Georgia, March 18-20, 1992.

Sagona, F.J., and Malone, D.L., *Middle Fork Holston River Watershed Aerial Inventory Data Report*, Tennessee Valley Authority, Chattanooga, Tennessee, 1989.

Sagona, F.J., and Phillips, C.G., Application of watershed index of pollution potential to aerial inventory of land uses and nonpoint pollution sources, in *Proceedings of WATERSHED '93, A National Conference on Watershed Management*, Alexandria, Virginia, March 21-23, 1993, US EPA Rep. 840-R-94-002.

Saylor, C.F., Hill, D.M., Ahlstedt, S.A. and Brown, A.M., *Middle Fork Holston River Watershed Biological Assessment--Summers of 1986 and 1987*, Rep. TVA/ONRED/AWR-88/20. Tennessee Valley Authority, Chattanooga, Tennessee, 1988.

EFFECTS OF OPEN-RANGE LIVESTOCK GRAZING ON STREAM COMMUNITIES

Christopher T. Robinson and G. Wayne Minshall

INTRODUCTION

For over a century, streams of the Intermountain West have been strongly influenced by various land use practices. For example, the relative intensity and extensiveness of open-range livestock grazing over time reflect historical socio-economic developments. Indeed, major increases in open-range livestock densities have resulted from extensive mining activities in the 1850's and 60's, war efforts in the 1940's, and population growth presently (Young and Sparks, 1985; Holechek, 1993).

Present livestock densities, however, are at an all time high (Ferguson and Ferguson, 1983) with grazing policy still representing special interest groups. Livestock grazing is permitted on 91% of federal lands comprising 48% of the land area in 11 western states (Ferguson and Ferguson, 1983). Open-range livestock grazing occurs on 69% of western rangeland covering 260 million ha (Platts and Nelson, 1989) with about 40% of this land in a state of degradation (Platts, 1991).

The economic benefits of open-range livestock grazing currently are being questioned in light of public environmental concerns. Historically, livestock grazing in the West has resulted in poor economic returns, averaging only 2-6% yield on investment, with over 8100 ha needed to produce a reasonable income (Holechek, 1993). For example, areas having less than 112 kg/ha of quality forage typically result in financial loss (Holechek, 1993). Public lands also are heavily subsidized, with only 37% of grazing costs recovered on Bureau of Land Management-managed lands and 30% from US Forest Service lands (Ferguson and Ferguson, 1983). Further, vital livestock production areas will decrease as human population and recreational demands increase.

Livestock grazing is considered the primary long-term cause of habitat degradation of Western stream ecosystems (Platts, 1991). Serious concerns towards range conditions and, in particular, stream habitats were evident in the 1920's, resulting in the formulation of the US Grazing Service (now the BLM), the US Soil Conservation Service in 1930, and the 1934 Taylor Grazing Act in hopes of reversing the degradation of public lands (Minshall et al., 1989; Platts,

1991). However, the Forest and Rangeland Resource Planning Act of 1974 continued to stress demand of rangeland for grazing livestock, thus forcing resource managers to develop management strategies compatible with grazing and fishery goals.

Because of the propensity of livestock to use riparian and stream habitats for shade, quality forage, drinking water, and the relatively gentle topography, these areas tend to be degraded. Documented effects of livestock grazing on lotic habitats include alteration of riparian vegetation, stream channel widening, water table lowering, increase in fine sediments, destabilized stream banks, and increased water temperatures (Platts, 1991). In addition, although livestock exclusion along isolated stream segments typically results in improved riparian, streambank, and channel conditions, little improvement is observed in the fishery because of upstream sediment inputs (Platts and Nelson, 1989), stressing the non-point nature of open-range livestock grazing.

The present chapter focuses on the effects of open-range livestock grazing on stream macroinvertebrate communities in southern Idaho. A rapid bioassessment approach was utililized to evaluate stream habitats and associated macroinvertebrate assemblages (Plafkin et al., 1989; Robinson and Minshall, 1994). Rapid bioassessment protocols are an important tool in assessing the biological integrity of freshwater ecosystems in a cost-effective manner (Plafkin et al., 1989; Karr et al., 1986; Karr, 1991). Rapid bioassessment, based on a strong theoretical framework in community and ecosystem ecology, attempts to combine aspects of water quality and physical habitat measures with that of the aquatic biota to more fully assess a system's biological integrity (references in Rosenberg and Resh, 1993). Our primary objective was to relate differences in macroinvertebrate assemblages among streams subjected to various degrees of watershed disturbance (i.e., livestock grazing) to observed differences in habitat conditions.

METHODS

Rapid bioassessment was completed on 60 streams, comprising 32 sites in the Northern Basin and Range (NBR) ecoregion and 28 sites in the Snake River Plain (SRP) ecoregion. Stream types analyzed included a range of nongrazed and grazed sites on 2nd- to 4th-order streams (Strahler, 1957). Representative degraded sites were lowland areas perturbed primarily by livestock grazing and agricultural inputs.

Habitats were assessed for stream slope (clinometer), elevation (topographic maps), width/depth ratio, mean width, percent canopy cover, land-use, vegetative characteristics, discharge (Platts et al., 1983), riparian conditions, substrate measures, water temperature, pH, specific conductance, alkalinity, hardness, nitrate, phosphorus, and turbidity using standard methods (APHA, 1989; Robinson and Minshall, 1994). Field surveys also included a tally of an overall habitat quality score based upon a qualitative ranking of 12 habitat categories

(Plafkin et al., 1989; Clark and Maret, 1993). The score for each category is weighted in respect to its importance in describing habitat quality.

Periphyton chlorophyll a and ash-free dry mass (AFDM)(n=5) and benthic organic matter (BOM) were quantified at each study site. BOM was estimated from material obtained with the benthic macroinvertebrate samples. Periphyton was collected by scraping a known area from the surface of a stone and transfering the material onto a Whatman GF/F filter (Robinson and Minshall, 1986). Chlorophyll a was quantified using a Gilford Model 2600 spectrophotometer (APHA, 1989). Periphyton AFDM of each sample was determined using the remaining material from chlorophyll a analysis.

Multiple discriminant analysis (MDA) and principal components analysis (PCA) were completed using the habitat measures and the habitat assessment categories in order to distinguish between ecoregions and stream types. Selected quantitative measures were scored by proportional scaling of measured values over an arbitrary range of 0 to 15 (maximum score = 15) to make them comparable with the habitat assessment categories. Scores for each measure (category) were summed for each site for an overall habitat score.

Benthic sampling followed protocols III and V of the Rapid Bioassessment Protocols recommended by the US Environmental Protection Agency (Plafkin et al., 1989; Resh and Jackson, 1993), being collected at five riffle/run habitats using a modified Hess net (250-μm mesh) (Platts et al., 1983). In the laboratory, a minimum 300-count subsample of invertebrates was systematically handpicked from each sample for metric analysis. All picked macroinvertebrates were identified to lowest feasible taxonomic unit (usually genus) and enumerated.

Eighteen biotic metrics were calculated from the macroinvertebrate data from each site as described in Winget and Magnum (1979), Platts et al. (1983), Plafkin et al. (1989), and Clark and Maret (1993). Additional analyses were completed using the 20 most abundant taxa from each site. Important biotic metrics to distinguish between ecoregions and among stream types were determined for invertebrates using MDA and PCA (Statsoft, 1991). Once important metrics were determined for macroinvertebrates, metric criteria scores were summed for each site and regressed against respective habitat assessment scores. Additional regressions were completed for summed metric scores against habitat assessment scores by ecoregion. ANOVA was used to test for differences between the summed criteria metric scores among stream types and ecoregions (Zar, 1984).

RESULTS

Average habitat assessment scores were greater in non-grazed sites than in grazed streams for both ecoregions (Figure 1). Principal components analysis revealed that inclusion of maximum water temperature and ionic concentrations (e.g., specific conductance, alkalinity, and total hardness) explained an

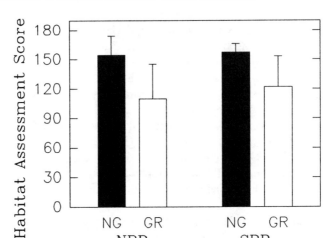

Figure 1. Habitat assessment scores for non-grazed (NG) and grazed (GR) streams within the NBR and SRP ecoregions in Idaho. Error bars equal one standard deviation from the mean.

additional 10% of the variation among sites for both ecoregions. Grazed sites displayed higher water temperatures and ionic concentrations than non-grazed sites.

Other habitat measures also showed average, but non-significant, differences among non-grazed and grazed sites because of high among-site variability. For example, mean chlorophyll a values were greater in grazed sites (NBR = 2.74, SRP = 2.13 µg/cm^2) relative to non-grazed sites (NBR = 0.50, SRP = 0.84 µg/cm^2). Average substrate size was lower and embeddedness values greater in grazed than in non-grazed sites. Further, the mean quantity of BOM was greater in degraded than in non-grazed streams.

Six of the 18 macroinvertebrate metrics calculated were found important (based on PCA results) for discriminating among stream types for either ecoregion: species richness, EPT richness, Hilsenhoff Biotic Index (HBI), %Dominance, Shannon's (H') diversity, and Simpson's index. PCA results also revealed that the metrics EPT/Chironomidae (CH), EPT/CH+Oligochaeta (O), %CH, and %CH+O were important for distinguishing among streams in the NBR ecoregion, while %Scrapers and the ratio of Scrapers/Filterers were important for the SRP ecoregion (Figure 2). The first two axes of the PCA explained 88% (NBR) and 78% (SRP) of the variation among sites within an ecoregion. MDA indicated that the EPT index was highly important for distinguishing among NBR sites (F = 34.3, $P < 0.0001$), and the HBI was important for distinguishing among sites in the SRP ecoregion (F = 13.7, $P < 0.001$). MDA also revealed that non-grazed sites were similar between NBR and SRP ecoregions, based on macroinvertebrate metrics, but that grazed sites were quite distinct between ecoregions (F = 3.53, $P < 0.001$). The difference between

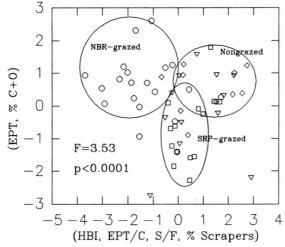

Figure 2. Scatterplot from MDA results based on the macroinvertebrate metrics. Circles fitted by eye to denote major groupings. Triangle = NBR-non-grazed, circle = NBR-grazed, diamond = SRP-non-grazed, and square = SRP-grazed.

grazed streams was primarily attributed to differences in the EPT index and %CH+O, a result also found by PCA.

A score of 5 indicated optimal values for each metric, based on 90% confidence limits around the average reference value, and resulted in a maximum summed score of 50 for the NBR ecoregion (10 metrics) and 40 for the SRP ecoregion (8 metrics). Scores ranged from 10 to 50 for NBR streams and from 7 to 31 for SRP sites (Figure 3). The biotic metric score displayed a positive regression against the habitat assessment score ($r^2 = 0.10$, NBR; $r^2 = 0.25$, SRP).

Multivariate statistics also were performed using the relative abundances of the 15-20 most abundant taxa collected among all sites. These taxa generally comprised over 80% of the macroinvertebrate assemblage at any one site. PCA results revealed that taxa of the Heptageniidae (e.g., the genera *Heptagenia, Epeorus, Cinygmula,* and *Rhithrogena*), Elmidae (e.g., the genera *Heterlimnius, Optioservus, Rhizelmis,* and *Cleptelmis*), and predacious Rhyacophilidae (all collected species except *Rhyacophila acropedes*) were more abundant in nongrazed than in grazed streams.

The first two axes of the PCA explained 61% (NBR) and 54% (SRP) of the variation among sites within ecoregions. Axis-1 for each ecoregion was described by Heptageniidae, Rhyacophilidae, *Capnia, Drunella,* and Turbellaria (Figure 4). Axis-2 had two taxonomic groups similar among ecoregions (*Brachycentrus* and *Hexatoma*), while the *Ephemerella* and the amphipod *Hyallela azteca* also were important for distinguishing NBR sites and the shredder *Zapada* was important for distinguishing among sites in the SRP ecoregion.

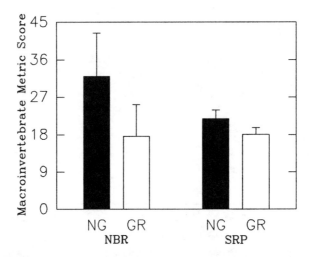

Figure 3. Mean macroinvertebrate metric scores for non-grazed (NG) and grazed (GR) streams in the NBR and SRP egoregions of Idaho. Bars represent +1SD from the mean.

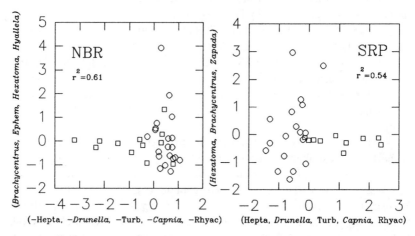

Figure 4. Scatterplot of PCA results based on the relative abundances of the 20 most abudant taxa for each ecoregion. Circles represent grazed and squares non-grazed sites

MDA results displayed patterns similar to the PCA, emphasizing the predominance of Heptageniidae and Rhyacophilidae in non-grazed sites for either ecoregion.

Based on the multivariate results (PCA and MDA), metric scores were derived for the Heptageniidae and Rhyacophilidae for each ecoregion and included with the invertebrate score. This modified score resulted in an average

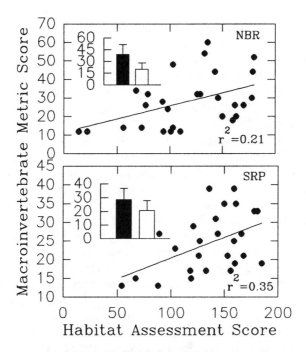

Figure 5. Regression of habitat assessment score against macroinvertebrate metric score for each ecoregion. Inset depicts mean macroinvertebrate score for non-grazed (solid bar) and grazed (open bar) sites within each ecoregion. "T" represents +1SD from the mean.

metric score of 20 for grazed and 40 for non-grazed sites in the NBR and scores of 21 and 28, respectively in the SRP (Figure 5). The addition of these taxonomic groups as metrics greatly improved the resolution of the macroinvertebrate metric score as evidenced by a major increase in correlation coefficients for the regression against the habitat assessment score.

DISCUSSION

Our results imply a strong integration between socio-economics and the eco-logical integrity of stream ecosystems in the Intermountain West. The present condition of aquatic resources is a consequence of over a century of extensive open-range livestock grazing. Even prior to the immigration of settlers, native Americans grazed large numbers of horses for trade purposes over large tracts of land (Robbins and Wolf, 1994). Most rangeland of these arid environs does not contain the foliage capacity to meet grazing demands at current stocking levels, thus augmenting the degradation of rangeland, riparian, and stream systems (Young and Sparks, 1985; Platts, 1991).

The results show grazed stream habitats to be substantially degraded with poor riparian conditions. Typically, streams influenced by grazing had reduced riparian cover, exposed stream banks, high sediment levels, greater water temperatures, and higher nutrient levels than non-grazed streams. However, even grazed streams that were exposed to minimal grazing pressure displayed degradation of some habitat parameters as a result of overgrazing in the headwaters. For example, streams heavily grazed in their headwaters typically showed accumulation of fine sediments, at times being over 30 cm in depth (C.T. Robinson, pers. observ.).

These changes in habitat also were reflected in the invertebrate assemblages with distinct communities being present among stream types. Heavily grazed streams generally were depauperate of cold-water taxa such as heptageniid mayflys and rhyacophilan caddisflies. These taxa also appear to prefer faster flowing water and cobble substrata; proportions of both are greatly reduced in degraded grazed streams. The alteration and degradation of stream habitats because of long-term livestock grazing has shifted the species assemblage to a predominance of more stress-tolerant taxa.

Our data also suggest a change in energy budgets and sources among stream types. The loss of elmid coleopteran taxa, consumers of detrital material, in livestock-degraded systems indicates differences in the quantity and/or quality of BOM among stream types. Further, autochthonous resources as an energy base may become predominant over allochthonous sources in lotic ecosystems in which the canopy coverage has been reduced (Minshall, 1978). However, functional processes have yet to be measured among stream types.

Open-range livestock grazing in the West has a long history of tradition and politics, e.g., over 90% of open-range livestock operations are controlled by less than 5% of the operators (Ferguson and Ferguson, 1983); thus change in management policies, although necessary, will be difficult. The Intermountain West contains some of the fastest growing areas in the USA with public recreational demands and environmental concerns keeping pace with this growth. Consequently, increased public scrutiny of management policies is imminent, and is already being seen with major river stretches being designated as wild and scenic or recreational. State and federal personnel must be held accountable for fostering the degradation of our streams and rivers as a result of overgrazing practices. Present legislation provides for the restoration and maintenance of the biotic integrity of our streams and rivers and needs to be put into practice by state and federal agencies.

ACKNOWLEDGMENTS

A number of individuals assisted in the successful completion of this project: D.M. Anderson, J.W. Check, T. Curzon, P.D. Dey, R. Gill, P. Koetsier, D.E. Lawrence, J.S. Mann, J. Mihuc, T.B. Mihuc, S.C. Minshall, G.C. Mladenka, D.C. Moser, C.A. Nelson, J.S. Nelson, M.A. Overfield, S.E. Relyea, T.V. Royer, K.W. Sant, S.A. Thomas, and J.T. Varricchione. Special thanks go to W.H. Clark, R.T. Litke, M. McMasters, D.W. Zaroban, and M. McIntyre of Idaho

Department of Health and Welfare, Division of Environmental Quality and T.R. Maret of the USGS for advice and assistance throughout the project. The project was funded by the Idaho Department of Health and Welfare, Division of Environmental Quality, Boise, Idaho.

REFERENCES

American Public Health Association (APHA), *Standard methods for the examination of water and wastewater*, American Public Health Association, American Water Works Association and Water Pollution Control Federation, Publ., Washington D.C., 1989.

Clark, W.H. and Maret, T.R., *Protocols for Assessment of Biotic Integrity (Macroinvertebrates) in Idaho Streams, Water Quality Monitoring Protocols*, Report 5, Idaho Department of Health and Welfare, Division of Environmental Quality, Boise, Idaho, 1993.

Ferguson, D. and Ferguson, N., *Sacred Cows at the Public Trough*, Maverick Publ., Bend, Oregon, 1983.

Holechek, J.L., Policy changes on federal rangelands: A perspective, *J. Soil and Water Cons.*, 48, 166, 1993.

Karr, J.R., Biological integrity: A long neglected aspect of water resource management, *Ecol. Appl.*, 1, 66, 1991.

Karr, J.R., Fausch, K.D., Angermeier, P.L., Yant, P.R., and Schlosser, I.J., *Assessing Biological Integrity in Running Waters: A Method and Its Rationale*, Special Publication 5, Illinois Natural History Survey, 1986.

Minshall, G.W., Autotrophy in stream ecosystems, *Bioscience*, 28, 767, 1978.

Minshall, G.W., Jensen, S.E., and Platts, W.S., *The Ecology of Stream and Riparian Habitats of the Great Basin Region: A Community Profile*, U.S. Fish Widl. Serv. Biol. Rep., 85(7.24), 1989.

Plafkin, J.L., Barbour, M.T., Porter, K.D., Gross, S.K., and Hughes, R.M., *Rapid Bioassessment Protocols for Use in Streams and Rivers: Benthic Macroinvertebrates and fish*, USEPA, EPA/444/4-89-001, 1989.

Platts, W.S., Livestock grazing, in *Influences of Forest and Rangeland Management on Salmonid Fishes and Their Habitats*, Am. Fish. Soc. Special Publ., 19, 389, 1991.

Platts, W.S., Megahan, W.F., and Minshall, G.W., *Methods for Evaluating Stream, Riparian, and Biotic Conditions*, Gen. Tech. Rep., INT-138, Ogden, Utah, USDA, Forest Service, Intermountain Forest and Range Experiment Station, 1983.

Platts, W.S., and Nelson., R.L., Characteristics of riparian plant communities and streambanks with respect to grazing in Northeastern Utah, in *Practical Approaches to Riparian Resource Management: An Educational Workshop*, Gresswell, B.E., Barton, B.A., and Kershner, J.L., eds., BLM-MT-PT-89-001-4351, 1989.

Resh, V.H., and Jackson, J.K., Rapid assessment approaches to biomonitoring using benthic macroinvertebrates, in *Freshwater Biomonitoring and Benthic Macroinvertebrates*, Rosenberg, D.M. and Resh, V.H., eds., Chapman-Hall, New York, 195, 1993.

Robbins, W.G., and Wolf, D.W., *Landscape and the Intermontane Northwest: An Environmental History*, USDA, Forest Serv., Gen. Tech. Rep., PNW-GTR-319, 1994.

Robinson, C.T., and Minshall, G.W., Effects of disturbance frequency on stream benthic community structure in relation to canopy cover and season, *J. N. Am. Benthol. Soc.*, 5, 237, 1986.

Robinson C.T., and Minshall, G.W., *Biological Metrics for Regional Biomonitoring and Assessment of Small Streams in Idaho*, Final Report submitted to Idaho Division of Environmental Quality, Boise, Idaho, 1994.

Rosenberg, D.M., and Resh, V.H., Introduction to freshwater biomonitoring and benthic macroinvertebrates, in *Freshwater Biomonitoring and Benthic Macroinvertebrates*, Rosenberg, D.M. and Resh, V.H., eds., Chapman-Hall, New York, 1993.

Statsoft, Statistica: Complete statistical system with data base management and graphics, Statsoft, Inc., Tulsa, Oklahoma, 1991.

Strahler, A.N., Quantitative analysis of watershed geomorphology, *Am. Geophys. Union Trans.,* 38, 913, 1957.

Winget, R.N., and Magnum, F.A., Biotic condition index: Integrated biological, physical, and chemical stream parameters for management, in *Aquatic Ecosystem Inventory: Macroinvertebrate Analysis*, USFS Intermountain Region Contract No. 40-84 M8-8-524, Brigham Young University, Provo, Utah, 1979.

Young, J.A., and Sparks, B.A., *Cattle in the Cold Desert*, Utah State University Press, Logan, Utah, 1985.

Zar, J.H., *Biostatistical Analysis*, 2nd edition. Prentice-Hall, Inc. Englewood Cliffs, New Jersey, 1984.

RELATIONSHIP OF FERTILIZATION WITH CHICKEN MANURE AND CONCENTRATIONS OF ESTROGENS IN SMALL STREAMS

L.S. Shore, D.L. Correll, and P.K Chakraborty

INTRODUCTION

Steroidal estrogens and testosterone from human and animal sources are constantly excreted into the environment. For example, the Eastern Shore of Maryland produces 200,000 metric tons/year of manure from broilers alone. Since this manure contains about 30 ng/g of steroidal estrogen (see below), approximately 6000 g/year of estrogen is produced, and much of it is used as fertilizer. However, virtually nothing is known of the fate of the gonadal steroids (Knights, 1980), other than some very early work some decades ago (Zondek and Sulman, 1943) indicating that no common soil or fecal bacteria can metabolize estrogen.

In addition to the steroidal estrogens, i.e., estradiol and estrone, many other chemically unrelated weakly estrogenic compounds, both natural (Shemesh and Shore, 1994) and synthetic (pesticides, herbicides; Colburn et al., 1993), are found in the environment. The most potent natural estrogens (phytoestrogens), found in legumes, have long been known to cause infertility and disrupt reproductive processes in cows, sheep, pigs, and chickens.

In alfalfa, the principal phytoestrogen is coumestrol, which is not normally found in significant amounts. However following exposure to a variety of environmental stimuli (fungal infection, U-V, ozone), coumestrol levels are highly elevated. Recently we found that alfalfa plants irrigated with treated sewage water had elevated coumestrol concentrations. This effect appeared to be related to the high concentrations of estrogen (see below) present in sewage water. These observations raised questions as to exactly what is the environmental fate of estrogens. We therefore undertook to determine (a) what is the prevalence and concentration of estrogen in water sources, (b) what are the possible sources, and (c) do estrogens exist in water sources in sufficient amounts to have a physiological effect on plants and animals?

PREVALENCE OF ESTROGENS IN SURFACE WATERS

Estrogen (E) and testosterone (T) were determined in water samples by specific radioimmunoassays (Shore et al., 1993a).

LAKE WATER

Estrogen was detected (20-22 ng/L) in the Sea of Galilee, which is the principal source of water in Israel. It was detected in the third year of an extreme drought when the lake was at a record low, and the estrogen fell to 4 ng/L after the drought broke. Of particular interest was the presence of estrogen detected in the drinking water even though it had been filtered and chlorinated (11-19 ng/L in the drought year vs < 0.5 ng/L in a normal year) (Shore et al., 1993a).

SMALL STREAMS

To determine the prevalence in small streams in the Chesapeake Bay watershed, 17 streams in the Conestoga River watershed, an area with high agricultural pollution, were studied. Four of these streams measured in May-June 1994 were found to have appreciable levels (> 0.5 ng/L) of T, ranging from 1.2 ± 0.1 SD (n = 5) to 4.1 ± 0.1 ng/L or E, 0.8 ± 0.4 to 2.9 ± 0.9 ng/L. Three of the sites were near fields fertilized with chicken manure and one below a sewage treatment plant. Estrogen was found in the streams during all of the months of the year except January and February. Testosterone was found from May to October, but from November to April, the levels were undetectable.

SOURCES OF ENVIRONMENTAL ESTROGENS

Although other sources can not be excluded, the three most logical sources of estrogen found in the streams were (a) sewage effluent; (b) chicken manure and (c) sludge.

SEWAGE EFFLUENT

Raw sewage contains large amounts of E (40 to 200 ng/L) and T (16 to 700 ng/L) that are highly variable depending on source and dilution (Shore et al., 1993a). Even in sewage originating from septic tanks with a fairly constant flow throughout the year, E levels ranged from 40 to 130 ng/L (78 ± 13; mean ± 1 standard error (SE); n = 6). Both E and T were readily detected in the effluent even after secondary and tertiary treatments (Table 1).

Similarly, Sumpter and coworkers (Jobling and Sumpter, 1993; Purdom et al., 1994) found a high level of estrogenicity in many rivers and streams that receive effluent from sewage treatment plants, and the activity was present many kilometers downstream from the plant. The estrogenicity was measured by exposing male trout to the effluent and measuring the level of vitellogenin in the blood. The source of the estrogenicity was not determined, but there were estrogenic alkyl phenols present in the effluent.

Table I. Comparison of concentrations of estrogen (E) and testosterone (T) in sewage water effluent after various treatments. Values are means ± SE for at least five samples made during times of moderate precipitation. Reprinted from Shore et al., 1993b.

Type of Treatment	E (ng/L)	T (ng/L)
sedimentation ponds	151 ± 6	74.1 ± 7.2
activated sludge	6.4 ± 1.6	6.7 ± 2.1
constructed wetlands	2.2 ± 0.4	1.6 ± 0.3
percolation through sand (3 mo.)	< 0.5	< 0.5

CHICKEN MANURE

Although other manures are used for fertilizer, chicken manure has much higher concentrations of hormones as birds have much higher concentrations of hormones than mammals. The effects of these high levels of hormones can be seen in heifers fed chicken manure silage (a common practice in Israel). The heifers may either develop premature udders (high estrogen) or have delayed puberty and enlarged clitoris (high testosterone), depending on the relative concentration of the hormone (Shemesh and Shore, 1994). The effect of origin on hormone content is shown in Table 2.

In order to evaluate the effect of the manure (more correctly called chicken litter in the US) on small streams, we studied a farm and two streams originating from it on the eastern shore of Maryland from August 1993 to April 1994. The field was manured in March 1993 and again in February 1994 (two weeks earlier than the surrounding farms). The fertilizer was found to contain 28 ng/g E and 34 ng/g T. The field runoff drained into an unused irrigation pond from which the Burnt Mill branch of Burnt Mill Creek originated. A second branch, the Ayalodyte, which joined the main branch about 15 Km from the farm, was also sampled. The sampled farm was the only one found on the Burnt Mill branch while there were a number of manured fields that drain into the Ayalodyte.

Estrogen was found in the pond throughout the time studied (Figure 1) but decreased from August till March 1994. There was virtually no precipitation in May-August 1993, which may account for the very high levels observed. The

Table 2. Concentration (dry weight) of estrogen (E) and testosterone (T) in poultry litter from 7-week-old broilers, 5-month-old layers, and 5-month-old roosters (from Shore, 1993b with permission).

Source	T	E	N
	(ng/g d.wt.)	(ng/g d.wt.)	
Immature broilers			
Females	133 ± 13	65 ± 7	10
Males	133 ± 12	14 ± 4	10
Laying Hens	254 ± 22	533 ± 40	17
Roosters	670 ± 95	93 ± 13	10

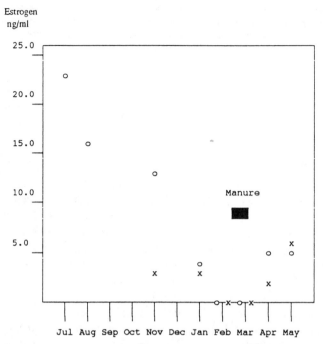

Figure 1. Estrogen concentrations in field runoff (open circles) and the primary drainage streams (x). Estrogen concentrations in the streams dropped from 3.7 ± 0.5 ng/L (mean ± SE, n = 6) to < 0.5 ng/L in February and March during the spring thaw. After application of chicken manure in late February, estrogens rose to 5.4 ± 1.2 ng/L in May.

runoff from the manured fields was reflected in the streams, and there was no difference between the two streams. In contrast, T concentrations did reflect the earlier application of manure (February instead of March) on the field leading to the Burnt Mill branch (Figure 2). Measurement of the six sites for T was less than 1 ng/L from November to March but rose to 9.9 ± 5.5 and to 15.3 ± 5.7 ng/L in May and April, respectively.

SLUDGE

No field work has been done on the effect of the wide use of sludge for fertilizer on the E and T content in the environment or on the plants themselves. However, aqueous extracts of sludge contain from 18 ng to 70 ng/L of E and 10 to 173 ng/L of T (Shore et al., 1993a) so sludge could be a substantial source of these hormones.

TRANSPORT AND ELIMINATION FROM THE ENVIRONMENT

There is preliminary evidence that T and E do not leave the environment by the same routes.

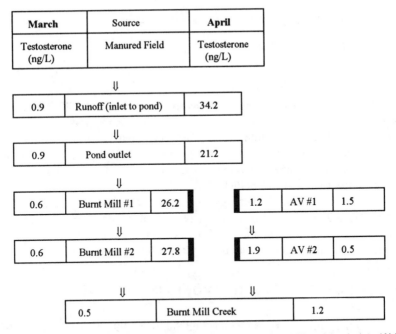

March	Source	April
Testosterone (ng/L)	Manured Field	Testosterone (ng/L)

⇓

| 0.9 | Runoff (inlet to pond) | 34.2 |

⇓

| 0.9 | Pond outlet | 21.2 |

⇓

| 0.6 | Burnt Mill #1 | 26.2 | | 1.2 | AV #1 | 1.5 |

⇓   ⇓

| 0.6 | Burnt Mill #2 | 27.8 | | 1.9 | AV #2 | 0.5 |

⇓   ⇓

| 0.5 | Burnt Mill Creek | 1.2 |

Figure 2. Comparison of testosterone concentrations in Burnt Mill and Ayalodyte (AV) Branches of Burnt Mill Creek 1 wk (March) and 5 weeks (April) following application of manure.

1. T is leached by aqueous solution from the soil more readily than E, probably due to the phenolic group of E binding more firmly to organic matter (Shore et al., 1993a);
2. T but not E was found in well water from a farm using chicken manure (T = 1.0 ± 0.2 ng/L versus E = < 0.10 ± 0.02 ng/L; n = 6);
3. Percolation of waste water through peat fields (but not emergent wetlands) is more effective in removing E than T.
4. Sub-surface samples of soil from manured fields contained T but not E. However, E was shown to be leached from the fields containing chicken manure and in puddles of water below the fields.

This would indicate that E reaches the streams by surface runoff while T reaches the streams by both ground water and surface water routes. Furthermore, finding estrogen at concentrations greater than 1 ng/L in the streams but not higher than 5 ng/L throughout the year indicates that the estrogen is constantly leached in small quantities from the soil. The ultimate fate of estrogen is not known, but early work indicates that it is destroyed by physicochemical processes rather than bacterial action (Zondek and Sulman, 1943).

PHYSIOLOGICAL EFFECTS OF LOW DOSES OF ESTROGENS

Both plants and animals have been reported to respond to low dose estrogen (10 ng to 1μg/L).

EFFECTS ON PLANTS

Steroidal estrogen in concentrations as low as 10 ng/L has been shown to significantly increase vegetative growth in alfalfa (Shore et al., 1992) and to decrease root growth in mung bean (Guan and Roddick, 1988). Estradiol is found in small concentrations (2-4 ng/g) in phaseolus (Young et al., 1978) and alfalfa (0.34 ng/g), but E and T do not enter the plant under normal physiological conditions (Jones and Roddick, 1988). However, we have found that when alfalfa plants are irrigated with treated sewage water, both E (1.30-2.42 ng/g) and T (1.58-2.74 ng/g; normally < 0.1 ng/g) concentrations in the plant were elevated. In addition, irrigation with treated sewage water increased the endogenous phytoestrogen content of the plant to levels that can affect fertility in cattle.

EFFECT OF STEROIDAL HORMONES IN THE ENVIRONMENT ON FISH

Sumpter and coworkers (Jobling and Sumpter, 1993; Purdom et al., 1994) have shown that exposing fish to effluent from sewage plants resulted in increased vitellogenin production. This was similar to the effects they observed when male and female sturgeons ate feed containing high level of phytoestrogen (soya) (Pelissero and Sumpter, 1992). The substances in the water that cause this effect are not known but are believed to be degradation products of detergents. However, estrogen alone was effective in the bioassay at 10 ng/L.

The effects of higher concentrations of E or T ingested in the feed or added to the water have been extensively documented as causing sex reversal and having anabolic effects in a wide variety of fish (for review see Donaldson et al., 1979). Although most of these experiments added hormones to the feed at fairly high concentrations, Ashby (1957) using trout alevins, showed that at a relatively low dose of E or T (50 μg/L) there was growth retardation and a powerful inhibitory effect on gonadal development. Others (Woo et al., 1993) have shown a wide variety of disruptions in blood values and enzyme levels in fish fed 3 μg/g of E or T for 30 days.

DISCUSSION

The purpose of the present work was to determine if steroidal estrogen is present in sufficient quantities in the environment to affect plants and animals. It was found that estrogen is normally found in raw sewage from septic tanks (73 ± 18 ng/L; n = 5) and summer runoff from field fertilized with chicken manure (14-20 ng/L). However, there seems to be an upper limit of about 5 ng/L in freely flowing streams. Since 10 ng/L has been shown to affect both alfalfa and trout, the potential for environmental effects is present. Finally, considerable attention has recently been drawn to evidence indicating that the environment is under an estrogen "load." Specifically, environmental estrogens have been implicated in the drastic reduction in sperm counts among Western men (Sharpe and Skakkebaek, 1993) and widespread reproductive disorders in a variety of wildlife (Colburn et al., 1993) and particularly in birds, alligators, and turtles. There is evidence that some of these reproductive disorders are due to ingestion of phytoestrogens or exposure to the numerous weakly estrogenic xenobiotics (DDT, methoxychlor). However, the effect of these compounds is probably additive to the level of estradiol and estrone already in the environment, and this should be taken into account before evaluating any additional effects of the xenobiotics.

REFERENCES

Ashby, K.R., The effect of steroid hormones on the brown trout (*Salmo trutta*, L.) during the period of gonadal differentiation, *J. Embryol. Exp. Morph.*, 5, 225, 1957.

Colburn, T., vom Saal, F.S., and Soto, A.M., Developmental effects of endocrine-disrupting chemicals in wildlife and humans., *Environ. Health Prospect*, 101, 378, 1993.

Donaldson, E.M., Fagerlund, U.H.M., Higgs, D.A., and McBride, J.R., Hormonal enhancement of growth, in *Fish Physiology*, Vol. 7, Hoar, W.S., Randall, D.J., and Brett, J.R., eds., Academic Press, New York, 455, 1979.

Guan, M., and Roddick, J.G., Comparison of the effects of epibrassinolide and steroidal estrogens on adventitious root growth and early shoot development in mung bean cuttings, *Physiol. Plant.*, 73, 426, 1988.

Jobling, S., and Sumpter, J.P., Detergent components in sewage effluent are weakly oestrogenic to fish: An *in vitro* study using rainbow trout (*Oncorhynchus mykiss*) hepatocytes, *Aquat. Toxicol.*, 27, 361, 1993.

Jones, J.L., and Roddick, J.G., Steroidal estrogens and androgens in relation to reproductive development in higher plants, *J. Plant Physiol.*, 133, 510, 1988.

Knights, W.M., Estrogens administered to food-producing animals: Environmental considerations, in *Estrogens in the Environment*, McLachlan, J.A., ed., North Holland, Amsterdam, 391, 1980.

Pelissero, C., and Sumpter, J.P., Steroids and "steroid-like" substances in fish diets, *Aquaculture*, 107, 283, 1992.

Purdom, E.C., Hardiman, P.A., Bye, V.J., Eno, N.C., Tyler, C.R., and Sumpter, J.P., Estrogenic effects of effluents from sewage treatment works, *Chem. Ecol.*, 8, 275, 1994.

Sharpe, R.M., and Skakkebaek, N.E., Are oestrogens involved in falling sperm count and disorders of the male reproductive tract? *Lancet*, 341, 1392, 1993.

Shemesh, M., and Shore, L.S., Effects of hormones in the environment on reproduction in cattle, in *Factors Affecting Net Calf Crops*, Fields, M.J. and Sand, R.S., eds., CRC Press, Boca Raton, Florida, 289, 1994.

Shore, L., Gurevich, M., and Shemesh, M., Estrogen as an environmental pollutant. *Bull. Environ. Contam. Toxicol.*, 51, 361, 1993b.

Shore, L.S., Harel-Markowitz, E., Gurevich, M., and Shemesh, M., Factors affecting the concentration of testosterone in poultry litter, *J. Environ. Sci. Health*, A78, 1737, 1993a.

Shore, L., Kapulnik, Y., Ben-Dov, B., Fridman, Y., Weninger, S., and Shemesh, M., Effects of estrone and 17ß-estradiol on vegetative growth of *Medicago sativa, Physiol. Plant.*, 84, 217, 1992.

Young, J., Hillman, J., and Knights, B.A., Endogenous estradiol 17ß in *Phaseolus vulgaris, Z. Pflanzerphysiol.*, 90, 45, 1978.

Woo, N.Y.S., Chung, A.S.B., and Ng, T.B., Influence of oral adminstration of estradiol-17ß and testosterone on growth, digestion, food conversion and metabolism in the underyearling red sea bream, *Chrysophrys major, Fish Physiol. Biochem.*, 10, 377, 1993.

Zondek, B., and Sulman, F., Inactivation of estrone and diethylstilbestrol by microorganisms, *Endocrinology*, 33, 204, 1943.

THE ROLE OF WETLANDS, PONDS, AND SHALLOW LAKES IN IMPROVING WATER QUALITY

Dennis F. Whigham

INTRODUCTION

Wetlands and shallow aquatic ecosystems (lakes and ponds) occur in almost all landscapes (Patten, 1990; Brinson, 1993; Whigham et al., 1993; Mitsch and Gosselink, 1993), and in many instances they have been shown to improve the quality of incoming water. Water quality improvement has been shown to occur at the landscape level where groups of aquatic and wetland ecosystems are hydrologically connected and at the level of individual systems. At the landscape level, one example of the influence that groups of wetlands and shallow aquatic ecosystems have on water quality comes from research on beaver (*Castor canadensis*). Naiman et al. (1994) have shown that beaver influence water quality over large areas of boreal landscapes and that the impacts are dynamic and long lasting. Beaver alter hydrologic patterns and create shallow impoundments and wetlands that provide conditions that are ideal for altering water quality, particularly through their influence on nitrogen and carbon dynamics (Cirmol and Driscoll, 1993). In some parts of the world (i.e., most countries in Central Europe), large portions of landscapes have been hydrologically modified to increase the number of shallow ponds, lakes, and associated wetlands for aquaculture and water quality management (Uhlmann and Recknagel, 1982; Kvet et al., 1990; Szumiec and Szumiec, 1993; Kub et al., 1994). At a smaller landscape scale, the quality of runoff from agricultural fields has been shown to improve when it passes through wetlands and shallow aquatic ecosystems before reaching streams and larger bodies of water (e.g., Karr and Schlosser, 1978; Schlosser and Karr, 1981; Peterjohn and Correll, 1984; Hill, 1991; Lowrance, 1992; Gilliam, 1994). Individually, wetland and shallow aquatic ecosystems have also been shown to improve water quality, and engineered systems have been used successfully to treat waste water (Godfrey et al., 1985; Carpenter et al., 1976; McNabb, 1976; Tourbier et al., 1976; Hammer, 1989).

It has been demonstrated that wetlands and shallow aquatic ecosystems can be used to treat animal wastes (e.g., Culley et al., 1990; Tanner et al., 1995a,b). The question is whether or not wetlands and shallow ponds and lakes can be used to treat animal wastes without degrading water quality in downstream aquatic and/or ground water systems. It has been shown, for example, that the addition of nutrients to shallow ponds and lakes can result in hyper-trophic conditions, leading to severe degradation of water quality (Moss, 1988). Waste water inputs to natural wetlands have also been shown to result in unde-sirable changes such as the loss of biodiversity and deterioration of downstream water quality (Kadlec, 1983).

In this chapter the processes that control the movement of nitrogen (N) and phosphorus (P), two primary components of waste water, through wet-lands and shallow aquatic ecosystems are discussed first, followed by a discus-sion of the potential limitations on the amounts of N and P that can be treated in wetlands and shallow aquatic ecosystems. The chapter concludes with the suggestion that, whenever possible, treatment of animal wastes should be ef-fected in constructed wetland/pond systems and that natural ecosystems should be used only when they are associated with suitable constructed wetland/pond systems or under circumstances in which downstream water quality is not the subject of primary interest and/or where some form of aquaculture can be prac-ticed.

NITROGEN AND PHOSPHORUS IN WETLANDS AND SHALLOW LAKES/PONDS

Considerable attention has been given to the issue of whether aquatic eco-systems, including wetlands and shallow lakes and ponds, are sources, sinks, or transformers of nutrients (Howard-Williams, 1985; Ryding and Rast, 1989; Mitsch and Gosselink, 1993). In most situations, it is not desirable to use shal-low aquatic ecosystems to treat waste water unless the influent is pre-treated before it enters the pond/lake system (Ryding and Rast, 1989). In contrast, it has been shown that waste water can be effectively treated by many types of wetlands under a wide range of conditions (Godfrey et al., 1985). Richardson (1990) reviewed the literature on wetlands, and his conclusions are relevant today; those related to water quality are listed here.

1. Wetlands function as effective transformers of nitrogen (N) and phos-phorus (P).
2. Wetlands maintain biogeochemical processes that are responsible for transforming and releasing significant quantities of dinitrogen.
3. Phosphorus is either adsorbed onto amorphous aluminum and iron in soil, stored in peat, or is taken up by microbes and plants in small quan-tities and recycled annually.
4. Wetland types differ in terms of the magnitude and form of N and P released to output waters.
5. N and P retention by wetlands varies among seasons, and each element has to be analyzed separately in terms of seasonal and annual retention patterns.

6. Wetlands can function as either sinks or sources for N and P, depending on wetland type, level of N and P loading into the wetland, the season of year, and whether or not the ecosystem is aggrading. Wetlands do not retain P as efficiently as terrestrial ecosystems.

While some of Richardson's (1990) conclusions are very specific, most are general statements that clearly suggest that it is necessary to consider each wetland separately before decisions are made about its ability to effectively treat waste water (e.g., animal wastes) that contain large amounts of nutrients, organic matter, and potentially harmful viruses and bacteria. There is little doubt, however, that almost every wetland transforms materials and at least changes the forms of N and P.

Most nitrogen enters wetlands and aquatic ecosystems through atmospheric or surface and sub-surface hydrologic inputs (Mitsch and Gosselink, 1993; Koerselman and Verhoeven, 1992), and the total loadings control whether the wetland will be nutrient poor (oligotrophic), nutrient rich (eutrophic), or intermediate (mesotrophic). Once nitrogen has entered a wetland, it can follow numerous pathways before it leaves in runoff or leaching, is returned to the atmosphere through denitrification or volatilization, or is stored in the wetland or transported from it in the form of biomass of plants and animals, or stored in litter and/or soil (Figure 1). Nitrogen cycling in ponds and shallow lakes follows the same pathways as shown in Figure 1 except that plankton and fila-

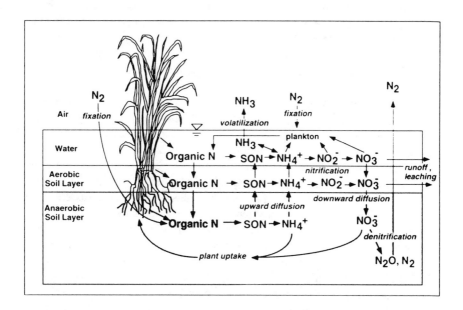

Figure 1. Nitrogen cycling in wetlands. SON = soluble organic nitrogen. (From Mitsch and Gosselink, 1993, with permission.)

mentous algae would be more important than macrophytes in nitrogen fixation and denitrification (Likens, 1975).

To be effective in treating animal wastes, wetlands and shallow aquatic ecosystems would have to provide long-term storage of more nitrogen than entered and/or have a high capacity to intercept nitrogen and return large amounts to the atmosphere as oxides of nitrogen. Wetlands that have high rates of denitrification have an oxidized surface soil-litter layer lying over a deeper layer of sediment that is anaerobic (Gambrell and Patrick, 1978), a continuous external (i.e., the addition of animal wastes) or internal (i.e. high ammonification rates) source of nitrate, and a source of carbon that is used as an energy source by microbes involved in transforming nitrogen.

Riparian forests are examples of wetland ecosystems that have a high capacity to remove nitrogen and return it to the atmosphere in various gaseous forms (e.g., Peterjohn and Correll, 1984; Gilliam, 1994). They occur at the boundary between terrestrial and aquatic ecosystems, they receive high inputs of nitrate from upslope agricultural fields, the soils often have an oxidized surficial layer and a deeper anaerobic zone, and the forests are highly productive and produce large amounts of organic matter.

Many wetlands, however, have lower denitrification rates than riparian forests because most of the nitrate forms and cycles internally and nitrate availability often limits denitrification rates (Verhoeven et al., 1994). Similar to riparian forests, denitrification rates increase in wetlands where nitrate inputs have increased (Koerselman and Verhoeven, 1992). Because animal wastes are high in both organic and inorganic forms of nitrogen, it can be expected that most wetlands would be able to remove nitrogen components associated with water quality. It needs to be emphasized, however, that other factors (e.g., soil acidity and temperature) are important, and most wetlands have a limited capacity to process nitrogen, as will be discussed in a later section of this chapter. Efficient nutrient cycling in shallow aquatic ecosystems can also improve water quality, but these systems often become eutrophic when waste water is added (Kub et al., 1994) because mechanisms for nutrient removal or long-term storage are not as great as those that occur in wetlands.

The number of pathways that phosphorus can cycle through (Figure 2) is less than those associated with nitrogen because there is not any significant atmospheric component to the phosphorus cycle. Once phosphorus enters a wetland or shallow aquatic ecosystem, it is transported through it or it is retained in the substrate or in the biomass. Within the substrate, the cycling of phosphorus is quite complex and depends on chemical conditions and the types and forms of sediments and minerals that they contain (Richardson, 1985; Richardson and Marshall, 1986; Barbanti et al., 1994). If a wetland or shallow pond/lake system receives and retains sediments, especially fine clays and silts that are high in phosphorus, then it is very likely that P will be removed from the water and water quality will be improved. Aquatic ecosystems with highly

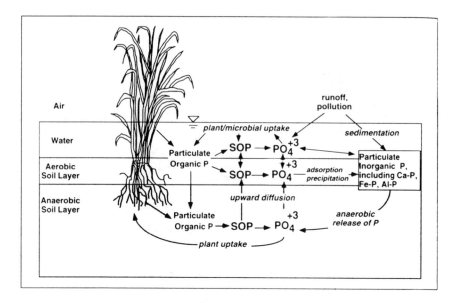

Figure 2. Phosphorus cycling in wetlands. SOP = soluble organic phosphorus. (From Mitsch and Gosselink, 1993, with permission.)

organic substrates (i.e., peat) can also efficiently remove phosphorus, but the absorption capacity is limited, the systems can ultimately become P saturated, and water quality improvement will cease. Phosphorus can effectively be stored in the substrates in shallow aquatic ecosystems, but it is often returned to the water column during periods of anoxia or when the substrates are disturbed by winds, bioturbation, etc.

LIMITATIONS ON THE ABILITY OF WETLANDS AND SHALLOW PONDS/LAKES TO IMPROVE WATER QUALITY

All aquatic ecosystems have the ability to change the quality of water but the levels of nutrients that are acceptable in a wetland or in downstream aquatic ecosystem is determined by social values and by legislation such as Section 303 of the Clean Water Act. In the US it is, in general, no longer acceptable to degrade water quality through the discharge of point sources of potential contaminants. Much of the current efforts to improve water quality in water of the US has, in fact, moved to efforts to effectively treat non-point source runoff (US EPA, 1986; Baker, 1992). Using existing wetlands and aquatic ecosystems to treat animal wastes would, therefore, be allowed only if the discharges had already been treated (e.g., see examples of pre-treatments in Ryding and Rast, 1989) or if it could be demonstrated that the natural systems would improve water quality while not deteriorating downstream aquatic ecosystems.

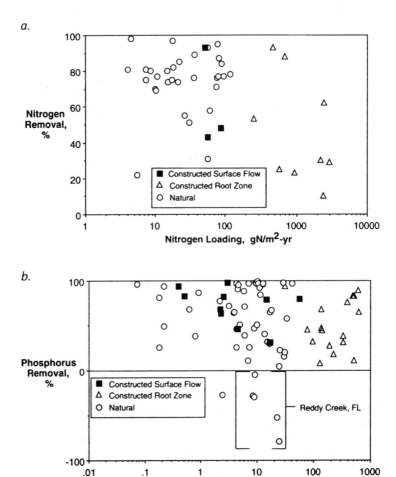

Figure 3. Nutrient retention in wetlands receiving wastewater or river water as a function of loading for a) nitrogen and b) phosphorus. (From Mitsch and Gosselink, J.G., 1993, with permission.)

What factors limit the ability of a wetland or shallow aquatic ecosystem to improve water quality in downstream areas? This issue has been addressed by a number of authors in recent years (e.g., Tourbier and Pierson, 1976; Ryding and Rast, 1989; Richardson, 1990; Verhoeven and Van der Toorn, 1990). Mitsch and Gosselink (1993) summarized a paper by Olson (1992) that provides useful guidance for managing wetlands for water quality improvement (Table 1). Olson's comments make it clear that no waste water should be discharged into a wetland (and by extension to shallow aquatic ponds and lakes) before its ability to improve water quality has been assessed, along with other functions.

Table 1. Questions that should be answered when consideration is being given to using a wetland or aquatic ecosystem for purposes of treating animal wastes. Source: Olson, 1992, as summarized by Mitsch and Gosselink (1993).

Technical Considerations

1. Are other values associated with the wetland (e.g., wildlife habitat, recreational value, long-term storage of water) more important?
2. What nutrient loading rates are within the natural ability of the system to treat waste water?
3. How would the addition of animal wastes impact exiting ecological conditions within the wetland (e.g., would a diverse emergent vegetation be replaced by monotypic stands of cattail)?
4. What is the hydrology of the wetland? For example, wetlands that are hydrologically isolated from other aquatic ecosystems are likely to retain nutrients added in waste water and are likely to undergo significant biotic and substrate changes.

Institutional Considerations

1. Are there any conflicts among agencies in the use of the wetland for treating animal wastes?
2. Will the use of the wetland for treating animal wastes result in positive benefits in functions and values?
3. Are there any legal constraints (i.e., liabilities) associated with using the site for treating animal wastes?
4. Are there local, state, and/or federal regulations that would need to be modified before the wetland could be used for treatment of animal wastes?

In general, shallow aquatic ecosystems (ponds and lakes) have a limited ability to improve water quality (see Giussani and Callieri (1993), which describes many of the problems associated with lakes) and the addition of waste water to aquatic ecosystems most often results in eutrophication (Ryding and Rast, 1989). Once ponds and shallow lakes have reached a eutrophic stage, the condition can be reversed only when the nutrient loading is eliminated or greatly decreased, or by regulation of the types of organisms in the aquatic food web. Excellent examples of this condition can be found in the long-term patterns of water quality deterioration followed by recovery in the Great Lakes and in Lake Washington near Seattle.

Wetlands have a greater potential to treat waste water than ponds and shallow lakes, but they also have limited ability. Figure 3 shows that the ability of natural wetlands to remove N and P varies greatly and that the removal efficiency of both nutrients decreases as loading rates increase. While data compilations such as those shown in Figure 3 have limited value due to natural variability in ecosystems, they provide a cautionary note that should not be ignored.

In conclusion, there have been numerous studies that have demonstrated that wetlands, ponds, and shallow lakes can improve water quality. It has also been demonstrated that a limited number of aquatic ecosystems can be used to effectively process waste water. I believe that most wetland ecologists would support the use of constructed wetlands-pond systems rather than natural systems to treat waste water. It seems unlikely that natural aquatic systems can be used to effectively treat animal wastes and a variety of engineered systems that use constructed wetlands have been developed that could be used for that purpose (e.g., see Table 17-1 in Mitsch and Gosselink, 1993).

REFERENCES

Baker, L.A., Introduction to nonpoint source pollution in the United States and prospects for wetland use, *Ecological Engineering,* 1, 1, 1992.

Barbanti, A., Bergamini, M.C., Frascari, F., Miserocchi, S., and Rosso, G., Critical aspects of sedimentary phosphorus chemical fractionation, *Journal of Environmental Quality,* 23, 1093, 1994.

Brinson, M.M., *A Hydrogeomorphic Classification for Wetlands,* Wetlands Research Program Technical Report WRP-DE-4, US Army Corps of Engineers, Washington, D.C., 1993.

Carpenter, R.L., Coleman, M.S., and Jarman, R., Aquaculture as an alternative waste water treatment system, in *Biological Control of Water Pollution,* Tourbier, J., and R.W. Pierson, Jr., eds., University of Pennsylvania Press, Philadelphia, 1976.

Cirmol, C.R., and Driscoll, C.T., Beaver pond biogeochemistry: Acid neutralizing capacity generation in a headwater wetland, *Wetlands,* 13, 277, 1993.

Culley, D.D., Kvet, J. and Dykyjová, D., Agricultural waste management, in *Wetlands and Shallow Continental Water Bodies, Volume 1, Natural and Human Relationships,* Patten, B.C., ed., SPB Academic Publishing, The Hague, 1990.

Gambrell, R.O., and Patrick, W.H., Jr., Chemical and microbiological properties of anaerobic soils and sediments, in *Plant Life in Anaerobic Environments,* D.D. Hook and R.M.M. Crawford, eds., Ann Arbor Scientific Publishers, Inc. Ann Arbor, Michigan, 1978.

Gilliam, J.W., Riparian wetlands and water quality, *Journal of Environmental Quality,* 23, 896, 1994.

Giussani, G., and Callieri, C., eds., *Strategies for Lake Ecosystems Beyond 2000,* Stresa (Italy), 1993.

Godfrey, P.J., Kaynor, E.R., and Pelczarski, S., eds., *Ecological Considerations in Wetlands Treatment of Municipal Wastewaters,* Van Nostrand Reinhold Company, New York, New York, 1985.

Hammer, D.A., ed., *Constructed Wetlands for Waste Water Treatment: Municipal, Industrial and Agricultural,* Lewis Publishers, Chelsea, Michigan, 1989.

Hill, A.R., A ground water nitrogen budget for a headwater swamp in an area of permanent ground water discharge, *Biogeochemistry,* 14, 209, 1991.

Howard-Williams, C., Cycling and retention of nitrogen and phosphorus in wetlands: A theoretical and applied perspective, *Freshwater Biology,* 15, 391, 1985.

Kadlec, R.H., The Bellaire Wetland: Waste water alteration and recovery, *Wetlands,* 3, 44, 1983.

Karr, J.R., and Schlosser I.J., Water resources and the land-water interface, *Science,* 201, 229, 1978.

Koerselman, W., and Verhoeven, J.T.A., Nutrient dynamics in mires of various trophic status: Nutrient inputs and outputs and the internal nutrient cycle, in *Fens and Bogs in the Netherlands: Vegetation, History, Nutrient Dynamics and Conservation,* Verhoeven, J.T.A., ed., Kluwer Academic Publishers, Dordrecht, 1992.

Kub, F., Kvet, J., and Hejný, S., Fishpond management, in *Wetlands and Shallow Continental Water Bodies, Volume 2. Case Studies,* Patten, B.C., ed., SPB Academic Publishing, The Hague, 1994.

Kvet, J., Pamprommin, Ch., Svirezhev, Y.M., and Wecomme, R.L., Fish and other aquaculture, in *Wetlands and Shallow Continental Water Bodies, Volume 1, Natural and Human Relationships,* Patten, B.C., ed., SPB Academic Publishing, The Hague, 1990.

Likens, G.E., Nutrient flux and cycling in freshwater ecosystems, in *Mineral Cycling in Southeastern Ecosystems*, Howell, F.G., Gentry, J.B., and Smith, M.H., eds., US Energy Research and Development Administration, Springfield, Virginia, 1975.

Lowrance, R., Groundwater nitrate and denitrification in a coastal plain riparian forest, *Journal of Environmental Quality*, 21, 401, 1992.

McNabb, C.D., Jr., The potential of submersed vascular plants for reclamation of waste water in Temperate Zone ponds, in *Biological Control of Water Pollution*, Tourbier, J., and R.W. Pierson, Jr., eds., University of Pennsylvania Press, Philadelphia, 1976.

Mitsch, W.J., and Gosselink, J.G., *Wetlands*, Academic Press, New York, 1993.

Moss, B., *Ecology of Fresh Waters: Man and Medium*, Blackwell Scientific Publications, Oxford, 1988.

Naiman, R.J., Pinay, G., Johnston, C.A., and Pastor, J., Beaver influences on the long-term biogeochemical characteristics of boreal forest drainage networks, *Ecology*, 75, 905, 1994.

Olson, R.K., ed., The role of created and natural wetlands in controlling nonpoint source pollution, *Ecological Engineering*, 1, 1, 1992.

Patten, B.C., ed., *Wetlands and Shallow Continental Water Bodies, Volume 1, Natural and Human Relationships*, SPB Academic Publishing, The Hague, 1990.

Peterjohn, W.T., and Correll D.L., Nutrient dynamics in an agricultural watershed: Observations of the role of a riparian forest, *Ecology*, 65, 1466, 1984.

Richardson, C.J., Mechanisms controlling phosphorus retention capacity in freshwater wetlands, *Science*, 228, 1424, 1985.

Richardson, C.J., Biogeochemical cycles: Regional, in *Wetlands and Shallow Continental Water Bodies, Volume 1, Natural and Human Relationships*, Patten, B.C., ed., SPB Academic Publishing, The Hague, 1990.

Richardson, C.J., and Marshall, P.E., Processes controlling movement, storage and export of phosphorus in a fen peatland, *Ecological Monographs*, 56, 279, 1986.

Ryding, S.-O. and Rast, W., eds., *The Control of Eutrophication of Lakes and Reservoirs*, Unesco, Paris, France, 1989.

Schlosser, I.J., and Karr, J.R., Riparian vegetation and channel morphology impact on spatial patterns of water quality in agricultural watersheds, *Environmental Management*, 5, 233, 1981.

Szumiec, M.A., and Szumiec, J., Fish ponds: A way of better utilization, creation and management of wetlands, in *Wetlands and Ecotones: Studies on Land-Water Interactions*, Gopal, B., Hillbricht-llkowska, A. and Wetzel, R.G., eds., International Scientific Publications, New Delhi, 1993.

Tanner, C.C., Clayton, J.S., and Upsdell, M.P., Effect of loading rate and planting on treatment of dairy farm waste waters in constructed wetlands-I, Removal of oxygen demand, suspended solids and faecal coliforms, *Water Research*, 29, 17, 1995a.

Tanner, C.C., Clayton, J.S., and Upsdell, M.P., Effect of loading rate and planting on treatment of dairy farm waste waters in constructed wetlands-II, Removal of nitrogen and phosphorus, *Water Research*, 29, 27, 1995b.

Tourbier, J., and Pierson, R.W., Jr., eds., *Biological Control of Water Pollution*, University of Pennsylvania Press, Philadelphia, 1976.

Uhlman, D., and Recknagel, F., Dimensioning of ponds for biological waste treatment, in *Ecosystem Dynamics in Freshwater Wetlands and Shallow Water Bodies*, Centre of International Projects GKNT, Moscow, U.S.S.R., 1982.

US Environmental Protection Agency, *National Water Quality Inventory: Report to Congress*, US Environmental Protection Agency, Washington, D.C., 1986.

Verhoeven, J.T.A., and Van der Toorn, J., Marsh eutrophication and waste water treatment, in *Wetlands and Shallow Continental Water Bodies, Volume 1, Natural and Human Relationships*, Patten, B.C., ed., SPB Academic Publishing, The Hague, 1990.

Verhoeven, J.T.A., Whigham, D.F., van Kerkhoven, M., and O'Neill, J., A comparative study of nutrient-related processes in geographically separated wetlands: Towards a science base for functional assessment procedures, in *Wetlands of the World: Biogeochemistry, Ecological Engineering, Modelling and Management*, Mitsch, W.J., ed., Elsevier, Amsterdam, 1994.

Whigham, D.F., Dykyjová, D., and Hejný, S., *Wetlands of the World I: Inventory, Ecology and Management*, Kluwer Academic Publishers, Dordrecht, The Netherlands, 1993.

DESIGN MODELS FOR NUTRIENT REMOVAL
IN CONSTRUCTED WETLANDS

Robert H. Kadlec

INTRODUCTION

T he intentional use of wetlands, both natural and constructed, for waste water treatment began in Europe in the early 1950s, and has grown ever since. There are now thousands of treatment wetlands, in all climatic regions of the world. Most of these are constructed on previous upland. Treatment wetlands may be classified by type and by water source (Table 1). The two most common are the constructed emergent marsh, also called a surface flow wetland (SF); and the constructed gravel bed, also called a sub-surface flow wetland (SSF). Sub-surface systems are better in terms of prevention of mosquitoes, odors, and human contact. Surface systems are better in terms of ancillary wildlife habitat and cost. This chapter deals with constructed and natural emergent marshes and SSF wetlands. Forested and other natural wetlands, submerged aquatic beds, and wetlands with floating leaved plants are not considered here.

Wetlands treatment has grown to the point of being considered a technology of its own and is being used for treatment of waste water from animal operations. It falls into the general classification of ecological engineering, in the subset known as natural systems for waste water treatment. In turn, natural treatment systems include terrestrial (land treatment) and aquatic (lagoons),

Table 1. Types and applications of treatment wetlands.

Types	Applications
Natural	
Marshes, swamps and bogs	Municipal waste water
Constructed	Mine drainage
Surface flow (marsh)	Urban stormwater
Densely vegetated	Rivers, lakes & reservoirs
Pond and island	Agricultural runoff
Submerged aquatic bed	Livestock runoff
Floating leaved	Industrial
Subsurface flow	Leachate
Gravel bed	Sludge drying
Soil based	

with wetlands occupying the middle ground.

Several component wetland processes combine to provide the observed overall treatments. These processes all lead to the transformation and transfer of a "removed" pollutant either to the atmosphere or to the wetland sediments and soils. The vegetation is extremely important for nutrient transformations and transfers, because it plays a key role in the cycling and temporary storage of nitrogen and phosphorus.

This chapter deals with nutrient removal in selected types of wetlands. Hydraulics are also important, and there are other pollutants of concern in many situations. Regulatory considerations and habitat values often dictate decision making just as much as treatment efficiency. However, those topics are outside the scope of this work.

The literature is replete with strongly suggestive terms that describe wetland functions. Descriptive words such as "net source," "net sink," and "net exporter" should be interpreted with great care, because there is not a common understanding of their meaning. The preferred method of description is the mass balance for a particular substance: the amount entering must leave by one or more routes, be chemically converted, or be stored in the wetland.

Several decades of extensive research on wetland processes have delineated the great complexity of the contributing internal processes, but very few have been quantified in terms of field-validated predictive models. Consequently, the technology must rely upon performance data from complete ecosystems and simplistic models that quantify that data. That approach carries a responsibility not to extrapolate outside the range of the contributing experience.

THE DATABASE

Performance has been measured and documented for hundreds of wetland treatment systems. Most of the information describes the overall performance of an entire ecosystem, and not the details of component processes. The parent form of the data is typically to be found in data reports, issued by the agencies responsible for research and monitoring at a particular wetland site. Several wetland treatment sites have each generated over a thousand pages of such reports. Depending on the motivations of the investigators, some small portion of this "dark gray literature" may sometimes be condensed into descriptive papers to be found in the "light gray" literature, comprised of lightly reviewed conference proceedings papers (Table 2). Only a fraction of the information eventually finds its way into the peer-reviewed journals.

The North American Database (NADB) (Knight et al., 1993), although terminated by US EPA before completion, contains information for 323 wetland cells of several types at 178 locations. The Danish summary (Schierup et al., 1990) contains data on 109 reed bed wetlands. It is clear that there is an enormous amount of data from which to obtain a definition of expected pollutant removals.

Table 2. Major literature sources on constructed wetlands in pollution control.

10	Moshiri, G.A., ed., *Constructed Wetlands for Water Quality Improvement*, Proceedings of an International Symposium, Lewis Publishers, Chelsea, Michigan, 1993.
9	Knight, R.L., Ruble, R.W., Kadlec, R.H., and Reed, S.C., *Database: North American Wetlands for Water Quality Treatment*, Phase II Report, prepared for US EPA, September, 1993.
8	Wetland Systems in Water Pollution Control, *Science and Technology*, 29, 1994.
7	Strecker, E.W., Kersnar, J.M., Driscoll, E.D., and Horner, R.R., *The Use of Wetlands for Controlling Stormwater Pollution*, distributed by the Terrene Institute, 1700 K Street, NW, Suite 1005, Washington, D.C. 20006, 1992.
6	Schierup, H.H., Brix, H. and Lorenzen, B., *Spildevandsrensning i Rodzonanlaeg (Waste water Purification in Root Zone Systems)*, Botanical Institute, Aarhus University, Nordlandsvej 68, DK-8420, Risskov, Denmark, 1990.
5	Cooper, P.F., and Findlater, B.C., eds., *Constructed Wetlands in Water Pollution Control*, Pergamon Press, London, 1990.
4	Water Pollution Control Federation, *Manual of Practice: Natural Systems, Wetlands Chapter*, MOP FD-16 WPCF, 1990.
3	Hammer, D.A., ed., *Constructed Wetlands for Waste water Treatment*, Conference Proceedings, Lewis Publishers, Chelsea, Michigan, 1989.
2	US EPA, *Design Manual: Constructed Wetlands and Aquatic Plant Systems for Municipal Waste water Treatment*, US EPA 625/1-88/022, 1988.
1	Reddy, K.R., and Smith, W.H., eds., *Aquatic Plants for Water Treatment and Resource Recovery*, Magnolia Press, Orlando, Florida, 1987.

THE BITS AND PIECES

In addition to the system data described in the preceding section, there is a large amount of literature pertaining to individual wetland processes. It is typically derived from research on microcosms and mesocosms, which do not generally include the full suite of wetland variables. The interactions with weather and animals are usually missing, and the time scales are too short to include either seasonal effects or the long-term (years-long) response of a complete ecosystem to its altered driving forces of flows and nutrients. It is, therefore, very dangerous to draw conclusions about full-scale, long-term wetland performance from highly specialized "more scientific" studies.

Other difficulties with small-scale studies involve mesocosm edge effects: the plants hang over the edge, creating abnormal vegetation density, the container precludes lateral advection, and hydrologic regime is artificially contrived.

At this point in time, wetland science is not advanced enough to allow synthesis of short-term, small-scale, highly specialized studies into an accurate prediction of the character or function of a fully developed and stabilized treatment wetland ecosystem. Calibration of a synthesized compartment model requires more data than is presently available.

For these reasons, it is necessary to use full- or pilot-scale operating data to project full-scale results for new systems and to use specialized small-scale studies to build our understanding of the component processes.

METHODS OF ANALYSIS

The principal variables associated with nutrient removal are the concentrations, depths, and flows; the size, type, and age of the wetland; and the climatic conditions in which it operates. Data sets usually include this information for inflows and outflows but often do not provide data on either the internal gradients in concentration or the status of soils and vegetation. There are several ways to utilize the available data, and some are more useful than others. The most simplistic are single-number, input/output rules, which have great appeal because they are quickly applied and easily understood. However, it is preferable to utilize internal mass balances coupled with causal removal equations.

SINGLE NUMBERS: DEFINITION OF TERMS

It is very easy to compare the amounts of a pollutant in the inlet and outlet streams of a wetland, and compute the percentage difference. Two such calculations are in use: percentage concentration reduction and percentage mass reduction. The literature is replete with review papers that tabulate percentage reductions or removals for a selected spectrum of wetlands (for example, Watson et al., 1989; Johnston, 1993; Strecker et al., 1992). The implication is that wetlands of a similar type will achieve a similar reduction. Such groups of data begin to elucidate the bounds of performance, but the effects of size, loading, flow patterns, depth, and other design variables cannot be deduced.

The idea that more time in the wetland is good for water quality is intuitively very appealing. Depth is one primary controlling factor for nominal detention time; wetland area is the other:

$$\tau = \frac{\varepsilon A H}{Q} \qquad (1)$$

where A = wetland area, m^2
  H = water depth, m
  ε = water column void fraction
  τ = nominal detention time, d

The activity of the wetland in pollutant removal is associated with the immersed sediments and biota. As a consequence, the rate of removal is strongly dependent on vegetation density: a bare soil, shallow pond has the minimum efficiency; a densely vegetated, fully littered wetland of the same depth has a higher efficiency. Most surface flow (SF) treatment wetlands operate at about 30 cm; most sub-surface flow (SSF) wetlands operate at 50 cm of media.

If the detention time is increased by deeper submergence of these active components, at constant wetland area, little further removal activity is engendered. In contrast, increasing the area of the wetland at constant depth does in fact increase the biotic material in contact with the water, and act to provide more detention time.

For the reasons stated above, hydraulic loading rate is the most relevant parameter for the design of SF wetlands. The relation between detention time and hydraulic loading rate is:

$$q = \frac{Q}{A} = \frac{\varepsilon H}{\tau} \tag{2}$$

where q = hydraulic loading rate, m/d

The water volume fraction is normally quite high for SF wetlands, with values typically in the range of 0.90 - 0.95; values are typically about 0.30-0.40 for SSF wetlands. The depth chosen for SF wetlands is usually in the range 0.15 - 0.45 meters. The seven- to ten-day detention time range for SF systems, therefore, translates to a typical range of hydraulic loading rates from 1.5 to 6.5 cm/day, with a central tendency of about 3 cm/day for marshes. Sub-surface wetlands are more heavily loaded, ranging from 5-40 cm/day.

MULTIVARIATE ANALYSIS

Three techniques may be easily used to improve upon the single number quantification: graphical displays of data sets, regression equations, and mass balance equations combined with rate equations. The first two are self-explanatory. The graphs display large scatter, and the regressions display poor correlation coefficients. Effects of flow, temperature, biota, and chemistry lead to the scatter.

There are several features of the observed wetland data that can be accommodated in relatively simple mass balance models. Those ideas are:

* Nutrients typically show a gradual decline from their inlet values, along a line parallel to flow, to the values at the wetland outlet.
* Wetlands with more area, and less water flow, are more efficient.
* Nutrients are not reduced to zero in wetlands.
* Flow patterns in wetlands are intermediate between complete mix (CM) and plug flow (PF).

The simplest models that embody these observations are the k-C* models for the extreme cases, shown here for the case of negligible rain and evaporative effects:

PLUG FLOW k-C* Model $\qquad \dfrac{C - C^*}{C_i - C^*} = \exp\left[-\dfrac{k}{q}\, y\right]$ $\qquad$ (3)

Mth CM Unit in a Series of N $\qquad \dfrac{C - C^*}{C_i - C^*} = \left[1 + \dfrac{k}{Nq}\right]^{-M}$ $\qquad$ (4)

where C = pollutant concentration, at y or M, gm/m^3
$\qquad C_i$ = inlet pollutant concentration, gm/m^3

C* = background pollutant concentration, gm/m^3
k = first order areal rate constant, m/d
N = number of complete mix units in series
M = number of a complete mix unit, from inlet
y = fractional distance through wetland

The concentration at the wetland outlet is calculated using y = 1 or M = N, respectively. Equation (3) is the limiting form of equation (4) for an infinite number of CM units in series; it is practically achieved when N>10. These models both show the necessary decreasing concentrations through the wetland and embody the correct flow and area dependence. They both are based on a return of the pollutant from the ecosystem to the water, calculated as:

$$J_r = kC* \qquad (5)$$

where J_r = pollutant return flux, gm/m^2/d

Design for non-ideal mixing is accomplished by appropriate selection of the number of well mixed units.

An examination of the model shows that it requires three parameters: a background concentration, C*; a rate constant, k; and a mixing descriptor, N. In general, a tracer study is required to determine the value of N for a given wetland, and that has been done in a limited number of instances. Typically, N ≈ 3 provides a description of wetland mixing. In the absence of data from which to determine N, the conservative data analysis assumes plug flow (N = ∞), since that yields the minimum value of the rate constant that can be derived from a particular concentration and flow data set. In design calculations, other values of N may be more appropriate, especially if hydraulics cannot be optimized for the system being designed.

RESULTS FOR PHOSPHORUS

The mechanisms of phosphorus removal are well understood. In SF wetlands, soil sorption may provide storage for a short time (a few months), but that capacity is small in the context of treatment wetlands. The media in an SSF wetland may be selected to provide a much greater potential (Jenssen et al., 1994). Particulate phosphorus is settled and trapped in the litter (SF) or media (SSF). Soluble phosphorus may precipitate or co-precipitate with iron or calcium minerals. Soluble reactive phosphorus is utilized by bacteria, algae and the macrophytes, which grow, die, and decompose but leave an undecomposed residual as new sediments and soils.

Phosphorus is found in several forms in wetlands: dissolved, particulate, available, mineral, organic, etc. In the discussion that follows, only total phosphorus will be considered.

The return flux of phosphorus from the static ecosystem components is quite small, and consequently C* is negligible compared to commonly encountered phosphorus concentrations. It appears that $C^* < \approx 0.02$ mg/L, and, therefore, will be taken as zero in analysis. There remains only the areal rate constant k to be determined. Values for emergent marshes, determined from quarterly averages of data (Table 3), center on approximately 12 m/year. The average value for 65 Danish SSF soil base wetlands, determined from project lifetime average data, was k = 9.8 m/year.

These model parameters are considered to apply on an average annual basis and are known to inadequately describe higher frequency phenomena. The ratio of the maximum monthly concentration to the average annual concentration for the NADB was 1.8 for 43 wetlands of all types. This variability is associated with random and deterministic phenomena that cannot be described at the k-C* level of modeling.

The expectation would be for lower rate constants in winter at cold temperatures. However, wetland data do not support this intuitively appealing view; there is not a temperature effect in performance data.

Stable operation of a treatment wetland is not achieved quickly; one to two years is required in southern climates, and two to four years in northern climates. A constructed system starts with sparse or no vegetation to assist in the phosphorus cycle and must develop the biotic components of the ecosystem in order to reach full potential. During this period, phosphorus sorption adjusts

Table 3. First order phosphorus rate constant for marshes.

Site	No. of Wetlands	Data Years	HLR cm/day	TP In mg/L	TP Out mg/L	k Value m/year
Des Plaines, IL	4	6	4.77	0.10	0.02	23.7
Jackson Bottoms, OR	17	2	6.34	7.51	4.14	14.2
Pembroke, KY	2	2	0.77	3.01	0.11	9.3
Great Meadows, MA	1	1	0.95	2.00	0.51	5.7
Fontanges, QUE	1	2	5.60	4.15	2.40	11.2
Houghton Lake, MI	1	16	0.44	2.98	0.10	11.0
Cobalt, ONT	1	2	7.71	1.68	0.77	20.9
Brookhaven, NY	1	3	1.50	11.08	2.33	8.9
Leaf River, MS	3	5	11.68	5.17	3.96	11.2
Sea Pines, SC	1	8	20.20	3.94	3.36	11.7
Benton, KY	2	2	4.72	4.54	4.10	2.4
Listowel, ONT	5	4	2.41	1.91	0.72	8.2
Humboldt, SAS	5	3	3.04	10.16	3.24	12.8
Tarrant County, TX	9	1	9.44	0.29	0.16	20.1
Iron Bridge, FL	16	7	2.69	0.43	0.10	13.5
Boney Marsh, FL	1	11	2.21	0.05	0.02	14.2
WCA2A, FL	1	14	0.93	0.12	0.02	10.2
OCESA, FL	4	6	0.83	0.27	0.16	6.4
Kis-Balaton, HUNG	1	5	3.3	0.54	0.23	8.6
					Average	12

itself to the conditions of the overlying waters. That may involve either release or uptake, depending on the antecedent condition of the soil.

The largest treatment wetlands in the world are operating for phosphorus removal, notably the 1,800-ha Kis-Balaton project in Hungary (Pomogyi, 1993), and the 1,500-ha Everglades Nutrient Removal project (Newman et al., 1993).

RESULTS FOR NITROGEN

Nitrogen compounds interconvert in the wetland environment (Figure 1), including the gasification processes of ammonia volatilization and denitrification. As a result, removal may be to the atmosphere as well as to newly formed sediments. There is soil sorption potential for ammonium nitrogen, but nitrate does not sorb. The macrophytes utilize a large amount of ammonium nitrogen for growth, some of which is returned to the water as organic nitrogen. Data interpretation is made difficult by the superposition of nitrogen speciation on the inter-compartment transfers in the wetland ecosystem.

Microbial conversion processes require the presence of appropriate bacterial populations, as well as energy supplies. Decomposition of biomass can provide the necessary carbon in an SF wetland. But that carbon may not reach the water in an SSF wetland, thus limiting the denitrifier populations (Gersberg et al., 1984). The conversion of ammonium to nitrite and nitrate (nitrification) requires an oxygen supply. Wetland plants bring oxygen to their root zone via porous stems to supply respiratory needs, and this mechanism may provide oxygen for nitrification adjacent to the roots. However, the majority of research results indicate that the excess oxygen available for nitrification is quite small (Brix and Schierup, 1990). A sequential chemical reaction model of the interconversions in Figure 1 is required to describe ammonium and nitrate concentrations for most conditions.

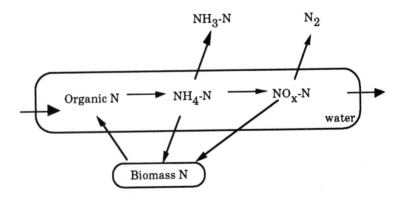

Figure 1. The nitrogen transfers and conversions in wetlands.

Startup periods for nitrogen removal may be just as long as for phosphorus removal if the loading is low enough that burial and biomass expansion can extract a significant portion of the added nutrients. But for heavily loaded wetlands, the preponderance of activity is in the microbial community, which responds rapidly to changing water quality conditions

TOTAL NITROGEN

Despite the complexity of the species, storages, and transfers, there is a common observation for treatment wetlands: total nitrogen is removed from incoming waste waters. Outlet concentrations increase with increasing loading rates for both SF and SSF wetlands. SSF and SF wetlands do not appear to differ when viewed in terms of these input-output variables; but SSF wetlands are loaded more heavily. The scatter is large, as for phosphorus, and probably for the same reasons.

The k-C* model may be calibrated to operating data for total nitrogen, based on the conservative plug flow premise discussed above. Corrected to 20°C, k = 22 m/year for SF wetlands, and 27 m/year for SSF wetlands. Temperature is an important determinant of the k value for TN. Based on limited information, it appears that the effect of temperature may be described by:

$$K_T = k_{20}\theta^{(T-20)} \tag{6}$$

where T = temperature, °C
θ = temperature factor, =1.05

C* is not zero for total nitrogen. Return processes, perhaps combined with a refractory form of organic nitrogen, lead to a minimum summer TN of about 1.5 mg/L. In northern wetlands, under-ice processes can raise that value to over 5 mg/L.

The k-C* model parameters above are for annual averages. Maximum monthly values of TN concentrations average 1.6 times the annual average.

OXIDIZED (NITRATE) NITROGEN

The interconversions of nitrogen species often confuse the analysis of wetland data. If the incoming water contains significant ammonium or organic nitrogen, a production of nitrate is possible by nitrification. The rate of denitrification is high compared to its formation reaction (nitrification), and hence only small and variable amounts of nitrate are usually measured in the wetland or at its exit. Input-output data are, therefore, often inadequate to determine rate constants.

There have been a number of studies that have dealt with the introduction of nitrate only. This species does not readily convert back to ammonium in a wetland (van Oostrom and Russell, 1994); therefore, its removal rate may be studied independently. The SF data of Crumpton et al. (1993) and the SSF data of Stengel et al. (1987) suggest that a first order areal model with $C^* = 0$ is appropriate. The k value at 20°C is approximately 35 m/year for the SF mesocosms and about 50 m/year for SSF systems that are not carbon limited (Gersberg et al., 1984). The absence of a carbon source can drop the k value to less than 10 m/year in an SSF wetland (Gersberg et al., 1984). This rate constant is temperature sensitive in wetlands, with $\theta = 1.09$. It is extremely important to note that an isolated calculation of nitrate removal via the k-C^* model is incorrect in those systems that contain a significant quantity of other nitrogen species.

Maximum monthly values of nitrate concentrations average 2.5 times the annual average.

ORGANIC NITROGEN

This species is produced by biomass decomposition and is mineralized to ammonium nitrogen. Regression produces a k value at 20°C of approximately 17 m/year for SF wetlands and about 35 m/year for SSF wetlands. These rate constants are temperature dependent, with $\theta = 1.05$. $C^* = 1.5$ for this nitrogen species.

AMMONIUM NITROGEN

Ammonium is the preferred nutrient for plant growth and is formed from mineralization. Microbial nitrification converts ammonium to nitrite and nitrate. Wetland parameter estimation must take these processes into account, as must the subsequent design calculations. The mass balances are complex, and lead to k values at 20°C of approximately 18 m/year for SF wetlands and about 34 m/year for SSF wetlands. These rate constants are temperature dependent, with $\theta = 1.04$.

DISCUSSION

Wetlands are a feasible alternative to other treatment technologies in a number of circumstances. An enormous amount of data exists upon which to base a rational design for the removal of nutrients and other common pollutants. Regression equations and scatterplots can help to determine initial feasibility, but mass balance models are a more rational technique. At the present time, only very simple k-C^* removal equations can be calibrated to accompany the mass balance. Total phosphorus, total nitrogen, and organic nitrogen may be calculated in this way. The interconversions of nitrogen compounds in fully speciated situations require a sequential model to determine nitrogen speciation, particularly the partitioning of TKN into organic and ammonium nitrogen.

The role of the plants is to cycle nutrients and to provide carbon for the microbes that conduct denitrification and other pollutant removal reactions. The decomposition step in the cycle returns some nutrients to the water, leading to the necessity for including a return rate in reduction models, embodied in a background concentration. Not all nutrients are returned to the water; some fraction is retained in new sediments.

Data from operating wetland treatment systems do not always support preconceived notions based on other ecosystems and other waste treatment technologies. Phosphorus removal does not necessarily slow down in winter. Wetlands do not drive nutrients to zero values at all times of the year. The plants are very important in nutrient cycling but probably provide little root zone oxygenation. There are not "big spring flushes" of nutrients from treatment wetlands. In contrast to the aquatic experience, phosphorus removal is just as efficient at low redox potentials as at high.

The practitioner is also advised to pay close attention to the factual basis for equations that may be found in the literature on wetland design. At present, there is rapid growth of available data, with a doubling time of only a year or two. There is a publication lag time of the same magnitude for issuance of result papers and a second publication lag time associated with incorporation into the design literature. Design equations that were the best available five years ago may not come close to explaining today's data because the author did not have the benefit of most of it.

The models presented above are deterministic in character and carry the implication of a well-ordered result in an operating wetland. That impression should be discarded: there is large inherent variability around the averages discussed and modeled above. Coefficients of variation are large, often ranging from factors of two to five, the larger being at shorter sampling intervals, such as weekly.

The equations in this chapter were calibrated to data averaged over periods ranging from three to 12 months and longer. Whenever possible, the effects of system startup were not included, and so these equations should not be expected to apply during that period. There are good reasons for a long averaging period: detention times are long, and there is a significant associated response lag for nutrient removal. Temporary storages and releases are smoothed by such averaging, and the inherent stochastic variability becomes less dominant when averages are used. The equations presented above, therefore, should not be used in short-time step calculations: they will not produce accurate results.

Treatment wetlands are user friendly; when properly designed they perform the intended functions and fit pleasingly into the landscape.

REFERENCES (See also Table 2)

Brix, H., and Schierup, H.H., Soil oxygenation in constructed reed beds: The role of the macrophyte and soil-atmosphere interface oxygen transport, in *Constructed Wetlands in Water Pollution Control*, P.F. Cooper and B.C. Findlater, eds., Pergamon Press, Oxford, UK, 1990.

Crumpton, W.G., Isenhart, T.M., and Fisher, S.W., Fate of non-point source nitrate loads in freshwater wetlands results from experimental wetland mesocosms, in *Constructed Wetlands for Water Quality Improvement*, G.A. Moshiri, ed., Lewis Publishers, Chelsea, Michigan, 1993.

Gersberg, R.M., Elkins, B.V., and Goldman, C.R., The use of artificial wetlands to remove nitrogen from wastewater, *Jour. of the Water Pollution Control Federation*, 56(2), 152, 1984.

Jenssen, P. D., Maehlum, T., and Krogstad, T., Adapting constructed wetlands for waste water treatment to northern environments, in *Global Wetlands: Old World and New*, W.J. Mitsch, ed., Elsevier, Amsterdam, The Netherlands, 1994.

Johnston, C.A., Mechanisms of wetland-water quality interaction, in *Constructed Wetlands for Water Quality Improvement*, G.A. Moshiri, ed., Lewis Publishers, Chelsea, Michigan, 1993.

Newman, S., Roy, J. and Obeysekera, J., The Florida Everglades Nutrient Removal Project, in *Hydraulic Engineering '93*, H.W. Shen, S.T. Su and F. Wen, eds., ASCE, New York, 1993.

Pomogyi, P., Nutrient retention by the Kis-Balaton water protection system, *Hydrobiogia*, 251, 309, 1993.

Stengel, E., Carduck, W., and Jebsen, C., Evidence for denitrification in artificial wetlands, in *Aquatic Plants for Water Treatment and Resource Recovery*, K.R. Reddy and W.H. Smith, eds., Magnolia, 1987.

van Oostrom, A.J., and Russell, J.M., Denitrification in constructed waste water wetlands receiving high concentrations of nitrate, *Water Science and Technology*, 29(4), 7, 1994.

Watson, J.T., Reed, S.C., Kadlec, R.H., Knight, R.L., and Whitehouse, A.E., Performance expectations and loading rates for constructed wetlands, in *Constructed Wetlands for Wastewater Treatment*, D.A. Hammer, ed., Lewis Publishers, Chelsea, Michigan, 1989.

IMPACTS OF NON-POINT SOURCE RUNOFF FROM AGRICULTURAL OPERATIONS ON LAKE OKEECHOBEE, FLORIDA

Nicholas G. Aumen, Alan D. Steinman, and Karl E. Havens

INTRODUCTION

Accelerated eutrophication of aquatic ecosystems from agriculturally derived non-point source runoff is a pervasive problem. In the United States, an estimated 54% of land surface is managed for agriculture (Bureau of the Census, 1990). Non-point sources of nutrient enrichment in these agricultural areas are difficult and expensive to correct due to their non-localized nature. Indeed, most affected aquatic ecosystems are still in decline or have not shown significant response after large expenditures on corrective actions (National Research Council, 1992)

The main sources of excess nutrients from agricultural operations are animal waste and fertilizers applied to food and forage crops. Animal waste is relatively high in soluble inorganic and organic nutrients, which are readily transported into surface and ground water. Agricultural activities that concentrate animals in relatively small areas increase the likelihood of negative impacts from animal waste. In addition, application of fertilizers to fields typically exceeds the amount assimilated by vegetation or stored in soils, resulting in transport of nutrients to water bodies.

This chapter describes agriculture activities in rapidly growing southern Florida and presents a case study of accelerated eutrophication in a highly valued lake ecosystem caused, in part, by phosphorus (P) inputs from agricultural animal waste. The following sections document a chronology of environmental deterioration, public and governmental response, monitoring and research, corrective management, and ecosystem responses. Lessons learned should be applied to future resource management at the land-water interface, particularly with respect to P enrichment of aquatic ecosystems.

AGRICULTURAL ACTIVITY IN SOUTH FLORIDA AND NON-POINT SOURCE RUNOFF

Within the natural watershed boundaries of the South Florida Water Management District (SFWMD), from Orlando to the Florida Keys, agricultural land occupies approximately 1.2 X 10^6 ha. The agricultural economy is based primarily on sugar cane, winter vegetables, citrus, dairy, and beef cattle farming. Dairy and beef cattle ranching dominate agricultural land use north of the lake. As of the late 1980s, 43 dairy operations occupying 25,000 ha of land were located north of Lake Okeechobee (Gunsalus et al., 1992).

Dairy farming is characterized by relatively large herds located on small land areas, resulting in concentrated animal waste deposition. In contrast, beef cattle ranching on improved pastures results in fewer cattle per unit area than dairies. However, the overall contributions of dairy and beef operations to non-point source nutrient runoff are estimated to be of the same order of magnitude because approximately 10 times more land area is in improved pasture for beef cattle than for dairies (Flaig and Havens, 1995).

Phosphorus imports to the approximately 12,000 km^2 Lake Okeechobee basin are primarily in the form of animal feed and fertilizers. Net P imports in dairy feed and fertilizer are estimated to be 1,200 and 1,500 tons P/year, respectively, during the late 1980s. Approximately 90% of the annual net imports of P from fertilizer application, animal waste, and other sources is retained in upland soils and wetland sediments in the basin, and 10% is discharged into Lake Okeechobee through non-point source runoff.

KISSIMMEE-OKEECHOBEE-EVERGLADES ECOSYSTEM

South Florida's landscape is dominated by the interconnected Kissimmee River-Lake Okeechobee-Everglades ecosystem. Historically, sheet flow of water occurred from Lake Okeechobee through the Everglades to Florida Bay. However, alterations of surface water flow have occurred for water supply and flood control purposes. More than 200 water control structures and 2,200 km of canals intersect the southerly flow, storing and conveying water for agriculture and urban use.

A central feature of southern Florida is Lake Okeechobee (Figure 1), a large (1730 km^2), shallow (2.7 m mean depth), subtropical lake (26° 58' N, 80° 50' W). The lake is surrounded by a flood-control levee, and almost all surface water inflow and outflow are regulated through more than 30 water control structures (James et al., 1995a). The lake features a large (400 km^2), diverse littoral zone of submergent and emergent plants and is highly influenced by wind-driven sediment resuspension (Maceina and Soballe, 1990). Despite Lake Okeechobee's historically high biological productivity (Gleason and Stone, 1975), its biota may have been limited primarily by P (Havens, 1995).

EUTROPHICATION OF LAKE OKEECHOBEE

A monitoring program begun by the SFWMD in 1973 documented declining lake water quality. Nutrient loads to the lake increased from the early 1970s to the early 1980s (James et al., 1995a) and were positively related to P imports to the basin. Total P and N concentrations in the lake doubled during the same

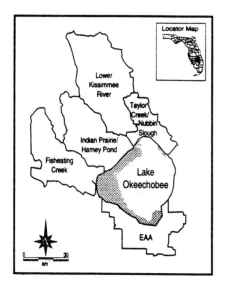

Figure 1. Lake Okeechobee and its basin. Shaded area represents the lake's littoral zone.

time period, with total P concentrations increasing from approximately 0.05 to 0.1 mg P/L (Figure 2) (James et al., 1995b). Increased P input results in more than 400 metric tons of P being added to the lake sediments annually (James et al., 1995a).

Reduced inputs of N-rich agricultural water from the Everglades Agricultural Area (EAA) south of the lake were correlated with a shift in water quality trends between 1981 and 1982 (James et al., 1995b). In the lake, total N concentrations declined while total P concentrations remained relatively constant, resulting in a drop in TN:TP ratios (Smith et al., 1995). Laboratory experiments using lake sediment cores suggest that P flux between the sediments and the water column is large (Reddy, 1993), and the sediments probably serve as a P source to the water column.

The declines in water column TN:TP ratios have led to conditions favorable to nuisance cyanobacteria (Smith et al., 1995). Analysis of water quality monitoring data and *in vitro* nutrient enrichment bioassays suggest that Lake Okeechobee has changed from P to N limitation (Havens, 1995) and that N limitation of algae is common (Aldridge et al., 1995) and favors blooms of N-fixing cyanobacteria (Havens et al., 1995). Extensive cyanobacterial blooms occurred in 1986, covering as much as 700 km^2, or 42%, of the lake's surface (Jones, 1987). The large extent and aesthetically unpleasant nature of the blooms triggered intense public pressure to commit money and other resources to elucidate the causal factors (Aumen, 1995).

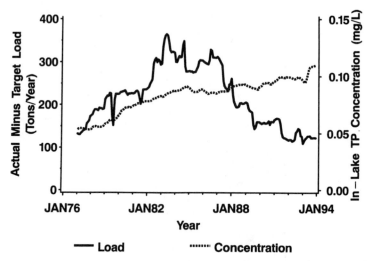

Figure 2. Lake Okeechobee phosphorus (P) loading (five-year moving average) and monthly average in-lake P concentrations.

Ecosystem-level research has documented other shifts in Lake Okeechobee's biota that accompanied declining water quality. Although historical data are limited, samples collected during the early 1970s indicate a phytoplankton community dominated by diatoms (Marshall, 1977). More recent studies reveal that cyanobacteria are now dominant (Cichra et al., 1995). The zooplankton community also has changed since the late 1970s, with increases in rotifer and copepod numbers (Crisman et al., 1995). Changes in the benthic macroinvertebrate community over a 24-year period include a dramatic increase in the relative proportion of oligochaetes--organisms indicative of enriched organic sediments--relative to other taxa (Warren et al., 1995).

LAKE OKEECHOBEE BASIN PHOSPHORUS CONTROL PRACTICES

Public pressure and a lake ecosystem in decline led to management activities designed to reduce the amount of P in non-point source runoff. Best Management Practices (BMPs) were implemented, including land application of animal waste as a nutrient source to forage crops, fencing of cattle from waterways, and waste water storage and treatment lagoons. The SFWMD contributed to non-point source runoff reductions in 1979 by reducing the pumping of N-rich agricultural runoff from the EAA to the lake.

Florida promulgated the Dairy Rule in 1987 (Ch. 17-670, F.A.C.), which required that dairies implement improved animal waste management systems to reduce P inputs to the lake (SFWMD, 1993). The intent of the Rule was to implement BMPs on dairies to collect manure and waste water from areas of high-intensity animal activity, to process that waste water in anaerobic lagoons, and to apply processed water to forage crops. This design encourages P recycling on-site and decreases the potential for off-site runoff.

Even with cost-sharing provided by the state, requirements of the Dairy Rule created financial hardships. The state responded with a dairy buy-out program in which dairy farmers could cease dairy farming in return for reimbursement for cows taken out of production. Forty percent of the dairies participated, reducing herds from 54,000 to 33,000 milk cows between 1987 and 1992 (Flaig and Havens, 1995).

Phosphorus concentration limits were established in 1989 for runoff from all non-dairy land uses greater than 0.24 ha (Flaig and Havens, 1995). The limits are concentration-based, with an upper limit of 1.2 mg P/L for intensive land use such as feed lots, and a limit of 0.35 mg P/L for improved pastures. Other land uses are required to maintain or decrease runoff P concentrations that existed prior to Rule implementation. A P loading target for Lake Okeechobee of 397 tons/year (annual average) was established in the 1987 Surface Water Improvement and Management Act (Chapter 373.451 and 373.454, F.S.). This target was developed using a modified Vollenweider model and varies temporally based on changes in lake volume and water residence time (Federico et al., 1981).

One of the most successful dairy BMPs was the design and implementation of confinement practices. The approach of confining large numbers of cows in small land areas facilitated the collection and management of nutrient-enriched runoff. A typical confinement dairy operation was based on the construction of confinement barns located next to milking barns (Figure 3). Dairy herds are confined to these barns the majority of time, and manure is removed by wash-down systems and scraping. Runoff is collected in a drainage system,

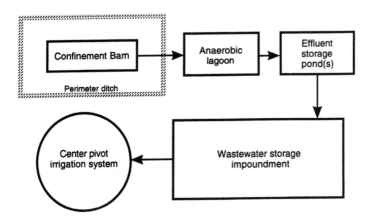

Figure 3. Schematic design of waste water management in a typical dairy in the Lake Okeechobee basin.

which empties into adjacent storage lagoons. Effluent is then sprayed on irrigation fields, where P assimilation by forage crops and soils encourages on-site P retention.

Confinement dairy operations and reductions in feed P content have coincided with some dramatic reductions in P concentrations in dairy runoff. In one case, implementation of this BMP resulted in a P reduction from over 10 mg P/L to less than the 1.2 mg P/L limit (Figure 4). In addition to reductions in off-site P transport, confinement practices in at least one dairy led to increased efficiency in milk production. Despite government cost-sharing, construction of BMPs and unfavorable economic conditions have created a financial burden for the dairy industry, and several dairies have gone out of business since the buy-out program was implemented.

ECOSYSTEM RESPONSES TO PHOSPHORUS CONTROL PRACTICES

Reductions in nutrient inputs from agricultural areas south of the lake, reductions of P in cattle feed, reductions in the application of fertilizers, and implementation of dairy BMPs have resulted in the downward trend in P loading to Lake Okeechobee since the early 1980s (Figure 2). However, P loads still are approximately 100 tons above the average annual target of 397 tons/year.

Failure to meet the P loading target may stem from the relatively high capacity of basin soils and sediments to retain, and gradually release, excess P from agricultural operations over the last several decades. Although the majority of the P inputs to the soil profile become part of relatively stable complexes,

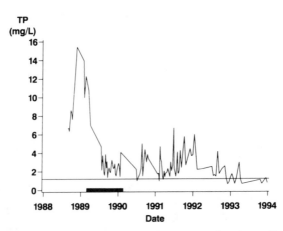

Figure 4. Monthly average total phosphorus (P) concentrations in runoff from a dairy pre- and post-BMP (best management practice) implementation. Bar represents period of BMP construction. Horizontal line is at 1.2 mg P/L, the present concentration limit for dairy runoff.

the less-stable fractions can serve as a source of P to overlying water (Reddy et al., 1995). This release mechanism can result in excess P being exported from sites previously impacted by agricultural activities, even with BMPs in place.

Additional P exports to Lake Okeechobee occur from agricultural land uses where BMPs are not required or from ineffective pasture management on some dairies. For example, heifer operations and beef cattle ranches are not required to have waste water management systems. Improvements in dairy pasture management, such as ensuring that cattle are excluded from wetlands and creek banks, are needed. Also, P export from high-intensity animal use areas on abandoned dairies may contribute to excess P loads to the lake.

The achieved reductions in P loading have not led to corresponding reductions of lake P concentrations (Figure 2), which is the ultimate goal of BMPs. There is an indication of stabilizing P concentrations since the early 1980s, but the high degree of variability makes trend determination difficult. The lack of immediate response in P concentrations to reductions in P loadings most likely is related to the lake sediments' role as a source of water column P.

COMPARISONS WITH OTHER SHALLOW LAKES

An impressive number of studies of shallow lakes in The Netherlands and Denmark have shown that the in-lake P concentration does not decline following reductions in external P loads (e.g., Jeppesen et al., 1991; Van Liere and Janse, 1992). In shallow lakes, sediment resuspension plays a critical role in supplying P to the water column, thereby counteracting the positive effects of BMPs. Thus, further reduction of external P loading may not result in lake water P concentration reductions for decades because of compensatory P release from sediments.

Numerous lake restoration measures have been attempted to reduce internal nutrient loading (Cooke et al., 1993), and it is essential that these procedures be coordinated with reduction of external nutrient inputs. In particular, flushing with unpolluted water (Jagtman et al., 1992), aeration (Jaeger, 1994), dredging (Van der Does et al., 1992), biomanipulation (Hosper and Jagtman, 1990), and P inactivation (usually with an Al, Fe, or Ca salt; Welch and Schrieve, 1994; Van der Veen et al., 1987) all have been attempted and met with varying degrees of success.

No similar measures have been attempted in Lake Okeechobee, where the sheer size of the lake limits their economic feasibility. Interestingly, two natural biological developments may result in enhanced restoration without intervention. First, large *Daphnia lumholtzii* recently have been observed in Lake Okeechobee. Large zooplankton can graze prodigious amounts of phytoplankton, thereby having the potential to reduce turbidity in lakes (Hosper and Jagtman, 1990). However, because the lake's turbidity derives largely from suspended sediments and not phytoplankton, the role of *Daphnia* may be restricted. Second, dense beds of benthic algae in the class Characeae (with some areas exceeding 50 g/m²) have been observed in the southern end of the lake. *Chara* has been observed as a pioneer species in the recovery stages of other shallow lakes (Hofstra and Van Liere, 1992; Scheffer et al., 1992). Characean growth may serve a dual purpose in reducing P concentrations: (1) sediment stabilization, thereby reducing desorption of P; and (2) uptake of P for growth, thereby removing P from the overlying water.

RELEVANCE TO WATERSHED AND LAKE MANAGEMENT

Research has suggested specific locations within the lake that will respond first to reductions in non-point source runoff. A spatially intensive five-year study of the relationship between physical/chemical parameters and phytoplankton indicate that the lake functions as a series of distinct ecological zones (Phlips et al., 1993, 1995). Phytoplankton located at the periphery of the limnetic region, particularly in the southern part of the lake, may respond first to P load reductions (Phlips et al., 1995). The central limnetic region is unlikely to exhibit a response to P load reductions because of its location over unconsolidated muddy sediments and the effect of wind-driven sediment resuspension.

Our increased knowledge of the processes responsible for transport and retention of P in the basin lead to better estimates of the time frame for recovery. Public perception is that rapid responses should occur in return for vast expenditures of time and money on BMPs. Instead, research indicates that response times on the order of decades may be more realistic.

Finally, management action must be taken to address all potential sources of agriculturally derived P to Lake Okeechobee to ensure favorable ecosystem response. Dairies were the target of initial action because they were the locations of concentrated animal activity and high P levels in runoff. Because of P concentration reductions in dairy runoff, beef cattle pastures have increased in their relative importance to total P input to Lake Okeechobee. Research is underway using field-scale experimentation to design pasture fertilization rates, grazing rotational schemes, and stocking rates that result in lower levels of P in surface and sub-surface flows. As with efforts directed at dairies, research must strive toward developing agricultural practices in southern Florida that are both economically and environmentally sustainable.

ACKNOWLEDGMENTS

The authors are grateful to: Tom James for calculating P loading rates in Lake Okeechobee; Eric Flaig, Susan Gray, and Barry Rosen for their helpful comments on early versions of this manuscript; and Tom James and JoAnn Hyres in the preparation of graphics.

REFERENCES

Aldridge, F.J., Phlips, E.J., and Schelske, C.L., The use of nutrient enrichment bioassays to test for spatial and temporal distribution of limiting factors affecting phytoplankton dynamics in Lake Okeechobee, Florida, in *Ecological Studies of the Littoral and Pelagic Systems of Lake Okeechobee, Florida (USA)*, Aumen, N.G. and Wetzel R.G., eds., *Archiv für Hydrobiologie Beiheft Ergebnisse der Limnologie*, 45, 177, 1995.

Aumen, N.G., The history of human impacts, lake management, and limnological research of Lake Okeechobee, Florida (U.S.A.), in *Ecological Studies of the Littoral and Pelagic Systems of Lake Okeechobee, Florida (USA)*, Aumen, N.G. and Wetzel R.G., eds., *Archiv für Hydrobiologie Beiheft Ergebnisse der Limnologie*, 45, 1, 1995.

Bureau of the Census, *Statistical Abstract of the United States, 1990*: The National Data Book, US Department of Commerce, US Government Printing Office, Washington, D.C., 1990.

Cichra, M.F., Badylak, S., Henderson, N., Rueter, B.H., and Phlips, E.J., Phytoplankton community structure in the open water zone of a shallow sub-tropical lake (Lake Okeechobee, Florida, U.S.A.), in *Ecological Studies of the Littoral and Pelagic Systems of Lake Okeechobee, Florida (USA)*, Aumen, N.G. and Wetzel R.G., eds., *Archiv für Hydrobiologie Beiheft Ergebnisse der Limnologie*, 45, 157, 1995.

Cooke, G.D., Welch, E.B., Peterson, S.A., and Newroth, P.R., *Restoration and Management of Lakes and Reservoirs*, Lewis Publ., Boca Raton, Florida, 1993.

Crisman, T.L., Phlips, E.J., and Beaver, J.R., Zooplankton seasonality and trophic state relationships in Lake Okeechobee, Florida, in *Ecological Studies of the Littoral and Pelagic Systems of Lake Okeechobee, Florida (USA)*, Aumen, N.G. and Wetzel R.G., eds., *Archiv für Hydrobiologie Beiheft Ergebnisse der Limnologie*, 45, 213, 1995.

Federico, A.C., Dickson, K.G., Kratzer, C.R., and Davis, F.E., *Lake Okeechobee Water Quality Studies and Eutrophication Assessment*, Technical Publication 81-2, South Florida Water Management District, West Palm Beach, Florida, 1981.

Flaig, E.G., and Havens, K.E., Historical trends in the Lake Okeechobee ecosystem I, Land use and nutrient loading, *Archiv für Hydrobiologie/Suppl. 107*, 1, 1, 1995.

Gleason, P.J., and Stone, P.A., *Prehistoric Trophic Level Status and Possible Cultural Influences on the Enrichment of Lake Okeechobee*, Central and Southern Florida Flood Control District, now South Florida Water Management District, West Palm Beach, FL., 1975.

Gunsalus, B., Flaig, E.G., and Ritter, G., Effectiveness of agricultural best management practices implemented in the Taylor Creek/Nubbin Slough and the lower Kissimmee River basins, in *Proceedings of the National Rural Clean Water Program Symposium*, 1992.

Havens, K.E., Secondary nitrogen limitation in a subtropical lake impacted by non-point source agricultural pollution, *Environmental Pollution*, in press, 1995.

Havens, K.E., Hanlon, C., and James, R.T., Historical trends in the Lake Okeechobee ecosystem, V, Algal blooms, *Archiv für Hydrobiologie/Suppl. 107*, 1, 87, 1995

Hofstra, J.J., and Van Liere, L., The state of the environment of the Loosdrecht lakes, *Hydrobiologia*, 233, 11, 1992.

Hosper, S.H., and E. Jagtman, E., Biomanipulation additional to nutrient control for restoration of shallow lakes in The Netherlands, *Hydrobiologia*, 200/201, 523, 1990.

Jaeger, D., Effects of hypolimnetic water aeration and iron-phosphate precipitation on the trophic level of Lake Krupunder, *Hydrobiologia*, 275/276, 433, 1994.

Jagtman, E., Van der Molen, D.T., and Vermij, S., The influence of flushing on nutrient dynamics, composition, and densities of algae and transparency in Veluwemeer, The Netherlands, *Hydrobiologia*, 233, 187, 1992.

James, R.T., Jones, B.L., and Smith, V.H., Historical trends in the Lake Okeechobee ecosystem. II. Nutrient budgets, *Archiv für Hydrobiologie/Suppl. 107,* 1, 25, 1995a.

James, R.T., Smith, V.H., and Jones, B.L., Historical trends in the Lake Okeechobee ecosystem. III. Water quality, *Archiv für Hydrobiologie/Suppl. 107,* 1, 49, 1995b.

Jeppesen, E., Kristensen, P., Jensen, J.P., Sondergaard, M., Mortensen, E., and Lauridsen, T., Recovery resilience following a reduction in external phosphorus loading of shallow eutrophic Danish lakes: Duration, regulating factors and methods for overcoming resilience, *Mem. Ist. ital. Idrobiol*, 48, 127, 1991.

Jones, B.L., Lake Okeechobee eutrophication research and management, *Aquatics,* 9, 21, 1987.

Maceina, M.J., and Soballe, D.M., Wind-related limnological variation in Lake Okeechobee, Florida, *Lake and Reservoir Management,* 6, 93, 1990.

Marshall, M.L., *Phytoplankton and Primary Productivity Studies in Lake Okeechobee during 1974,* Technical Publication 77-2, South Florida Water Management District, West Palm Beach, Florida, 1977.

National Research Council, *Restoration of Aquatic Ecosystems,* National Academy Press, Washington, D.C., 1992.

Phlips, E.J., Aldridge, F.J., and Hansen, P., Patterns of water chemistry, physical and biological parameters in a shallow subtropical lake (Lake Okeechobee, Florida, U.S.A.), in *Ecological Studies of the Littoral and Pelagic Systems of Lake Okeechobee, Florida (USA),* Aumen, N.G. and Wetzel R.G., eds., *Archiv für Hydrobiologie Beiheft Ergebnisse der Limnologie,* 45, 117, 1995.

Phlips, E.J., Aldridge, F.J., Hansen, P., Zimba, P.V., Ihnat, J., Conroy, M., and Ritter, P., Spatial and temporal variability of trophic state parameters in a shallow, subtropical lake (Lake Okeechobee, Florida, U.S.A.), *Archiv für Hydrobiologie,* 128, 437, 1993.

Reddy, K.R., *Lake Okeechobee Phosphorus Dynamics Study: Biogeochemical Processes in the Sediments,* Report to the South Florida Water Management District, West Palm Beach, Florida, 1993.

Reddy, K.R., Flaig, E.G., and Graetz, D.A., Phosphorus storage capacity of uplands, wetlands and streams of the Lake Okeechobee basin, *Environmental Management,* in press, 1995.

Scheffer, M., de Redelijkheid, M.R., and Noppert, F., Distribution and dynamics of submerged vegetation in a chain of shallow eutrophic lakes, *Aquatic Botany,* 42, 199, 1992.

Smith, V.H., Bierman, V.J., Jr., Jones, B.L., and Havens, K.E., Historical trends in the Lake Okeechobee ecosystem. IV. Nitrogen:phosphorus ratios, cyanobacterial dominance, and nitrogen fixation potential, *Archiv für Hydrobiologie/Suppl. 107,* 1, 69, 1995.

SFWMD, *Surface Water Improvement and Management (SWIM) Plan for Lake Okeechobee,* South Florida Water Management District, West Palm Beach, Florida, 1993.

Van der Does, J., Verstraelen, P.J.T., Boers, P.C.M., Van Roestel, J., Roijackers, R., and Moser, G., Lake restoration with and without dredging of phosphorus-enriched upper sediment layers, *Hydrobiologia,* 233, 197, 1992.

Van der Veen, C.A., Graveland, A., and Kats, W., Coagulation of two different kinds of surface water before inlet into lakes to improve the self-purification process, *Water Science Technology,* 19, 803, 1987.

Van Liere, L., and Janse, J.H., Restoration and resilience to recovery of the Lake Loosdrecht ecosystem in relation to phosphorus flow, *Hydrobiologia,* 233, 95, 1992.

Warren, G.L., Vogel, M.J., and Fox, D.D., Trophic and distributional dynamics of Lake Okeechobee sublittoral benthic invertebrate communities, in *Ecological Studies of the Littoral and Pelagic Systems of Lake Okeechobee, Florida (USA),* Aumen, N.G. and Wetzel R.G., eds., *Archiv für Hydrobiologie Beiheft Ergebnisse der Limnologie,* 45, 317, 1995.

Welch, E.B. and Schrieve. G.D., Alum treatment effectiveness and longevity in shallow lakes, *Hydrobiologia,* 275/276, 423, 1994.

DESIGN OF CONSTRUCTED WETLANDS FOR DAIRY WASTE WATER TREATMENT IN LOUISIANA

Shulin Chen, Gianna M. Cothren, H. Alan DeRamus, Stephen Langlinais, Jay V. Huner, and Ronald F. Malone

INTRODUCTION

A nimal production processes usually generate significant waste effluent. A variety of systems have been used to treat agricultural wastewaters to reduce their environmental impacts. For most dairy operations in Louisiana, two-stage lagoon systems are typically used for waste stabilization and waste water treatment. In a two-stage dairy lagoon system, the
first lagoon is typically anaerobic, and the second is aerobic or facultative.
Although this type of treatment system was designed to retain the waste water
on the farm, actual systems often overflow to the surrounding environment due
to insufficient holding capacity of either the waste water generated from the
operation or the rainfall runoff water. Additionally, because the lagoon systems
were designed for zero discharge, the level of pollutant treatment within the
lagoon is limited. As a result, when discharges occur, the effluent quality does
not usually meet the discharge standard, primarily because of the high organic
loadings and algae growth in the systems.

The waste water treatment process at the dairy farm at the University of
Southwestern Louisiana (USL) is accomplished through such a two-stage lagoon system. This facility was originally designed to handle a 150-cow herd
capacity. A flush tank delivers 32,000 gal (121,120 L) of water to carry manure, bedding, and debris from the free stall holding and feeding slab area into
a concrete drain and a slurry pit where large solids are removed by a separator.
The waste water is then sent to two lagoons, each with a surface area of 1.2
acres (4,856 m²). The typical water discharge rate from the lagoon system is 19
L/min. Since the effluent quality fluctuates and suffers from the effects of algae
growth and high organic loading, the discharge water of the lagoon system
does not consistently meet the secondary effluent standards.

To address similar problems, the Louisiana Department of Environmental
Quality has supported several projects to identify cost-effective processes for
dairy lagoon effluent polishing and possible water reuse. One such effort is the
project conducted at the USL Dairy Farm to compare the effectiveness and
feasibility of using different constructed wetlands for upgrading the dairy la-

goon effluent. The purpose of this chapter is to provide a basic description of these wetlands, their design features, and system characteristics.

SYSTEM DESIGN

DESIGN RATIONALE

The two major types of constructed wetlands, sub-surface flow (SF) and free water surface flow (FWS), were designed for this project. The philosophy behind the design of the SF wetland is to use short, wide beds to optimize solids capture capacity and reduce costs (Chen et al., 1993). The design rationale for the FWS system is to maximize the shading effects of the plants so that light is blocked, thus accelerating algae die off and subsequent settling. Sizing of the wetland cells was accomplished by referencing the currently available procedures (WPCF, 1990). The design requires that the effluent be maintained below 30 mg/L five-day biological oxygen demand (BOD_5) and total suspended solids (TSS).

CONFIGURATION

The water and waste water handling and treatment components in the dairy farm are divided into five separate processes: clean water storage, waste water pretreatment, stabilization, polishing, and final water storage/retention.

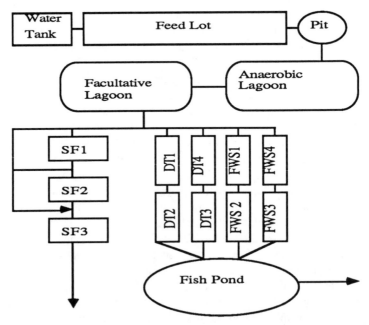

Figure 1. University of Southwestern Louisiana dairy farm waste water treatment system configuration. (SF- Sub-surface Flow system, FWS - Free Water Surface system, DT - Deep Trench system.)

The configuration of these processes is illustrated in Figure 1. During operation, well water stored in the water tank is used to wash the feed lot, resulting in waste water flow into the solids/liquid separation pit. This pit discharges to the anaerobic lagoon, which overflows to the facultative lagoon.

The waste water is pumped from the facultative lagoon to the newly constructed effluent polishing system which consists of three types of tertiary treatment systems. The first is a sub-surface flow constructed wetland (SF) which contains rock media. There are three of these SF wetland cells which can be operated in parallel or series. The second type is the free water surface system (FWS) comprised of shallow ponds planted with vegetation or crops. And the third type of treatment system incorporates deep narrow trenches (DT) bordered by tall plants. The FWS and DT systems consist of four cells each. The dimensions of each system are listed in Table 1. Due to the research and demonstration nature of this project, each of the treatment systems was oversized to provide flexibility for evaluating system performance with different hydraulic detention times. A reservoir or fish pond used to store the treated waste water for discharge or possible reuse is the final water containment structure.

Considering the fact that the design of constructed wetlands has been the topic of many references (EPA, 1988; WPCF, 1990; Reed et al., 1995), the following description will focus on the unique features of this system.

FLOW DISTRIBUTION AND MEASUREMENT

In order to be measured and then further polished, the effluent from the facultative lagoon is pumped into a constant head device designed to maintain a stable flow rate. Connected to the constant head device are two 1.5-in. PVC pipes directing the flow to two 0.6 m x 0.6 m x 0.76 m measurement boxes. The excess water pumped to the constant head device is returned back to the lagoon via a bypass pipe. A manual valve controls the flow to each measurement box. Within each measurement box, a sump pump pumps the water from the box to the constructed wetland systems. Since constant head is maintained

Table 1. Dimensions and aspect ratios of different constructed wetland cells.

Cell Number	Length	Width	Aspect Ratio	Depth
	m	m	length/width	m
DT1	16.8	3.0	5.6	1.2
DT2	16.8	2.7	6.2	1.2
DT3	18.2	3.6	5.1	1.2
DT4	21.3	2.7	7.9	1.2
FWS1	19.1	5.0	3.8	0.6
FWS2	17.1	5.9	2.9	0.6
FWS3	12.5	5.7	2.2	0.6
FWS4	15.2	3.8	4.0	0.6
SF1	5.2	8.8	0.6	0.6
SF2	4.5	8.5	0.5	0.6
SF3	5.2	6.9	0.8	0.6

within the measurement box, the flow to the wetland systems is also maintained at a constant rate.

SUB-SURFACE SYSTEM

Mechanisms contributing to the upgrade of lagoon effluent by a SF wetland entails direct physical filtration, sedimentation of algae and other solids, and microbiological decay of organics. The primary design objective of a SF constructed wetland is to determine the surface area, depth, and slope of the rock bed. Currently, the design of a SF constructed wetland is based on first order BOD_5 removal kinetics, Darcy's law and the assumption of plug flow. According to this design theory, the total surface area needed for a given hydraulic detention time is determined by the required removal efficiency and the total loading. The bed slope and aspect ratio are not independent variables; therefore, the selection of one will determine the other. There are several different approaches available to determine the slope and the aspect ratio (US EPA, 1988; Reed et al., 1995). The design of this system observed the approach suggested by Chen et al. (1993). From the perspective of minimizing excavation costs, Chen et al. (1993) have demonstrated that a SF wetland with a smaller aspect ratio costs less than a wetland with a larger aspect ratio. An additional, significant advantage of a short, wide bed is the uniform deposition of suspended solids within the whole bed. With the use of a long narrow bed, deposition of solids tends to occur near the inlet.

Selection of media for a SF wetland is based on the cost, unit surface area, and clogging potential of the media. It is generally agreed that large media has less unit surface area for biological activity but less potential for clogging. In contrast, smaller-sized media has a greater unit surface area but can be more easily clogged. Although there is no general consensus on media sizes, 5 cm was recommended as a proper media size after balancing the major related factors (Reed and Brown, 1992). To compare the impact of media size on system behavior, three types of media, No. 2 limestone, No. 57 limestone, and river gravel, were used in SF1, SF2, and SF3, respectively. According to the specifications, 30-65% of the No. 2 limestone is larger than 2 in. (5.08 cm) and 40-75% of the No. 57 limestone is larger than 0.5 in. (1.27 cm).

A common concern is the clogging of SF constructed wetlands. The hydraulic conductivity of the rock bed declines over time due to the accumulation of solids within the pores formed between the media. As this occurs, the same bed can no longer handle the same flow rate. Consequently, ponding and surface flow may occur, greatly deteriorating the effluent quality. Therefore, it is important to monitor solids accumulation within the wetlands and to predict their serviceable lives. For this purpose, appropriate mechanisms have been incorporated into the design of the SF wetlands for measuring the *in situ* solids accumulation and hydraulic conductivity changes. This was done by embedding 0.25 in. (0.635 mm) hardware cloth basket assemblies and perforated

PVC cylinders in the media during construction. The basket assembly consists of a 10 cm x 10 cm x 61 cm basket centered inside a 20 cm x 20 cm x 61 cm basket. Both baskets were filled with the rock media. The layout is designed so that one basket assembly can be retrieved every 3 months. The undisturbed gravel inside the smaller basket will be placed in an apparatus for measuring the hydraulic conductivity. In addition to the basket assembly, PVC cylinders of 6-in. diameter were also embedded. These cylinders will be retrieved for measuring solids accumulation at the same time interval as that for the basket assembly. In each of the three cells, 12 baskets and 12 cylinders were embedded. The information on hydraulic conductivity changes and solids accumulation over time can be used for predicting the serviceable life of these constructed wetlands.

FREE WATER SURFACE SYSTEM

The major mechanism for upgrading the lagoon effluent in a FWS wetland is sedimentation of algae after die off. An advantage of the free water surface system is that clogging is avoided. However, a major disadvantage when compared to the SF system is the reduction in efficiency due to the lack of media surfaces, thus decreasing the physical filtration potential.

Two types of free water surface systems were designed and constructed at the experimental site. The first type, similar to that described by other researchers (WPCF, 1990; Reed, 1991), consists of channels with a relatively shallow depth with vegetation growing within. The straightforward design of the free water surface wetland chiefly involves the determination of hydraulic detention time and water depth. The water depth depends on the plant type and is controlled by varying the height of the standpipe at the outlet.

The second type of free water surface system is the shaded deep trench designed to reduce light exposure to the flowing waste water. The shading was achieved by planting tall plants around the border of the trenches. Since the shaded cells are relatively narrow, the light is blocked by the plants on the trench banks. Additionally, because the trenches are relatively deep, light penetration into the water column is further limited. Therefore, as the waste water from the facultative lagoon flows through the cell, algae will die off due to insufficient light. Compared to the typical free water surface system, the shaded trench is more easily maintained as a plug flow system and requires a smaller area.

POROSITY AND HYDRAULIC CONDUCTIVITY
OF THE SUB-SURFACE FLOW WETLANDS

Porosity and hydraulic conductivity of a SF wetland impact both the system design and operation. The inter-related components, porosity and hydraulic conductivity, determine the serviceable life and flow capacity of the bed. Therefore, it is important to know the initial and subsequent change in porosity and hydraulic conductivity in the beds. The porosity of the SF wetlands was determined by measuring the water volume that filled the void space within the media. This was performed both in the laboratory and in the field. In the laboratory, water was added to a media-filled container of known volume, thus permeating the pore space. In the field, water was pumped to the actual rock bed at a given flow rate. The porosity was then determined based on the water volume needed to fill the pore space of the bed and the total volume of the bed. It was observed that the initial porosity values obtained in the laboratory and in the field are very close.

Hydraulic conductivity was determined by conducting a constant head test that is widely used in soil research (Whitlow, 1990). The water flow rate though a volume of media confined within a container with a fixed cross sectional area, and length was recorded. Also, the pressure head difference across the media column was recorded. The hydraulic conductivity was then calculated in accordance with Darcy's law.

Both porosity and hydraulic conductivity tests were conducted for each of the three types of media used in the SF constructed wetlands. The results are given in Table 2. The results presented in Table 2 indicate that the porosity increased as the media size increased. However, the hydraulic conductivity did not change significantly. This suggests that size distribution of the media is a very important factor. Although the average size of the No. 2 limestone is much larger than that of No. 57, the actual hydraulic conductivity did not increase significantly, possibly because of the portion of the No. 2 limestone that has smaller sizes. This suggests that the particle size at the smaller end size range controls the hydraulic conductivity.

PERFORMANCE DATA

Water samples have been collected from the effluent of each system and analyzed on a bi-weekly schedule since 11 August 1994. Tables 3 and 4 contain the BOD_5 and TSS data for the SF systems and the FWS and DT systems. The flow rates during the operation ranged about 1.5-5 gpm (7.6-18.9 L/M)

SUMMARY

This chapter presents a description of constructed wetland systems used to upgrade dairy lagoon effluent. It is obvious from the above discussion that the system design has three distinct features: 1) use of a wide bed for the sub-

Table 2. Characterization of the sub-surface flow wetland.

Cell Number	Media Type	Porosity	Hydraulic Conductivity (m^3/m^2-d)
SF1	#2 limestone	0.446	308
SF2	#57 limestone	0.415	280
SF3	River Gravel	0.353	283

Table 3. Five-day biochemical oxygen demand (BOD₅) and total suspended solids (TSS) data for SF1, SF2, and SF3

Date	System	BOD$_5$(mg/l)		TSS(mg/l)	
		Inf.	Eff.	Inf.	Eff.
8/11/94	SF1	35.7	12.0	61.0	36.0
8/24/94	SF1	6.9	2.1	32.0	12.0
	SF2	6.9	4.5	32.0	16.0
9/7/94	SF1	9.4	8.2	36.0	19.0
9/21/94	SF1	8.4	4.5	41.0	20.0
	SF2	8.4	3.6	41.0	15.0
10/19/94	SF1	12.2	9.2	36.0	24.0
	SF2	12.2	8.0	36.0	12.0
11/2/94	SF1	10.5	6.3	73.0	30.0
	SF2	10.5	6.6	73.0	27.0
11/16/94	SF1	17.4	3.9	120.0	8.0
	SF3	17.4	18.3	120.0	54.0
11/30/94	SF1	24.9	7.0	82.0	16.0
	SF2	24.9	5.2	82.0	12.0
	SF3	24.9	16	82.0	44.0
12/4/94	SF1	21.6	1.8	101.0	14.0
	SF3	21.6	5.7	101.0	35.0
12/28/94	SF1	23.0	7.8	130.0	32.0
	SF2	23.0	3.6	130.0	19.0

Table 4. Five-day biochemical oxygen demand (BOD₅) and total suspended solids (TSS) data for free water surface system (FWS1) and deep trench system (DT1).

Date	System	BOD$_5$(mg/l)		TSS(mg/l)	
		Inf.	Eff.	Inf.	Eff.
8/24/94	DT1	6.9	10.5	32.0	59.0
9/7/94	FWS1	9.4	4.9	36.0	31.0
	DT1	9.4	8.95	36.0	14.0
9/21/94	FWS1	8.4	0.3	41.0	13.0
	DT1	8.4	6.95	41.0	30.0
10/5/94	FWS1	13.7	3.45	74.0	41.0
	DT1	13.7	9.0	74.0	49.0
10/19/94	FWS1	12.2	3.5	36.0	28.0
	DT1	12.2	10.5	36.0	21.0
11/2/94	FWS1	10.5	7.2	73.0	45.0
	DT1	10.5	10.8	73.0	44.0
11/16/94	FWS1	17.4	6.0	120.0	142.0
	DT1	17.4	3.3	120.0	124.0
11/30/94	FWS1	24.9	17.2	82.0	92.0
	DT1	24.9	16.6	82.0	60.0
12/4/94	FWS1	21.6	21.3	101.0	77.0
	DT1	21.6	21.3	101.0	84.0
12/28/94	DT1	23.0	11.0	130.0	115.0

surface flow system, 2) *in situ* baskets and cylinders for hydraulic conductivity and clogging monitoring, and 3) use of deep trenches for algae removal. The preliminary results indicate that even with shorter hydraulic detention times, the SF wetlands were more effective than FWS wetlands in terms of TSS removal.

ACKNOWLEDGMENTS

Funding for this project is provided by Louisiana Department of Environmental Quality through contract No. 24400-93-19. Mr. Larry Fall was involved in some of the early work.

REFERENCES

Chen, S., Malone, R.F., and Fall, L., A theoretical approach for minimization of excavation and media costs of constructed wetland for BOD$_5$ removal, *Trans ASAE*, 36(6), 1625, 1993.

Reed, S.C., Constructed wetlands for wastewater treatment, *BioCycle,* 32, 4449, 1991.

Reed, S.C., and Brown, D.S., Constructed wetland design--the first generation, *Water Environment Research Journal,* 64(6),776, 1992.

Reed, S.C., Middlebrooks, E.J., and Crites, R.W., *Natural Systems for Waste Management and Treatment,* 2nd. Edition, McGraw Hill, New York, New York, 1995.

US EPA, *Design Manual: Constructed Wetlands and Aquatic Plant Systems for Municipal Wastewater Treatment,* U.S. Environmental Protection Agency, EPA/625-88/022, 1988.

Whitlow, R., *Basic Soil Mechanics,* 2nd edition, Longman Scientific & Technical, 1990.

WPCF, *Natural Systems for Wastewater Treatment, Manual of Practice,* Water Pollution Control Federation, FD-16, 1990.

AN OVERVIEW OF PHOSPHORUS BEHAVIOR IN WETLANDS WITH IMPLICATIONS FOR AGRICULTURE

P.M. Gale and K.R. Reddy

OVERVIEW OF WETLANDS

Wetland ecosystems function as transitional areas between aquatic and terrestrial systems and subsequently buffer the interactions of these two environments. Wetland definitions and delineation are legally and emotionally volatile issues, and as yet no one definition or delineation scheme has been accepted by everyone. For our purposes the definition proposed by the Soil Conservation Service (USDA SCS, 1987) will be used in which wetlands are defined as areas that are inundated or saturated for long enough periods to produce hydric soils and support hydrophytic vegetation. This definition encompasses the unique chemistry of wetland environments, which is the subject of this chapter.

As buffer zones in the landscape, wetlands have developed in response to the movement of water. Sediment and nutrient movement through runoff from terrestrial environments during rain events can become trapped and assimilated within the wetland as the water flows through it. The presence of water in wetlands is often easily controlled due to its shallow and temporary nature. Drained and cleared wetlands have provided prime agricultural land that is relatively level and high in organic matter and available nutrients. Once a wetland is drained and put into production, its ability to serve as a buffer between the aquatic and terrestrial environments is lost. More than half of the original wetlands in the conterminous United States have been drained for other land uses (Dahl, 1990).

The ability of wetlands to remove and retain nutrients is the concept behind using wetland areas as on-site treatment options for agricultural and municipal waste waters (Boyt et al., 1977; Fetter et al., 1978). Processes of nutrient removal under anaerobic conditions (i.e., denitrification) are enhanced and more rapid in wetland environments; thus wetlands can be relied upon heavily as tools for permanent removal of nitrogen from waste waters. The extended presence of water in a wetland results in the exclusion of oxygen from the soil environment. Without oxygen to serve as the electron acceptor for aerobic respiration, alternative pathways must be sought. These alternative pathways depend upon the availability of other electron acceptors such as nitrate, manganic manganese, ferric iron, sulfate, and carbon dioxide (Reddy and D'Angelo, 1994). The efficiency of cellular respiration using alternative electron acceptors is much lower than that of aerobic respiration (Gale and Gilmour, 1988). Thus, organic matter accumulates in the wetland environment.

The ability of a wetland to assimilate and retain nutrients will depend upon the form (either organic or inorganic) and concentration of the nutrient and the chemistry of the soil and water column of a wetland. The chemistry of the wetland will depend upon the type of wetland, hydrology, and management (Mitsch and Gosselink, 1993). This chapter will present an overview of the ability of wetlands to retain the nutrient phosphorus and explain how this ability is influenced by agricultural impacts on wetland systems.

AGRICULTURAL IMPACTS ON WETLANDS

Former wetland areas have provided the US with prime agricultural land for many years. If the hydrology of the system can be controlled, then the farmer is provided with level, easily cultivated fields that are high in organic matter and available nutrients. Drainage of wetlands for agricultural production has resulted in the most extensive loss of this resource (Dahl, 1990). This loss of wetlands has also resulted in a loss of the cleansing nature of this landscape component and the nutrient removal mechanisms associated with these ecosystems.

The high organic matter content of wetland soils becomes subject to oxidation and loss once the wetland is drained. This process, which is one form of subsidence, can be quite dramatic in organic soils, resulting in losses of organic matter (Snyder, 1987). Although the C is lost through oxidation, the nutrients remain. Reflooding of organic soils after periods of drainage can result in the release of nutrients, especially P, into the flood waters. Nutrient flushing is a common phenomena of reclaimed wetlands.

Irrigated agriculture and high-density animal systems often generate quantities of nutrient-laden waters that require treatment before they are discharged into the surface water systems of the area. The flux of P associated with runoff from an agricultural watershed was found to be 8 to 10 times higher than that from a similar, but forested, landscape (Vaithiyanathan and Correll, 1992). This and similar studies support the conclusions of Gilliam (1994) that riparian buffers are the most important factor influencing non-point source pollutants in streams. Constructed and natural wetlands are being looked upon as a means of polishing these waters with a natural filtering mechanism. The idea behind using this practice is good, but the capacity of each system needs to be taken into account (Wetzel, 1993). Hydraulic retention time, flow path, presence and kinds of vegetation, microbial populations, and chemical properties of the soil will all contribute to the effectiveness of nutrient removal in these systems.

Recently fertilized or manured fields are subject to nutrient losses through runoff during rainfall events. Sharpley et al. (1994) state that the amount of bioavailable P in runoff waters can be significant and possibly detrimental to downstream water bodies. This has been demonstrated by Mozaffari and Sims (1994) and Graetz and Nair (1995), who have shown that long-term manuring greatly enhanced P availability in the upper 0-5 cm of the soil profile. Although direct leaching of P is significant, high surface P concentrations greatly enhance the potential for P losses in runoff and drainage waters. Retention of the runoff water in wetland areas can help to reduce the nutrient load of these runoff waters.

Agricultural impacts associated with P discharged from adjacent dairies in the Lake Okeechobee Basin, Florida, are clearly evident through accumulation of P in wetland soils and stream sediments. Continuous P loading from these dairies has resulted in increased soil or sediment EPC (equilibrium P concentration), suggesting a decrease in P retention capacity (Graetz and Nair, 1995). Similarly, continuous discharge of drainage water from the Everglades Agricultural Area (EAA) into adjacent wetlands in the northern Everglades has resulted in elevated soil P and alteration in plant communities (DeBusk et al., 1994).

Incorporation of wetland areas into an overall farm management scheme may be the BMP (Best Management Practice) of the future. This management scheme provides an area for nutrient dissipation off the field, thus protecting surface (and possibly) ground water systems. It provides a wildlife corridor for the stream system where it is located and a high water buffer for the agricultural fields. In other words, loss of crops and damage to buildings can be avoided during the regular flooding periods by providing an area for the flood waters to dissipate into--a wetland.

PHOSPHORUS CHEMISTRY IN WETLANDS

Phosphorus is an essential plant nutrient and is often limiting in crop production systems, resulting in a need for regular additions of P fertilizers (Sharpley and Menzel, 1987). In aquatic systems P is often the element that limits primary productivity in the water column (Levine and Schindler, 1992). Cultural eutrophication of lakes and streams has been found to be associated with increased non-point source nutrient loading from the surrounding watershed. Agriculture is often implicated as the activity contributing the majority of this nutrient load, although inadequately treated municipal and industrial waste waters can also be significant contributors. Wetlands, however, can reduce P loadings to aquatic systems. This phenomena has been used as the basis for using natural wetlands to treat waste waters (Nichols, 1983).

In wetland soils several processes influence the concentration of P in the overlying flood water. These processes are biological, physical, and chemical in nature. Physical processes of P retention are associated with the movement of water. Flowing waters entering a wetland are slowed, resulting in the settling, or sedimentation, of particles and the P associated with these particles. The slowing of the flowing water also provides longer contact times between the water and sediments or biota, thus influencing the removal of the nutrient by these species.

Biological processes of P removal vary with the plant and microorganisms that are found in the wetland. Plant uptake and storage of P was shown to be a function of species, with herbaceous species removing P at a much faster rate, whereas woody species provided more permanent storage (Reddy and DeBusk, 1987). Variations in the ability of plants to uptake P is one of the bases by which species compete. As such, P enrichment can result in the development of monocultures of highly efficient plant species and a reduction in the environmental quality with a reduction in species diversity (Koch and Reddy, 1992). Biological uptake can be a significant short-term storage mechanism, but this sink is not permanent, since senescence and decomposition can release the nutrients back into the water column. Only the processes of woody tissue accumulation and sediment accretion are long-term biological storage mechanisms in wetland environments.

Biological uptake is not limited to the macrophyte community. Periphyton and phytoplankton in the water column can strongly influence the chemistry of the flood water. In poorly buffered water columns, with high phytoplankton densities, photosynthesis can result in high dissolved oxygen concentrations and high pH during the day, which can dramatically drop during the night. Richardson and Marshall (1986) have shown that microorganisms play a significant role in the initial uptake of added dissolved inorganic P. They concluded that the soil and microorganisms controlled the water column inorganic P concentrations and the amount of plant-available P. Vadstein et al. (1993) demonstrated that planktonic bacteria were efficient scavengers of available P in the water column and contributed to soluble P removal through death and sedimentation. This phenomena was described by Simmons and Cheng (1985) as a two-step mechanism of removal. However, the microbial community serves as a temporary storage for P, and studies by Mitsch and Cronk (1993) found P storage in this compartment to be only 2 to 6% of the total P being retained by the system.

Chemical removal mechanisms within the soil are important (and possibly the most permanent) means of retaining P in wetlands. The processes of diffusion, exchange, adsorption, and precipitation all combine to determine the EPC (equilibrium P concentration) of a given wetland environment (Richardson, 1985). Concentration gradients between the overlying flood water and the soil column can drive diffusion-related movement of P (Berkheiser et al., 1980). This process can be enhanced by the physical process of the movement of water. Exchange processes are driven by the concentration of competing ions in the soil solution, which are often organic in nature. These organic ions are important to wetlands in that their lifetimes are greatly enhanced under reduced conditions. Thus, although organic ions are produced under a variety of conditions, they tend to persist in flooded soils.

Adsorption and precipitation reactions are most likely the major removal mechanisms for P in a wetland environment. These processes are related to the concentration of cations and coprecipitates along with the presence of surfaces for sorption and will be limited by the concentration of these species. Studies of P sorption in wetland soils have shown that sorption is a function of the presence of Fe and Al oxides, along with organic matter and available Ca (Patrick and Khalid, 1974). Richardson (1985) suggested that P saturation will occur in fresh water wetlands when the sorption capacity is reached with the subsequent loss of the wetlands ability to retain P. However, Gale et al. (1994) showed that soil P retention involved both chemical and biological pathways and estimates of P retention based on soil chemical properties alone would underestimate potential P removal.

In a flooded soil environment, oxygen is consumed and becomes limiting. Without molecular oxygen to drive cellular respiration, alternative electron acceptors must be found. This process results in the reduction of the redox potential and alters the chemistry of the system. Although P is not directly involved in redox-sensitive reactions, the changes in the soil solution can alter the P chemistry. Patrick and Khalid (1974) found that under reduced conditions, soils released P to flood waters low in inorganic P and sorbed P from flood waters high in inorganic P. These changes in P solubility upon flooding are likely due to reduction of ferric oxyhydroxide and ferric phosphate compounds to the more soluble ferrous iron forms and hydrolysis of polyphosphate species (Reddy and D'Angelo, 1994). Reddy and Reddy (1993) and Amer et al. (1991) have found the P sorption capacity of sediments to increase under anaerobic conditions. Clausen and Johnson (1990) found that greater nutrient removal efficiencies in wetlands were associated with higher water levels. Thus the ability to retain P is a function of the chemistry of the system, and changes in pH and Eh will influence the chemical P retaining processes (Moore et al., 1994).

INFLUENCE OF IMPACTS ON PHOSPHORUS CHEMISTRY

Agricultural impacts on P chemistry are a function of management schemes. Excess P may be associated with applied soil amendments (fertilizers) or as the result of land being used as a disposal site for farm-generated wastes (animal manures) (Sharpley and Menzel, 1987). High-density animal production in feedlots presents massive problems due to the volumes and concentrations of wastes generated. Due to the high concentrations of nutrients in manures, leaching and runoff from fields where these wastes are over applied is highly likely. The pollutants that leach or runoff (most likely N and P) can contaminate ground and surface waters. Their presence can result in the early aging or accelerated eutrophication of surface water systems and other natural systems. Reddy et al. (1993) found that increased nutrient inputs in the Everglades have resulted in increased peat accumulation in the affected areas.

Another impact of agriculture on wetlands involves the degradation of organic matter. Wetlands are usually organic matter-accreting systems. Drainage and clearing of wetlands for crop production results in enhanced degradation of the soil organic matter. This enhanced degradation rate is what provides the increased nutrient availability of these soils and thus their enhanced fertility. However, it also means that the chemistry of the system is drastically altered, and reclamation at a later time may be a success or failure. The slow release of available nutrients to the soil during organic matter decomposition may become a flush of nutrient laden water upon flooding. Diaz et al. (1993) demonstrated the enhancement of P availability as a result of periodic draining and flooding of Florida histosols. However, this loss mechanism is not limited to the humid subtropics as similar observations have been made in Canada (Longabucco and Rafferty, 1989) and Michigan (Sloey et al., 1978).

These released nutrients can then be lost to the surface and ground waters that drain the effected area, thus increasing the nutrient loads into these systems. This problem is nowhere more evident than in the Everglades Agricultural Area region of southern Florida in which drainage waters from muck farms are the major input of nutrients into the surrounding watershed. However, the problem is not limited to this region as any area where organic soils are drained and cultivated can decompose. Decomposition of organic soils can be slowed through management practices that flood the fields during times of non-production. This drainage water is often nutrient enriched and must be treated before being released into surrounding surface waters. Studies conducted in Florida have shown that constructed wetlands can be used to polish these nutrient laden waters and as such enhance the quality of the drainage waters (Kadlec and Newman, 1992).

IMPLICATIONS AND RECOMMENDATIONS

Wetlands can also be incorporated into a farm management system. State natural resource managers are beginning to establish guidelines and provide workshops to encourage the establishment and maintenance of wetlands by private landowners (US EPA, 1992). Small riparian systems can be added as buffer zones between streams and fields. Large shallow water areas can be used for drainage water disposal and treatment prior to its release into surrounding surface waters. Constructed wetlands can be used to provide treatment of animal waste waters generated by animal production. Biogeochemical benefits from such systems would include enhanced denitrification and plant and microbial uptake of labile P. Ecological and aesthetic benefits would also be reaped through the provision of wildlife habitat and greenspace.

Wetlands have provided agriculture with prime farmland and can continue to serve agriculture through reduction of non-point source runoff. Set aside programs are currently being implemented along the Missouri River as a buy-out option for farmers recently devastated by the 1993 floods (Columbia Daily Tribune, 1994). This option provides the land owner with compensation for not using environmentally sensitive land, with the added benefits of water quality enhancement and improvement. Restoration of riparian flood plains in areas where channelization has been tried and failed can help to ensure reduced sediment and nutrient loading into major streams and rivers. Once reestablished, these floodplain areas can serve as dissipation areas for future high water events.

All areas are not conducive to wetlands, nor should they be. However, in the humid areas of the United States where wetlands have historically been part of the landscape, it would be wise to incorporate them back into a watershed management plan. Wetlands form as a response to excess water in the terrestrial landscape. This response needs to be kept in mind as movement towards managing the environment for the 21st century is made. Designation of all bottom lands as set aside for wetlands overlooks the prime agricultural land that these areas often are. Farming practices in the bottom land areas can and should be encouraged but with the stipulation that when high waters again approach, these areas can be flooded. This is not a new idea as ancient Egypt developed its agriculture in response to the flooding Nile. By using agricultural land as flood water dissipation areas during major floods, threats to urban areas could be reduced.

REFERENCES

Amer, F., Saleh, M.E., and Mostafa, H.E., Phosphate behavior in submerged calcareous soils, *Soil Sci.*, 151, 306, 1991.

Berkheiser, V.E., Street, J.J., Rao, P.S.C., and Yuan, T.L., Partitioning of inorganic orthophosphate in soil-water systems, in *CRC Critical Reviews in Environmental Control*, 1980.

Boyt, F.L., Gayley, S.E., and Zoltek, J. Jr., Removal of nutrients from treated municipal wastewater vegetation, *J. Water Pollut. Control Fed.*, 49, 789, 1977.

Clausen, J.C., and Johnson, G.D., Lake level influences on sediment and nutrient retention in a lakeside wetland, *J. Environ. Qual.*, 19, 83, 1990.

Columbia Daily Tribune, U.S. proposes Big Muddy Refuge, Sect. A-12, May 27, 1994.

Dahl, T.E., *Wetlands Losses in the United States 1780's to 1980's*, US Department of the Interior, US Fish and Wildlife Service, Washington, D.C., 1990.

DeBusk, W.F., Reddy, K.R., Koch, M.S., and Wang Y., Spatial distribution of soil nutrients in a Northern Everglades Marsh: Water Conservation Area 2A, *Soil Sci. Soc. Am. J.*, 58, 543, 1994.

Diaz, O.A., Anderson, D.L., and Hanlon, E.A., Phosphorus mineralization from Histosols of the Everglades Agricultural Area, *Soil Sci.*, 156, 178, 1993.

Fetter, C.R., Jr., Sloey, W.E., and Spangler, F.L., Biogeochemical studies of a polluted Wisconsin marsh, *J. Water Pollut. Control Fed.*, 50, 290, 1978.

Gale, P.M., and Gilmour, J.T., Net mineralization of carbon and nitrogen under aerobic and anaerobic conditions, *Soil Sci. Soc. Am. J.,* 52, 1006, 1988.

Gale, P.M., Reddy, K.R., and Graetz, D.A., Phosphorus retention by wetland soils used for treated wastewater disposal, *J. Environ. Qual.,* 23, 370, 1994.

Gilliam, J.W., Riparian wetlands and water quality, *J. Environ. Qual.,* 23, 896, 1994.

Graetz, D.A., and Nair, V.D., Fate of phosphorus in Florida spodosols contaminated with cattle manure, *Ecol. Engrg.,* in press, 1995.

Kadlec, R.H., and Newman, S., *Phosphorus Removal in Wetland Treatment Areas,* Report for South Florida Water Management District, 1992.

Koch, M.S., and Reddy, K.R., Distribution of soil and plant nutrients along a trophic gradient in the Florida Everglades, *Soil Sci. Soc. Am. J.,* 56, 1492, 1992.

Levine, S.N., and Schindler, D.W., Modification of the N:P ratio in lakes by in situ processes, *Limnol. Oceanogr.,* 37, 917, 1992.

Longabucco, P., and Rafferty, M.R., Delivery of nonpoint-source phosphorus from cultivated mucklands to Lake Ontario, *J. Environ. Qual.,* 18, 157, 1989.

Mitsch, W.J., and Cronk, J.K., *Phosphorus Retention and Ecosystem Productivity of Constructed Wetlands at the Des Plaines River Wetland Demonstration Project,* Final Report, U.S. Army Corps of Eng., Contract No. DACW39-91-C0071, 1993.

Mitsch, W.J., and Gosselink, J.G., *Wetlands,* 2nd Edition, Van Nostrand Reinhold, 1993.

Moore, P.A. Jr., and Reddy, K.R., Role of Eh and pH on phosphorus geochemistry in sediments of Lake Okeechobee, *Florida, J. Environ. Qual.,* 23, 362, 1994.

Mozaffari, M., and Sims, J.T., Phosphorus availability and sorption in an Atlantic coastal plain watershed dominated by animal-based agriculture, *Soil Science,* 157, 97, 1994.

Nichols, D.S., Capacity of natural wetlands to remove nutrients from wastewater, *J. Water Pollut. Control Fed.,* 55, 495, 1983.

Patrick, W.H., Jr., and Khalid, R.A., Phosphate release and sorption by soils and sediments: Effect of aerobic and anaerobic conditions, *Science,* 186, 53, 1974.

Reddy, G.B., and Reddy, K.R., Phosphorus removal by ponds receiving polluted water from nonpoint sources, *Wetland Ecology and Management,* 2, 171, 1993.

Reddy, K.R., and D'Angelo, E.M., Soil processes regulating water quality in wetlands, in *Global Wetlands: Old World and New,* Mitsch, W.J., ed. Elsevier Science, 1994.

Reddy, K.R., and DeBusk, W.F., Nutrient storage capabilities of aquatic and wetland plants, in *Aquatic Plants for Water Treatment and Resource Recovery,* Reddy, K.R., and Smith, W.H., eds., Magnolia Publ., 1987.

Reddy, K.R., DeLaune, R.D., DeBusk, W.F., and Koch, M.S., Long-term nutrient accumulation rates in the Everglades, *Soil Sci. Soc. Am. J.,* 57, 1147, 1993.

Richardson, C.J., Mechanisms controlling phosphorus retention capacity in freshwater wetlands, *Science,* 228, 1424, 1985.

Richardson, C.J., and Marshall, P.E., Processes controlling movement, storage and export of phosphorus in a fen peatland, *Ecol. Monogr.,* 56, 279, 1986.

Sharpley, A.N., and Menzel, R.G., The impact of soil and fertilizer phosphorus on the environment, *Adv. Agron.,* 41, 297, 1987.

Sharpley, A.N., Chapra, S.C., Wedepohl, R., Sims, J.T., Daniel, T.C., and Reddy, K.R., Managing agricultural phosphorus for protection of surface waters: Issues and options, *J. Environ. Qual.,* 23, 437, 1994.

Simmons, B.L., and Cheng, D.M.H., Rate and pathways of phosphorus assimilation in the Nepean river at Camden, New South Wales, *Wat. Res.,* 19, 1089, 1985.

Sloey, W.E., Spangler, F.L., and Fetter, C.W., Jr., Management of freshwater wetlands for nutrient assimilation, in *Freshwater Wetlands: Ecological Processes and Management Potential,* Good, R.G., Whigam, D.F., and Simpson, R.L., eds., Academic Press, New York, 1978.

Snyder, G.H., *Agricultural Flooding of Organic Soils,* Bulletin 870, Florida Agricultural Experiment Station, 1987.

USDA Soil Conservation Service, *Hydric Soils of the United States,* USDA, Washington, D.C., 1987.

US Environmental Protection Agency, *The Private Landowner's Wetlands Assistance Guide: Voluntary Options for Wetlands Stewardship in Maryland,* US EPA, 1992.

Vadstein, O., Olsen, Y., and Reinertson, H., The role of planktonic bacteria in phosphorus cycling in lakes - sink and link, *Limnol. Oceanogr.,* 38, 1539, 1993.

Vaithiyanathan, P., and Correll, D.L., The Rhode River watershed: Phosphorus distribution and export in forest and agricultural soils, *J. Environ. Qual.,* 21, 280, 1992.

Wetzel, R.G., Constructed wetlands: Scientific foundations are critical, in *Constructed Wetlands for Water Quality Improvement,* Moshiri, G.A. ed., Lewis Pub., 1993.

USING EUTROMOD WITH A GIS FOR ESTABLISHING TOTAL MAXIMUM DAILY LOADS TO WISTER LAKE, OKLAHOMA

W.C. Hession, D.E. Storm, S.L. Burks, M.D. Smolen,
R. Lakshminarayanan, and C.T. Haan

INTRODUCTION

Wister Lake, located in southeastern Oklahoma, is the sole water supply for the majority of residents in LeFlore County and three adjacent counties. In addition, the lake and related recreational activities are important to the economy of the area. Wister Lake has been considered eutrophic since it was first surveyed by US EPA in 1974 (US EPA, 1977). Oklahoma's 1990 Water Quality Assessment Report identifies Wister Lake as being eutrophic and having turbidity problems. In addition, the Wister Lake watershed has been targeted in Oklahoma's Section 319 Nonpoint Source (NPS) Management Plan.

Lakes are often classified as oligotrophic, mesotrophic, or eutrophic based on their primary productivity and other attributes. Oligotrophic lakes tend to be geologically young, low-productivity lakes; eutrophic lakes are older, highly productive ecosystems. Eutrophication of surface waters can be accelerated by an increased input of nutrients, which can limit water use for fisheries, recreation, industry, or drinking. Although nitrogen and carbon are associated with eutrophication, most attention has focused on phosphorus inputs because of the difficulty in controlling the exchange of nitrogen and carbon between the atmosphere and water and fixation of atmospheric nitrogen by some blue-green algae. Thus, phosphorus often limits eutrophication, and its control is of prime importance in decreasing accelerated eutrophication (Sharpley, 1993). Of the major nutrients, phosphorus is the most effectively controlled using existing engineering technology and land use management (Reckhow et al., 1980).

Under the Clean Water Act of 1987, states are required to compute total maximum daily loads (TMDLs) for their priority water bodies. A TMDL determines the maximum pollutant loading from both point and non-point sources that a receiving water can accept without exceeding water quality standards (US EPA, 1991). The TMDL process has become an important and required portion of US EPA's water quality initiatives.

The objectives of this study were 1) to determine the TMDL for total phosphorus (TP) to Wister Lake; 2) to investigate alternative management scenarios; and 3) to make recommendations on management alternatives to achieve specific water quality objectives. The size and complexity of the Wister Lake watershed and variability of the natural system required the use of state-of-the-art tools and techniques. Specifically, a geographic information system (GIS) and computer model were used to complete this project.

PROJECT AREA DESCRIPTION

Wister Lake, located in the Arkansas River Basin on the Poteau River, is a reservoir completed by the US Army Corps of Engineers in 1949 to provide flood control, water supply, low flow augmentation, and water conservation. Wister Lake has a surface area of 2,970 ha, a shoreline length of 185 km, a mean depth of 2.3 m, and a maximum depth of 13.4 m at normal pool elevation of 146 m. Wister Lake's watershed covers approximately 260,000 ha, of which two thirds is in Oklahoma and one third is in Arkansas (Figure 1).

The lake receives inputs from a variety of pollutant sources, both point and non-point. There are nine permitted waste water treatment plants in Wister Lake's watershed. Non-point pollution contributing to the lake includes agricultural, forestry, resource exploration and extraction, and urban sources. A major source of nutrients in the watershed originates from the poultry industry in the region, which is expected to double in the coming years.

The Wister Lake watershed includes portions of the Ouachita Mountains and the Arkansas Valley ecoregions. Land use in the watershed is approximately three-fourths forest and one-fourth pasture, with small amounts of cropland, urban, and disturbed land. The topography ranges from level flood plains along Fourche Maline Creek and the Poteau River to gently sloping uplands to

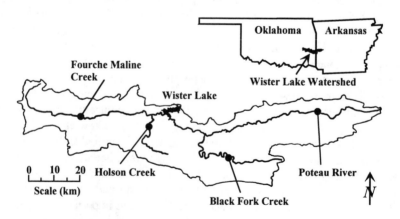

Figure 1. Location of Wister Lake watershed with major tributaries.

steep mountainous areas. The relief ranges from the lake's normal pool elevation to the 817-m peak of Rich Mountain in Arkansas.

TOTAL MAXIMUM DAILY LOAD DEVELOPMENT

A standard TMDL analysis includes the following activities: (1) determine pollutant of interest; (2) estimate the water's assimilative capacity; (3) quantify pollutant loading from all sources; (4) determine total allowable pollutant load; and (5) allocate the allowable loads among different pollutant sources (US EPA, 1991). Computing a TMDL is difficult for a combination of point and non-point pollution sources because of the fundamentally different nature of the two sources. Point source (PS) loadings are essentially continuous in time, while most NPS loadings occur only intermittently (Rossman, 1991). In reality, the TMDL varies from day to day as a receiving water's capacity to assimilate pollutant loads varies. However, an operational TMDL, where a constant daily load is defined, can be useful in terms of management.

The TMDL can be interpreted as the sum of the long-term average loadings from each source category that achieves water quality standards (Rossman, 1991). Due to the importance and manageability of phosphorus, a TMDL for TP loading to Wister Lake was developed. A nutrient loading and lake response model, EUTROMOD (Reckhow et al., 1992), was used to estimate Wister Lake's assimilative capacity, to quantify TP loading from all sources, to determine total allowable pollutant load, and to allocate these loads among the different sources.

Phytoplankton population or algal biomass has been related to nutrient loading (Vollenweider, 1968) and is often used as an indicator of primary productivity or trophic state of water bodies. Chlorophyll a (chl. a), as the dominant photosynthetic pigment in phytoplankton, is often measured as an indicator of phytoplankton biomass. Since neither Oklahoma nor Arkansas has set water quality standards for nutrients, we used in-lake chl. a estimates as indicators of whether or not water quality goals were met. This assumes that excessive growth of aquatic plants interferes with desirable water uses.

We used a chl. a concentration of 10 µg/L, which US EPA's National Eutrophication Survey indicated as the breakpoint between mesotrophic and eutrophic lakes (Gakstatter et al., 1974), as our endpoint or water quality goal and for determining a TMDL for TP. Wister Lake was one of 812 lakes and reservoirs included in the survey (US EPA, 1977).

EUTROMOD

EUTROMOD is a computer model developed to provide guidance and information for managing eutrophication in lakes and reservoirs (Reckhow et al., 1992). It is a collection of spreadsheet-based nutrient loading and lake response models that may be used to relate water quality goals to allowable nutrient inputs. The model, thereby, provides information concerning the appropriate mix of PS discharges, land use, and land management controls that result in acceptable water quality.

EUTROMOD predicts lake-wide, growing season average conditions as a function of annual nutrient loadings. The annual loadings are simulated with a simple, lumped watershed modeling procedure that includes the Rational Equation's runoff coefficient for surface runoff, the Universal Soil Loss Equation (USLE) for estimating soil loss, loading functions for nutrient export from NPSs, and user-provided PS information. Lake response is predicted by a "robust" set of non-linear regression equations from multi-lake regional data sets in the US. These regression equations are used to estimate lake nutrient levels (mg/L), chl. a (μg/L), and Secchi Disk depth (m). Estimates of the prediction uncertainty of these equations are provided as standard model output.

Data required for simulating basin loadings and lake response include information about climate, watershed characteristics, and lake morphometry. Climate parameters include precipitation and lake evaporation estimates. Several parameters are needed to describe the watershed in terms of land use, soils, and topography. Lake morphometry is described using surface area and mean depth.

We used EUTROMOD to simulate annual TP loading from point and nonpoint sources as well as resulting lake response (chl. a). The model utilizes an annual mean precipitation and corresponding coefficient of variation to account for hydrologic variability or uncertainty. The lake response results are presented as means bounded by the 5th and 95th percentile estimates. Utilizing this year-to-year variability estimate, we included a margin of safety in our TMDL by requiring the 95th percentile estimate of mean chl. a (CA_{95}) to be equal to or less than 10 μg/L.

RESULTS AND DISCUSSION

We executed EUTROMOD for natural, current, and expected future watershed conditions. The future conditions included increased human populations and resulting increased waste water loads (point sources) as well as increased poultry production. In addition, we explored a variety of management options to satisfy the TMDL. These management alternatives included upgrading waste water treatment facilities to include phosphorus removal as well as NPS reductions. Watershed-level input parameters were area-weighted by land use category utilizing soil, land use, and topographic digital data layers in the Geographic Resources Analysis Support System (GRASS) GIS developed by the US Army Corps of Engineers. The land use percentages for each condition are provided in Table 1.

The model results for current conditions are summarized and compared with monitored or previously computed values in Table 2. These comparisons were by no means adequate for validation but provided some confidence in the

Table 1. Watershed land use percentages for simulated conditions.

	Land Use Category					
Condition	Cropland	Pasture	Manured Pasture	Forest	Urban	Disturbed
	--%--					
Natural	0	0	0	100	0	0
Current	<1	18	5	75	1	<1
Future	<1	14	9	75	1	<1

Table 2. Summary of model results and previous estimates.

Parameter	EUTROMOD	Previous Estimate	Source
Runoff:			
Volume (m³*10⁶)	1030	1230	Smolen et al. (1993)
Depth (m)	0.40	0.38-0.51	Pettyjohn et al. (1983)
Sediment (t/year)	53,200	47,400	Hession et al. (1993)
Nutrient Loads:			
Phosphorus (kg/year)	117,000	94,000	Smolen et al. (1993)
Nitrogen (kg/year)	921,000	664,000	US EPA (1977)
In-Lake Parameters:			
Phosphorus (mg/L)	0.07	0.06-0.12	US EPA (1977)
Nitrogen (mg/L)	0.73	none available	--
Chl. a (µg/L)	10.4 (mean)	4.8	US EPA (1977)
		13.6-37.0	OWRB (1990)
Secchi Depth (m)	0.42	0.30-0.90	US EPA (1977)
		0.23-0.38	OWRB (1990)

simulation process. The model results compared well with existing monitoring values and/or estimates.

The CA_{95} concentration for current conditions was estimated to be 10.7 µg/L (Table 3). To determine the maximum allowable annual TP load, we adjusted the annual load through trial and error until the CA_{95} was 10.0 µg/L. An annual TP load of 97,000 kg/year resulted in this acceptable lake trophic level. TMDLs for lakes are established by dividing the annual load to the lake by the number of days in a year (365) (Griffin et al., 1991). For current conditions, the long-term average maximum daily load to keep the lake at a mesotrophic state is 266 kg of TP.

Natural loads and lake conditions were estimated by simulating the watershed as 100% forest (Table 3). Even though these conditions never existed (the reservoir was not created until 1949, at which time urban and agricultural areas already existed), these results provide loads and conditions that are natural and, therefore, not due to anthropogenic influences. By definition, natural loads are not pollution and do not have to be mitigated (Griffin et al., 1991). It is interesting to note that, even under these theoretically "pristine" conditions, the lake is predicted to be borderline oligotrophic/mesotrophic based on US EPA's trophic level classification system (Gakstatter et al., 1974). This is likely due to the fact that Wister Lake is relatively shallow, and shallow lakes

Table 3. Model results for natural, current, and future conditions.

Condition	Total Phosphorus		Chlorophyll a		Total Phosphorus by Source		
	Mean Load	Mean In-lake Concentration	Mean	95th Percentile	Back-ground	Point Source	Nonpoint Source
	kg/year	mg/L	-------µg/L--------		----------------%---------------		
Natural	11700	0.01	3.9	4.0	100	0	0
Current	117000	0.07	10.4	10.7	10	14	76
Future	132000	0.08	10.9	11.2	9	14	77

tend to be more biologically productive. In fact, many shallow man-made lakes are naturally eutrophic when initially filled (NALMS, 1988).

Future loads and lake trophic state were also simulated with EUTROMOD (Table 3). The PSs were increased based on year-2000 population estimates. In addition, we assumed a doubling of the poultry industry in the watershed. This was modeled by doubling the amount of manure-spread pasture acreage from current conditions. The future TP load increased to 132,000 kg/year, a 13% increase, which resulted in a CA_{95} concentration estimate of 11.2 µg/L. The percentages of annual TP load by source are shown in Table 3. The relative contributions do not change significantly from the current to future scenarios. These percentages indicated that NPS pollution is by far the most significant source of TP in the watershed.

We simulated two management scenarios for current and future conditions using a variety of PS and NPS control combinations in order to meet the TMDL (Table 4). These simulations allocated the allowable loads among different pollutant sources and investigated alternative management options to meet water quality goals.

Scenario 1 was performed for the current and future conditions by assuming the implementation of phosphorus removal processes at all waste water treatment plants, thereby lowering the mean nutrient concentrations. The percentage reductions listed in Table 4 for PSs were computed from reductions in annual loadings from PSs due to the lower nutrient concentrations. In addition, NPS loads were incrementally reduced until the TMDL and, therefore, the CA_{95} requirement of 10 µg/L was met.

Scenario 2 was used to determine the percent reduction in NPSs needed to achieve the TMDL assuming no PS controls. We assumed that the NPS reductions were obtainable from pasture, crop, urban, and disturbed lands only through the implementation of best management practices (BMPs). Therefore, the percent reductions were simulated on those lands only, not from the entire NPS load.

The results of the two scenarios are provided in Table 4 for natural, current, and future conditions. The natural condition met our water quality goals and, therefore, needed no controls. However, even with PS phosphorus removal, reductions in NPSs of 10 and 27% would be needed under current and future conditions, respectively, to meet our TMDL. With no PS controls, BMPs must

Table 4. Alternatives for meeting the total maximum daily load
for total phosphorus.

| | Percent Reductions Required | | | |
| | Scenario 1 | | Scenario 2 | |
Condition	Point Source	Non-point Source	Point Source	Non-point Source
Natural	0	0	0	0
Current	72	10	0	22
Future	73	27	0	40

be implemented to reduce the NPS load by 22 and 40% for current and future conditions, respectively. Since NPSs contribute the majority of the phosphorus loads, it appears that BMPs should play a major role in achieving the TMDL currently and in the future. Most importantly, as the poultry industry expands, effective management of poultry litter and BMP implementation will be essential.

Many different combinations of point and non-point source controls can be generated to meet water quality goals. Generally, employing NPS controls (BMPs) is cheaper than upgrading or adding waste water treatment. A cost analysis can be used to determine the most cost-effective combination of controls resulting in watershed-scale pollution control optimization. One innovative management technique is to allow municipalities and utilities to trade NPS control for PS control (Griffin et al., 1991). If trading is allowed, and NPS control is significantly cheaper than additional PS treatment, water quality goals may be more obtainable. There are many different types of NPS controls (BMPs) that can be implemented to achieve water quality goals. However, cost analysis and actual BMP recommendations are beyond the scope of this study.

SUMMARY AND CONCLUSIONS

Wister Lake, important as a water supply and for recreational activities, has been considered eutrophic since 1974. A total maximum daily load (TMDL) for phosphorus was estimated using a nutrient loading and lake response model, EUTROMOD. Model input parameters were evaluated using digital data layers within the GRASS GIS. The TMDL was set such that in-lake chlorophyll a concentration estimates remained at levels considered mesotrophic by US EPA (10 µg/L).

The estimated TMDL was 266 kg of total phosphorus, which was computed as a long-term average load. Such a TMDL, which results from a combination of point and non-point sources, should not be taken literally but should be used operationally for management purposes. Additional work is needed to evaluate cost-effective alternative management options and to improve or update model input parameter estimates.

Uncertainty analyses should be a routine part of any modeling exercise. EUTROMOD allows for minimal uncertainty analyses by providing estimates of model error and hydrologic variability. We utilized the hydrologic variability component to incorporate a margin of safety in our TMDL analysis. However, there are many other uncertainties due to the lack of knowledge and system stochasticity that are not accounted for in the model. Presently we are working to incorporate more extensive uncertainty analyses into EUTROMOD.

REFERENCES

Gakstatter, J.H., Allum, M.O., and Omernik, J.M., Lake eutrophication: Results from the National Eutrophication Survey, in *Water Quality Criteria Research of the US EPA*, Proc. of an EPA-sponsored Symposium on Marine, Estuarine and Freshwater Quality, EPA-600/3-76-079, US Environmental Protection Agency, Washington, D.C., 1974.

Griffin, M., Kreutzberger, W., and Binney, P., Research needs for nonpoint source impacts, *Water Environment & Technology*, 3(6), 60, 1991.

Hession, W.C., Storm, D.E., Smolen, M.D., Lakshminarayanan, R., and Haan, C.T., *Analysis of Sediment Loading to Wister Lake*, ASAE Paper No. 93-2073, American Society of Agricultural Engineers, St. Joseph, Michigan, 1993.

NALMS, *Lake and Reservoir Restoration Guidance Manual*, North American Lake Management Society, EPA 440/5-88-002, US Environmental Protection Agency, Washington, D.C., 1988.

OWRB, *Trophic Classification of Oklahoma Reservoirs*, Lake Water Quality Assessment Final Report, Oklahoma Water Resources Board, Oklahoma City, Oklahoma, 1990.

Pettyjohn, W.A., White, H., and Dunn, S., *Water Atlas of Oklahoma*, University Center for Water Research, Oklahoma State University, Stillwater, Oklahoma, 1983.

Reckhow, K.H., Beaulac, M.N., and Simpson, J.T., *Modeling Phosphorus Loading and Lake Response Under Uncertainty: A Manual and Compilation of Export Coefficients*, EPA 440/5-80-011, US Environmental Protection Agency, Washington, D.C., 1980.

Reckhow, K.H., Coffey, S., Henning, M.H., Smith, K., and Banting, R., *EUTROMOD: Technical Guidance and Spreadsheet Models for Nutrient Loading and Lake Eutrophication*, Draft Report, School of the Environment, Duke University, Durham, North Carolina, 1992.

Rossman, L.A., *Computing TMDLs for Urban Runoff and Other Pollutant Sources*, Water and Hazardous Waste Treatment Research Division, Risk Reduction Engineering Laboratory, US Environmental Protection Agency, Cincinnati, Ohio, 1991.

Sharpley, A.N., *Phosphorus Management for Agriculture and Water Quality,* paper presented at 1993 Research Conference: Focus on Phosphorus, Arkansas Water Resources Center, Fayetteville, Arkansas, April 6, 1993.

Smolen, M.D., Lakshminarayanan, R., Hession, W.C., and Storm, D.E., *Estimating Total Maximum Daily Load (TMDL) for Wister Lake*, ASAE Paper No. 93-2072, American Society of Agricultural Engineers, St. Joseph, MI, 1993.

US EPA, *Report on Wister Reservoir LeFlore County, Oklahoma*, EPA Region VI Working Paper No. 595, National Eutrophication Survey Working Paper Series, US Environmental Protection Agency, Washington, D.C., 1977.

US EPA, *Guidance for Water Quality-Based Decisions: The TMDL Process*, Assessment and Watershed Protection Division, US Environmental Protection Agency, Washington, D.C., 1991.

Vollenweider, R.A., *Scientific Fundamentals of the Eutrophication of Lakes and Flowing Waters, with Particular Reference to Nitrogen and Phosphorus as Factors in Eutrophication*, Technical Report DAS/CSI/68.27, OECD, Paris, 1968.

EXPERIENCES WITH TWO CONSTRUCTED WETLANDS FOR TREATING MILKING CENTER WASTE WATER IN A COLD CLIMATE

Brian J. Holmes, Billi Jo Doll, Chet A. Rock, Gary D. Bubenzer,
R. Kostinec, and Leonard R. Massie

INTRODUCTION

Waste water from milking centers contains milk, detergents, acid rinses, and sanitizers and can contain manure, feed, antibiotics, and soil particles. Milk and manure contribute a significant organic load. Biochemical oxygen demand (BOD_5) can be greater than 1000 mg/L. Cleaning solutions contribute phosphorus. Milk, manure, and feed contribute phosphorus and nitrogen. Sanitizers and manure contribute chlorides. If allowed to enter surface water, this high-strength waste water can reduce dissolved oxygen and enhance eutrophication.

The best method to reduce the pollution potential of milking center waste water is by wide dispersal onto agricultural land under conditions that enhance soil infiltration. Methods for doing this include storage, transport by tank or pump to fields, and then spreading or injection. Dairymen handling solid manure seldom have manure storage or liquid hauling equipment, thus making it more costly to apply milking center waste water to fields. Intense land application methods for milking center waste water encounter difficulties of soil fouling with organic biosolids, a particular problem under the anaerobic conditions found in drain fields. Constructed wetlands offer a hope for treating milking center waste water.

Under temperate conditions, organic matter decomposition by microbes in a constructed wetland can be expected to reduce the BOD_5 strength of the waste water. During low temperature conditions, this process will be severely reduced or halted. Phosphorus precipitation may occur within the wetland by solids settling and by forming settleable compounds through chemical reactions. Chlorine removal is not expected to be significant. Nitrogen may be transformed to nitrate forms with little removal expected.

The success of constructed wetlands for improving water quality has been realized in temperate areas of North America and Europe. Wetlands to treat dairy wastes with a variety of aquatic plants are used in Mississippi (Ulmer et al., 1992). European farmers used constructed wetlands successfully for treatment of dairy waste water (Biddlestone et al., 1991).

This chapter reports on research being conducted on wetlands constructed to treat milkhouse waste water in the cold climates of Maine and Wisconsin.

DESIGN AND CONSTRUCTION OF TWO WETLANDS

WISCONSIN WETLAND

Waste water from the milkhouse of a 50-cow dairy is directed to the constructed wetland. The dairyman uses exceptional source control practices to limit the waste milk and water load leaving the milkhouse. Non-salable milk is fed to calves or is delivered to his manure handling system. Milk in the pipeline after milking is placed in the bulk tank. Due to a limited-capacity well, water conserving practices limit daily average water use to 200 gal/day (760 L/day). The milkhouse waste discharge contains milk from the milkline and bulk tank wash, washing and sanitizing chemicals, water softener chemicals, and solids trafficked into the milkhouse.

The milkhouse waste water is pumped from a pump pit to a flow diverter at the head end of the wetland (Figure 1). Half of the flow is diverted through a settling tank for pretreatment. A tipping bucket splits the flow before delivery to the wetland cells. The untreated portion of the flow is directed to another two series of wetland cells. Each cell is 76 ft (23.2 m) long with a bottom width of 4 ft (1.2 m). The cells are 18 in. (0.5 m) deep with 2:1 side slopes.

U-shaped steel channels convey the waste water into and out of the 12-cell series of wetland cells. V-notched weirs allow 4 in. (0.1 m) of water to accumulate in the cells and assist in measuring flow through the channels. Design and construction of the wetland is described in Holmes et al. (1992, 1994).

MAINE WETLAND

The project in Maine was implemented on a 330-cow dairy milking three times a day. The original milkhouse waste storage capacity was inadequate. A portion of the resulting 500-gal/day (1900 L/day) overflow reached the river. In this study, the constructed wetland cell system treats the overflow.

Hammer (1989) recommends a loading rate of 89 lb BOD_5/acre-day (100 kg/ha-day) if freezing conditions are expected. The wetland was designed for a loading rate of 65 lb BOD_5/acre-day (73 kg/ha-day) for the four discrete 20 x 50-ft (6.1 m x 15.2 m) cells connected in series (Figure 2) with 2:1 side slopes (Figure 3). Wooden weirs with stop-lock gates divide the cells. Stop-lock gates can be used to alter water depth, but a minimum depth of 4 in. (0.1 m) is always maintained. Retention time is approximately 5 days/cell, giving a total detention time of almost 20 days (Doll et al., 1994). The farmer provided the equipment and labor to build the ponds. An attempt was made to keep construction costs low so other dairymen can duplicate this affordable alternative for waste water treatment.

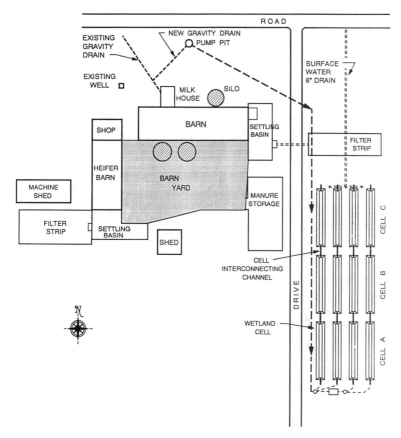

Figure 1. Partial farmstead layout with Wisconsin constructed wetland.

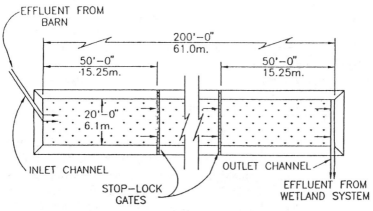

Figure 2. Plan view of Maine four-cell constructed wetland.

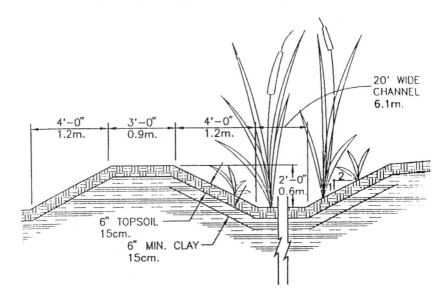

Figure 3. Cross-section of Maine constructed wetland cell.

WETLAND PERFORMANCE EXPERIENCE

WISCONSIN EXPERIENCE

Three wetland plant species (river bulrush, giant burreed, softstem bulrush) were planted in long rows in each wetland cell in the fall of 1992. Liquid levels in the cells were established at 4 in. (0.1 m) in the fall. The 1992/1993 winter was mild with minimal snow cover. In the fall of 1993, the liquid level in each of the eight downstream cells was established at 8 in. (0.2 m), while that in the four southernmost cells was established at 4 in. (0.1 m). The 1993/1994 winter was extremely cold for extended periods with good snow cover.

Each spring, the cells were totally drained to encourage the soil to warm and the plants to break dormancy. The spring of 1993 is remembered for the extended rain that caused the Midwest floods. This precipitation helped the plants get started. Later in the summer, the milkhouse waste did not provide sufficient flow to keep the eight downstream cells sufficiently moist. The reduced plant height compared to the southernmost cells was visibly apparent.

The spring and early summer of 1994 was particularly dry. Consequently, the milkhouse waste water seldom flowed to the downstream cells. During the summer of 1994, the population density of wetland plants declined in the downstream cells, and the remaining plants had desiccated tops. Grass species have invaded these cells. Plants in the upstream (southernmost) cells were tall and vigorous. The effects of winter on these plants appear to be minimal, while dry conditions are extremely detrimental.

Water Quality Parameters--1993

Throughout the summer and fall of 1993, grab samples of water (when available) were taken at the inlet to the first tier of cells and at the discharge of each set of cells. The sampling frequency was approximately two weeks. Average values for some of the water quality parameters are presented in Tables 1-4 (reported in Holmes et al., 1994). Interpreting these data requires care. Since values are concentrations, they can be influenced by precipitation or the lack thereof. The decreasing number of samples with increasing distance from the

Table 1. Average chemical oxygen demand concentration
at different locations within the wetland.

Location	No Treatment (mg/L)	(SD)	No. of Samples	Settling/Flotation Treatment (mg/L)	(SD)	No. of Samples
Inlet	488	(526)	15	290	(231)	19
First Cell Discharge	117	(15)	17	117	(24)	12
Second Cell Discharge	143	(26)	3	200	(0)	1
Third Cell Discharge	114	(41)	5	86	(6)	3

Table 2. Average biochemical oxygen demand concentration
at different locations within the wetland.

Location	No Treatment (mg/L)	(SD)	No. of Samples[†]	Settling/Flotation Treatment (mg/L)	(SD)	No. of Samples[†]
Inlet	168	(231)	13	97	(121)	17
First Cell Discharge	14	(5)	13	14	(4)	10
Second Cell Discharge	10	(0)	1	N/A	(N/A)	0
Third Cell Discharge	15	(11)	4	17	(12)	3

[†]Some samples deleted from averages because of quality control difficulties in analysis.

Table 3. Average total phosphorus concentration
at different locations within the wetland.

Location	No Treatment (mg/L)	(SD)	No. of Samples	Settling/Flotation Treatment (mg/L)	(SD)	No. of Samples
Inlet	16.9	(5.9)	19	13.5	(4.8)	21
First Cell Discharge	8.1	(3.4)	17	7.7	(2.5)	12
Second Cell Discharge	5.8	(0.15)	3	10.2	(0.0)	1
Third Cell Discharge	2.8	(0.8)	5	2.4	(0.8)	3

Table 4. Average total Kjeldahl nitrogen concentration
at different locations within the wetland.

Location	No Treatment (mg/L)	(SD)	No. of Samples	Settling/Flotation Treatment (mg/L)	(SD)	No. of Samples
Inlet	19.8	(15.9)	19	14.7	(8.8)	21
First Cell Discharge	6.5	(2.1)	17	7.2	(2.8)	12
Second Cell Discharge	6.8	(1.1)	3	15.0	(0.0)	1
Third Cell Discharge	5.2	(1.4)	4	4.4	(0.45)	3

inlet to the wetland is an indication of how often water was available to sample.

BOD$_5$ and chemical oxygen demand (COD) values at the inlet are lower than normally expected (1500-3000 mg/L) for milkhouse waste water because the dairyman makes a special effort to keep milk from entering the floor drain.

Successes and Difficulties

One difficulty experienced was delivering waste water during the first winter. A construction error, completion of construction in December 1992, and inadequate winterization caused the system to freeze downstream of the flow distributer during the mild winter of 1993. Better winterization of the system and the good snow cover of the 1993/1994 winter allowed the system to continue flowing without difficulty.

Controlling water flow from one wetland cell to the next has been difficult. Initially tar was used to seal joints between the weirs and channels. This was never fully effective, and slow leaks let water drain the cells. This may have had benefits for some of the downstream cells receiving water in dry periods. However, it was not beneficial when those same cells were drained by leaks. In the spring of 1994, the tar was replaced with a silicone calk.

Automatic flow meters (12) and samplers (8) were installed in the spring and summer of 1994. The goal is to obtain flow-proportioned samples that can be tested in the laboratory for pollutant concentration. From these data, treatment effectiveness of the wetland cells can be analyzed.

MAINE EXPERIENCE

Plant Survival

After the cells were constructed in the fall of 1993, common cattails (*Typha latifolia*) were planted. Cattail marshes are one of the simplest types of wetland systems to create due to the cattail's monotypic structure, aggressive and productive growth habits, and wide tolerance to hydrologic and edaphic conditions (Dobberteen and Nickerson, 1991). Mass balance studies demonstrate that *Typha* has the capacity to assimilate 90% of the nutrients loaded into a waste water wetland (Busnardo et al., 1992).

The cells in this study were planted with rhizomes and seed. In the spring of 1994, there was a 35% survival rate of plants derived from the rhizomes. The winter of 1993/1994 was one of the worst on record for Maine, which may have contributed to plant mortality. However, the cattails from rhizomes have responded to the nutrient-rich waters with vigorous growth. Over 200 new seedlings have emerged from the seeds. Growth of plants from seed has been slower than that for plants from rhizomes. Milkhouse waste was not diverted to the wetland until mid-July 1994.

Successes and Difficulties

The system was relatively easy to construct due to the presence of a layer of marine clay that served as an excellent liner for the system. Cattail transplanting was labor-intensive, using mature plants and rhizomes. However, these plants were capable of receiving waste water sooner than plants generated from seeds.

SUMMARY

Constructed wetlands have been found to reduce pollutants in milkhouse wastes in cold climates. Use of constructed wetlands in cold climates to augment the reduction of nutrients from dairy wastes can be a viable complement to traditional storage facilities and spreading wastes as fertilizer. Such an approach to dairy herd wastes may significantly reduce non-point source pollution.

ACKNOWLEDGMENTS

The authors thank the following for their contributions to the success of this project: USDA-Soil Conservation Service; Department of Civil and Environmental Engineering, University of Maine-Orono; United States Geological Survey; Richard Varnum, Dover-Foxcraft, Maine; Doug Rohn, Bitterroot Design and Drafting, Bozeman, Montana; Water Quality Demonstration Project - East River, Brown County, Wisconsin; Department of Agricultural Engineering, University of Wisconsin- Madison; Wisconsin Department of Agriculture, Trade and Consumer Protection; Wisconsin Department of Natural Resources; Fox Valley Technical College, Appleton, Wisconsin.

REFERENCES

Biddlestone, A.J., Gray, K.R., and Job, G.D., Treatment of dairy farm wastewaters in engineered reed beds, *Biochemistry*, 26, 265, 1991.

Busnardo, Max, Gersberg, Richard M., Langis, Rene, Sinicrope, Theresa L., and Zedler, Joy B., Nitrogen and phosphorus removal by wetland mesocosms subjected to different hydroperiods, *Ecological Engineering*, 1, 287, 1992.

Dobberteen, R.A., and Nickerson, N.H., Use of created cattail (*Typha*) wetlands in mitigation strategies, *Environmental Management*, 15(6), 797, 1991.

Doll, B., Rock, C., and Kostinec, R., Dairy waste treatment: A look at wetland resources, (submitted) Water Environment Federation, Proc. of Ann. Conf., Oct. 15-19, 1994.

Hammer, Donald A., Designing constructed wetland systems to treat agricultural nonpoint source pollution, *Ecological Engineering*, 49, 1989.

Holmes, B.J., Massie, L.R., Bubenzer, G.D., and Hines, G., *Design and Construction of a Wetland to Treat Milkhouse Wastewater*, ASAE Paper 924524 presented at Int. Winter Meet. Amer. Soc. of Agricultural Engineers, Nashville, TN, Dec. 15-18, 1992.

Holmes, B.J., Bubenzer, G.D., Massie, L.R., and Hines, G., A constructed wetland for treating milkhouse wastewater in a cold climate, *Proc. Workshop on Constructed Wetlands for Animal Waste Management*, Purdue Univ., West Lafayette, Indiana, 1994.

Ulmer, Ross, Strong, Lon, Cathcart, Thomas, Pote, Jonathan, and Davis, Sam, *Constructed Wetland Site Design and Installation*, ASAE Paper 924528 presented at Int. Winter Meet. Amer. Soc. of Agricultural Engineers, Nashville, Tennessee, Dec. 15-18, 1992.

EVALUATING THE EFFECTIVENESS OF FLOOD WATER RETENTION RESERVOIRS AS INTEGRATORS OF WATER QUALITY

Anne M.S. McFarland and Larry Hauck

INTRODUCTION

Erath County, located in North-Central Texas, is the number one milk producing county in Texas. County-wide, nearly 200 dairies with approximately 60,000 cows present potential sources of non-point source pollution from the active disposal of liquid and solid animal wastes on agricultural fields. Peanuts, pecan orchards, and hay fields represent other significant agriculture in the area.

Ubiquitous features of the county's watersheds are flood-retardation reservoirs built by the US Department of Agriculture Soil Conservation Service (SCS), now the Natural Resources Conservation Service. These relatively small reservoirs, constructed during the 1950s and 60s under funding from Public Law 566 (PL-566), have storage capacities of 247,000 m^3 or less at the principal spillway elevation (though temporary flood storage is typically many times larger). Collectively, however, these PL-566 reservoirs have a significant impact on the hydrology of area streams and rivers. Forty flood-retardation reservoirs are located in the North Bosque River Basin along with approximately half of Erath County's dairies and cows.

To evaluate the impact of PL-566 reservoirs on the water quality in the North Bosque River, two PL-566 reservoirs and the streams in their watersheds were intensively monitored in a paired watershed study. While these watersheds have similar morphological characteristics and are within 10 km of one another, they exhibit conspicuous differences in vegetative cover and agricultural land-use activities (Table 1). The South Fork Watershed is impacted to a lesser degree by intensive agricultural practices (e.g., dairy waste application fields) than the North Fork Watershed. The areas of the watersheds are roughly comparable with similar ratios of reservoir volume to contributing drainage area. Given natural variability constraints, these two study areas represent comparable systems with the exception of the desired variation in intensive agricultural land use.

Previous studies have indicated significant differences in water quality between these two watersheds (Hauck et al., 1994). The North Fork generally

231

Table 1. Watershed and reservoir characteristics.

Characteristic	North Fork	South Fork
Drainage Area	1,553 ha	913 ha
Average Basin Slope	11.4 m/1,000 m	9.8 m/1,000 m
Number of Active Dairies	3	0
Number of Permitted Cows	2,050	0
Intensive Agricultural Land-Use Coverage	53%	4%
Reservoir Pool Surface Area	18 ha	6 ha
Reservoir Pool Storage Capacity	246,000 m^3	93,480 m^3
Reservoir Maximum Depth	3 m	3 m

has higher nutrient levels, attributed, at least in part, to the intensive agriculture in the watershed. The objective of this chapter is to evaluate reservoir-induced removal of water-borne constituents by comparing inflow and outflow relationships for each reservoir. Four storm events with varying levels of precipitation occurring between October 1993 and May 1994 were chosen for this analysis.

METHODS

An extensive monitoring program using ISCO automatic samplers was established in both watersheds to provide characterization of storm water run-off as tributary inflow into each reservoir and outflow through the principal spillway from each reservoir. A rise in water level activates the samplers into a programmed collection sequence. Samples were taken at hourly intervals for the first three hours, at two-hour intervals until the eleventh hour, and then at six-hour intervals until the end of the storm event. Total suspended solids (TSS), total Kjeldahl nitrogen (TKN), ammonia-nitrogen (NH_3-N) nitrate-nitrogen (NO_3-N), and orthophosphate-phosphorus (OPO_4-P) are the major water-borne constituents discussed in this chapter. All samples were collected and analyzed following a US EPA-approved Quality Assurance Project Plan (Texas Institute for Applied Environmental Research, 1992, 1993) based on US EPA guidelines (Kopp and McKee, 1983).

The automatic samplers also measure water level at five-minute intervals throughout the storm event. Manual measurements of water flow are routinely conducted on the tributaries to allow development of site-specific relationships of flow to water level. Stage-discharge relationships based on hydraulic calculations for the riser-outlet pipe structure of each reservoir's principal spillway were provided by the SCS (Goertz, 1993).

The water quality of the main body of each reservoir is monitored with monthly grab samples taken the first week of each month at the deepest point near the principal spillway. Measurements are taken 0.3 m below the surface, mid-depth, and 0.3 m above the bottom of the reservoir. The monthly reservoir analyses include the same constituents as the storm water sampling. The highly intermittent nature of stream flow effectively restricts routine sampling to the permanent reservoir sites, since the stream sites generally do not flow between storm events.

Four storm events between October 1993 and May 1994 were chosen for detailed analysis of inflow/outflow relationships based on the completeness and availability of relevant information on water quality and flow variables. Loadings were calculated by combining the instantaneous discharge record with the water quality data. Discharge was divided into intervals based on the date and time when water quality samples were taken using a midpoint rectangular method between water quality samples (Stein, 1977). Constant flow was assumed between each five-minute discharge measurement to estimate volume. Mass for each five-minute interval was calculated as volume times the associated water quality constituent. Storm event masses were then calculated by summing the mass associated with each five-minute interval over the entire storm event. The average constituent concentration by storm event was calculated as a weighted average based on the volume of flow associated with each measurement. The inflow water quantity and constituent mass associated with the ungauged portion of each watershed were estimated based on simple proportions of total drainage area (gauged and ungauged) to the gauged drainage areas. The direct contribution by rainfall was not included. Hydraulic detention time for storm water in each reservoir was estimated based on the period of outflow, i.e., the time span between when outflow started and ended, the outflow volume, and the reservoir pool storage capacity of each reservoir.

RESULTS

Similar amounts of precipitation occurred on the North and South Fork Watersheds during each storm event, although longer detention times and larger outflow volumes generally occurred in the North Fork reservoir than in the South Fork reservoir (Table 2).

Table 2. Storm and reservoir outflow characteristics.

Watershed	Storm	Precipitation	Period of Outflow	Volume of Outflow	Hydraulic Detention Time
		mm	days	m^3	days
North Fork	1	177	10.6	644,593	4.1
	2	45	8.1	37,491	53
	3	12	0.9	508	436
	4	173	14.8	686,639	5.3
South Fork	1	168	9.6	355,605	2.5
	2	42	8.8	26,153	31.5
	3	17	3.3	7,235	42.6
	4	183	11.0	419,993	2.5

Table 3. Mass and volume weighted mean and standard deviation of water quality constituents for four storm events on the North Fork watershed and reservoir water quality prior to each storm event.

Storm	Vol. (m³)	NH_3-N (kg)	NH_3-N (mg/L)	NO_3-N (kg)	NO_3-N (mg/L)	TKN (kg)	TKN (mg/L)	OPO_4-P (kg)	OPO_4-P (mg/L)	TSS (kg)	TSS (mg/L)
Storm 1 (Oct. 12–24)											
Total Inflow (n†=17)	745848	63	0.08±0.07	428	0.57±0.29	2773	3.72±1.81	671	0.90±0.42	1806233	2422±1913
Outflow (n=24)	644593	73	0.11±0.08	148	0.23±0.05	1376	2.14±0.69	506	0.78±0.12	121406	188±260
Retention	14%	-16%		65%		50%		25%		93%	
Reservoir (Oct. 6, n=3)			0.10±0.03		0.03±0.04		1.50±0.27		0.69±0.01		14±3
Storm 2 (Jan. 22– Feb. 1)											
Inflow (n=12)	36654	30	0.82±1.41	105	2.86±2.23	126	3.44±1.01	28	0.78±0.33	11606	317±266
Outflow (n=14)	37491	2	0.05±0.01	11	0.28±0.22	58	1.54±0.09	8	0.21±0.02	763	20±3
Retention	-2%	93%		90%		54%		73%		93%	
Reservoir (Jan. 6, n=3)			0.02±0.01		0.13±0.01		2.30±0.18		0.28±0.01		20±1
Storm 3 (March 26–28)											
Inflow (n=7)	1409	0.50	0.35±0.51	0.80	0.57±0.14	2.24	1.59±0.28	0.11	0.07±0.01	133	94±81
Outflow (n=4)	508	0.05	0.09±0.01	0.07	0.14±0.00	1.47	2.90±1.15	0.07	0.13±0.01	16	31±4
Retention	64%	91%		91%		34%		38%		88%	
Reservoir (March 3, n=3)			0.11±0.03		0.28±0.01		2.02±0.09		0.11±0.00		23±1
Storm 4 (May 9–24)											
Inflow (n=24)	730856	315	0.43±0.37	628	0.86±0.37	2746	3.76±1.56	807	1.10±0.49	959763	1313±1016
Outflow (n=26)	686639	241	0.35±0.20	328	0.48±0.10	1893	2.76±0.35	543	0.79±0.63	95963	140±178
Retention	6%	24%		48%		31%		33%		90%	
Reservoir (May 5, n=3)			0.03±0.03		0.04±0.01		2.34±0.21		0.19±0.03		25±1

†'n' refers to the number of water quality samples taken during a storm event or monthly reservoir sampling period.

NORTH FORK

Inflow and outflow relationships for the North Fork for each storm event are presented in Table 3 along with the constituent concentrations of the reservoir prior to each storm.

For the two smaller storm events (Storms 2 and 3), outflow concentrations were generally much lower than inflow concentrations and more similar to the monthly reservoir concentrations. Average reservoir concentrations were greater than outflow concentrations for TKN in Storm 2 and for NO_3-N in Storm 3. This was likely a function of changes in reservoir water quality between the time of the monthly reservoir sample and the occurrence of the storm event. Percent retention of NH_3-N, NO_3-N, and TSS mass was fairly high for these two storm events, while percent retention of TKN and OPO_4-P occurred at lower levels.

For the larger storm events (Storms 1 and 4), the water quality of the outflow was generally more similar to that of the inflow than to the water quality within the reservoir since the large storm volume effectively replaces the reservoir pool capacity several times over. The final inflow and outflow volumes for Storms 1 and 4 were quite similar. The constituent mass and concentrations associated with the inflow and outflow for all water quality variables, except NH_3-N and TSS, were also fairly similar for both these storm events. In Storm 1, the outflow mass of NH_3-N was greater than the inflow mass, indicating the possible generation of NH_3-N within the reservoir, although more likely this is a function of the accuracy of the laboratory analyses at these low concentrations. The inflow mass of TSS from Storm 4 was about half that from Storm 1, which may be a function of seasonality. The percent retention of nutrients generally decreased with increasing storm volume, although the percent retention of TSS was similar for all four storm events. This is probably a function of decreasing detention time with increasing outflow volume (Table 2).

SOUTH FORK

The mass of constituents in the inflow and outflow was noticeably smaller on the South Fork than on the North Fork, presumably due to smaller storm runoff volumes and lower constituent concentrations on the South Fork. The volume of inflow and outflow for the South Fork was 30% to 50% less than for the North Fork for Storms 1, 2, and 4 (Tables 3 and 4), which roughly corresponds to the 41% difference in watershed drainage areas (Table 1). During Storm 3, inflow and outflow volumes of the South Fork greatly exceeded those of the North Fork. Although precipitation for Storm 3 on the South Fork was only 5 mm above the 12 mm received on the North Fork, the USDA-SCS Curve Number Method for calculating runoff (McCuen, 1982) demonstrates that even small differences in precipitation may significantly impact the volume of runoff, especially for small rainfall events.

The relative differences between inflow and outflow mass and concentrations for the different storm events was generally very small for NH_3-N, NO_3-

Table 4. Mass and volume weighted mean and standard deviation of water quality constituents for four storm events on the South Fork watershed and reservoir water quality prior to each storm event.

Storm	Vol. (m³)	NH_3-N (kg)	NH_3-N (mg/L)	NO_3-N (kg)	NO_3-N (mg/L)	TKN (kg)	TKN (mg/L)	OPO_4-P (kg)	OPO_4-P (mg/L)	TSS (kg)	TSS (mg/L)
Storm 1 (Oct. 12-16)											
Inflow (n†=24)	375597	6.5	0.02±0.03	31	0.08±0.06	710	1.89±1.37	39	0.10±0.07	453588	1208±2505
Outflow (n=24)	355605	13.9	0.04±0.08	33	0.09±0.06	391	1.10±0.35	18	0.05±0.03	29314	82±72
Retention	5%	-115%		-7%		45%		52%		94%	
Reservoir (Oct. 6, n=3)			0.05±0.00		0.01±0.00		0.98±0.27		0.02±0.01		13±1
Storm 2 (Jan. 22-31)											
Inflow (n=14)	25492	0.60	0.02±0.02	5.4	0.21±0.07	30	1.18±0.31	1.25	0.05±0.02	3909	153±153
Outflow (n=15)	26153	0.82	0.03±0.01	5.6	0.21±0.05	19	0.73±0.11	0.49	0.02±0.02	435	17±5
Retention	-3%	-37%		-3%		37%		61%		89%	
Reservoir (Jan. 6, n=3)			0.14±0.11		0.21±0.01		0.79±0.07		0.01±0.00		10±4
Storm 3 (March 26-29)											
Inflow (n=12)	6614	0.27	0.04±0.03	0.17	0.03±0.02	7.6	1.15±0.52	0.04	0.01±0.00	741	112±134
Outflow (n=14)	7235	0.42	0.06±0.04	0.05	0.01±0.00	3.6	0.50±0.10	0.10	0.01±0.02	59	8±2
Retention	-9%	-54%		74%		52%		-150%		92%	
Reservoir (March 3, n=3)			0.12±0.03		0.13±0.10		0.57±0.09		0.02±0.01		3±1
Storm 4 (May 9-20)											
Inflow (n=25)	480635	37	0.08±0.04	28	0.06±0.04	509	1.06±0.39	18	0.04±0.02	71337	148±114
Outflow (n=29)	419993	32	0.08±0.04	31	0.07±0.03	423	1.01±0.35	27	0.07±0.06	34556	82±71
Retention	13%	12%		-12%		17%		-54%		52%	
Reservoir (May 5, n=3)			0.01±0.00		0.01±0.00		0.36±0.18		0.01±0.00		4±2

†'n' refers to the number of water quality samples taken during a storm event or monthly reservoir sampling period.

N, and OPO$_4$-P. In several instances outflow mass even exceeded inflow mass for several constituents having very low concentrations. The accuracy of the laboratory procedures at low concentrations could be responsible for these results. The high retention rate of TSS was similar to that of the North Fork except for Storm 4 when 52% retention occurred. The large volume of storm water runoff and the relatively short reservoir detention time of this event may have provided inadequate time for settling to occur.

As with the North Fork, outflow constituent concentrations generally showed a closer association to inflow concentrations than to reservoir concentrations on the large storm events, and vice versa on the small storm events, although these relationships were not as clear as on the North Fork. The inflow, outflow, and reservoir nutrient concentrations were consistently very low on the South Fork for all storm events.

DISCUSSION

In the North Fork reservoir, the high removal efficiencies or retention of solids and, to a lesser degree, organic nutrients exhibited in the four storm events can be attributed primarily to the physical process of sedimentation, which occurs in any relatively quiescent body of water. Smaller storm events, which are typically accompanied by longer detention time, may have a greater impact on the accumulation of nutrients in these reservoirs than do larger storm events in which complete flushing of the reservoir can occur. While high solids retention occurred in the South Fork reservoir, nutrient removal was not consistently demonstrated. The relatively low nutrient concentrations of inflow into the South Fork reservoir probably can be attributed to this lower removal rate.

The substantial aquatic plant populations in each reservoir should result in nutrient uptake of retained nutrients, especially of inorganic nitrogen, since the reservoirs are nitrogen limited (Hauck et al., 1994). The North Fork reservoir exhibited large reductions in inorganic nutrients from storm events. These reductions were generally greater for smaller storm events, which provided greater detention time for nutrient uptake than for larger storm events. Because of the consistently low levels of dissolved, inorganic nutrients in the South Fork reservoir, plant uptake was probably not a major factor in nutrient removal in the South Fork reservoir.

While the PL-566 reservoirs were designed for flood control, they appear to perform an important role in the attenuation of water pollutants to delay and/or reduce downstream pollution loads in watersheds impacted by non-point source pollution, especially from smaller storm events. The appropriateness of this role is a question that should be debated in evaluating the controls for non-point source pollution.

ACKNOWLEDGMENTS

The Texas Institute for Applied Environmental Research would like to acknowledge funding provided by the State of Texas, the Soil Conservation Service, and the Environmental Protection Agency. In addition, we would like to acknowledge the support and understanding of cooperating land owners who made this study possible.

REFERENCES

Hauck, L., Jones, T., and Coan, T.L., *Report on the Role of Two PL-566 Reservoirs in Agricultural Pollution Control: North Bosque River Basin Erath County, Texas,* Texas Institute for Environmental Research Report prepared for the Texas State Soil and Water Conservation Board, July 1994.

Goertz, L., Computer generated rating provided to TIAER in correspondence, USDA-SCS, Temple, Texas, May 11, 1993.

Kopp, J.F., and McKee, G.D., *Methods for Chemical Analysis of Water and Wastes,* Revised, EPA-600/4-79-020, March 1983.

McCuen, R.H., *A Guide to Hydrologic Analysis Using SCS Methods,* Prentice-Hall, Englewood Cliffs, New Jersey, 1982.

Stein, S.K., *Calculus and Analytic Geometry,* second edition, McGraw-Hill Book Company, New York, 1977.

Texas Institute for Applied Environmental Research, *Quality Assurance Project Plan for the National Pilot Project,* Tarleton State University, Stephenville, Texas, 1993.

Texas Institute for Applied Environmental Research, *Quality Assurance Project Plan for the Texas State Soil and Water Conservation Board for Environmental Monitoring and Measurement Activities Relating to Nonpoint Source Pollution in the Upper North Bosque River Watershed,* Tarleton State University, Stephenville, Texas, 1992.

IMPACT OF MANURE ON ACCUMULATION AND LEACHING OF PHOSPHATE IN AREAS OF INTENSIVE LIVESTOCK FARMING

A. Breeuwsma, J.G.A. Reijerink, and O.F. Schoumans

INTRODUCTION

In the past decade, livestock densities have increased substantially in Western Europe and, particularly, the Netherlands. The animal feeds required for the increasing livestock production generally have been imported from abroad. Thus manure production often exceeds crop requirements on a farm scale and, due to concentration of livestock holdings in certain areas, on a regional scale (Breeuwsma and Silva, 1992). The manure surpluses not only added to the general environmental load by modern agriculture but also introduced a more specific problem related to phosphate (P) accumulation in soils. Thus, long-term manure applications in areas of intensive livestock farming may enhance leaching of P as well as nitrate to ground water. Soil P accumulations were first recognized in the Netherlands some 15 years ago (Lexmond et al., 1982). Since that time, many Dutch studies have addressed sorption and desorption of P in soils, the degree of accumulation (saturation), the relationship with soil P level, and leaching. It is the purpose of this study to summarize the major results of those reports (mostly in Dutch).

ENVIRONMENTAL IMPACT OF PHOSPHORUS SURPLUSES

DIFFERENCES BETWEEN NITROGEN AND PHOSPHORUS

The environmental concerns regarding the use of manures as well as fertilizers often differ for N and P. Nitrogen is much more mobile than P, which is strongly adsorbed in most soils. Leaching to ground water is, therefore, not a problem except at high loads and shallow water tables (see below). N, on the other hand, is commonly present in the nitrate form, and this ion is only slightly adsorbed by most soils. The net surplus of N (input minus uptake and losses by volatilization, denitrification, and immobilization), therefore, directly leaves the root zone with the excess rainfall.

Secondly, environmental concerns for N and P may also differ because of the level of the critical concentrations (quality standards). For N in ground water, a drinking water limit of 11.3 mg/L NO_3-N is used by the European

Union. This is well above the background concentration of nitrate in ground water. The acceptable P level in ground water, as used in the definition of P-saturated soils in the Netherlands, is 0.10 mg/L of ortho-P (see below). Eutrophication standards for surface water vary mostly between 0.05 and 0.15 mg/L of total P, which is close to the background concentration in many soils. Therefore, ground water protection P standards are far more strict than N standards.

For surface waters, however, differences posed by N and P standards are much smaller. For example, in the Netherlands the critical concentration for N has been set at 2.2 mg/L NO_3-N. In percentages of the application rates, allowable N and P losses would then amount to about 2 and 1% respectively, based on application rates of 350 kg/ha/year of N and 40 kg/ha/year of P and a rainfall excess of 300 mm/year.

RUNOFF

Annual P losses by surface runoff from agricultural soils generally amount to 0.5-2.0 kg/ha/year of P, which is about 1-5% of the input (Chesters et al., 1978; Sherwood, 1990). On frozen grounds annual losses of up to 13% of manure input have been reported (Young and Mutchler, 1976). Most of the losses occur in periods with heavy rainfall shortly after application of fertilizer or manure on soils with low permeability or water storage capacity and slopes of more than a few percent. In addition, soil P status also plays a significant role (Sharpley et al., 1994).

LEACHING

Generally, P concentrations in ground water are more influenced by soil parent material than by inputs. For example, high concentrations of the order of 1-5 mg/L of ortho-P can be found in eutrophic peat soils. In oligotrophic peat, concentrations are usually less than 1 mg/L. Marine clay soils can also have a high P content. Mineral soils generally have low concentrations in the subsoil of the order of 0.02 - 0.06 mg/L of ortho-P.

In the mineral top soil of areas with intensive livestock farming, P concentrations can be much higher. For example, in the Netherlands shallow (within 50 cm below surface) ground water may occur during wet periods and reach P-enriched soil layers. The effects on ground water quality can be illustrated with data from an area with a high P surplus and sandy soils (Breeuwsma et al., 1989). More than 80% of the agricultural soils in the catchment area were saturated with P. During the winter of 1987-1988, concentrations in ground water exceeded the general quality goal for surface waters in the Netherlands (0.15 mg/L of total P) in 80% of the samples (Figure 1). About 40% of the samples had concentrations above 1.0 mg/L total P and 6% above 6 mg/L of total P. During the summer the ground water level was deeper, and P concentrations decreased considerably. However, the transport to surface waters primarily occurs in wet periods when phosphate concentrations are high. The

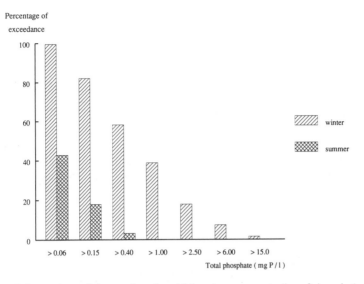

Figure 1. Percentage of observations for which a given concentration of phosphate (mg/L of total P) in ground water was exceeded in the catchment area of the Schuitenbeek, the Netherlands, in 1988.

annual flux-weighted concentration of total P in surface waters in this area was 1 mg/L and the mean leaching rate 2.5 kg/ha/year. The desorbable amount is of the order of 500-1000 kg/ha of P, indicating that in this case, enhanced phosphate leaching may last for a very long time.

PHOSPHATE SATURATION

Though strongly adsorbed, even phosphate can show enhanced leaching to ground water because the sorption capacity of the soil is not unlimited. Once a soil profile is saturated with P to the ground water table, P concentrations would reach very high levels as present in slurries (up to 100 mg/L, Gerritse and Zugec, 1977). Obviously, this situation should be prevented. Therefore, the concept of a critical degree of P saturation was introduced in Dutch environmental and manure policies.

The degree of phosphate saturation has been defined by Breeuwsma et al., 1989:

$$DPS = \frac{P_{ox}}{PSC} \times 100\% \qquad (1)$$

DPS = degree of phosphate saturation (%)
P_{ox} = sorbed P (extractable by oxalate) (mol/kg)
PSC = phosphate sorption capacity (mol/kg)

The critical value of DPS is defined as the saturation percentage that should not be exceeded to prevent adverse effects on ground water quality. The environmental goals have been specified as follows: the phosphate concentration in ground water should not exceed 0.10 mg/L of ortho-P at the level of the Mean High Water table (MHW). This MHW is defined as "the arithmetic mean during at least eight years of the three highest water levels per hydrological year (April 1 - March 31)." A critical DPS of 25% has been established (Van der Zee, 1988; Van der Zee et al., 1990). The surplus of phosphate that can be applied to the soil before saturation occurs varies with soil type and, particularly, the depth to the water table. For example, for poorly drained sandy soils, with a shallow MHW (20-40 cm below surface), this amounts to approximately 1000-2000 kg/ha of P_2O_5 (Table 1). Well-drained soils with a MHW of 100 cm below surface may get saturated at a surplus of 2000-4000 kg/ha of P_2O_5, depending on the soil type.

Table 1. Phosphate surplus (kg/ha of P_2O_5) that can be applied to the soil before saturation occurs, for five values of MHW (cm below surface).[†]

Soil type	Poorly drained			moderately drained	well drained
	20	30	40	50	100
Gleyic Podzol (Haplaquod)	810	1290	1810	2250	3630
Humic Gleysol (Typic Humaquept)	770	1200	1530	1750	1940

[†]MHW: Mean High Water table

ASSESSMENT OF PHOSPHORUS SATURATION ON A REGIONAL SCALE

The degree of P saturation in a given area has been assessed by measurements and model calculations. For economic reasons, measurements based on statistical sampling techniques were usually restricted to smaller areas of a few thousand hectares in the Netherlands (Breeuwsma et al., 1989; Hack-ten Broecke et al., 1990; Brus et al., 1992; Brus, 1993), and up to 100,000 ha in Belgium (de Smet et al., 1994; Lookman et al., 1994).

In the Netherlands model calculations were used to assess the P saturation status of areas with a manure surplus on a national scale. These areas are concentrated on sandy soils in the central, eastern, and southern part of the country. The area of P-saturated soils was estimated in various studies with increasing degree of detail. The tools comprise the following:
- soil data base with Al_{ox} and Fe_{ox} contents per soil type
- transfer function between $Al_{ox} + Fe_{ox}$ and the phosphate sorption capacity of the soil (Schoumans et al., 1987)
- soil map, scale 1:50,000

- land use map, scale 1:50,000 (Landsat images)
- data base on animal production and crops (based on annual registration per farm)

In a preliminary study, P saturation was examined only for land in maize using a DPS of 100% (Breeuwsma and Schoumans, 1987). A second study included all farmland and used a critical DPS of 25%, as mentioned earlier (Breeuwsma et al., 1990).

In the third, most recent, study, assessments were improved by using more detailed data on animal production, at a scale of 2.5 km x 2.5 km, updated data for the MHW and new estimates of the P loads prior to the strong development of intensive livestock holdings around 1970 (Reijerink & Breeuwsma, 1992). P inputs were calculated on a farm level by the DLO Agricultural Economics Research Institute and supplied as a frequency distribution for areas of 2.5 km x 2.5 km. For the P loads, alterations were prompted by the results of validation studies in three regions, which showed that the model underestimated the P content of the soil and/or the degree of P saturation (Breeuwsma et al., 1989; Hack-ten Broeke et al., 1990). A literature survey indicated a hitherto-neglected surplus of about 50 kg/ha/year of P_2O_5 between 1950 and 1970. Accounting for this additional input resulted in a better agreement between measured and calculated results for the P content of the soil and the area of P-saturated soils (± 10%).

From 1970-1990, the net annual load of P was of the order of 150-450 kg/ha/year of P_2O_5 for maize. For permanent pastures mean net loads are lower but still substantial (50-150 kg/ha/year of P_2O_5). The cumulative surplus of P in the soil often exceeds saturation values for the poorly drained soils and even for the well-drained soils with a MHW at 100 cm below surface (Table 1). Using the improved and geographically more detailed model input gave a new estimate of the percentage of P-saturated farmland of 70% (Figure 2). The percentages for maize and pastures are 88% and 69%, respectively. More than 21% of the farmland is strongly saturated, with a DPS above 50%, and 6% is very strongly saturated, with a DPS above 75%. The degree of P saturation varies significantly between regions, indicating sensitive areas that can be targeted for watershed control (Figure 3).

MANURE AND FERTILIZER MANAGEMENT

Can appropriate manure management prevent soil P saturation? At first sight, a simple solution would be to restrict applications to levels required for optimum crop yields. Fertilizer recommendations based on this objective are related to the P status of the soil as measured by soil testing. In the example given in Figure 4, application rates meet the removal by harvested crops at the P status "amply sufficient." Higher rates are recommended at lower P levels, and vice versa. In addition to this crop-based recommendation, there is also a

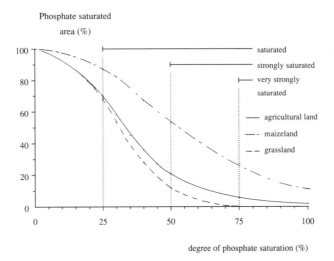

Figure 2. Percentage of phosphate-saturated soils as a function of the degree of phosphate saturation for manure surplus areas in the Netherlands in 1990.

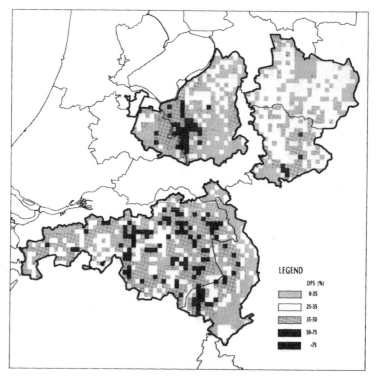

Figure 3. Degree of phosphorus (P) saturation of 50% of the farmland in areas of 2.5 km x 2.5 km.

soil-based recommendation that aims at raising the P status of the soil to the level "amply sufficient." Sandy soils with a low natural soil fertility level require high applications (Table 2) of the same order as those causing P saturation in poorly drained soils (Table 1). In other words, the aim of the farmer to attain an appropriate level of soil fertility may not be compatible with environmental objectives in these soils.

But even when the soil is not saturated by "restoration" of the soil fertility, as farmers call it, it is still possible that soils may eventually become saturated when soil fertility is maintained at the desired level. In the example given in Figure 4, controlling the P status at an "amply sufficient " level also requires an overdose of about 15 kg/ha/year of P_2O_5. A recent study in the Netherlands indicated that even higher surpluses of 25-70 kg/ha/year of P_2O_5 may be required under good agricultural practices (Oenema and Van Dijk, 1994). These surpluses are considered to be inevitable agricultural P losses. They are significantly higher than the P losses that are environmentally acceptable on a long term. For example, a quality goal of 0.15 mg/L of total P corresponds to an environmentally acceptable loss of about 1 kg/ha/year of P_2O_5 at an excess rainfall of 0.3 m/year.

In reality, the gap between "agriculturally inevitable" and "environmentally acceptable" losses may be somewhat smaller. The P losses mentioned above have been assessed at relatively high soil P levels (P_w-number: 30 - 60) to cover all agricultural crops, including highly P demanding vegetables. Target levels are usually lower. For example, the target level for a common crop rotation on arable land (potatoes, sugar beets, wheat) is P_w-number 30. But even if smaller P surpluses of 15-25 kg/ha/year of P_2O_5 would be possible, without reducing crop production, soils with a MHW shallower than 1 m below surface may eventually get saturated. A rough estimate of the time needed for saturation can be obtained from the data in Table 1, assuming an average loss of 20 kg/ha/year of P_2O_5, a "target application" of 1000 kg/ha of P_2O_5 (Table 2), and

Table 2. Recommended rates of phosphorus (P) fertilization (kg/ha of P_2O_5) to improve the P status of the soil to the target level in the Netherlands (P_w-number = 30; "amply sufficient") (Henkens, 1984).[†]

| Soil P level | P_w-number[‡] | Soil type | |
		Sand, löss, river clay	Marine clay
Very low	1	1710	1500
	5	1340	1130
	10	990	780
Low	15	700	490
	20	440	230
Sufficient	25	210	0

[†]Crop rotation: potatoes, sugar beet, barley/grass,wheat
[‡]Water extract (Van der Paauw, 1971; Sissingh, 1971)

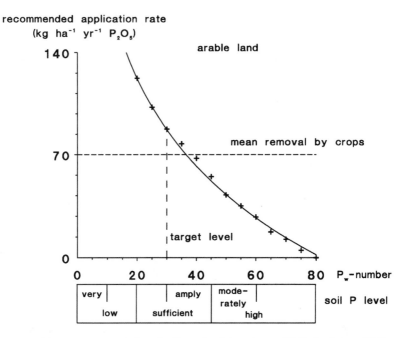

Figure 4. Mean recommended application rate of phosphorus (P) (kg/ha/year of P_2O_5 as a function of soil P level for arable land in the Netherlands. Crop rotation: potatoes, sugar beet, wheat (2x).

a travelling wave behavior of the phosphate front (Van der Zee and Gjaltema, 1992). In that case, residual P surpluses allowed amount to 200-2630 kg/ha/year of P_2O_5 for unsaturated soils, and the saturation time would be 10-130 years.

Obviously, more research is needed to quantify the P losses that are "agriculturally inevitable." Nevertheless, in many cases leaching losses from the tillage layer will most likely be above 1 kg/ha/year of P_2O_5. The use of this figure as a general quality goal also needs further consideration. One quality standard does not seem realistic in view of large variations in the natural water quality. Moreover, different uses of the water bodies, for example water supply, swimming and fishing water, do not have the same quality requirements.

Adverse effects of the use of manures on the environment can also be diminished by adjusting application times and methods. In the Netherlands, application times are limited to the period 1 February through 1 September, except for arable land on clay and peat soils. Sub-surface placement is required on grassland and ploughing-in on arable land. Furthermore, chemicals may be used to reduce the solubility of P, for example slaked lime or alum in poultry litter (Sharpley et al., 1995)

WATERSHED MANAGEMENT

Targeting of sensitive watersheds, or areas within watersheds, is one of the first steps in watershed control. Both qualitative and quantitative methods can be used to identify the most vulnerable areas. A qualitative approach is useful for a quick identification of "hot spots" and effective measures, as for example in the P indexing system, which relates potential P loss in runoff to site characteristics (Sharpley et al., 1995). In other cases, a more quantitative approach is required to assess whether quality standards or emission goals can be attained. The assessment of P saturation status of soils is a first step in this direction. Leaching of P from these soils is now being studied using the ANIMO model (Kroes et al., 1990) to quantify the effects of manure management on the water quality.

In the targeted areas measures can be promoted or regulations imposed to reduce or improve the use of animal waste and fertilizers ("source-oriented" measures) or to undertake site-specific actions that reduce the effects that remain even at a proper use. These effects are due to deleted losses of P (and N) caused by historical loads. In the Netherlands research is under way to study the reduction of losses from P-saturated soils by adding P binding materials and by hydrological measures. More generally, non-application zones and buffer strips can also be used to reduce losses to surface waters by runoff and leaching.

CONCLUSIONS

Concentrations of high livestock densities have led to manure surpluses on a regional scale and, in the Netherlands, even on a national scale. High application rates of manure may pose a threat to water quality by leaching of N and runoff of P. Less well known but equally important in soils with shallow water tables is the leaching of P caused by high accumulations of soil P. The breakthrough of P that is to be expected in these soils will substantially add to the eutrophication of sensitive surface waters. To prevent this the degree of saturation of the sorption capacity of soils should not exceed a critical limit. Applying the Dutch definition of a saturated soil indicates that P saturation is a widespread phenomenon in the Netherlands and Belgium. P-saturated soils would also be expected in other areas in Western Europe and the US with similar soils, drainage conditions, and livestock concentrations.

Another important conclusion is that P saturation cannot always be prevented by following fertilizer recommendations. Soils with a low natural fertility level require a substantial surplus of P to obtain the desired soil-P level. In addition, a small surplus is needed to control this level. This compensates for leaching losses, which are nevertheless substantially higher than what is acceptable environmentally. Therefore, in sensitive areas P concentrations in ground water and surface waters may increase (on a long term) even under good agricultural practices.

On a short term, proper manure management can significantly contribute to improving and controlling the water quality in a watershed. However, in P-saturated areas, additional measures may be needed to restore the quality of surface waters, for example additions of chemicals to the soil and hydrological measures.

REFERENCES

Breeuwsma, A., and Schoumans, O.F., Forecasting phosphate leaching from soils on a regional scale, in *Vulnerability of Soil and Ground Water to Pollutants,* Proc. and Information / TNO Committee on Hydrological Research, W. van Duijvenbooden and H.G. van Waegeningh, eds., 38, 973, 1987.

Breeuwsma, A., and Silva, S., *Phosphorus Fertilization and Environmental Effects in the Netherlands and the Po Region (Italy),* Wageningen, The Netherlands, DLO Winand Staring Centre for Integrated Land, Soil and Water Research, Report 57, 1992.

Breeuwsma, A., Reijerink, J.G.A., and Schoumans, O.F., *Phosphate Saturated Soils in the Eastern, Central and Southern Sand Districts* (in Dutch), DLO Winand Staring Centre for Integrated Land, Soil and Water Res., Rpt. 68, Wageningen, The Netherlands, 1990.

Breeuwsma, A., Reijerink, J.G.A., Schoumans, O.F., Brus, D.J., and van het Loo, H., *Phosphate Loads on Soil, Ground and Surface Waters in the Catchment Area of the Schuitenbeek* (in Dutch), Wageningen, DLO Winand Staring Centre for Integrated Land, Soil and Water Research, Rapport 10, Wageningen, The Netherlands, 1989.

Brus, D.J., *Incorporating Models of Spatial Variation in Sampling Strategies for Soil,* PhD thesis, Agricultural University, Wageningen,The Netherlands, 1993.

Brus, D.J., de Gruijter, J.J., and Breeuwsma, A., Strategies for updating soil survey information: A case study to estimate phosphate sorption characteristics, *J. Soil Sci.,* 43, 567, 1992.

Chesters, G., Stiefel, R., Bahr, T., Robinson, J., Ostry, R., Coote, D.R., and Whitt, D.M., *Pilot Watershed Studies: Summary Report,* International reference group on Great Lakes pollution from land use activities, Great Lakes Reg. Off., Ontario, Canada, 1978.

de Smet J., van Meirvenne, M., Scheldeman, K., Baert, L., Hofman, G., Vanderdeelen, J., Lookman, R., Schoeters, I., Vlassak, K., and Merckx, R., The P status of the soil in Flemish areas with concentrations of livestock production, 1, Sampling strategies and data collection, in *Phosphate Problems in Agriculture* (in Dutch), Technologisch Instituut, Kon. Vlaamse Ingenieurs Vereniging, Antwerpen, Belgium, 1994.

Gerritse, R.G., and Zugec, I., The phosphorus cycle in pig slurry measured from $^{32}PO_4$ distribution ratio, *J. Agric. Sci. Camb.,* 88, 101, 1977.

Hack-ten Broeke, M.J.D., Kleijer, H., Breeuwsma, A., Reijerink J.G.A., and Brus, D.J., *Phosphate Saturation of the Soils in Two Areas in the Province of Overijssel* (in Dutch), DLO Winand Staring Centre for Integrated Land, Soil and Water Research, Rapport 108, Wageningen, The Netherlands, 1990.

Henkens, P.L.C.M., Fertilizer recommendations for attaining and controlling soil P and K levels (in Dutch), *Bedrijfsontwikkeling* 15, 969, 1984.

Kroes, J.G., Roest, C.W.J., Rijtema, P.E., and Locht, L.J., *The Effect of Fertilization Scenarios on Nitrogen and Phosphorus Loads to Surface Waters in the Netherlands* (in Dutch), DLO Winand Staring Centre for Integrated Land, Soil and Water Research, Rapport 55, Wageningen, The Netherlands, 1990.

Lexmond, Th.M., van Riemsdijk, W.H., and de Haan, F.A.M., *Phosphate and copper in soils of areas with intensive livestock production* (in Dutch), Bodembeschermingsreeks nr. 9, Ministry of Housing, Physical Planning and Environment, Leidschendam, The Netherlands, 1982.

Lookman, R., Schoeters, I., Merckx, R., Vlassak, K., de Smet, J., Scheldeman, K., van Meirvenne, M., Vanderdeelen, J., Hofman, G., and Baert, L., The P status of the soil in Flemish areas with concentrations of livestock production, 2, Results, in *Phosphate Problems in Agriculture* (in Dutch), Technologisch Instituut, Kon. Vlaamse Ingenieurs Vereniging, Antwerpen, Belgium, 1994.

Oenema, O., and Van Dijk, T.A., eds., *P Losses and Surpluses in Dutch Agriculture, Report of the Technical P Desk Study* (in Dutch), Ministry of Agriculture, Nature and Fisheries, The Hague, 1994.

Paauw, F. van der, An effective water extraction method for the determination of plant-available soil phosphorus, *Plant and Soil,* 34, 467, 1971.

Reijerink, J.G.A., and Breeuwsma, A., *Spatial Distribution of Phosphate Saturated Soils in Areas with Manure Surpluses* (in Dutch), DLO Winand Staring Centre for Integrated Land, Soil and Water Research, Rapport 222, Wageningen, The Netherlands, 1992.

Schoumans, O.F., Breeuwsma, A., and de Vries, W., Use of soil survey information for assessing the phosphate sorption capacity of heavily manured soils, in *Vulnerability of Soil and Groundwater to Pollutants,* Proceedings and Information/TNO Committee on Hydrological Research, W. van Duijvenbooden and H.G. van Waegeningh, eds., 38, 1079, 1987.

Sharpley, A.N., Chapra, S.C., Wedepohl, R., Sims, J.T., Daniel, T.C., and Reddy, K.R., Managing agricultural phosphorus for protection of surface waters: Issues and options, *J. Environ. Qual.,* 23, 437, 1994.

Sharpley, A.N., Meisinger, J., Breeuwsma, A., Sims, J.T., Daniel, T., and Schepers, J.S., Impacts of annual manure management on ground and surface water quality, in *Effective Management of Animal Waste as a Soil Resource,* Proc. Workshop, Kansas City, Missouri, 1995.

Sherwood, M., Runoff of nutrients following landspreading of slurry, in *Environmental Impact of Landspreading of Wastes,* Proc. Seminar Johnstown Castle Centre, Wexford, Ireland, 111, 1990.

Sissingh, H.A., Analytical technique of the P_w method, used for the assessment of phosphate status of arable soils in the Netherlands, *Plant and Soil,* 34, 483, 1971.

Van der Zee, S.E.A.T.M., *Transport of Reactive Contaminants in Heterogeneous Soil Systems,* PhD thesis, Agricultural University, Wageningen, The Netherlands, 1988.

Van der Zee, S.E.A.T.M., and Gjaltema, A., Simulation of phosphate transport in soil colums, 1, Model development, *Geoderma,* 52, 87, 1992.

Van der Zee, S.E.A.T.M., van Riemsdijk, W.H., and de Haan, F.A.M., *Protocol for Phosphate Saturated Soils* (in Dutch), Department of Soil Science and Plant Nutrition, Agricultural University, Wageningen, The Netherlands, 1990.

Young, R.A., and Mutchler, C.K., Pollution potential of manure spread on frozen ground, *J. Environ. Qual.,* 5, 174, 1976.

NITRATE-NITROGEN LEVELS IN WELL WATER IN THE LITTLE RIVER/ROOTY CREEK AGRICULTURAL NON-POINT SOURCE HYDROLOGICAL UNIT AREA

M. Charles Gould

INTRODUCTION

The Little River Rooty Creek Agricultural Non-point Source Hydrological Unit Area is one of 74 five-year federal water quality projects across the nation. The overall purpose of the water quality project is to increase the voluntary farmer adoption of best management practices (BMPs) that will protect and improve surface and ground water quality while maintaining agricultural productivity and profitability.

BMPs are conservation techniques used to control or prevent agriculturally caused, non-point source pollution. BMPs implemented in this water quality project include sediment retention ponds, terracing, permanent pasture, nutrient management plans for manure and commercial fertilizer application, composting poultry mortality and litter, and pump-out of animal waste lagoons. Implementing BMPs specific to each farm's water quality problem(s) is expected to achieve a 65-75% reduction in agriculturally caused non-point source pollution at the completion of the project.

The 99,912 ha Little River/Rooty Creek watershed includes portions of Jasper, Morgan, Newton, Putnam, and Walton Counties in the Piedmont region of east-central Georgia. There is an estimated farm population of 790 and a rural non-farm population of 6,020.

Morgan and Putnam Counties lead the state in numbers of dairy cattle with a combined total of over 18,000 head. Morgan County leads the state in total number of cattle and calves with over 30,000 head (Georgia Agriculture Statistics Service, 1992). Within the watershed there are 80 dairies, 70 beef cattle farms, and over 3 million chickens and turkeys. These animals excrete an estimated 386,654 metric tons of manure annually (HUA Work Plan, 1991).

There is an estimated 5,123 ha of cropland in the Little River/Rooty Creek basin. The major agricultural crops grown include corn (*Zea mays* L.) for silage, cotton (*Gossypium hirsutum* L.), bermudagrass (*Cynodon dactylon* (L) Pers.) for pasture and hay production, wheat (*Triticum aestivum* L.), rye (*Secale cereale* L.), and ryegrass (*Lolium multiflorum* Lam.).

Surface and ground water contamination from agricultural activities is a concern in the project area. The watershed's streams are listed in Georgia's Non-point Assessment and Management Plan as being threatened to meet their designated "fishing stream" classification (Environmental Protection Division, 1989). The primary surface water problems across the watershed are caused by sediment and excess nutrients (nitrogen and phosphorus) (US Environmental Protection Agency, 1991).

At the onset of the project, the extent of ground water contamination from agriculture was not known. Very few homeowners throughout the project area routinely have mineral tests (including nitrate) performed on their water, so no base data on ground water quality existed. Furthermore, five areas totaling approximately 17,658 ha are significant recharge areas for localized aquifers. All of these recharge areas contain some potential agricultural pollutants. The majority of rural residents throughout the project depend on ground water for their domestic needs.

Excess nitrate-nitrogen (nitrate-N) in well water samples is an indication that animal waste, commercial fertilizer, or human waste is leaching into the ground water. Nitrate-N at levels above 3 mg/L are considered introduced by human activity (Nielson and Lee, 1987). The US Environmental Protection Agency (US EPA) has set the maximum contaminant level for nitrate-N in drinking water at 10 mg/L. When infants less than six months old and the very elderly drink water with levels of nitrate-N greater than 10 mg/L, methemoglobinemia or "blue-baby syndrome" may occur. Simply put, vital tissues such as the brain receive blood with less oxygen than normal. This may cause brain damage or even death (Nugent et al., 1988)

METHODS

One of the objectives of the water quality project is to increase land owner knowledge of the effect of agricultural activities on potable water quality. To achieve this objective, a goal was set to analyze 25% of participating farmers' private potable water supply for nitrate-N.

From 1 June 1991 through 30 September 1993, 236 well samples were taken in the project area and analyzed for nitrate-N. Sixty-four well samples came from non-farm wells. The remaining 172 samples represent 88% of the dairies and 58% of poultry operations in the project area. Also included are well samples from swine and beef operations.

Sampling procedures outlined by Tyson and Harrison (1989) were followed to ensure consistency. The water system at each well was purged to eliminate standing water and ensure that the sample taken was fresh ground water. The sampling site for each well was the spigot located closest to the well head.

RESULTS AND DISCUSSION

Well test results indicate that, as a whole, the ground water in the project area is relatively free of serious nitrate-N contamination (Table 1). However, there are "hot spots" of nitrate-N contamination in ground water throughout the project area.

These results are consistent with other major ground water quality studies conducted on aquifers across the United States by the US Geologic Survey (1988) and the US EPA (1990). These studies found that the majority of principal aquifers sampled had median nitrate-N levels well below 10 mg/L (US Department of Agriculture, 1991). However, the studies identified "hot spots" within some aquifers where nitrate-N levels exceeded the EPA's maximum contaminant level. Some of this contamination has been linked to agricultural activity.

Likewise, some of the nitrate-N contamination of ground water in the project area can be linked to agricultural activity. Twenty-eight of the 30 well samples in Table 1 exceeding US EPA's nitrate-N standard originated from farms. Two samples came from a poultry farm, while the remaining 26 came from dairy farms. A study of nine of these dairies to determine the sources of nitrate-N in well water found that unpaved loafing areas, rather than animal waste lagoons or septic systems, were the most likely source of ground water nitrate-N contamination (Drommerhausen et al., 1994).

The Piedmont region of Georgia is characterized as having deep clay soil on top of granite rock aquifers. The average depth of the clay soil is 18 m (Lineback, 1991). Wells dug or bored in the clay less than 18 m, not in the aquifer, are considered shallow wells. Typically these wells have a casing diameter of either 609 mm or 914 mm. Deep wells, on the other hand, are greater than 18 m deep and are drilled into the aquifer. These wells generally have 152-mm-diameter well casings. On the average, wells in the Piedmont flow between 3.7 - 37.8 L/minute. This is very slow compared to South Georgia wells that can flow over 7570 L/minute.

Of the 236 well samples taken in the project area, 110 (47%) came from wells greater than 18 m deep and had less than 10 mg/L nitrate-N concentration (Table 2). Thirty-four well samples (14%) were 18 m deep or less and had less than 10 mg/L nitrate-N concentration. It should be pointed out that 62 (26%) samples were taken from wells whose depth was not known. Therefore, 206 (87%) of the 236 well samples had nitrate-N concentrations of less than 10 mg/L.

Table 1. Nitrate-N levels of 236 wells sampled in The Little River/Rooty Creek Water Quality Project.

Wells	Nitrate-N levels-mg/L		
	< 3.0	3.01-9.99	> 10.0
Number (% of total)	145 (61%)	61 (26%)	30 (13%)

Table 2. The number of well samples with less than 10 mg/L nitrate-N
and corresponding well depths.

Samples	Well depth-meters		Unknown	Total
	< 18	> 18		
Number	34	110	62	206
(Percent of total)	(14)	(47)	(26)	(87)

Table 3. Farm and non-farm shallow and deep well samples
exceeding 10 mg/L nitrate-N level.

Wells	Well depth-meters		
	<18	18-61	>61
Non-farm	1	1	0
Poultry	2	0	0
Dairy	4	7	15
Total (% of total)	7 (23%)	8 (27%)	15 (50%)

The majority of wells with nitrate-N levels at or above 10 mg/L are deep wells (Table 3). These wells are on dairy farms. Visual inspection of these wells reveal they are in close proximity to cattle loafing areas or are influenced by surface water runoff. Upon questioning farmers about how old these wells are, each one indicated they were installed pre-1985 (before the Georgia legislature passed the "Water Well Standards Act of 1985"). Therefore, it may be that these wells were not properly installed, or because of age the casings have cracked, allowing nitrate-N contaminated water to mix with ground water.

CONCLUSION

Although there is a high animal density in the watershed, generally, the ground water is free of widespread nitrate-N contamination. There are, however, "hot spots" throughout the watershed where well water exceeds the maximum contaminant level for nitrate-N of 10 mg/L set by the US EPA. These wells are deep wells, found on dairy farms, and were installed before laws were passed in Georgia that regulated well installation. Visual inspection of these wells revealed that they are being influenced by surface water runoff from cattle loafing areas and, in some instances, septic effluent. Some wells are suspected as being contaminated from both sources.

This points to the need for an effective wellhead protection program. It is recommended that farmers with high nitrate-N in their well water first complete the Farm*A*Syst worksheets and follow the recommended wellhead protection measures. In addition to the worksheets, BMPs that control and eliminate ground water contamination are recommended. These practices would include using a nutrient management plan to spread manure and commercial fertilizer, rotational grazing, annual well testing, diverting surface water away from wells, and installing heavy-use fabric ("cow carpet") in heavily-used areas around cattle feeders, waterers, and loafing areas. Cost share money is available to farmers on some of these BMPs from Consolidated Farm Services Agency.

Follow up well water testing is planned during the 1994-95 year as farms with wells with high nitrate-N complete the Farm*A*Syst worksheets and/or implement appropriate BMPs.

REFERENCES

Drommerhausen, D.J., Radcliff, D.E., Brune, D.E., and Gunther, H.D,. *Assessing Dairy Lagoon Seepage Using Ground Electromagnetic Conductivity*, ERC 01-94, Department of Crop and Soil Sciences, The University of Georgia, Athens, Georgia, 1994.

Environmental Protection Division, *Georgia Nonpoint Source Assessment Report*, Georgia Department of Natural Resources, Atlanta, Georgia, September 1989.

Georgia Agricultural Statistics Service, *Georgia Agricultural Facts,* Athens, Georgia, 1992

HUA Work Plan, USDA ASCS, SCS, UGA CES, and Georgia SWCC, Athens, Georgia, 1991.

Lineback, Jerry, Personal Communication, 1991.

Nielson, E.G., and Lee, L.K., *The Magnitude and Costs of Groundwater Contamination From Agriculture Chemicals*, Staff report AGES87, Economic Research Service, US Dept. of Agriculture, Washington, D.C., 1987.

Nugent, M., Kamrin M., Wolfson, L, and D'Itri, F. M., *Community Assistance Program in Environmental Toxicology*, Center for Environmental Toxicology and The Institute of Water Research, Michigan State University, 1988.

Tyson, A., and Harrison, K., *Water Quality for Private Water Systems*, Bulletin 939, Cooperative Extension Service, The University of Georgia College Of Agriculture, Athens, Georgia, 1989.

US Department of Agriculture, *Nitrate Occurrence in U. S. Waters (and Related Questions), A Reference Summary of Published Sources From An Agricultural Perspective*, USDA Working Group on Water Quality, Washington, D.C. , 1991.

US Environmental Protection Agency, *The Quality Of Our Nations Water: A Summary of the 1988 National Water Quality Inventory*, Washington, D.C., 1990.

US Environmental Protection Agency, *1991 Water Quality Monitoring Results: Rooty Creek Watershed*, Athens, Georgia, 1991.

US Geological Survey, *State Summaries of Ground Water Quality, National Water Quality Summary 1986*, Water Supply Paper 2325, US Department of The Interior, Washington, D.C., 1988.

IMPACT OF ANIMAL WASTE ON WATER QUALITY IN AN EASTERN COASTAL PLAIN WATERSHED

P.G. Hunt, K.C. Stone, F.J. Humenik, and J.M. Rice

INTRODUCTION

As part of the USDA's Presidential Water Quality Initiative, a five-year water quality demonstration project was initiated in 1990 on a watershed located on the Cape Fear River Basin in Duplin County, North Carolina (Stone et al., 1995). Streams of the basin have been officially designated environmentally sensitive and are, thus, subject to the highest applicable North Carolina water quality standards. The 2,044-ha demonstration watershed, Herrings Marsh Run (HMR), has many characteristics typical of an intensive agricultural area in the eastern Coastal Plain of the USA (Hubbard and Sheridan, 1989). These include 1) intensive crop and animal production, 2) shallow ground waters that are used for drinking water, and 3) close connections of shallow ground waters, streams, and sensitive environmental areas.

Annual nutrient usage for crop production in the watershed is estimated at 145 metric tons of nitrogen (N), 64 metric tons of phosphorus (P), and 243 metric tons of potassium (Stone et al., 1994). Although swine and poultry operations produce sufficient quantities of waste to supply 62% of the N and 96% of the P, 90% of the nutrients applied to cropland are supplied by commercial fertilizers. Animal waste is applied as solid and liquid to both row and forage crops. The application of large quantities of commercial fertilizers coupled with the production of large quantities of animal waste provides a potential for N and P contamination of surface and ground water. The initial phase of the project evaluated the effect of existing agricultural management practices on stream and ground water quality within the watershed. The second phase of the project evaluated the impact of management changes and landscape modifications on water quality.

METHODS

WATERSHED CHARACTERISTICS

Agricultural practices in the watershed include 1,093 ha of cropland, 708 ha of woodlands, and 212 ha of farmsteads, poultry facilities, and swine facilities. The major agricultural crops on the watershed include corn (*Zea mays L.*)

415 ha, soybeans [*Glycine max* (L.) Merr.] 273 ha, tobacco (*Nicotiana tabacum* L.) 131 ha, wheat (*Triticum aestivum* L.) 121 ha, and vegetables 162 ha. The primary soil series in the watershed is Autryville (Loamy, siliceous, thermic Arenic Paleudults); secondary soil series are Norfolk (Fine-loamy, siliceous, thermic Typic Kandiudults), Marvyn-Gritney (Claey, mixed, thermic Typic Hapludults), and Blanton (Loamy siliceous, thermic Grossarenic Paleudults).

MASS BALANCE

A simple N and P mass balance was calculated using results of farm surveys (personal communication, S.W. Coffey, North Carolina State University, Raleigh, North Carolina, 1994). Estimates of land applications of N and P were made using conversion factors typical of the swine and poultry industry of North Carolina (Barker and Zublena, 1995); N and P inputs (fertilizer, livestock waste, legume residual) and outputs (crop and residue) were recorded in the farm surveys. The differences between the inputs and outputs were calculated, and values of excess N and P were obtained. The excess N and P as estimated from the surveys were then compared with the N and ortho-phosphate-P loading rates observed at the stream sampling stations.

STREAM WATER EVALUATION

Stream water sampling stations were established within the watershed in cooperation with the US Geological Survey during 1990 (Figure 1) (Stone et

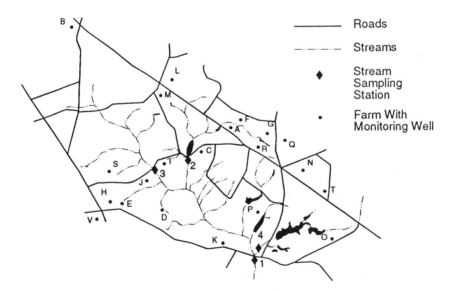

Figure 1. Location of stream sampling stations and farms with ground water monitoring wells on Herrings Marsh Run watershed.

al., 1995). Sub-watershed outlets are indicated in Figure 1 at the diamond-shaped stream sampling stations 1-4. Stations were located on tributaries as well as at the upper reach and at the watershed outlet of HMR. The upper reach of HMR (sampler 3) had extensive riparian buffer zones and relatively little poultry and swine production activities (257 ha of the total 537 ha were forested). The upper reach tributary (sampler 2) had extensive swine and crop production, but less extensive riparian buffer zones (146 ha of the total 425 ha were forested). Stream water samples were taken with auto samplers, and flow was measured at stream sampling stations as described in Stone et al. (1995). All water samples were acidified upon sampling and transported to the USDA-ARS research center in Florence, South Carolina, for analyses. Water samples were analyzed for nitrate-N, ammonium-N, total Kjeldahl-N, ortho-P, and total P using US EPA Methods 353.2, 350.1, 351.2, 365.1, and 365.4, respectively (US EPA, 1983). A TRAACS 800[1] Auto-Analyzer was used, and EPA-certified quality control samples were routinely analyzed to verify results.

GROUND WATER EVALUATION

Ground water monitoring wells were installed using a SIMCO 2800 trailer-mounted drill rig on 24 farms in the HMR watershed from 1991 through 1994 (Figure 1). Well bottoms were positioned on an impermeable layer or at a depth of 7.6 m if the impermeable soil layer could not be located above that depth (Stone et al., 1994). Water table depths in the watershed were generally 1.5 to 3 m below the soil surface. The farms exemplify the agricultural practices used in the watershed. The majority of the farms with monitoring wells were in row crops. There was a mix of farms with and without implemented nutrient management plans. Poultry litter and compost were the main sources of N on two row crop farms. Several farms applied swine lagoon effluent to pasture for hay production.

RIPARIAN ZONES

An overcut riparian zone of approximately 1.2 ha was contiguous to a 3.7-ha Coastal bermudagrass (*Cynodon dactylon* L.) field that was used for swine wastewater treatment. Approximately 2.2 and 1.0 Mg/ha/year of N and P, respectively, were applied in swine waste water. The portion of the field used for wastewater treatment was initially too small (about 1.1 ha), and overloading had occurred. During our study the treatment field was expanded, and hardwood trees were planted in the riparian zone. Starting from the field, the species planted were green ash (*Fraxinus pennsylvanica* Marsh.), red maple (*Acer rubrum* L.), sycamore (*Platanus occidentalis* L.), water oak (*Quercus nigra* L.), and cypress (*Taxodium distichum* (L.) Rich.). Ground water wells were

---

[1]Mention of trademark, proprietary product, or vendor does not constitute a guarantee or warranty of the product by the US Department of Agriculture and does not imply its approval to the exclusion of other products or vendors that may also be suitable.

established in the treatment field and in the riparian zone. Point in time "grab" samples were collected in the stream contiguous to two of the wells in the riparian zone. Denitrification potential in the soil of the spray field and riparian zone was assessed by use of the acetylene blockage method (Smith et al., 1978).

IN-STREAM WETLAND EVALUATION

The stream from sub-watershed 2 flowed through a wetland that existed in the bottom of an old pond with a breached dam. Beavers built a dam across the old breach in April 1993. The dam initially suffered substantial sidewall erosion, and it was necessary to reinforce the sidewalls and direct the flow over the center of the dam. The expanded wetland area upstream of station 2 was about 3.3 ha when the water level at the dam was approximately 3.1 m above the bottom of the stream bed. Stream water sampling stations have been installed upstream of the wetland. All statistical analysis of the data was accomplished using SAS version 6.07 (SAS, 1990).

RESULTS AND DISCUSSION

MASS BALANCE

The estimated total N and P applied as animal waste in sub-watershed 2 (425 ha) were 22.7 and 16.7 Mg/year, respectively (Table 1). Sub-watershed 3 (537 ha) had only swine waste, with 5.8 Mg N and 2.7 Mg P applied each year. The total watershed minus sub-watersheds 2 and 3 (1082 ha) received an estimated 47.4 Mg N and 42.0 Mg P in animal waste each year. The impact of animal waste on sub-watershed 2 was greater than on sub-watershed 3 or the whole watershed; the N and P would have averaged 53 and 39 kg/ha/year, respectively, if applied evenly to sub-watershed 2. A uniform application rate of animal waste in sub-watershed 3 would have resulted in 14 kg/ha N and 5 kg/ha P. Stone et al. (1995) estimated that the excess N applied to crops as animal waste and fertilizers was 85, 55, and 38 kg/ha/year for sub-watershed 2, sub-watershed 3, and the total watershed, respectively (Table 2). Excess P applications were estimated to be 57, 15, and 30 kg/ha/year in sub-watershed 2, sub-watershed 3, and the total watershed, respectively.

STREAM QUALITY

The geometric mean (antilog of the mean of the log of the data) of nitrate-N concentrations in the tributary from sub-watershed 2, the HMR from sub-watershed 3, and the HMR watershed outlet at station-1 were 5.4, 1.2, and 2.0 mg/L, respectively, from September 1991 to June 1993 (Table 3). Ammonia-N concentrations had the same relationship, but they were much smaller than nitrate-N concentrations. The stream flow from the upstream stations was less than one third the 0.147 m^3/s of the HMR watershed outlet. The mass flux of N from sub-watershed 3 was 4 kg/day. On the other hand, the N flux from sub-

Table 1. Nitrogen and phosphorus from manure
in the Herrings Marsh Run (HMR) watershed.

Waste Source	Location	N	P
		---------------Mg/yr---------------	
Swine	Sub-watershed 2	10.6	5.0
Poultry		12.1	11.7
Total		22.7	16.7
Swine	Sub-watershed 3	5.8	2.7
Poultry		0.0	0.0
Total		5.8	2.7
Swine	Main Watershed Minus Sub-watershed 2 & 3	11.4	5.4
Poultry		36.0	36.6
Total		47.4	42.0
Grand Total		75.9	61.4

Table 2. Excess applied nitrogen (N) and phosphorus (P) and stream loadings for the
Herrings Marsh Run watershed (modified from Stone et al., 1995).

	Sampling Station[†]		
Nutrient Loading	1	2	3
	-------------------kg/ha/year-------------------		
Excess N Applied	38	85	55
Stream N Loading	3	22	6
Excess N % Loss in Stream	8%	26%	11%
Excess P Applied	20	57	15
Stream Ortho-phosphate Loading	0.5	0.9	0.4
Excess P % Loss in Stream	3%	2%	3%

[†]Station 1 is located at the watershed outlet. Station 2 is the Herrings Marsh Run tributary.
Station 3 is the Herrings Marsh Run main and is used as a background reference.

Table 3. Mean daily nutrient concentrations and mass fluxes over the sampling
period for three stream monitoring stations in the Herrings Marsh Run watershed in
Duplin County, North Carolina (modified from Stone et al., 1994).

	Sampling Stations		
Nutrient Discharge	1	2	3
Concentration	----------------------mg/L----------------------		
Nitrate	2.01	5.34	1.18
Ammonia	0.15	0.42	0.08
o-phosphate	0.14	0.15	0.06
Mass Flux	---------------------kg/day---------------------		
Nitrate	22.17	19.61	3.56
Ammonia	2.08	1.34	0.28
o-phosphate	2.24	0.57	0.17
	----------------------m^3/s----------------------		
Stream Flow	0.147	0.041	0.034

watershed 2 (21 kg/day) was nearly as large as that from the watershed outlet (24 kg/day). Ortho-phosphate-P concentrations in the streams were generally less than 0.2 mg/L, and the mass flow differences among the streams resulting from the small concentration differences were of little environmental significance.

The impact of N and P inputs upon stream water quality was very different. Most of the P was bound by the soil, and very little of the excess (less than 3%) reached the streams. Nitrogen was much more mobile, and over 25% of the estimated excesses in sub-watershed 2 were lost in stream flow, predominantly as nitrate.

GROUND WATER QUALITY

Ortho-phosphate-P concentrations were less than 0.05 mg/L in all wells, and nitrate-N was less than 10.0 mg/L in wells on 19 of the 24 farms (Stone et al., 1994). In the five farms with wells that exceeded 10 mg/L of nitrate-N (Figure 1; farms A, B, C, F, and R), only farm A had wells that exceeded 20 mg/L nitrate-N. Farm A had mean concentrations of 20 and 83 mg/L of ammonia-N and nitrate-N, respectively, in wells in a bermudagrass field that had been overloaded with swine waste water prior to the Water Quality Demonstration Project. The waste water spray field has been expanded in area, but the ground water quality has not yet improved. It is anticipated that lower wastewater application rates, denitrification, and Coastal bermuda hay uptake of N will reclaim the site in time. Three of the other four high-nitrate-N farms were also located in sub-watershed 2. Thus, stream and ground water nitrate-N levels were highest in the portion of the watershed with the highest level of animal waste production.

HARDWOOD RIPARIAN ZONE

First year growth of the planted hardwood trees has been excellent in the reestablished riparian zone contiguous to the swine waste water disposal field of farm A. The trees will soon constitute a significant sink for the N that moves to the stream. Initial denitrification enzyme analyses indicated that the riparian zone had significant denitrification potential, particularly near the creek. The creek had mean ammonia-N and nitrate-N concentrations of 4.1 and 8.7 mg/L, respectively, substantially lower than the 13 and 59 mg/L of ammonia-N and nitrate-N, respectively, in the shallow ground water that flowed through the riparian zone. Thus, significant nitrate-N and ammonia-N were removed, but significant amounts also moved from the waste water disposal site through the riparian zone to the stream. In such instances, some form of in-stream treatment is desirable, and the need for stream clean-up suggested that an in-stream wetland would be desirable.

IN-STREAM WETLAND

Prior to the establishment of the 3.2-ha, in-stream wetland, the mean nitrate concentration from sub-watershed 2 at sampler 2 was about 5.5 mg/L, and the mean ammonia-N concentration was 0.42 mg/L (Table 2). The concentration of nitrate-N in the water entering the wetland a few hundred meters upstream of sampler 2 remained about the same as the pre-wetland concentrations (Table 4). After the wetland was established, the concentration of nitrate-N in the stream leaving the wetland at station 2 was 1 mg/L or less in the warmer months. However, the wetland was less effective in N removal during the cooler months. This seasonal effect was probably due to less plant growth and slower denitrification in the cooler months. Ammonia-N and ortho-phosphate-P were not altered greatly. They were generally less than 0.4 and 0.2 mg/L, respectively, before and after flow through the wetland.

SUMMARY

Nitrate-N in stream and ground water was highest in the portion of the HMR watershed with the highest concentration of swine and poultry production. Four of the five farms with high nitrate-N were in the sub-watershed with highest swine and poultry production density. However, only five of the 24 tested farms had mean ground water nitrate-N concentrations in excess of 10 mg/L, and only one of the farms had mean nitrate-N concentrations in excess of 20 mg/L. Ortho-phosphate-P in streams and ground waters was affected very little by animal waste applications, even when in close association with a heavily overloaded waste disposal site.

A riparian zone removed substantial amounts of nitrate-N and ammonia-N from the shallow ground water of an overloaded swine waste water disposal site. An in-stream wetland was very effective in the removal of nitrate-N from a stream that had nitrate-N in excess of 5 mg/L. Development and use of adequate nutrient management plans along with the creation, enhancement, and preservation of riparian and wetland landscape features offer the opportunity to minimize the adverse effects of animal waste disposal in the eastern Coastal Plain.

Table 4. Mean nitrate-N in stream water of an in-stream wetland in sub-watershed 2.

Flow	Fall 93	Time Period Winter 94	Spring 94
	------------------------mg/L------------------------		
In-flow	7.0 (1.1)	6.7 (1.1)	6.3 (1.4)
Out-flow	0.8 (1.0)	4.9 (0.9)	1.0 (1.1)

REFERENCES

Barker, J.C., and Zublena, J.P., *North Carolina Livestock Manure Production Characteristics and Nutrient Assessment,* Environmental, Biological, and Agricultural Engineering, North Carolina State Univ., Raleigh, North Carolina, manuscript, 1995 (*In press*).

Hubbard, R.K., and Sheridan, J.M., Nitrate movement to ground water in the southeastern Coastal Plain, *J. Soil and Water Cons.*, 44, 20, 1989.

Statistical Analysis System Institute, Inc., SAS©, Language: Reference, Version 6, First Edition, Cary, North Carolina, 1042, 1990.

Smith, S.M., Firestone, M.K., and Tiedje, J.M., The acetylene inhibition method for short-term measurement of soil denitrification and its evaluation using nitrogen-13, *Soil Sci. Soc. Am. J.*, 42, 611, 1978.

Stone, K.C., Hunt, P.G., Coffey, S.W., and Matheny, T.A., Water quality status of a USDA water quality demonstration project in the eastern Coastal Plain, *J. Soil and Water Cons.*, 1995 (*In Press*).

Stone, K.C., Hunt, P.G., Novak, J.M., and Matheny, T.A., Impact of BMP's on stream and ground water quality in a USDA demonstration watershed in the eastern Coastal Plain, in *Proc. of the Second Environmentally Sound Agriculture Conference*, Campbell, K.L., Graham, W.D., and Bottcher, A.B., eds., ASAE, St. Joseph, Michigan, p. 280, 1994.

US EPA. *Methods for Chemical Analysis of Water and Wastes,* US EPA-600/4-79-020, Kopp, J.F., and McKee, G.D., eds., Environmental Monitoring and Support Lab. Office of Research and Development. Cincinnati, Ohio, 1983.

NUTRIENT CYCLING FROM CATTLE FEEDLOT MANURE AND COMPOSTED MANURE APPLIED TO SOUTHERN HIGH PLAINS DRYLANDS

Ordie R. Jones, William M. Willis,
Samuel J. Smith, and Bobby A. Stewart

INTRODUCTION

Each of the 5.6 million head of cattle fed annually in confined operations on the Southern High Plains produces about one ton of manure. Thus disposal or utilization of feedlot wastes is a major agricultural and environmental concern (Smith et al., 1980). Stockpiled manure or composted manure is often applied to cropland. Normally, surface-applied manure is incorporated with disks or sweeps immediately upon application to prevent nitrogen (N) volatilization. However, on drylands, improved moisture management requires that most or all crop residues be maintained on the soil surface as a mulch to conserve soil water and protect against wind and water erosion. Manure application to drylands with little or no incorporation may result in increased nutrient content of rainfall runoff and increased loss of N due to volatilization.

Our research objective was to determine the nutrient cycling effects of applying cattle feedlot manure and composted manure to cultivated drylands using no-tillage management (wastes not incorporated) and stubblemulch management (wastes incorporated upon application).

MATERIALS AND METHODS

SITE CHARACTERISTICS

The research was conducted on dry-farmed cropland at the USDA Conservation and Production Research Laboratory at Bushland, Texas, in the Southern High Plains. The soil is slowly permeable Pullman clay loam with a 1% slope. Average annual precipitation is 465 mm, of which 75% occurs between 1 April and 30 September. The study was initiated in May 1993 on a level-terraced watershed that had been farmed for more than 30 years in a dryland stubblemulch-tilled (SM) three-year cropping sequence of winter wheat, grain sorghum, and fallow.

TREATMENT DESCRIPTIONS

No-tillage (NT) and stubblemulch (SM) tillage plots (12 x 36 m) were established with two replications during the fallow period between sorghum harvest and wheat planting. These plots were divided into four fertility sub-plots (9 x 12 m), which received the following treatments:

1) stockpiled manure from a commercial feedlot (manure),
2) commercially composted feedlot manure from the same feedlot from which manure was obtained (compost),
3) inorganic commercial fertilizer (18-46-0), and
4) untreated (check), (Table 1).

Fertilizer or animal waste applications were made by hand, and sub-plots on SM were immediately swept twice to incorporate. Applications on NT were not incorporated. Sub-plot borders were defined with aluminum garden edging material to control and collect runoff and prevent outside runoff from entering. Edging material was removed and replaced on SM plots when tillage or planting was performed. Weeds were controlled on NT plots with glyphosate. Winter wheat was sown on all plots using a NT drill in October 1993. Stubblemulch plots were swept prior to seeding.

SAMPLING

Runoff was measured and collected from one sub-plot in each treatment by using a runoff splitter to divert 10% of flow to a 400-L collection tank where volume was measured and samples obtained for laboratory analysis. Wheat was sampled for nutrient content and harvested in June 1994. Plots were incrementally soil sampled to a 1.5-m depth for N and P content prior to plot establishment, at wheat planting, and at wheat harvest.

LABORATORY ANALYSES

Chemical analyses of runoff samples for nitrate nitrogen (NO_3-N), ammonium nitrogen (NH_4-N), and total Kjeldahl nitrogen (TKN) were conducted using standard automated methods described in US EPA, 1979. Kjeldahl N represents primarily organic N but also includes any NH_4-N present. Soluble P

Table 1. Treatment description.

Treatment	Application Rate Mg/ha	Water Content % by weight	N %	P %	N kg/ha	P kg/ha	Notes
Check	0.0	0.0	0.0	0.0	0.0	0.0	No treatment applied
Compost	4.5	38.6	1.9	0.9	51.0	10.3	Commercially composted
Manure	13.4	51.4	0.7	0.3	48.2	8.7	Stockpiled manure
18-46-0	0.24	0.0	18.0	46.0	43.2	48.1	Granular fertilizer

(SP) was measured colormetrically on filtered runoff (Murphy and Riley, 1962). Digestion of unfiltered samples with perchloric acid determined total P (TP) (O'Connor and Syers, 1975). Bioavailable P (Bio-P) was measured by extraction of a 20-ml unfiltered sample with 180 ml of 0.1M NaOH for 17 h on an end-over-end shaker (Sharpley et al., 1991). Conductivity, pH, and dissolved solids were measured electrometrically. Sediment concentrations were determined from the total solids dried at 105°C.

Soil NO_3-N and NH_4-N were extracted using 2.0M KCl. TKN was digested with H_2SO_4 and HgO catalyst (ASA, 1982). KCl extractions and acid digestions were analyzed using automated analysis methods (Lachat Method No. 10-107-06-1-C, Method No. 10-107-04-1-C, and Method 10-107-06-2-D, respectively). Total P was determined using the previously described perchloric acid digestion.

RESULTS AND DISCUSSION

Treatment rates of application were designed to apply 40 to 50 kg/ha N (Table 1), the amount of N normally removed in grain with dryland cropping systems on dryland. Excessive available nitrogen with limited water supply results in good early-season vegetative growth but usually reduces grain yields due to moisture stress during grain filling. Thus, N application rates in excess of plant needs are not desirable on dryland except possibly for forage production.

Precipitation was near average for the 4.5-month period of fallow after the treatments were applied. More than three times as much runoff was measured from NT (5.4% of precipitation) as from SM plots (1.5% of precipitation), except for the manure treatment (Table 2). Normal runoff (10-year average), expressed as percentage of precipitation, is 4.9% for SM and 7.7% for NT (Jones et al., 1994). Runoff volume was considerably less than the long-term average due to small storm size. Most runoff occurs from large storms (> 75 mm) with high intensity. No-tillage had reduced infiltration rates and increased runoff because of soil crusting (Jones et al., 1994). With SM tillage, the soil crust was destroyed, the surface was roughened, and surface storage capacity was increased; thus greater infiltration occurred with SM management.

Sediment content (soil erosion) of runoff ranged from 35 to 218 kg for the 4.5-month period, much below the long-term annual erosion losses of 1300 and 600 kg/ha measured from SM and NT, respectively (Jones et al., 1995). As with runoff, storm size and intensity are more important in controlling erosion than is total seasonal precipitation.

With lower-than-normal runoff and erosion losses, the total amounts of nutrients (N and P) lost in runoff were very low for all treatments on both tillage systems (Table 2). Total loss of N or P was < 0.3 kg/ha for all treatments, regardless of tillage system. The higher losses of NO_3-N (0.37 kg/ha) and WS-P (0.19 kg/ha) were observed from the commercial fertilizer treatment

Table 2. Precipitation, runoff, and nutrient losses in runoff from May to September 1993.

Tillage	Treatment	Precip.	Runoff	Sediment	TKN	NO$_3$-N	NH$_4$-N	Total P	WS-P	Bio-P
		mm	mm	--kg/ha--						
SM†	Check	284	4.2	123.8	-	0.02	0.03	-	0.01	-
SM	Compost	284	4.8	217.6	0.09	0.01	0.03	0.03	0.02	0.01
SM	Manure	284	4.4	34.7	0.12	0.01	0.12	0.02	0.04	0.01
SM	18-46-0	284	4.2	61.5	-	0.02	0.01	-	0.01	-
NT	Check	284	17.8	125.8	0.08	0.02	0.11	0.03	0.04	0.01
NT	Compost	284	18.8	109.8	0.10	0.03	0.15	0.04	0.13	0.02
NT	Manure	284	6.7	140.6	0.12	0.02	0.06	0.04	0.06	0.02
NT	18-46-0	284	18.8	217.5	0.11	0.27	0.11	0.05	0.19	0.03

†SM=stubblemulch tilled; NT=no tillage.

on NT, indicating that for most efficient use of fertilizer it should be applied sub-surface when possible.

Concentrations of nitrate-N in runoff water were < 1.5 mg/L (Table 3), which is considered low. Ammonium-N was generally above the 0.5 mg/L level considered safe for human consumption. However, only runoff from the manure treatment had NH_4-N concentrations considerably greater than those measured on the check. Soluble P concentrations in runoff from manure treatment were also three times greater than from the check (Table 3). Soluble P concentrations ranged from 0.21 to 1.13 mg/L. Phosphorus concentrations > 0.05 mg/L are sufficient to promote increased aquatic activity. While the P concentrations in runoff that we measured are considered high, concentrations are in the same range as measured for 10 years from similarly cropped but unfertilized watersheds on an adjacent experiment (Jones et al., 1995). It is apparent that addition of manure resulted in increased concentrations of P in runoff from both NT and SM management. Electrical conductivity and pH measurements made on runoff are indicative of high-quality water.

Treatment and tillage effects on soil N and P concentrations at the end of fallow are shown in Table 4 for the 0- to 15-cm soil depth. The 18-46-0 treatment had significantly greater concentrations than the check treatment for NO_3-N and TP and was also greater than the manure treatment for TP. Nitrate-N levels tended to be greater for animal waste treatments than for the check, indicating that some accumulation of nitrate was occurring in the soil in response to treatment applications. Phosphorus concentrations were greatest on the 18-46-0 treatment because two to five times more P was added with fertilizer than was added with compost or manure. The high natural fertility of the Pullman soil and the nutrient mineralization process made it difficult to account for nutrients added to the soil with treatments.

There was no significant difference in any reported nutrient variable in the 0- to 15-cm soil depth due to tillage. Highest wheat grain and dry-matter yields were obtained with 18-46-0 (Table 5), probably due to favorable early-season nutrient supply, which enhanced tillering. This accounted for a greater number of heads/m^2, which is usually the dominant yield component. Improved plant stands and greater tillering with NT also resulted in significantly greater grain and dry-matter yields in comparison to SM.

CONCLUSIONS

The relatively small amounts of nutrients added to drylands via animal waste or fertilizer treatments did not appear to degrade surface water quality, even with unincorporated surface application. The amount of nutrients lost in runoff was low (losses <0.15 kg/ha N and <0.2 kg/ha P), and nitrogen concentrations were low. Phosphorus concentrations in runoff exceeded limits required for increased aquatic activity, but were in the same range as P concentrations measured in runoff from adjacent unfertilized watersheds.

Table 3. Runoff weighted concentrations of nitrogen (N) and phosphorus (P) in storm runoff, May to September 1993.

Tillage	Treatment	TKN	NO_3-N	NH_4-N	Total P	WS-P	Bio-P	pH	Cond.
				----mg/L----					mS/m
SM†	Check	-	0.5	0.8	-	0.33	-	7.9	4.6
SM	Compost	1.9	0.2	0.6	0.57	0.34	0.22	7.9	4.7
SM	Manure	2.7	0.3	2.6	0.50	1.01	0.32	7.7	6.9
SM	18-46-0	-	0.6	0.3	-	0.25	-	7.9	6.0
NT	Check	0.5	0.1	0.6	0.14	0.21	0.05	7.7	4.6
NT	Compost	0.5	0.2	0.8	0.20	0.70	0.11	7.7	3.3
NT	Manure	1.8	0.2	0.9	0.55	1.13	0.33	7.6	7.3
NT	18-46-0	0.6	1.4	0.6	0.27	0.99	0.17	7.2	2.8

†SM=stubblemulch tilled; NT=no tillage.

Table 4. Treatment and tillage effects on soil nitrogen (N) and phosphorus (P) forms and concentrations at the end of fallow (4.5 months after treatment application) for the 0- to 15-cm soil depth.

Treatment	TKN	NO$_3$-N	NH$_4$-N	TP
	----------------------------------mg/ kg----------------------------------			
18-46-0	798a[†]	36.7ab	6.1a	408a
Compost	784a	34.3ab	4.0a	403ab
Manure	828a	31.7a	4.3a	383b
Check	770a	24.5b	3.8a	382b
Tillage	TKN	NO$_3$-N	NH$_4$-N	TP
NT	783a	34.1a	4.8a	399a
SM	807a	29.5a	4.3a	389a

†Treatment or tillage means separated by Tukey's Studentized Range Test with α = 0.10. Means with different letters differ (P < 0.1).

Table 5. Wheat yield comparisons, 1994.

Treatment	Grain Yield	Total Dry matter	Head count
	--------------------Mg/ha---------------		no/m^2
18-46-0	1.71a[†]	4.27a	309a
Manure	1.68ab	4.13ab	283b
Compost	1.54ab	3.89b	278b
Check	1.51b	3.78b	277b
NT	1.71a	4.26a	308a
SM	1.61b	4.00b	279b

†Treatment or tillage means separated by Tukey's Studentized Range Test with α = 0.10. Means with different letters differ *(P* < 0.1).

Tillage did not appear to have a major effect on nutrient cycling, although most certainly volatilization of nitrogen did occur. High natural fertility and the ongoing microbial process of nutrient mineralization made it difficult to account for nutrients added to the soil with fertility treatments.

Our results suggest that the small amount of nutrients required to support dryland cropping can be surface applied as commercial fertilizer, manure, or composted manure without threatening surface or ground water supplies in this semiarid area.

REFERENCES

ASA, *Methods of Soil Analysis: Part 2 - Chemical and Microbiological Properties*, 2nd Edition, Page, A.L., ed., Am. Soc. Agron. Monograph No. 9, Madison, Wisconsin, 1982.

Jones, O.R., Hauser, V.L., and Popham, T.W., No-tillage effects on infiltration, runoff, and water conservation on dryland, *Trans. ASAE*, 37(2), 473, 1994.

Jones, O.R., Smith, S.J., Southwick, L.M., and Sharpley, A.N., Environmental impacts of dryland residue management systems in the Southern High Plains, *J. Environ. Qual.*, (in press), 1995.

Murphy, T.J., and Riley, J.P., A modified single solution method for determination of phosphate in natural waters, *Anal. Chem. Acta*, 27, 31, 1962.

O'Connor, P.W., and Syers, J.K., Comparisons of methods for the determination of total phosphorus in waters containing particulate material, *J. Environ. Qual.,* 4, 347, 1975.

Sharpley, A.N., Troeger, W.W., and Smith, S.J., The measurement of bioavailable phosphorus in agricultural runoff, *J. Environ. Qual.,* 20, 235, 1991.

Smith, S.J., Mathers, A.C., and Stewart, B.A., Distribution of nitrogen forms in soil receiving feedlot waste, *J. Environ. Qual.,* 9(2), 215, 1980.

US Environmental Protection Agency, *Methods for Chemical Analysis of Water and Wastes,* USEPA Rep. 600-4-79-020, Environmental Monitoring Support Lab., Cincinnati, Ohio, 1979.

WATERSHED MANAGEMENT AND CONTROL
OF AGRICULTURAL CRITICAL SOURCE AREAS

Gordon R. Stevenson

INTRODUCTION

I n the summer of 1993, legislation was passed in Wisconsin that provides for issuance of corrective orders to owners and operators of those critical sites on which non-point source water pollution abatement actions must occur in order to meet watershed-based water quality improvement goals. This legislation in Chapter 144 of Wisconsin Statutes, referred to as the Critical Site Provisions, is unprecedented. Prior to passage of this legislation, Wisconsin maintained separate voluntary and regulatory non-point source pollution abatement programs, the voluntary program seeking to produce watershed-wide water quality improvement goals and the regulatory program to abate acute water quality problems from discrete sources.

AGRICULTURAL POLLUTANTS AND WATER
QUALITY IN WISCONSIN

The problem of agriculturally generated pollutants is becoming more sharply focused in Wisconsin. The Wisconsin Department of Natural Resources (DNR) biennially publishes a comprehensive *Water Quality Assessment Report* (Turville-Heitz, 1992) to the federal Congress. The 1992 edition reported the following about agriculture in Wisconsin:

1) agriculture is the primary source of contamination to the state's lakes,
2) agriculture is responsible for major impacts on nearly 1500 miles (2500 km) of streams in the state, and
3) agriculture is the primary source of contaminants to ground water, eclipsing landfills and hazardous waste sources.

Lack of animal waste management on the part of a some farmers is among a group of environmental problems that contribute to degraded lakes, streams and ground water, a group that includes soil loss and stream sedimentation, misapplication of pesticides, and abuse of chemical fertilizers. While gains are occurring in soil erosion control, pesticide management, and other areas, the situation involving animal waste is not encouraging. Arguably, animal waste mismanagement may be the most severe water quality problem in the state. Wisconsin has the highest concentration of dairy farms of any state in the Union: 35,000. An additional 35,000 beef, poultry, swine, and sheep operations exist within the boundaries of the state (Pratt, 1991). The essential problem is that animal waste is perceived by some farmers as having small value. In terms of biochemical oxygen demand, a typical 100 cow-dairy herd may be compared to a community of 700-1,000 people without benefit of sewer service. Studies conducted by the University of Wisconsin Department of Rural Sociology indicate significant overapplication of crop nutrients due to lack of proper crediting of applied manure. In one survey, 68% of farmers responding reported that they were not crediting nitrogen (N) contained in manure applied. Of the remaining 32% of the respondents, only 2% were actually crediting N from manure properly (Shephard, 1993.) This leads to excessive nutrient application, some of which may infiltrate to ground water as nitrate and may also lead to polluted runoff to surface water.

Improperly constructed earthen manure pits are also indicated as a significant source of nitrate contamination to Wisconsin's aquifers. Of the hundreds of earthen manure pits within the state, only one pit out of three has been constructed with benefit of engineering assistance (Madison, 1988.) In 1988 and 1989, the Wisconsin Department of Agriculture selected a sample of 534 Grade A dairy farms to study nitrate concentrations in private water supply wells. Of this group, 48% exceeded the 2 mg/L ground water preventive action limit for nitrates as contained in state administrative codes (LeMasters and Doyle, 1989.)

Phosphorus (P)-enriched runoff to surface water is of equal, if not greater, concern. Agriculturally generated P is the primary contributor to accelerated eutrophication of Wisconsin's lakes (Turville-Heitz, 1992). In Wisconsin, animal agriculture alone generates more P than industry and municipalities, as indicated in Tables 1 and 2. Wisconsin recently promulgated Chapter NR 217, an administrative code for P control that restricts P concentrations from point source surface water discharges to 1 mg/L or less. Ironically, agriculture is effectively exempt from this code.

Biochemical pollution from feedlot runoff and over-application of animal waste appears to be contributory to loss or damage of some of Wisconsin's highest-quality fisheries. For instance, the world class smallmouth bass stream fishery of Southwestern Wisconsin is now a remnant of its former self. The loss of this fishery occurred simultaneously with the rise of more intensive livestock farming in that part of the state (Kerr, 1992). The advent of liquid manure storage systems has confronted the state with new hazards: spills. Last year, Department staff responded to several large storage structure failures, resulting in fish kills and property damage of several hundred thousand dollars.

Table 1. Estimated production of phosphorus from livestock in Wisconsin.

Livestock Type	Number of Head	Daily P Production	Daily P Production	Annual P Production
		kg/day/head	kg/day	kg/year
All cattle	4,170,000	0.0331	138,183	50,436,959
Swine	1,200,000	0.0100	11,977	4,371,757
Chickens	4,275,000	0.0004	1,592	580,925
Turkeys	7,000,000	0.0037	26,201	9,563,218
Total			177,953	64,952,858

Sources: Pratt, 1991; Midwest Plan Ser., 1985; Wisconsin Agric. Stat. Serv., 1992

Table 2. Estimated phosphorus production from industrial and municipal sources (Melby, 1992).

Industry	Number of Sources	Average Flow	Phosphorus Conc.	Phosphorus Production
		M liters/day	mg/L	kg/year
Dairy Processing (land application only)	20	0.3785	30.00	82,865
Dairy Processing (other discharge media)	180	0.0757	30.00	149,156
Pulp/Paper	20	26.4950	1.50	290,026
Canneries	40	0.7570	20.00	220,972
Meat Processing	4	0.7570	30.00	33,146
Publicly Owned Treatment Works	N/A	1,324.7500	4.00	1,933,509
Total				2,709,675

Water quality problems associated with animal waste are not limited to overapplication to land, feedlot runoff, and earthen storage practices. The list may also include stream sedimentation, transmission of pathogens, aesthetic damage, destruction of recreational opportunities, and others. Animal waste is a resource when managed properly; it is a pollutant when managed improperly. Unfortunately, improper animal waste management on the part of some operators in Wisconsin is causing measurable problems.

HISTORICAL BACKGROUND

The National Pollutant Discharge Elimination System (NPDES) was developed as a result of the 1972 Federal Water Pollution Control Amendments Act. The purpose of the NPDES program was to create a consistent means of regulating point source discharges within the United States through a national permit system. Federal water pollution legislation also provided the capability of delegating permit issuance authority to the states. In 1973, the Wisconsin legislature created the Wisconsin Pollutant Discharge Elimination System. Wisconsin was delegated permit issuance authority from the United States Environmental Protection Agency (US EPA) in 1974 and with the Wisconsin

Department of Natural Resources identified as the responsible agency. Subsequently, the Wisconsin Department of Natural Resources, in partnership with other federal, state, and local agencies, created essentially two programs that have sought to abate non-point source water pollution. They are:

1) Voluntary Non-point Source Water Pollution Abatement (NPS) Program,
2) The regulatory NR 243 Animal Waste Management Program (NR 243).

VOLUNTARY PROGRAMS

Prior to promulgation of the regulatory NR 243 Program, a number of federal, state, and county government grant programs existed that sought to protect Wisconsin's water resources through promotion of proper animal waste management and other agricultural practices. Virtually all of these programs were voluntary. From the standpoint of funding and staff, the flagship of agricultural pollution control efforts in Wisconsin has been and remains the NPS Program, administered by the Wisconsin Department of Natural Resources (DNR). Initially promulgated in the late 1970's, it is a broad and comprehensive program that seeks to meet pre-identified water quality improvement goals within selected watershed boundaries. A typical watershed project encompasses many square miles, all of which drain to a water body where water quality improvement goals have been pre-identified. Until the summer of 1993, it, too, was entirely voluntary. Watershed projects are selected on the basis of statewide ranking criteria. These criteria are established in accordance with relative conditions of streams, lakes, and ground water. Essentially, a comprehensive watershed plan is written that identifies impaired water bodies from non-point source pollution, establishes specific water quality improvement goals, and prescribes strategies for achievement of the established goals. The plan undergoes review and formal approval by various government bodies prior to its implementation. During implementation, generous cost-share grants are offered to largely agricultural operations within selected watershed boundaries to install and implement what have come to be called "Best Management Practices." A significant percentage of this program's activities have involved abatement of animal waste-related water pollution through installation of feedlot runoff control practices and construction of manure storage units meeting Soil Conservation Service standards or other recognized standards. Projects are very long-term, lasting 8-10 years.

As of the late summer of 1993, an enforcement component to the NPS Program was authorized by the Wisconsin Legislature in Chapter 144 of Wisconsin Statutes, providing for regulatory actions at sites within project boundaries whose participation is critical to achieving water quality improvement goals of projects.

REGULATORY PROGRAMS

The NR 243 Animal Waste Management Program is the only long-standing state regulatory program seeking to abate non-point source water pollution from animal waste-related operations. It views livestock operations in two distinct categories: those with greater than 1,000 animal units and those with less than 1,000 animal units. One animal unit equals a market weight steer or an equivalent number of other livestock. NR 243 provides for issuance of discharge permits to the larger facilities. This is a Wisconsin Pollutant Discharge Elimination System permit (WPDES), the same category of permit that is issued to pulp and paper mills and publicly-owned treatment works. Discharge permits are the centerpiece of the Federal Clean Water Act. Out of the approximately 2500 WPDES permits issued to all facilities in Wisconsin, about 50 have been issued to livestock operations.

Smaller farms are also subject to NR 243 through a system of complaints and citations. The Department receives about 100 complaints per year alleging water pollution from animal waste. Complaints are investigated, typically in company of someone from either county or state agricultural agencies. If complaints are substantiated, DNR issues a corrective order called a "notice of discharge," containing a timetable for compliance. Non-compliance with notices of discharge may result in issuance of a WPDES permit that could be enforced by the Wisconsin Department of Justice. Notably, farmers receiving notices of discharge qualify for generous cost-share grants. Rather than being triggered by pre-defined water quality improvement needs within defined boundaries, NR 243 activities may be initiated anywhere in the state by either the potential presence of a large livestock operation or in reaction to a complaint. Such programs are described as "unplanned intervention" (Frarey and Jones, 1993).

RELATIVE EFFECTIVENESS OF WISCONSIN'S PROGRAMS

The initial voluntary NPS projects that were begun in the 1970's fell short of their pollutant reduction goals. Voluntary participation rates were below 40% of the level needed for goal attainment.While current projects are showing better rates of participation, the program is still falling short of its goals. One simple reason is that agricultural facilities causing the most severe water quality impacts appear to be the very same individuals who are least likely to participate voluntarily in programs such as this. The NPS Program has fostered significant progress in non-point source water pollution control efforts. It has been recognized nationally for its pioneering efforts and has disseminated information to the public about non-point source pollution control. However, improved water quality success is desired by program administering agencies, the State Legislature and the public. There were three problems that motivated the Department of Natural Resources to promulgate the NR 243 Rule in 1984:

1) A long-standing awareness that animal waste-related water pollution was a serious problem in Wisconsin,
2) A need to issue Wisconsin Pollutant Discharge Elimination System (WPDES) permits for confined animal feeding operations under the Federal Clean Water Act, and

3) A lack of an administrative mechanism to respond to over 100 citizen complaints received annually by the Department alleging animal waste-related water pollution.

The NR 243 Rule provided some solutions to these problems. Over the life of the program, approximately 50 individual WPDES permits to large livestock operations have been issued, and nearly 400 notices of discharge have been issued. Through these activities, some large operations improved environmental performance and prevented the potentially severe local impacts from large livestock developments that some other states report. On smaller farms, a significant number of serious, localized impacts from animal waste mismanagement have been resolved. The NR 243 Program may also claim a number of other, less obvious accomplishments. The mere fact that the program exists in Wisconsin has kept out a number of operators who elected to locate their operations elsewhere after becoming aware of the NR 243 Rule and were unwilling to comply with the Rule. Through the Department's information and education efforts as well as media coverage, the NR 243 Program has attempted to elevate the public's awareness of the role of agriculture in environmental conditions. And at least to some degree, it has begun acclimating the agricultural community to the reality of environmental regulation.

Whether or not the NPS or NR 243 Programs accomplish their objectives as envisioned may not be as relevant as another broad question: Are the programs fostering overall water quality improvement on a statewide basis? Despite discrete and localized successes of both programs, they need improvement. The scale of animal waste-related water pollution, and agriculturally-generated pollution in general, in Wisconsin simply surpasses capabilities of these programs as administered in the past.

PROBLEMS CONFRONTING PROGRAMS

1) The NR 243 Program and NPS Program currently are not integrated and at times duplicate effort. Due to its complaint-driven configuration, NR 243 has been without a pro-active and systematic means of addressing this large-scale problem even though it has regulatory capabilities. The NPS Program, which Frarey and Jones described in 1993 as "planned non-intervention," has had a preplanned mission in place but without regulatory backing for full implementation.

2) Provisions for enforcement against recalcitrant notice of discharge recipients under NR 243 are cumbersome and sluggish. Due to the nature of the NR 243 Program, a body of unresolved notice of discharge cases continues to expand with each year. This situation will not be sustainable indefinitely.

3) Agricultural behaviors that cause water quality degradation are pervasive and have not yet been successfully addressed.

FUTURE DIRECTIONS

The Critical Site Legislation, which was passed in 1993, integrates an enforcement component into the NPS Program. It authorizes regulatory actions at sites within project boundaries whose participation is critical to achieving water quality improvement goals of projects. The Critical Site Provisions may be summarized as follows:

1) Critical sites are those sites where best management practices must be implemented to obtain a reasonable likelihood of achieving specific water quality improvement goals in Priority Watershed Projects.

2) Critical sites must be designated at the beginning of projects within Priority Watershed plans. Owners and operators of Critical Sites must be informed of designations. Designation of additional critical sites after formal plan approval is discouraged by the legislation.

3) Critical Site designees are allowed three years to voluntarily participate in Priority Watershed Projects. At the end of three years, designees become subject to corrective orders if the level of participation is inadequate to meet pre-identified water quality improvement goals of the project as determined by runoff models and other means.

4) Any non-point source may be designated as a Critical Site, provided its abatement is necessary to achieve the water quality improvement goals of the projects. However, the legislation does not provide uniform enforcement provisions. In the interest of preserving standing regulatory authority and simplifying drafting of the Critical Site legislation, enforcement activity associated with animal waste uses the NR 243 Rule. Enforcement actions involving sediments delivered to streams and other non-point sources are regulated by procedures established within the legislation.

5) Critical site activity may be conducted within ongoing and past Priority Watershed Project areas, provided all affected governing authorities approve.

6) Owners of agricultural critical sites may appeal critical site designations and orders to local elected officials in the county in which the non-point source is located. County officials have veto authority over designations and orders.

7) Owners of critical sites may further appeal designations and orders to higher governing bodies as well.

8) Owners of critical sites are afforded full due process of law; they may request contested case hearings and appeal further through the court system.

9) Parallel appeal rights are established for the Department of Natural Resources.

280 ANIMAL WASTE AND THE LAND-WATER INTERFACE

The Critical Site legislation is broad-based; the definition of Critical Sites suggests that relatively large numbers of land owners could be affected, not just those with the most severe problems. It was important to the authors of this legislation that landowners would be informed early in the project to allow sufficient time for them to voluntarily comply. Enforcement is bifurcated in a legal sense; i.e., animal waste cases will be regulated under the NR 243 Program, which itself will require revision in order to integrate with the Critical Site Provisions. Other non-point sources, such as sediments delivered to water resources, are subject to stepwise enforcement that includes issuance of a Notice of Intent to issue an order, an additional one year timetable to comply and, if necessary, issuance of an Administrative Order. Reasonable penalties may accompany the Order.

In light of the fact that the NPS Program has existed since 1978, there are a number of completed watershed projects. The Critical Site Provisions allow reopening of past projects for Critical Site activity. However, local county governments as well as other political bodies must approve. Additionally, elected local officials may veto Critical Site designation or orders; they may also uphold them. While it could be said that this feature of the legislation places technical decision-making into the hands of politicians, it can also be said that the legislation provides an opportunity for local "buy-in" to the program. This aspect is being keenly observed as the Critical Site program matures.

Environmental management and regulation of agriculture remains politically volatile. Agriculture is now among the primary sources of contamination to Wisconsin's water resources. Unlike other economic endeavors, agriculture has little capability to pass increased costs to consumers. The State of Wisconsin is now acknowledging the environmental impacts from agriculture. The adoption of the Critical Site Provisions signals significant changes in attitude: the long-standing idea that water quality improvement from non-point sources can be achieved on solely a voluntary basis has now been supplanted with the idea that regulatory back-up is needed in addition to voluntary opportunity. Further, previous non-point source regulatory activity had been designed to correct discrete, single-source problems on a limited scale. The Critical Site Provisions now allow such activity to be used to achieve broader water quality improvement goals.

Design and implementation of a practical Critical Site Program is just beginning. It is hoped that once this program is operational, it will be a model for *planned intervention* for non-point source water pollution abatement.

REFERENCES

Frarey, L. and Jones R., *Dimensions of Planned Intervention*, Texas Institute for Applied Environmental Research, Stephensville, Texas, 1993.

Kerr, R., personal communication, Bureau of Fisheries Management, Wisconsin Department of Natural Resources, Madison, Wisconsin, 1992.

LeMasters, G., and Doyle D., *Grade A Dairy Farm Well Water Quality Survey*, Wisconsin Department of Agriculture, Trade and Consumer Protection, and Wisconsin Agricultural Statistics Service, Madison, Wisconsin, 1989.

Madison, F., personal communication, Department of Soil Science, University of Wisconsin, Madison, Wisconsin, 1988.

Melby, J., personal communication, Bureau of Wastewater Management, Wisconsin Department of Natural Resources, Madison, Wisconsin, 1992.

Midwest Plan Service, *Livestock Waste Facilities Handbook*, Publication No. MWPS-18, Second Edition, Midwest Plan Service, Ames, Iowa, 1985.

Pratt, L. H., *Wisconsin 1991 Agricultural Statistics*, Wisconsin Agricultural Statistics Service, Madison, Wisconsin, 1991.

Shephard, R., *Beyond Superficial Targeting-Designed Strategies for Water Quality Education*, Ph.D. Thesis, Department of Rural Sociology, University of Wisconsin, Madison, Wisconsin, 1993.

Turville-Heitz, M., *1992 Wisconsin Water Quality Assessment Report To Congress*, Wisconsin Department of Natural Resources, Madison, Wisconsin, 1992.

Wisconsin Agricultural Statistics Service, personal communication, 1992.

OPTIMAL FARM-LEVEL USE AND VALUE
OF BROILER LITTER

Feng Xu and Tony Prato

INTRODUCTION

As broiler production becomes more concentrated, the amount of litter disposed in a given area increases. Broiler litter application to land is a major use of broiler litter. More broiler litter is generated than can be utilized in many concentrated production regions in the United States. As a result, litter over-application to land in litter-concentrated areas is a common practice. Increasing concern about water quality has focused attention on litter management in areas that have high litter concentration. Water quality problems occur when nutrients in commercial fertilizer or broiler litter are applied to farmland in amounts that exceed crop requirements. Southwestern Missouri is a region with high litter concentration and karst topography that makes ground water in the region highly vulnerable to pollution from nutrients in litter. Litter utilization must balance farm profit and water quality protection. Production options that consider both profitability and water quality have been studied (Norris and Shabman, 1992; Prato et al., 1991; Xu et al., 1993).

This study analyzes economic factors affecting litter utilization decisions at the farm level. Emphasis is placed on analyzing the value of litter and grower's decisions on litter utilization. Enterprise budgets for crops based on different fertilizer sources, broiler litter cleanout schedules, and litter management practices are used in an optimization model to determine litter application areas and rates that maximize the net return to litter and land management for a representative broiler farm. Water quality protection is considered in this study by basing litter application rates on soil type and crop requirements for nutrients. Economic profitability is approached by maximizing net returns to litter and land management.

PROCEDURES AND METHODS

The Shoal Creek watershed is located in Barry County, Missouri. This watershed has a total area of 17,232 ha (42,564 acres). Land in grass and pasture accounts for more than 80% of the watershed. Litter is applied to land in hay and pasture either for convenience or for its nutrient value. Broilers, cattle (cow-calf), and dairy are the major livestock activities.The karst features of soils in the area make it vulnerable to water contamination from broiler litter. There are 513 individual land owners in the watershed, of which 48 are broiler growers with 125 broiler houses. While forages are generally less susceptible to runoff and leaching than cultivated crops, the karst topography and high litter concentration increase the likelihood that surface contaminants will reach ground water.

An ARC/INFO Geographic Information System (GIS) is used to store and analyze data on watershed boundaries, land use, soil type, farm size, location and number of broiler houses, and other relevant information. Based on GIS information and other data sources, a representative broiler farm is constructed. The farm has three houses on 41 ha (102.4 acres). Each house has an area of 149 m^2 (16,000 ft^2) with a capacity of approximately 20,000 birds. Available area for forage production is 35 ha (85.51 acres). The prevalent soil type in the watershed, Scholten gravelly silt loam, is assumed to be on the farm.

A litter management practice (LMP) refers to the rate of litter application, litter cleanout frequency, and litter selling. Litter cleanout is the removal of accumulated broiler manure and litter from the house. Frequency of litter cleanout varies from every three flocks to every six flocks. About 85% of broiler growers clean out every six to seven flocks, or about once a year in the four south-central states of Missouri, Arkansas, Texas, and Oklahoma, and 92% of litter is applied to the land (Missouri Farming Planning Handbook, 1989). Cleanout is usually contracted out to custom operators. Broilers are marketed at an age of 6-8 weeks. The choice of LMPs affects farm profitability.

Information on nutrient content of broiler litter in Shoal Creek watershed was obtained from a study that sampled and analyzed litter from nine broiler houses in Barry county (Prato et al., 1991). Compared to Alabama (Payne and Donald, 1991) and Virginia (Bosch and Napit, 1991), Missouri litter has a higher phosphate content but a similar nitrogen and potassium content. Data show that both quantities and nitrogen (N) content of litter vary with cleanout frequency. Phosphate (P) and potash (K) content do not vary significantly with cleanout frequency (Table 1). N has been the primary nutrient of interest in terms of fertility and water quality (Payne and Donald, 1991).

Crop nutrient requirements for N, P, and K were estimated for fescue hay and pasture for each of four yield goals: low, moderately low, moderately high, and high. These requirements can be supplied by either commercial fertilizer or broiler litter. Based on crop nutrient requirements and the nutrient content of broiler litter, crop requirements for broiler litter were estimated on a N, P, and K basis. Detailed procedures for estimating crop nutrient requirements are provided in Buchholz (1990). *Nitrogen requirements* are estimated as NR = γYG for fescue hay and pasture where: NR is total nitrogen requirement (lb/acre), γ is pounds of nitrogen per unit of yield, and YG is yield goal (ton/acre for fescue hay and animal unit month (aum)/acre for fescue pasture), and γ =

Table 1. Cost for cleanout of broiler houses and nutrient content
for various cleanout frequencies.

Category	3-flock	4-flock	5-flock	6-flock
Litter per cleanout per house (tons)	77.36	91.20	105.12	118.96
Litter per house annually (tons)[†]	141.83	125.40	115.63	109.05
Variable cost @ $35/ton	386.80	456.00	525.60	594.80
Total cost = fixed cost ($545) + variable cost	931.80	1001.00	1070.60	1139.80
Cleanout cost w/o $60 compensation	12.04	10.98	10.18	9.58
Cost above base (base=6 flock w/o compensation)	2.46	1.40	0.60	0.00
Cleanout cost w/ $60 compensation	11.27	10.32	9.61	9.08
Cost above base (base=6 flock w/o compensation)	1.60	0.74	0.03	-0.50
Total annual cost w/o compensation ($)	1707.63	1376.89	1177.11	1044.70
Total annual cost with $60 compensation	1598.42	1294.13	1111.20	990.17
N after 25% loss (lbs/ton)	43.39	49.40	54.39	59.39
P_2O_5 (lb/ton)	82.14	82.14	82.14	82.14
K_2O_5 (lb/ton)	39.19	39.19	39.19	39.19
Value of litter based on nutrients ($/ton)	33.77	34.92	36.07	37.22

[†]Assumes an average of 5.5 flocks a year.

(40 18)' for hay and pasture, respectively. *Phosphate requirements* are estimated as $PR = (110/t)*(DP^{.5}-OP^{.5}) + øYG$, where PR is total phosphate requirement (lb/acre); t is number of years to increase soil test to desired level (t=8); DP is desired level (lb) = 40; OP is the observed level (lb); ø is phosphate removal (lb/acre). Estimation equations are: $PR = 13.75(40^{.5}-OP^{.5}) + 9YG$ for fescue hay, and $PR = 13.75(40^{.5}-OP^{.5}) + .15YG$ for fescue pasture. *Potassium requirements* are estimated as $KR = (75.5/t)*(DK^{.5}-OK^{.5}) + \lambda YG$, where KR = total potassium requirement (lb/acre); t is number of years to increase soil test to desired level (t=8); DK is the desired level (lb); OK is the observed level (lb); λ is potassium removal (lb/acre). Estimation equations are: $KR = 9.4375(DK^{.5}-OK^{.5}) + 34YG$ for fescue hay, and $KR = 9.4375(DK^{.5}-OK^{.5}) + 5.1YG$ for fescue pasture, where DK = 160+5(CEC) and CEC is a measure for cation exchange capacity. Soil parameters for Scholten soil used to estimate nutrient requirements are: organic matter = 1.8%, pH = 5.0, OP = 5, OK = 160, and CEC = 10.

Enterprise budgets were constructed for hay and pasture at the four yield goals. Production costs were assumed to be the same for all items except fertilizer, labor, and machinery, which are directly related to yield. Budgets were developed for fescue hay and pasture using cost and returns data from the Missouri Farm Planning Handbook (1989).

The optimization model is presented in (1)-(9) below. Variables and parameters in the model are defined according to subscripts such that: crop j has elements for fescue hay and pasture; fertilizer source k has elements for commercial fertilizer and litter for different cleanout schedules; yield goal (and therefore fertilizer level) l has four elements: low, moderately low, moderately high, and high; and nutrient m has elements of N, P, and K. Therefore, X_{jkl} is area in crop j with fertilizer k at level l, which is the decision variable to be determined by the optimization model. BL_{jl} is amount of broiler litter associated with each X_{jkl}, and CF_{jlm} is the amount of mth nutrient of commercial fertilizer associated with X_{jkl}. C_{jkl} is the cost of producing crop j with fertilizer k at level l, excluding fertilizer cost. SIZE is the number of acres on the farm that are suitable for forage production. BL is quantity of broiler litter available. Y_{jkl} is yield of crop j with fertilizer k at level l. P_j, PBL and P_m are prices of crop j, broiler litter, and m^{th} nutrient, respectively. α_m is the amount of mth nutrient in litter. β_{jlm} is crop j's requirements for m^{th} nutrient with yield goal l. $STCF_{jlm}$ is the amount of mth nutrient of commercial fertilizer that is used to supplement for crop j's nutrient requirement at yield goal l after litter application rate is determined on a particular nutrient basis. $WTBL_{jlm}$ is the amount of mth nutrient of litter that exceeds crop j's nutrient requirement at yield goal l after litter application rate is determined on a particular nutrient basis. BLS is the amount of litter that can be sold. Input parameters include P_j, Y_{jkl}, C_{jkl}, PBL, P_m, SIZE, α_m, β_{jlm}, and BL. Crop yields and costs (Y_{jkl} and C_{jkl}) are differentiated for each j, k, l.

$$\text{Maximize: } Z = \sum_{j=1}^{j} (P_j Y_{ijk} - C_{ijk}) X_{ijk} - \sum_{m=1}^{M} P_m (CF_m + STCF_m) + PBL \times BLS \tag{1}$$

$$\text{Subject to: } \sum_{j=1}^{J} \sum_{k=1}^{K} \sum_{l=1}^{L} X_{jkl} \leq SIZE \tag{2}$$

$$\alpha_m BL_{ji} - \beta_{jlm} X_{jk='BL'l} + STCF_{jlm} - WTBL_{jlm} = 0 \quad \forall \; j,l,m \tag{3}$$

$$CF_{jlm} - \beta_{jlm} X_{jk='CF'l} \geq 0 \quad \forall \; j,l,m \tag{4}$$

$$BLS + \sum_{j=1}^{J} \sum_{l=1}^{L} BL_{jl} \leq BL \tag{5}$$

$$-CF_m + \sum_{j=1}^{J} \sum_{l=1}^{L} CF_{jlm} = 0 \quad \forall \; m \tag{6}$$

$$-STCF_m + \sum_{j=1}^{J} \sum_{l=1}^{L} STCF_{jlm} = 0 \quad \forall \; m \tag{7}$$

$$\sum_{k=1}^{K} \sum_{l=1}^{L} X_{j='HAY'kl} \; / \; \sum_{k=1}^{K} \sum_{l=1}^{L} X_{j='PASTURE'kl} = \frac{1}{3} \tag{8}$$

$$X_{jkl}, CF_m, STCF_m \geq 0 \quad \forall \; j,k,l,m \tag{9}$$

As presented in the LP model, the representative farm efficiently utilizes the annual litter load for that farm. The objective function of the LP model (1) is the net return to land and management from utilizing litter. Costs of commercial fertilizer and broiler litter are explicitly stated in the objective function. The LP model chooses a crop mix that maximizes net return from applying broiler litter or commercial fertilizer to land. Constraint (2) represents the acreage restriction, namely, the sum of the acreage for forage activities should not exceed the area available for forage. Constraints (3) and (4) represent the crop nutrient requirements based on broiler litter and commercial fertilizer, respectively. Specific quantities of commercial fertilizer and/or broiler litter are required to achieve particular yield goals. Parameter α_m is the quantity of the mth nutrient per unit of broiler litter expressed in commercial fertilizer equivalents. β_{jlm} is the crop requirement for the mth nutrient by crop j for yield goal l (or at fertilizer level l). Crop nutrient requirements are met using either commercial fertilizer and/or litter.

Three scenarios are considered: N-basis, P-basis, and K-basis. In the N-basis scenario, litter application rates are estimated based on crop N requirement and N content of litter. Any shortages in P and K ($STCF_p$ and $STCF_k$) after using litter to meet N requirements are made up from commercial fertilizer. No credit is given for excess P and K in litter ($WTBL_p$ and $WTBL_k$). Scenarios for applying litter on a P or K basis are also analyzed. The assumption that certain excess nutrients in broiler litter are freely disposed may be practical from the perspective of farm management since nutrients in litter cannot be easily separated. Constraint (5) represents the availability of litter. Constraints (6) and (7) are the accounting constraints for the mth commercial fertilizer and supplemental commercial fertilizer, respectively. Constraint (8) restricts the ratio of hay to pasture to 1:3. This ratio conforms with the cow/calf feeding ratio in the Missouri Farm Planning Handbook (1989). Constraint (9) ensures that decision variables are non-negative.

Variations of the LP model are used to evaluate several litter utilization scenarios. Effects of changes in litter prices on utilization and net farm returns were examined by parametrically changing litter price in the LP model. Changes in net farm returns are examined for different optimal litter utilization alternatives. In this study, only cost differences were considered because cleanout is a necessary activity in broiler production.

Other factors that can be incorporated in the LP model include timing of broiler house cleanout, litter storage, time of application; water quality impacts of cattle manure; feeding litter to cattle; exporting litter to other areas; leaching of nutrients through the root zone; risk tolerance levels for water pollutants; and accounting for variability in the nutrient content of litter. Since no process model that is specific for broiler application and adequately handles litter on pasture was available at the time of this analysis, the risk of ground water contamination from use of broiler litter can not be demonstrated. As a result, nutrient leachate and/or runoff cannot be incorporated explicitly in the optimization model. A rule of thumb for water quality protection is to determine the litter application rate needed to meet crop nutrient requirements. While zero water contamination cannot be guaranteed, the risk of water contamination would be minimal.

RESULTS AND ANALYSIS

LITTER UTILIZATION

A profit-maximizing grower would use litter to produce forage on-farm because it is an inexpensive substitute for commercial fertilizer. Excess litter is sold to other farmers. Litter should be applied on an N basis because net returns are higher than on a P or K basis (Table 2). The grower should choose a 6-flock cleanout schedule because it provides the highest net returns.

About 327 tons of litter is generated annually by a three-house farm using a 6-flock cleanout schedule. For this farm, 125 tons (or 38%) is applied to the grower's 41 ha of land because the value of litter is higher than the market price. The remaining 202 tons (or 62%) of surplus litter is sold to other farmers. Current market price for litter is $10.83/ton. As litter price increases, the price advantage of litter relative to commercial fertilizer decreases. As number of houses per farm and litter availability increases, additional income is earned, provided the additional litter can be sold. As commercial fertilizer prices decrease, the incentive to apply litter to land decreases.

Land requirements for the three nutrient scenarios are different. Land area required to apply the available litter (327 tons) is higher on a P basis (180 ha or 444 acres) than on a N or K basis (90 ha or 223 acres). Land area is greatest for the P scenario because litter is relatively rich in P and hay and pasture do not require much P. Economic return is higher on an N basis than on a P or K basis. Net returns for the three scenarios vary due to differences in the price of each commercial fertilizer nutrient. Compared with over application of litter to land, applying litter to land based on its N content should reduce nitrate runoff and nitrate concentrations in ground water. However, excess P may contribute to pollution of streams and lakes receiving runoff from land-applied litter.

LITTER VALUE

Per-unit value of surplus litter equals the market price. To estimate the value of litter in forage production, a forage producer is modeled by dropping irrelevant litter activities and constraints from the optimization model. A forage producer can buy litter or commercial fertilizer but would prefer litter to commercial fertilizer because of its lower price. The optimal amount of litter to purchase is estimated. Shadow price is used to measure the contribution of additional unit of litter to farm profit when the optimal amount of litter is used, holding others variables constant (assuming no sale activity). Under current market conditions, the shadow price of litter is estimated to be $21.12, $23.41, $25.22, and $27.54/ton of litter for 3-, 4-, 5-, and 6-flock cleanout on an N basis, respectively (Table 3). Since the current market price for litter is below the shadow prices, forage producers have an incentive to use litter. Shadow prices of litter are higher when application rate is determined based on P. However, the total net return to litter when applied on a P basis is lower than when applied on an N or K basis because considerable N and K must be added to meet crop requirements. The value of litter ranges from $33.77 to $37.22/ton if the N, P, and K in litter are valued at current market prices. Since litter cannot be separated into its nutrient components, these litter values exceed the shadow prices, which range from $21.12 to 27.54/ton.

If the price of litter increases from its market price of $10.83 to its maximum N-basis shadow price of $27.54/ton, farmers would be indifferent between using litter (from 6-flock cleanout) and commercial fertilizer. For market prices above this level, commercial fertilizer becomes less expensive than litter. As long as market price for broiler litter does not exceed $27.54/ton, litter will be used in forage production. When market price of litter exceeds this upper limit, there is no incentive to use broiler litter in forage production.

When broiler growers view litter as a waste product rather than a valuable resource in forage production, they tend to minimize litter handling costs rather than maximize returns to litter. Hence, over-application of litter to land is likely in the absence of guidelines for litter disposal. The likelihood of over-application of litter is reduced when litter is applied to meet forage requirements for nutrients.

EFFECT OF CLEANOUT COMPENSATION

Even though the cleanout schedule is primarily determined by the broiler production plan, growers tend to clean out houses less frequently due to the high cost of cleanout. However, when the fertilizer value of litter is considered,

Table 2. Annual net returns by nutrient basis, with and without compensation for litter cleanout.

Clean-out schedule	Net return ($) without compensation			Net return ($) with compensation		
	N-basis	P-basis	K-basis	N-basis	P-basis	K-basis
3-flock	13,260	13,020	13,252	13,587	13,348	13,580
4-flock	13,397	13,079	13,311	13,646	13,327	13,559
5-flock	13,523	13,152	13,384	13,721	13,349	13,582
6-flock	13,621	13,218	13,451	13,784	13,382	13,615

Table 3. Shadow price ($/t) for litter in forage production by nutrient basis and cleanout schedule.

Cleanout schedule	N-basis	P-basis	K-basis
3-flock	21.12	33.44	21.58
4-flock	23.41	34.59	21.58
5-flock	25.22	35.74	21.58
6-flock	27.54	36.89	21.58

it may be profitable for growers to change their cleanout schedule. Broiler integrators can provide growers with a financial incentive to do more frequent cleanout in order to enhance the quality of broilers. For example, some integrators are willing to pay growers $60 for each cleanout. The effects of this incentive are analyzed using the optimization model. Cost and return information suggest that growers will choose a 6-flock cleanout. When growers receive $60/cleanout from integrators, they would switch to 4-flock cleanout with little change in net return (Table 2). Hence, cleanout will occur more often when growers are compensated.

Litter quality varies with cleanout schedule. Ranking of litter quality by cleanout frequency is 6-flock, 5-flock, 4-flock, and 3-flock cleanout. Litter application results in economic benefits because litter has a high nutrient content, making it an excellent fertilizer. To balance these benefits against the potential negative impacts on water quality, litter should be applied to land at a rate that does not exceed crop nutrient requirements.

SUMMARY AND CONCLUSIONS

This chapter examines optimal utilization of broiler litter at the farm level in Missouri's Shoal Creek watershed. A farm-level optimization model was used to analyze a broiler grower's choice of litter utilization and management practices. The model maximizes net returns to land and litter management and was used to estimate the fertilizer value of litter. The integrated model was used to evaluate changes in net farm returns from increased availability of broiler litter as well as economic incentives for increasing the frequency of litter cleanout. Total net returns and litter value varied, being higher on an N basis than on a P or K basis. Shadow prices were $21.58, $27.54, and $36.89/ton based on K, N, and P, respectively, for a 6-flock cleanout frequency.

Litter is a valuable resource whose rich nutrients and low cost make it an excellent substitute for commercial fertilizer in crop production. Litter is also a potential water pollutant. Such pollution can be reduced and minimized by basing litter application rates on crop nutrient requirements. Broiler growers typically have excess litter. Small farms with many houses have a large quantity of excess litter, which can be sold. Over-application of litter to land is likely to occur when buyers cannot be found or high transportation costs make shipping to other areas prohibitive.

ACKNOWLEDGMENTS

Authors thank J.R. Brown and D.D. Buchholz for their assistance in estimating crop nutrient requirements, M. Jenner for his comments on an earlier version of this chapter, and three reviewers for their helpful comments. This research was partially supported by the Department of the Interior, US Geological Survey, through the Missouri Water Resources Research Center, and the Missouri Agricultural Experiment Station, Columbia, Missouri.

REFERENCES

Bosch, D.J., and Napit, K.B., *The Economic Potential for More Effective Poultry Litter Use in Virginia*, SP-91-11, Department of Agricultural Economics, VPI & State U., Blacksburg, Virginia, 1991.

Buchholz, D.D., *Soil Test Interpretations and Recommendations Handbook,* Department of Agronomy, University of Missouri-Columbia, Missouri, 1990.

Missouri Farm Planning Handbook, University of Missouri-Columbia, Missouri, 1989.

Norris, P.E., and Shabman, L.A., Economic and environmental considerations for nitrogen management in the Mid-Atlantic coastal plain, *Amer. J. of Alternative Ag.,* 7(4), 148, 1992.

Payne, V.W.E., and Donald, J.O., *Poultry Waste Management and Environmental Protection Manual,* Circular ANR-580, The Alabama Cooperative Extension Service, Auburn Univ., 1991.

Prato, T., Vandepopuliere, J., Fulhage, C., Haithcoat, T., Xu, F., Fulcher, C., and Jenner, M., *Managing Land Application of Broiler Litter to Optimize Economic Value and Water Quality,* Report #G-1572-04, Missouri Water Resources Research Center, Univ. of Missouri-Columbia, Missouri, 1991.

Xu, F., Prato, T., and Fulcher, Chris, Broiler litter application to land in an agricultural watershed: A GIS approach, *Water Science and Technology,* 28(3-5), 111, 1993.

ECOSYSTEM-BASED ASSISTANCE AS A CORNERSTONE TO WATERSHED MANAGEMENT

David C. Moffitt

INTRODUCTION

Among the topics of conversation at most conservation agencies are "Watershed scale planning" and "watershed management." Managers discuss program delivery at the watershed level, and technicians are gearing up to prepare watershed plans. But what is a watershed? How big is it? How do we know when we have a watershed plan that will truly lead to "watershed management." All good questions.

The phrase "ecosystem-based assistance" also generates several questions. First of all, what is "ecosystem-based assistance?" What is the scale of an "ecosystem?" How is animal waste management included? Is ecosystem-based assistance something new or a recast of existing methodology?

This chapter will discuss these questions and outline the Natural Resources Conservation Service (NRCS) planning process, including the role of tools such as pollutant loading models. The use of these tools in determining the impact or effect of alternative management options (animal waste management for example) on the five resources (soil, water, air, plant, and animal), and on humans, will also be discussed.

WATERSHED

What is a watershed? *Webster's New World Dictionary of American English* (1988) defines a watershed in part as that area drained by a stream or river system. Other dictionaries offer slightly different wording, but the shared concept is that a watershed is a drainage area contributing to a water body that concentrates the water and removes it from the drainage area.

The offered definition does not provide much of a clue to the scale of a watershed except in the use of such terms as "river" or "stream." In the simplest form, a watershed can be a concentrated flow area in a field with its contributing area, possibly less than an acre in size. On the other end of the spectrum, consider as a watershed a river system such as the Mississippi that has a drainage area of millions of acres. The most common scale of watersheds addressed by the NRCS is in the range of tens to thousands of acres.

Another way of looking at watershed scale is through an understanding of stream orders after Strahler (1950). The concentrated flow area in the simple field sized watershed can be considered as a first order stream (or contributing directly to a first order stream). Two first order streams join to form a second order stream; two second order streams join to form a third order stream; and so forth. The watershed area associated with each stream order varies because of many factors including geology and topography; however, the typical NRCS watershed consists of relatively few stream orders.

Watershed planning or watershed management, then, normally references actions with a land unit of tens to thousands of acres, the potential for multiple land uses, and possibly many land owners and operators.

ECOSYSTEM-BASED ASSISTANCE

The terms "ecosystem" and "ecosystem-based assistance" are not new concepts; however, the use of these terms has become more popular in recent times. Hugh Hammond Bennett, the first Chief of the Soil Conservation Service, recognized the need for ecosystem considerations. In a talk to agency personnel in 1938 he stated;

> Animal life not only is intimately interrelated with plant life, but with the soil itself, and our knowledge of ecology is still insufficient for us to assume that we can afford to eliminate any species completely from our fauna or flora. It is only the part of common sense, therefore, to try to maintain the best biological balance that may be attained under agricultural conditions.

In a similar manner, Eugene Odum (1959) stated:

> The soil conservation movement in the United States, following as it has (and none to soon) an era of widespread destructive exploitation by one-crop systems, is based on the sound principle of ecosystem and of the cyclic use and renewal of the soil. The soil conservation program is also an outstanding example of cooperation between local people and the Federal government. In each of the soil conservation districts the program is run by the local people with technical help but not dictation from the central government.

The above are given as illustrations that planning based on ecological principals and concepts has been the basis of NRCS assistance in farm and ranch planning since its inception. The NRCS staffs have been and are composed of professional people with some form of ecology training. The traditional job of the ecological sciences specialist has been to train and assist field office staff personnel in applying ecological principles in their work. In recent years staffing of ecological specialists has been more limited. With the renewed emphasis on ecosystems, this is changing.

What are ecosystems? Ecosystems are the basic functional units in ecology including both a biological community (living organisms) and an abiotic (non-living) environment. The abiotic environment consists of inorganic and organic compounds, such as water, oxygen, salts, and nutrients, that are essential for the survival of the biological community. When ecosystems are in balance, their biological communities and abiotic environments are in equilibrium. Is there a given size for an ecosystem?

Again quoting from Odum (1959):

> The concept of the ecosystem is and should be a broad one, its main function in ecological thought being to emphasize obligatory relationship, interdependence and causal relationships. Ecosystems may be conceived and studied in various sizes. A pond, a lake, a tract of forest or even a small aquarium could provide a convenient unit of study. As long as the major components are present and operate together to achieve some sort of functional stability, even if for only a short time, the entity may be considered an ecosystem. A temporary pond, for example, is a definite ecosystem with characteristic organisms and processes even though its active existence is limited to a short period of time.

A watershed then, after Odum (1959), can be viewed as an ecosystem in which the organisms and their abiotic environment should be in balance. The diversity, both in organisms and their environment, present in the watershed defines the complexity of the ecosystem.

Ecosystem-based assistance in terms of the NRCS, then, can be broadly defined as assistance provided to land owners and operators that considers maintaining or improving the equilibrium between the organisms and their abiotic environment.

The ecosystem-based assistance process can be viewed in terms of its impact on all levels of our society. People are a dominant guiding force in our ecological system. They express at the local, state, and regional level their vision of needs for their ecosystem. These ecosystem visions are transmitted to the national level through congressional representatives, lobbyists, and individuals. People form national coalitions, and their concerns emerge in the media as conservation concerns. These concerns are also expressed into laws that cumulatively become national strategic plans. *Strategic plans are not driven by programs, but by visions.*

A national strategic plan provides directives or incentives for a regional coalition to develop a plan supporting or complementing the national plan. A regional strategic plan filters out those items that do not apply and adds other items of regional concern not addressed by the national coalition. These concerns also emerge in the media and other public forums.

This process continues down through the various levels until an area specific strategic plan is reached. A watershed plan is an example of this level of strategic plan. The watershed plan develops the detail that determines how the conservation programs will be delivered to the land locally. This is reflected in the NRCS Field Office Technical Guide (USDA-SCS, 1990) and in other conservation partners' guidance documents.

The final step is the site-specific application of conservation practices as identified in the watershed plan and formed into resource management systems that address ecosystem needs such as for animal waste management, the land owner's objectives, and society's concerns or requirements. Ecosystem needs are met when the soil, water, air, plant, and animal resources and human impacts are considered and one resource does not appreciably degrade another.

Implementation of a site-specific conservation plan begins the process of meeting a national strategic plan. One plan is combined with several to form a group of plans that meet watershed strategic plan objectives. Watershed plans are joined to accomplish county, then area, and then state requirements until all the pieces are together to accomplish the whole. The process begins again.

ECOSYSTEM-BASED ASSISTANCE TOOLS

Just how are ecosystem relationships addressed? How are diverse concepts such as increased (or more efficient) farm production combined with ecological concerns, such as importing nutrients in feeds for confined animals? How does one evaluate the effects of alternative management practices on the watershed ecosystem, including outputs? Daryl Simons (1980) answers these questions:

> The increasing interest in water resource and land-use planning has stimulated the development of particular and general watershed and river system models for predicting response from ecological systems. The models, whether physical process simulation or conceptual, are intended to be used to estimate physical quantities that describe the major ecosystem responses to precipitation such as water yield, sediment yield, changes of land and river morphology, and transport of pollutants. Methods to estimate water, sediment, and other pollutant transport yields are needed for analyzing the economic feasibility and trade-offs of any proposed water resources or land-use development in watershed and river systems and for predicting possible adverse environmental impacts associated with the proposed development.

Simons (1980) goes on to say, in part, that mathematical models used to simulate the effects of management activities on ecosystem responses can be classified as one of three types: regression, "black box" simulation, and physical process simulation. Where models fall into at least two or all three of these categories, they are classified according to the dominant traits of the model as a whole.

A general weakness of regression models available for use in water and land resources management is that the variables representing water and land uses and conditions are not specific enough to reflect the effects of many individual management activities on the ecosystem. In addition, the regression models usually require sufficient (and unavailable) observed data to correlate meaningful relations. Furthermore, time- and space-dependent processes are very difficult to predict using regression equations.

The lumped parameter of "black box" or "simulation programming" type of model interprets input-output relations using oversimplified forms that may or may not have physical significance. All processes related to movement of the water and sediment through the watershed are lumped together into several coefficients. The classic example of a lumped parameter model is the rational formula for estimating peak discharge, i.e., $Q = CIA$ where Q is the peak discharge, I is the rainfall input, A is the drainage area, and C is the runoff coefficient that represents the major hydrologic processes in the watershed. This model is easy to use but has limited physical meaning and can often be inaccurate. In most cases, it is impossible to predict the effects of alternative mixes and sequences of management activities occurring on complex upland watersheds utilizing lumped parameter models (i.e., how do we define C in the above equation?).

Physical process simulation models, however, avoid the "lumping" of physically significant variables by decomposing components such as infiltration and sediment detachment from raindrop splash. By simulating the selected phenomena into its separate components, each individual process can be analyzed and refined or altered to meet the needs of the user. Consequently, as each process component is upgraded, the model becomes more representative of the ecosystem.

Use of component process models also allows input of variables that have physical significance to the user and the field situation. All of the above characteristics of component process models allow greater flexibility than other types of models. Advantages of physical process component models over other types are numerous. In general, physical process simulation models are superior to regression type models or "black box" type mathematical models. The input variables to process models are physically significant as they indicate system response caused by changing one or more physically significant values. The physical process simulation models are "dynamic simulation systems." They are not assumed stationary in either time or space; therefore, with very little calibration (although some question this statement), they can be used for predicting the future response of the system in real time and space. Furthermore, because these models are formulated according to physical processes, they are applicable to areas in which the governing physical processes are the same.

Physical process simulation models include those models relating to the physical processes, such as surface runoff, sub-surface flow, raindrop splash erosion, overland flow erosion, channel bank erosion, fate of nutrients (some include nutrients from animal waste), and pesticide degradation. All of these processes receive a great deal of attention in ecosystem considerations.

The NRCS uses several physical process simulation models with names such as GLEAMS (Leonard et al., 1987; Knisel, 1993), EPIC (Sharpley and Williams, 1990), NLEAP (Follett et al., 1991), SWRRB (Williams et al., 1985), and AGNPS (Young et al., 1989). These are abundantly described in published literature, and the citations offered in "References" are included for the convenience of the reader. These models are often referenced as water quality models, but are in fact pollutant loading models. Each of these models is designed to represent portions of an ecosystem and can provide limited answers to impacts resulting from changes to the environment. Application of four of these models to answer water quality related questions will be discussed below.

The advent of the use of geographic information systems (GIS) has increased the planner's capability to collect and assemble data for physical process models, such as those mentioned above. The Geographic Resources Analysis Support System (GRASS) (Van Warren et al., 1988) is the GIS selected by the NRCS. GRASS provides the capability to enter, store, manage, and display digital data depicting spatial features such as field boundaries, roads, streams, soil maps, hydrologic units, and many other types of information. Because of its flexibility, GRASS can assist in analyzing and solving many important natural resource problems.

The discussion that follows summarizes two tools utilizing GRASS that are being used by the NRCS. Following the discussion of the tools will be a short summary of the NRCS planning process describing how physical process models can add to our knowledge as we provide ecosystem-based assistance.

HUMUS

The Hydrologic Unit Model for the United States (HUMUS) (Stinivasen et al., 1993) project was initiated for making national assessments of the status and conditions of the water resources in the United States under current and projected alternative conditions of the use and management, or mismanagement, of those resources. This project directly supports the Soil and Water Resources Conservation Act of 1977 (RCA), which was enacted to further "the conservation, protection, and enhancement of the Nation's soil, water and related resources for sustained use and other purposes."

The HUMUS project is a cooperative effort between two agencies within the United States Department of Agriculture (USDA), the Agricultural Research Service (ARS) and the NRCS. The Texas Agricultural Experiment Station is also a partner. The HUMUS project has three major components: computer simulation models, data bases, and a geographical information system.

The simulation models used in the HUMUS project are an adaptation of the Simulator of Water Resources in Rural Basins (SWRRB) model and the Environmental Policy with Integrated Climate (EPIC). The backbone of the HUMUS project is a distributed parameter, continuous time model adapted from SWRRB to assist water resource experts in assessing water supplies and non-point source pollution on watersheds and large river basins. EPIC is a field-scale model that is used to simulate pollutant loading (sediment, nutrients, and pesticides) leaving an agricultural field in runoff and percolate.

The most critical component of the HUMUS project is collection of the spatial and relational data to drive the models. The spatial data requirements include watershed boundaries, soil properties, topography, land use, political boundaries, weather station locations, aquifer maps, stream network, and stream gauge stations. Relational requirements include the NRCS National Resource Inventory, agricultural statistics, weather parameters, soil survey data, streamflow and reservoir operation data, and agricultural census data. GRASS is also needed to extract data from geographic information system data bases for use in the model and to graphically display model output.

HU/WQ TOOL

The Hydrologic Unit Water Quality (HU/WQ) Tool (USDA-SCS, 1994) provides the user with an integrated environment consisting of a database management system, geographic information system, and pollutant loading models and data bases. Much of the discussion that follows was edited from the 1994 reference. The HU/WQ Tool integrates four models previously mentioned; (a) Agricultural Non-point Source Pollution Model (AGNPS), (b) SWRRB, (c) EPIC, and (d) Groundwater Loading Effects of Agricultural Management Systems (GLEAMS). AGNPS and SWRRB are designed for watershed-scale analysis--collections of several to many fields within a watershed. EPIC and GLEAMS are designed for field-scale analysis--a single field for each analysis.

The HU/WQ Tool uses two levels for analysis. The first level is called project level. The next level is the simulation level. A project represents a geographic area such as a hydrologic unit. Simulations are grouped within a project to represent a set of conditions. A baseline simulation would be the conditions of the project area in the future if the same management practices take place. The models are designed to inform the user of possible future conditions. They should be used in conjunction with other information about the area before making planning decisions.

Users collect spatial and tabular data in one defined format for all the models according to a set of guidelines. The collected data are preprocessed into parameters required by, and into a format acceptable to, a selected water quality model. After the model is run, its output is presented to the user in a dynamic and flexible format for evaluation and generation of new requirements. This process is repeated to provide information to support the planning process.

The HU/WQ Tool addresses the entire process including (1) entering and revising collected data; (2) running the water quality models; (3) selecting the proper output, format, and reports; (4) analyzing output results; and (5) interpreting and displaying output results.

The following describes each of the models within HU/WQ Tools and a discussion of how the model can integrate factors within the ecosystem.

AGNPS was developed at USDA-ARS, North Central Soil Conservation Laboratory, Morris, Minnesota. It is a single-event-based computer model. It was originally developed to simulate sediment and nutrient transport from agricultural watersheds in Minnesota, but the principles are not limited exclusively to that state.

The basic components of the model are hydrology, erosion, sediment transport, nutrient transport, and chemical oxygen demand. The model works on a cell basis. The cells are uniform square areas that collectively represent the watershed. Contaminants are routed from the headwaters of the watershed to the outlet in a stepwise fashion so that flow through any cell may be examined.

AGNPS can be used to evaluate non-point source pollution from agricultural watersheds. It can compare the effects of implementing various conservation alternatives within the watershed. Cropping systems, fertilizer application rates and timing, point source loads, and the effect of terraced fields can be modeled.

The model partitions soluble nitrogen and phosphorus between surface runoff and infiltration. Chemical oxygen demand and nutrient contributions from manure sources are assumed to be soluble and transported with runoff. Once the soluble pollutants reach concentrated flow, they are conservative and accumulate in the flow. Sediment-transported nitrogen and phosphorus are determined. An erosion process, adjusting for slope shape, is used to predict local sediment yield within the originating cell. Sediment and runoff routing through impoundment terrace systems is also simulated.

SWRRB (also referenced as SWRRBWQ to denote water quality implications) was developed at USDA-ARS and Temple Agricultural Experiment Station, Temple, Texas, and the National Soil Erosion Research Laboratory, Purdue University, West Lafayette, Indiana. It is designed for simulating hydrologic and related processes in rural basins. The objective of the model is to predict the effect of management decisions on water, sediment, nutrient, and pesticide yields at the sub-basin or basin outlet. SWRRB is a comprehensive, continuous simulation model covering aspects of the hydrologic cycle, pond and reservoir storage, sedimentation, crop growth, nutrient yield, and pesticide fate. A basin can be divided into a maximum of 10 sub-basins to account for differences in soils, land use, crops, topography, vegetation, or weather. SWRRB allows for simultaneous computation on each sub-basin and routes the water, sediment, and chemicals from the sub-basin outlets to the basin outlet. It also has a lake water quality component that tracks the fate of pesticides and phosphorus from their initial application on the land to their final deposition in a lake.

SWRRB can model the effect of farm-level management systems such as crop rotations, tillage, planting date, irrigation scheduling, and fertilizer, and pesticide application rates and timing. The effect these management systems have on the movement of contaminants can be tracked from the farm to the watershed outlet.

EPIC was developed at USDA-ARS and Texas Agricultural Experiment Station, Temple, Texas. EPIC was originally developed to estimate the long-term relationship between soil erosion and soil productivity and was titled Erosion Productivity Impact Calculator. EPIC is a continuous-simulation, field-scale computer model that can be used to determine the effect of management strategies on crop yield. The drainage area considered by EPIC is generally a field-sized area, up to 250 acres, in which weather, soils, and management systems are assumed to be homogeneous. The major components in EPIC are weather simulation, hydrology, erosion-sedimentation, nutrient cycling, pesticide fate, plant growth, soil temperature, tillage, economics, and plant environment control.

EPIC was designed to compare management systems and their effects on nitrate, phosphorus, pesticides, and sediment. The management components that can be changed are crop rotations, tillage operations, irrigation scheduling, and nutrient and pesticide application rates and timing.

The model partitions nitrate losses among surface runoff, sub-surface lateral flow, leaching below the root zone, crop uptake, denitrification, and transport by soil evaporation. Other nitrogen processes modeled include organic N transport with sediment and mineralization. Soluble phosphorus in surface runoff and phosphorus suspended with sediment are also determined. Daily accounting of pesticide fate is modeled to include pesticide transported with surface runoff and sediment, leached below the root zone, and lost through degradation. Estimates of soil loss by water erosion or sediment yield are also calculated.

Automatic irrigation and fertilization options exist in EPIC. The time and quantity of irrigation water applications can be automatically scheduled to maintain the root zone water content at field capacity. The fertilization option applies N and P at planting and continues the nutrient applications throughout the growing season to satisfy crop needs. Animal manures can be used as a nutrient source.

GLEAMS was developed at USDA-ARS, Tifton, Georgia. This continuous-simulation, field-scale model was developed as an extension of the Chemicals, Runoff and Erosion from Agricultural Management Systems (CREAMS) (Knisel, 1980) model. GLEAMS assumes that a field has homogeneous land use, soils, and precipitation. It consists of four major components: hydrology, erosion/sediment yield, nutrient transformation and transport, and pesticide transport. GLEAMS was developed to evaluate the impact of management practices on potential agrichemical leaching within, through, and below the root zone. Nutrients from animal manures are considered separately from inorganic fertilizers. It also estimates surface runoff and sediment losses from the field. GLEAMS was not developed as an absolute predictor of pollutant loading. It is a tool for comparative analysis of complex nutrient and pesticide chemistry, soil properties, and climate. GLEAMS can be used to assess the effect of farm-level management decisions on water quality.

GLEAMS can provide estimates of the impact that management systems, such as planting dates, cropping systems, irrigation scheduling, and tillage operations, have on the potential for agrichemical movement. The model also accounts for varying soils and weather in determining leaching potential and can be useful in long-term simulations for pesticide screening of soil/management systems and is often used in that context.

One use of the HU/WQ Tool and imbedded models is to support the NRCS nine-step planning process (Stinivasen et al., 1993) in integrating ecosystem concerns to address water quality issues on a watershed basis.

The nine steps in the planning process are as follows:

1. Identify the Problem
2. Determine the Objectives
3. Inventory the Resources
4. Analyze the Resource Data
5. Formulate Alternative Solutions
6. Evaluate Alternative Solutions
7. Determine a Course of Action
8. Implement the Plan
9. Evaluate the Results

Of the nine steps, the HU/WQ Tool provides the most support for three steps: inventorying resources, analyzing the resources, and evaluating alternative solutions.

To support step 3, Inventory the Resources, the HU/WQ Tool provides an automated means to keep track of large amounts of collected and reference data and provides easy retrieval methods. The tool basically organizes data into a project associated with a particular watershed. This project may have one or more simulations that correspond to unique runs of a water quality model.

To support step 4, Analyze the Resource Data, the capabilities of the HU/WQ Tool allow for quick and efficient methods of generating graphic and tabular results. With the integration of the water quality models and geographic information system, data that could previously be seen only in a tabular format can now be examined spatially. The user-friendly menu system provides viewing and comparison capabilities of project and simulation data.

Each model used with the HU/WQ Tool has limits on generating quantitative estimates because they were developed for specific purposes. For example, results generated by AGNPS are for single storm events and should not be compared directly to averages from SWRRBWQ. Results from GLEAMS and EPIC are similar; however, GLEAMS looks at a whole field-sized watershed whereas EPIC focuses more at a point.

Alternative solutions are analyzed and compared to determine if they meet the client's objectives, comply with NRCS policy, and satisfy technical and legal requirements of the ecosystem. Using the HU/WQ Tool, simulations can help determine results of alternative plans of action before they are actually implemented. In this way, the most feasible solutions can be examined.

HUMUS also has application in the planning process. Although the present HUMUS data base is national in scope, the tool can be used to identify portions of larger watersheds with the most contribution of water quality pollutants. Identification of these "hot spots" within a larger ecosystem allows the planner to zero in on a more detailed study that will identify the critical interactions within the smaller ecosystem.

CONCLUSION

As discussed above, tools are available to incorporate ecosystem concerns into watershed concepts in an orderly planning process. Introduction of animal manures from confined livestock and poultry add a level of complexity to the planning process, but can be addressed with existing tools. What does the future hold? Undoubtedly more sophisticated adaptations of existing ecosystem related tools as well as new tools will be incorporated into our planning process. There certainly does not appear to be any future for a retreat from watershed management based on ecologically sound principles.

REFERENCES

Bennett, Hugh H., Wildlife and the Soil Conservation Service Program, talk to the Soil Conservation Service regional biologists, Washington, D.C., 1938.

Follett, R.F., Keeney, D.R., and Cruse, R.M., eds., *Managing Nitrogen for Groundwater Quality and Farm Profitability*, Soil Science Society of America, Inc., Madison, Wisconsin, 1991.

Knisel, W.G., ed., *CREAMS: A Field Scale Model for Chemicals, Runoff, and Erosion from Agricultural Management Systems*, Conservation Resource Report No. 26, USDA - Science and Education Administration, Washington, DC, 1980.

Knisel, W.G., Ed., *GLEAMS: Groundwater Loading Effects of Agricultural Management Systems,* University of Georgia Coastal Plains Experiment Station, Biological and Agricultural Engineering Department Publication No. 5, 1993.

Leonard, R.A., Knisel, W.G., and Still, D.A., GLEAMS: Groundwater loading effects of agricultural management systems, *Trans ASAE,* 30(5), 1403, 1987

Odum, Eugene, *Fundamentals of Ecology,* 2nd edition, Philadelphia: W.B. Saunders Company, 1959.

Sharpley, A.N., and Williams, J.R., eds., *EPIC -- Erosion Productivity Impact Calculator, 1, Documentation,* USDA Tech. Bulletin # 1768, 1990.

Simons, D.B., Overview of Stream Mechanics and River Systems Analysis - SCS Short Course Handout, 1980.

Stinivasen, R., Arnold, J., Muttiah, R.S., Walker, C., and Dyke, P.T., *Hydrologic Unit Model for the United States. Advances in Hydro-Science and Engineering,* Volume 1, Part A., 1993

Strahler, A.N., Hypsometric analysis of erosional topography, *Bulletin of the Geological Society of America,* 63, 1120, 1950.

USDA-Soil Conservation Service, *Field Office Technical Guide, General Manual,* GM 450, Part 401, Washington, D.C., February 1990.

USDA-Soil Conservation Service, *National Planning Procedures Handbook,* Part 600.2, Washington, DC, September 1993.

USDA-Soil Conservation Service, *Hydrologic Unit Water Quality Tool (Beta Version 1.0) User's Guide,* Fort Collins, Colorado, 1994.

Van Warren, L., Shapiro, M., Westervelt, J., *Geographical Resources Analysis Support System (GRASS),* US Army Construction Support Laboratory, Champaign, Illinois, September 1988.

Webster's New World Dictionary of American English, Third college edition, Neufeldt, Victoria, Editor in chief, Simon and Schuster, Inc., New York, Page 1510, 1988.

Williams, J.R., Nicks, A.D., Arnold, J.G., Simulator for water resources in rural basins, *Journal of Hydraulic Engineering,* ASCE 111(6), 970, 1985

Young, R.A., Onstad, C.A., Bosch, D.D., and Anderson, W.P., AGNPS: A nonpoint-source pollution model for evaluating agricultural watersheds, *Journal of Soil and Water Conservation,* 168, 1989.

NUTRIENT ANALYSIS OF THE MALIBU WATERSHED USING THE AGNPS MODEL

V.L. Finney, M.A. Cocke, S.T. Moorhead, and Jay Klug

INTRODUCTION

The AGNPS (Agricultural Non-point Source) watershed model (Young et al., 1987) is widely used as a planning tool. In this chapter, version 4.02 of AGNPS is used to track soluble nitrogen and soluble phosphorus movement through the Malibu Creek Watershed for a 2.2-in., six-hour duration precipitation. Malibu Creek Watershed, with a 109-mi² drainage area, discharges into Santa Monica Bay, California.

A comparison is made between AGNPS calculations of baseline nutrient conditions, the addition of fertilizers and animal waste, and the addition of tertiary treatment point sources.

BASIN MODEL

The Malibu Watershed was divided into eight sub-watersheds (Figure 1) to preserve the integrity of the model AGNPS and to facilitate data management. AGNPS runs were made for each sub-watershed for a 2.2-in., six-hour duration precipitation. Lake samples of nitrogen (N) and phosphorus (P) taken in November of 1993 were used to determine soluble N and P concentrations of waters discharging into downstream sub-watersheds. The Soil Conservation Service (SCS) TR-20 (SCS, Rev. 1982) model was used to calculate discharges into these sub-watersheds. For those sub-watersheds with no lakes at their outlets, AGNPS calculations were used for waters and soluble nutrients discharging into downstream sub-watersheds.

To more accurately depict watershed conditions, model default values were replaced where better data existed. Non-point source nutrients were entered as fertilizer and point source nutrients as point sources.

CHANGED MODEL DEFAULTS

The default rainfall N concentration used in the model AGNPS is 0.8 mg/L. A calculated mean annual concentration of 0.35 mg/L was used. The calculation is based on an annual loading of 1.36 kg/ha/year (Chapin and Uttormark, 1973) and a mean rainfall of 39.40 cm/year.

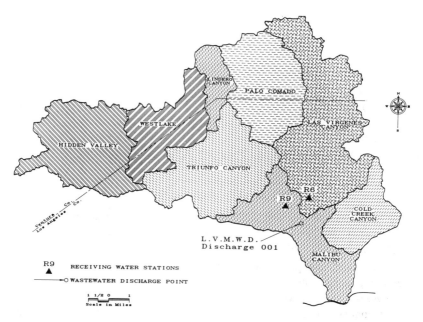

Figure 1. AGNPS Sub-watersheds.

AGNPS default values for soil pore nitrogen, soil pore phosphorus, organic matter, total phosphorus, and total nitrogen were replaced. Replacement values were based on analysis of 20 soil samples collected by SCS soil scientists and tested by the SCS Midwest National Soils Laboratory in Lincoln, Nebraska, for particle size, organic C, Bray P, total N, total P, and water-soluble P.

The 20 soil samples were assigned to AGNPS textural classes (Figure 2). Average values by textural class were determined for soil pore nitrogen, soil pore phosphorus, organic matter, total phosphorus, and total nitrogen. Concentration of soluble P and soluble N in surface soil layers was limited to an upper level of 5 mg/L (Frere et al., 1980). Water-soluble N was not analyzed; total N was used to rank the soil samples for soluble N within the range 2 to 5 mg/L.

NON-POINT SOURCE NUTRIENT INPUTS

Fertilizer inputs for the model were developed for agricultural land uses based on the crop grown. The principal crops were pasture and hay. Additional nutrients were applied from the animals that graze the harvested hay fields.

Nutrient contributions from grazing animals were developed from tabular data on the characteristics of animal waste per pound of animal (SCS, 1992). Animals were counted and their weight estimated.

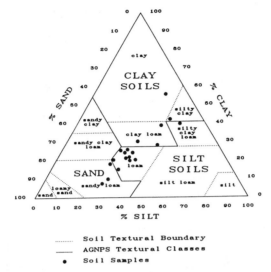

Figure 2. Plotted soil samples.

The amount of fertilizer applied to other identified land uses was calculated using typical application values for the different land uses (SCS, 1993). The acreage and type of fertilized area was developed from 1989 false color infra-red aerial photography. The density of the lawn and landscaped areas was estimated. Irrigation water contributions were developed based on two samples taken in May and August of 1993 in a 285-unit subdivision.

POINT SOURCE NUTRIENT INPUTS

The 1990 Census data was used to develop the density of septic tank systems in the study area. The amount of daily septage was calculated for the number of households in each sub-watershed. The average household contribution to the system is 150 gal/day. This rate was converted to cubic feet per second for each sub-watershed. Septic tanks provide primary treatment of household wastes. The liquid portion goes to the leach field where the portion lost to evapotranspiration was estimated to be 50%. The resulting net flow rate was entered as a point source at the outlet of each sub-watershed. Chemical concentrations of soluble N, 18 mg/L, and soluble P, 0.5 mg/L, were derived from tabular data sets (Culp/Wesner/Culp, 1986; SCS, 1992).

The largest point source is the Tapia waste water treatment plant. A discharge of 7 cfs and soluble N and P concentrations of 24 and 7 mg/L, respectively, were entered into the cell downstream of the Tapia waste water treatment plant. These values are at the higher range of reported data for 1991 and 1992 (LVMWD, 1991 and 1992).

LAKE NITROGEN AND PHOSPHORUS TESTING

AGNPS does not model lakes, and no lake model has been proven in the study area. Therefore, lake samples were taken and tested to provide values for lake nutrients in modelling soluble nitrogen and soluble phosphorus movement through the Malibu Creek Watershed. Due to time constraints, one-time grab samples were taken. The assumption was made that inflowing waters from a two-year frequency event would displace an equivalent volume of lake water. Surface concentrations of soluble N and P samples (range of soluble N 0.10 to 0.30 mg/L, range of soluble P 0.05 to 0.45 mg/L) taken on 15 and 16 November 1993 were input to AGNPS downstream sub-watersheds as point sources. The point source discharge used was that calculated for a two-year frequency event using the SCS TR-20 model.

STREAM NITROGEN AND PHOSPHORUS TESTING

Two stream gage sampling stations run by the Las Virgenes Municipal Water District (LVMWD) were used to check the model AGNPS for reasonability. Soluble N and soluble P for water years 1991 and 1992 at stream gages R-6 and R-9 (Figure 1) are plotted in Figures 3 and 4. A linear regression line relating N concentration to discharge was plotted. The slope of this line was tested using the t-statistic to determine if it was signifiantly different from zero. The computed t-value was 0.08, indicating that it was not. Similar results were obtained for P.. Since no relationship was found between discharge and concentrations, the calculation of means of the data sets was used for statistical comparisons. The calculated soluble N and P means, for 1991-1992, at a 95% confidence interval for stream samples taken by the LVMWD are shown in Table 1.

No quantifiable data exist for point sources above gages R-6 and R-9. AGNPS calculated N and P concentrations for a two-year frequency storm are also plotted on Figures 3 and 4.

MODEL RESULTS

Initial runs with the AGNPS decay default functions proved inadequate. The effect of additions of fertilizer and point sources were measurable at the outlets of cells where applied but were not measurable, to two significant figures, at the cell outlet of the next downstream cell. Also, AGNPS calculations of soluble N and P at the cell outlets of the two cells used as a comparison to stream gage data were non-measurable to two significant figures. These results were not acceptable. AGNPS 4.02 does not allow the user to completely shut off the decay functions. Thus, AGNPS runs were made at user input decay rates of 1 and 10%. The 1% decay rate gave AGNPS calculations at the same order of magnitude as the stream gage means. The calculated AGNPS values compared to the 1991-1992 means at LVMWD stream gages R-6 and R-9 are shown in Table 2.

Table 1. Soluble nitrogen (N) and soluble phosphorus (P) stream gage data

Stream Gages	Soluble N mg/L	Soluble P mg/L
R-6	2.3+/-0.3	0.4+/-0.1
R-9	0.5+/-0.3	0.1+/-0.2

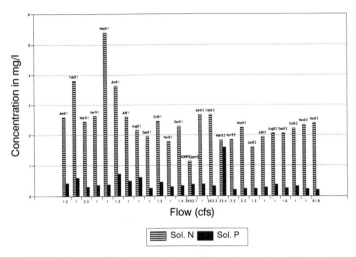

Figure 3. Soluble nitrogen (N) and phosphorus (P) grab samples at stream site R6 and AGNPS calculation.

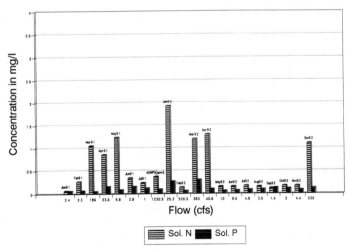

Figure 4. Soluble nitrogen (N) and phosphorus (P) grab samples at stream site R9 and AGNPS calculation.

Table 2. Stream gage data compared to AGNPS calculations.

Station	Soluble N and P Station Means	Soluble N and P AGNPS Calculations
	----------------------mg/L--------------------	
R-6		
N	2.3	1.14
P	0.4	0.4
R-9		
N	0.5	0.4
P	0.1	0.1

AGNPS calculated soluble nutrient loadings to watershed lakes, sub-watershed outlets, and Malibu Lagoon are shown in Table 3. Based on these calculations, the addition of fertilizer and animal waste accounts for a 43.8% increase in water soluble N, and a 85.7% increase in water soluble P delivered to Malibu Lagoon. The addition of Tapia's treated effluent accounts for most of the AGNPS calculated additional point source increases of 10.9% in water soluble N, and 7.7% in water soluble P. Septic tank effluent as a point source was insignificant.

CONCLUSIONS

The decay equations added to versions AGNPS 4.0+ did not improve the application of the model in this study. Dissolved oxygen of the sampled water was higher than 1.0 mg/l. No denitrification should be expected in the flowing waters of Malibu Creek and Las Virgenes Creek (Metcalf and Eddy, 1991). At gage R-6 for a 2-year frequency, 6-hour-duration storm,TR-20 calculated a peak discharge of 459 cfs at 9.8 hours with the end of the hydrograph coming at 30.5 hours. At gage R-9 for the same storm, the calculated peak discharge

Table 3. AGNPS nitrogen (N) and phosphorus (P) loadings to sub-watershed lakes and outlets for a 2.2-in., 6-hour duration storm.

Sub-watershed Outlets	Baseline Water Soluble		Fertilizer and Animal Waste Water Soluble		Fertilizer, Animal Waste, Point Sources Water Soluble	
	Nitrogen	Phosphorus	Nitrogen	Phosphorus	Nitrogen	Phosphorus
	----------------------------------mg/L----------------------------------					
Hidden Valley 1[†]	0.38	0.07	4.17	0.7	4.17	0.7
Westlake 1	0.23	0.07	2.3	0.87	2.3	0.87
Lindero 1	0.34	0.08	3.03	1.1	3.03	1.1
Palo Comado 2	0.34	0.08	2.31	0.85	No point sources	
Triunfo Canyon 1	0.3	0.08	1.32	0.47	1.32	0.47
Las Virgenes Cyn. 2	0.34	0.07	1.03	0.34	1.03	0.34
Cold Creek Cyn. 2	0.37	0.08	1.05	0.34	1.05	0.34
Malibu Canyon 2	0.32	0.07	0.46	0.13	0.51	0.14

[†]1 = Total stream loadings to lake; 2 = stream loadings to sub-watershed outlets.

was 2307 cfs at 13.7 hours with the end of the hydrograph coming at 42.6 hours.

Using a 1% decay, reasonable average values for soluble N and P were calculated by AGNPS. The calculated percent contributions of soluble N and P from fertilizer and animal waste and effluent from the Tapia waste treatment plant seem reasonable.

The largest discrepancy between AGNPS calculations and calculated stream measurement means was the soluble N at gage R-6. This discrepancy may be due to the lack of quantification of point sources above gage R-6.

ACKNOWLEDGMENTS

The authors wish to thank Natural Resource Conservation Service, formerly SCS, employee Ken Oster for collecting soil samples, Tom Share for preparation of Figures 1 and 2, and Susan Southard for coordinating the field collection and lab analysis of soil samples by the SCS National Soil Survey Center in Lincoln, Nebraska. The authors also wish to thank Rene McGaugh for making the TR-20 runs. Special thanks are owed to Jacqy Gamble of the Las Virgenes Municipal Water District for the analyses of lake samples and providing the stream data and Tapia discharge data.

REFERENCES

Chapin, J.D., and Uttormark, P.D., *Atmospheric Contributions Of Nitrogen And Phosphorus*, Technical Report Wis-Wrc-73-2, The University of Wisconsin Water Resources Center, Hydraulic and Sanitary Laboratory, Madison, Wisconsin 53706, 35, 1973.

Culp/Wesner/Culp, *Handbook of Public Water Systems*, B. Williams and Gorden L. Culp, eds., Van Nostrand Reinhold, New York, 1986.

Frere, M.H., Ross, J.D., and Lane, L.J., *The nutrient submodel. CREAMS, A Field Scale Model for Chemicals, Runoff, and Erosion from Agricultural Management Systems*, Cons. Res. Rpt. No. 26. Agr. Res. Serv., US Dept. Agr., Washington, 1980.

Las Virgenes Municipal Water District (LVMWD), *1991 NPDES Annual Report*, NPDES Permit CA0056014, File C1 4760, Order 89-076, Effluent Discharge To Malibu Creek, LVMWD Report # 1861, Las Virgenes Municipal Water District (LVMWD), Tapia Water Reclamation Facility, Laboratory Assurance Program, 4232 Las Virgenes Road, Calabasas, California, 91302, 1991.

LVMWD, Personal Communication, Tapia Water Reclamation Facility, Laboratory Assurance Program, 4232 Las Virgenes Road, Calabasas, California, 91302, 1992.

SCS, *Computer Program For Project Formulation - Hydrology*, Technical Release Number 20 (TR-20), United States Department of Agriculture, Soil Conservation Service (SCS), Engineering Division, Washington, D.C., 308. Rev. 1982.

SCS, *Agricultural Waste Management Field Handbook*, National Engineering Handbook, Part 651, United States Department of Agriculture, Soil Conservation Service (SCS), Engineering Division, Washington, D.C., 638, 1992.

SCS, *Escondido Creek Hydrologic Area, Project Report*, San Diego County California, Water Resources Planning Staff, United States Department of Agriculture, Soil Conservation Service (SCS), Davis, CA., 27, 1993.

Metcalf and Eddy, Inc., *Wastewater Engineering - Treatment, Disposal, and Reuse*, Revised by George Tchobanoglous and Franklin L. Burton, Mc Graw Hill Inc., 1991.

Young, R.A., Onstad, C.A., Bosch, D.D., and Andersons, W.P., *AGNPS, Agricultural Nonpoint Source Pollution Model: A Watershed Analysis Tool*, Cons. Res. Rpt. 35, Agr. Res. Serv., US Dept. Agr., Washington, D.C., 1987.

MANAGEMENT STRATEGIES FOR LAND-BASED DISPOSAL OF ANIMAL WASTES: HYDROLOGIC IMPLICATIONS

W.J. Gburek and H.B. Pionke

INTRODUCTION

Shuyler (1994) stated, "Nutrient management should reduce soluble nutrient transport and (yet) provide enough nutrients to produce a realistic crop yield." However, nutrient management can also be oriented toward animal waste *disposal*, where wastes are applied to the land surface such that nutrients are in excess of crop needs. Under these conditions, it becomes critical that we have the capability to predict nutrient movement within and from a watershed flow system.

Development of an animal waste disposal strategy must address down gradient water quality impacts. It should integrate effects at the local scale where management is implemented (i.e., the field), with the scale of the logical management unit (i.e., the farm), and finally with the larger scale at which results of the strategy are sampled and evaluated (i.e., the watershed). A management strategy developed in this way must incorporate definition of what we call "critical source areas" (CSAs), identifiable zones which are dominant controls on export of nutrients from the watershed.

The single most important factor needed to define these CSAs is knowledge of water movement within the watershed, i.e., the hydrologic cycle. Water can translocate contaminants from source-areas of application (typically on or within the soil zone), to or through zones of reaction and sinks within the watershed (in either the surface or sub-surface), and finally to positions where they can be removed from the watershed (generally by stream flow but possibly via ground water). Thus, the key to understanding contaminant transport within and from a watershed is detailed understanding of pathways of flow.

In this chapter, we consider hydrologic implications for development of land-based animal waste disposal strategies. Emphasis is on humid-climate upland watershed conditions, since such watersheds are the dominant source of water and chemicals to larger downstream flow systems, such as rivers, reservoirs, lakes, or estuaries. Research results from an upland agricultural watershed in east-central Pennsylvania are presented as case studies.

NUTRIENTS AND FLOW COMPONENTS OF CONCERN

To determine flow components of the hydrologic cycle critical to the problem of animal waste disposal, we must define the important nutrients and their dominant interactions with the natural flow system. Nutrients within animal waste that can be transported by a watershed's flow system to affect down gradient positions are nitrogen (N) and phosphorus (P). We are not concerned with application methodologies, rates, or method of spreading or incorporating here; rather, we address only the transport and fate of "excess" N and/or P introduced into the flow system.

Hydrologic controls on the transport of N and P within the watershed are different; in some ways they are opposed. In the case of P, surface runoff, with its associated water-borne sediment, is the flow component of most concern. In the case of N, though (concern is usually with nitrate, a soluble species), the sub-surface flow system is important since nitrate is generally moved into the sub-surface by infiltrating water and ultimately exported from the watershed, primarily via sub-surface flow.

To complicate the issue, all components of the hydrologic cycle are closely connected within the watershed conditions considered. Rainfall, soil moisture, evapotranspiration, ground water recharge, surface runoff, and ground water discharge to the stream respond at the same time scales, event-based and seasonal. Further, surface and sub-surface waters are both important within and from these watersheds. The rural population of the East relies almost entirely on ground water for water supply, so effects of animal waste disposal on ground water quality is of concern. This same ground water is also the dominant source of stream flow, up to 70-80% of annual flow (Gburek et al., 1986). The sub-surface-derived flow, combined with surface runoff, provides the water for the larger rivers, impoundments, and estuaries important to fisheries, recreation, and water supply.

WATERSHED HYDROLOGY - RESPONSE VERSUS FLOW PATHS

Present views of hydrology, and consequently of how contaminants move through a watershed, remain slanted by tradition. Watershed hydrology was founded on the concept of defining watershed-scale response to inputs. When rain occurs, the watershed produces a storm hydrograph; when a well is pumped, the ground water table drops. Because water quality was not originally of concern, hydrologists gave minimal consideration to pathways of water movement when quantifying these responses.

Our concern with water quality, beginning in the late 1960s, required an alternative view. Quantifying contaminant transport requires knowledge of pathways of water movement, not simply watershed response. We must specify where contaminants are introduced, what CSAs they move through and how fast, and, finally, where they exit the watershed. All these require definition of flow paths from all positions over the watershed land surface to the watershed outlet, within both the surface and sub-surface systems.

A HYDROLOGIC BASIS FOR ANIMAL WASTE DISPOSAL

Management strategies for land-based disposal of animal wastes can be developed in context of important aspects of upland watershed hydrology-- layering and fracturing within shallow aquifers and variable-source-area watershed response. They must also incorporate effects of the riparian zone and describe all relevant interactions at the watershed scale.

At the most basic, the intersection of surface runoff and ground water recharge source-areas with areas of animal waste disposal over the landscape is what creates the initial CSAs controlling export of P and N, respectively, from a watershed. Complications arise when pathways of flow from these CSAs are considered; they may interact with down gradient CSAs that alter concentration and/or mass of a contaminant by dilution, reaction, or sink types of processes, adding an additional control on contaminant export.

GROUND WATER

Limited amounts of nitrate are moved to the stream in surface runoff, but most environmentally related problems occur when nitrate from fertilizers or animal wastes is infiltrated and escapes from the root zone to enter the subsurface. We will not consider the unsaturated zone--basically, it provides a control on timing to the movement of water and nitrate from the root zone to the water table. At the watershed scale, we typically assume that unsaturated flow is one-dimensional vertically, so contaminants introduced thereto by a particular land use initially impact ground water directly below. Only after they reach the water table do we become concerned with their movement laterally and with depth throughout the ground water body.

Ground water flow systems have traditionally been analyzed in terms of deeper water supply aquifers and homogeneous media. Flow paths developed for these analyses extend deep into the sub-surface and are regular in form, while travel times are at scales of years, decades, and even centuries. However, concerns with non-point source pollution from agriculture have required us to become more familiar with the characteristics of the shallow surficial aquifers, those most easily and more likely affected by land use. This leads directly to a focus on effects of layering and fracturing on patterns of flow and contaminant transport, as well as the shallow aquifer's interactions with the surface water flow system.

AQUIFER LAYERING

The land surface is typically underlain by relatively conductive layers of soil and rock; these are, in turn, underlain by less permeable strata. The more permeable layers forming the surficial aquifer can range in thickness from a few meters to tens of meters. Ground water moving within the highly conductive surficial layers generally forms an unconfined aquifer, i.e., an aquifer in which the water table is the upper boundary of the saturated zone.

Water quality in the shallower zones of surficial aquifers tends to be a reflection of the immediately overlying and up gradient land uses. The degree to which the aquifer is affected deeper is a direct result of the severity of layering of conductivity in the vertical (generally decreasing with depth). The more extreme this layering, the more the flow paths remain in the shallower parts of the aquifer, and the less the deeper aquifer is affected by overlying and up gradient land uses.

BEDROCK FRACTURING

Heath (1984) showed that nearly all ground water regions of the eastern US have some degree of fracture control on ground water flow within the shallow bedrock. The shallow zone of fracturing may be overlain by relatively thick glacial deposits or regolith, or as in the case study described subsequently, by only a thin soil. In all cases, though, ground water within the shallow fracture zones is affected directly by overlying land use.

Bedrock fracturing exaggerates the characteristics of the aquifer, producing extremes in the properties governing flow and transport--high hydraulic conductivity (compared to matrix properties alone) and relatively low specific yield (considered approximately the effective porosity). When combined with layered conditions discussed previously, fracturing is a major influence on patterns of ground water quality, and as such, must be considered in development of waste disposal strategies.

CASE STUDY

The EPIC root zone model, MODFLOW ground water model, and a generic contaminant transport model were used to examine effects of layered and fractured aquifer geometry on patterns of nitrate contamination in ground water (Gburek, 1993). Flow modeling was based on aquifer properties previously developed, and a simulated corn land use provided high nitrate inputs to the ground water. These can be considered analogous to excess nitrate inputs from disposal of animal wastes. Land use over the remainder of the cross section provided minimal inputs of N.

Figure 1 shows concentration patterns resulting from the distribution of high N inputs shown by the corn icons. With the corn land use positioned directly on the divide at the south, the entire depth of the aquifer is contaminated with a nitrate-N concentration of 16.0 mg/L from the corn percolate. Down slope of the corn, however, ground water in the shallow fracture zone exhibits continuing dilution of this higher concentration by recharge from the unfertilized land use. In the north, where corn land use is adjacent to the stream, only the shallowest portions of the flow system are affected. The 16.0 mg/L nitrate-N input is diluted by low-concentration ground water from the land use

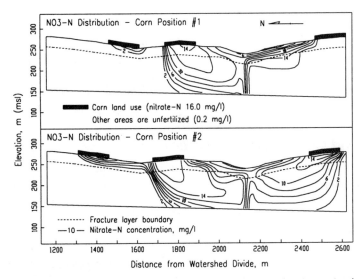

Figure 1. Effects of layered fractured aquifer geometry on ground water contamination.

up slope moving laterally through the shallow fracture zone. Lastly, the corn land use in the middle of the cross section is directly over the topographic divide, yet its effects are seen only on one side of this divide. The ground water divide is over 100 m to the north of the topographic divide, so land use within that distance of the topographic divide affects only ground water to the south.

The lower panel of Figure 1 shows changes resulting from moving the corn inputs only short distances. Moving the corn field 30 m off the southern divide reduces nitrate-N concentrations within the deeper aquifer to less than 10 mg/L and moderately reduces nitrate concentrations in all down slope portions within the fracture layer. This illustrates the very sensitive nature of the recharge zone to positioning of input sources in strongly layered hydrogeology. The corn land use adjacent to the stream was moved approximately midway up slope. There is still no major contamination of the deeper aquifer; only the zone of fracturing continues to be impacted, and even this is minor. The middle field of corn was moved entirely to the north side of the topographic divide, but it still impacts ground water only to the south of the divide, illustrating more dramatically the fact that ground water and topographic divides can be offset from one another.

Finally, nitrate concentrations of the two streams draining the section are identical under both land use configurations. This is because these are steady-state simulations with the same aggregate inputs of percolate and nitrate. This leads directly to the conclusion that for closed systems and non-reactive contaminants, positioning of land use does not affect long-term concentrations and/or loads of contaminants leaving the watershed in base flow as long as relative percentages of each land use remain constant.

SURFACE RUNOFF

Surface runoff is the direct result of rainfall impacting the land surface. It was traditionally thought to be a soil-controlled phenomenon occurring over the entire watershed, and techniques developed to predict storm runoff (e.g., curve number) were based on this assumption. Recently, however, partial-area hydrology, which then evolved to variable-source-area hydrology, has become accepted as a descriptor of watershed response to precipitation.

VARIABLE-SOURCE-AREA HYDROLOGY

In the humid east, variable-source-area (VSA) hydrology controls runoff production (Ward, 1984). The basic premise of the VSA hydrologic concept is that there is a contributing sub-watershed within the topographically defined watershed that varies in time; it expands and contracts rapidly during a storm as a function of precipitation, soils, topography, ground water level, and moisture status. VSA runoff is dominated by saturation overland flow and rapidly responding sub-surface flow. The remainder of the watershed provides no runoff, only infiltration and ground water recharge.

CASE STUDY

A methodology for identifying the P-exporting areas of a watershed has recently been developed using a VSA-based storm runoff model integrated within the GRASS GIS framework (Zollweg et al., 1995). Data layers representing hydrologic processes and P availability over the watershed are required. Early simulation results compare favorably to observed P output data from a wide range of storms on a 26-ha agricultural watershed.

Figure 2 shows the result of applying the model to a 21-mm storm from April 1992. Simulated runoff distribution over the watershed is shown as the top layer and Bray P (representing P availability) on the middle layer. On the bottom, simulated soluble P loss over the watershed is shown as the result of runoff intersecting areas of available P. Runoff is generated primarily in near-stream zones, but other locations at distance from the stream also generate runoff; these tend to be at slope breaks, in areas of converging sub-surface flow, or where slopes are generally shallow. Phosphorus availability is low in the near-stream areas of the lower part of the watershed (the left); while runoff amounts are high, P loss per unit area is low. Phosphorus loss is much higher over upper watershed positions where storm runoff originates within cropped fields having higher P availability. Losses generally reflect P availability where runoff occurred--higher where there was high runoff and significant available P, and lower where there was low runoff and/or low P.

RIPARIAN ZONE

The riparian zone (RZ) is that portion of the watershed where sub-surface flow intersects the land surface, usually in a near-stream position, causing high

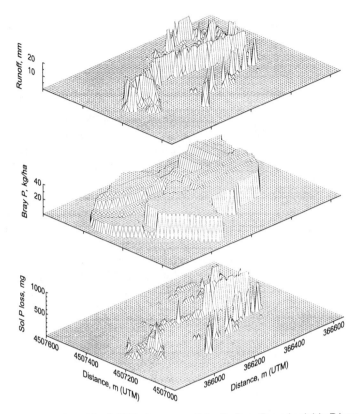

Figure 2. Simulation of April 1992 storm; runoff depth, Bray P, and soluble P loss.

water tables and moisture levels. Elevated levels of nitrate are expected in ground water below agricultural land use, but these may be attenuated enroute to becoming stream flow if the ground water passes through a RZ.

HYDROLOGIC CONSIDERATIONS

Cooper (1990) concluded that catchment hydrology, particularly flow paths through the RZ, determine the control of near-stream environments on pollutant flux. Whether riparian zone processes can substantially reduce nitrogen discharge to a stream, however, depends on both the areal extent of the RZ and the hydrologic linkages between the up gradient nitrate sources and the RZ. A variety of linkages with varying degrees of complexity are possible on a watershed, ranging from landscape configurations with short, shallow, direct paths between cropped areas and riparian zones, to those with deep flow paths through RZs of limited extent. When flow paths are shallow, nearly all up gradient drainage passes laterally through the RZ before discharging to the stream. In contrast, when flow paths are less constricted, more ground water by-passes the RZ, either as deeper flow before converging to the channel or as loss to some deeper ground water system. Flow path definition becomes critical to describe delivery of recharge from the agricultural land use to the RZ.

CASE STUDY

Figure 3 shows flow paths in a near-stream zone of the cross section from Figure 1 (Gburek et al., 1994). Figure 3A represents springtime high recharge conditions; 3B represents the late-summer. Flow paths are labeled with percentages of the total ground water moving above them. A "riparian zone" 20 m wide and 3 m deep is also shown; it represents an area potentially active in removal of nitrates from ground water.

It is easy to evaluate potential for the RZ to remove nitrates from up gradient land uses in this portrayal since entry of specific flow paths into the RZ is clearly defined. The high-recharge configuration shows that 95% of ground water flow through the section, and consequently 95% of the land surface area, is potentially affected by RZ processes. Under low-flow conditions, though, recharge from only about 80% of the land surface is fully influenced by the RZ. In evaluating potential RZ effects, the hydrologic status of the watershed associated with the two figures must also be considered. The high-recharge configuration represents springtime conditions when denitrification processes are less effective because of cooler temperatures and residence times within the RZ are shorter because of higher flow rates. Conversely, when potential for deni-

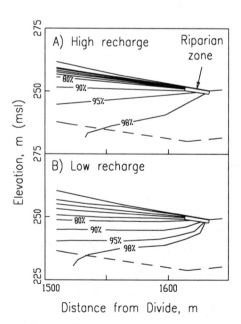

Figure 3. Flow paths in the riparian zone.

trification is greater (summer temperatures and longer residence times), more of the ground water bypasses the RZ. Thus, RZ processes must be integrated with hydrologic conditions to fully evaluate the potential for a landscape to attenuate nitrate inputs to the ground water.

IMPLICATIONS FOR ANIMAL WASTE DISPOSAL

We remain far removed from development of a single animal waste disposal strategy that can account for both N and P as well as all hydrologic implications. Modeling tools and field data are simply not available to integrate all aspects of the hydrologic cycle from the flow perspective alone, much less from that of water quality. However, we can draw conclusions based on what has been shown.

Animal wastes are relatively high in P compared to N when plant needs are considered. If animal wastes are applied to the land surface at rates of N sufficient for plant growth, there will be an excess of P at the soil surface. Hydrologic implications are that to control P loss, we control P application and build-up, primarily in near-stream zones. Levels in the soil at distance from the stream are of minimal concern since there is only a limited chance of runoff occurring to move the P to the stream. Thus, the most obvious control from the P point of view is to dispose of animal wastes on landscape positions at distance from the stream--the further the better.

Concern with nitrate is different though--we must control N in areas of ground water recharge. The ideal animal waste disposal strategy should provide for N in the soil at amounts needed by the plants, both in timing and areal distribution. If excess wastes are applied though, nitrate escaping crop uptake will move with any water available to move it. In this case, the management strategy becomes *disposal* (i.e., excess N) limited to those portions of the landscape that do not affect ground water zones critical to water supply. Lastly, if we can identify those flow paths that pass through the RZ, we can apply the excess N to watershed areas having the best chance to be affected by RZ processes, thereby realizing maximum reduction of input concentrations.

SUMMARY

There is no single strategy currently available for land-based disposal of animal wastes that minimizes impact of both N and P. Optimum disposal zones on the landscape are different for each. We do have an opportunity to isolate loss of P in animal wastes, since wastes can be disposed of at distance from the stream where transport mechanisms are not available. However, in these same positions, we must account for the N component and what parts of the subsurface will be affected by its disposal.

When all flow components of the hydrologic cycle are intimately connected, as in humid-climate upland watersheds, we must consider interactions among these components when developing effective animal waste disposal strat-

egies. Management must account for interactions among areas of waste disposal over the landscape, source areas of ground water recharge, patterns of
ground water movement and associated contaminant transport, surface runoff
source areas and associated transport, and riparian zone controls. To do this,
we must accurately portray the hydrologic flow system. External and internal
boundaries, controlling geometries, hydraulic properties, major flow path patterns, and areas of recharge, sub-surface discharge, and surface runoff must all
be defined and integrated into any disposal strategy.

Finally, the most critical problem of all may be interfacing this objective
and technical approach to animal waste disposal with the interests of land
owners. A purely hydrologic analysis implies that different levels of management for different parts of the watershed may be the most efficient approach to
minimize contamination of ground water and/or surface runoff. Thus, there
may be differing management practices suggested from farm to farm. A major
challenge in implementing an animal waste disposal management strategy
derived from these hydrologic implications may be simply to demonstrate to
the land owners that the results will be of sufficient benefit to override the
apparent inequities associated with its application over the watershed.

ACKNOWLEDGMENTS

Contribution from the Pasture Systems and Watershed Management Research Laboratory, US Department of Agriculture, Agricultural Research Service, in cooperation with the Pennsylvania Agricultural Experiment Station,
the Pennsylvania State University, University Park, Pennsylvania.

REFERENCES

Cooper, A.B., Nitrate depletion in the riparian zone and stream channel of a small headwater catchment, *Hydrobiologia*, 202, 13, 1990.

Gburek, W.J., Land use effects on ground water within a layered fractured aquifer, *Proc. ASAE Inter. Symp. on Integrated Resource Management and Landscape Modification for Environental Protection*, J.K. Mitchell, ed., ASAE Publication 13-93, 176, 1993.

Gburek, W.J., Folmar, G.J., and Schnabel, R.R., Ground water controls on hydrology and water quality within rural upland watersheds of the Chesapeake Bay Basin, *Proc. 1994 Chesapeake Res. Conf.*: accepted for publication, 1994.

Gburek, W.J., Urban, J.B., and Schnabel, R.R., Nitrate contamination of ground water in an upland Pennsylvania watershed, *Proc. Agric. Impacts on Ground Water, Nat. Water Well Assn.*, 352, 1986.

Heath, R.C., *Ground-Water Regions of the United States*, USGS Water Supply Paper 2242, U.S. Government Printing Office, Washington, DC, 1984.

Shuyler, L.R., Why nutrient management? in *Nutrient Mangement*, supplement to *J. Soil and Water Cons.* 94(2), 3, 1994.

Ward, R.C., On the response to precipitation of headwater streams in humid areas. *J. Hydrol.* 74, 171, 1984.

Zollweg, J.A., Gburek, W.J., Pionke, H.B., and Sharpley, A.W., GIS-based delineation of source areas of phosphorous within northeastern agricultural watersheds, *Proc. IAHS Symp. on Modelling and Management of Sustainable Basin-Scale Water Resource Systems*: accepted for publication, 1995.

NLEAP APPLICATION FOR DEVELOPING MUNICIPAL WELLHEAD PROTECTION STRATEGIES IN THE CENTRAL WISCONSIN SAND PLAIN

J. D. Kaap, W. Ebert, G. Kraft, and M.K. Brodahl

STUDY AREA

The Stevens Point, Whiting, Plover (SWP) Wellhead Protection Hydrologic Unit Project focuses on non-point source pollution in a 108-mi^2 basin located in the Central Wisconsin Sand Plain of Portage County (Figure 1). This basin is the ground water recharge area for municipal wells serving over 40,000 people in and near the communities of Stevens Point, Whiting, and Plover. The project's main goal is to encourage and facilitate voluntary adoption of Best Management Practices (BMPs) to reduce the risk of nitrate and pesticide contamination of ground water. The project focuses on the areas's diverse irrigated and dryland vegetable and field crops and dairy farming. The Nitrogen Leaching and Economic Analysis Package (NLEAP) was used to model nitrogen movement.

The basin is sensitive to ground water pollution because sandy soils overlie a sandy and gravel aquifer. Ground water is typically within 10 to 20 ft of the surface (Ebert et al., 1993). The area has a long history of ground water problems. In 1979, Whiting was forced to shut down its municipal well because the water exceeded the 10 mg/L NO_3-N standard (Hagan, 1992). In 1980, the pesticide aldicarb was detected in numerous domestic wells, two years after initial use (WDATCP, 1989). About 25% of rural domestic wells now exceed the nitrate standard (CWGC, 1994).

Land use (Figure 1, Table 1) consists of 38% cropland, 36% woods and wetland areas, and 26% commercial, roads, residential, pasture, idle land or surface water (Ebert and Moberg, 1992). About 38% of the cropland, or 10,000 acres, is sprinkler irrigated, mostly with center pivots. In a typical year, vegetables make up 7,100 acres (71%) of irrigated cropland. Potato is grown on approximately 3,600 acres, which is 5% of the basin area and accounts for 5% of the total Wisconsin potato acreage.

Land use within the SWP area is not uniform. Irrigated agriculture is concentrated in the southern part of the SWP (Figure 1). Some center pivot systems are near urban areas and municipal wells. Stevens Point's recharge area is mostly forest, wetlands, and dryland crops; but Whiting's and Plover's re-

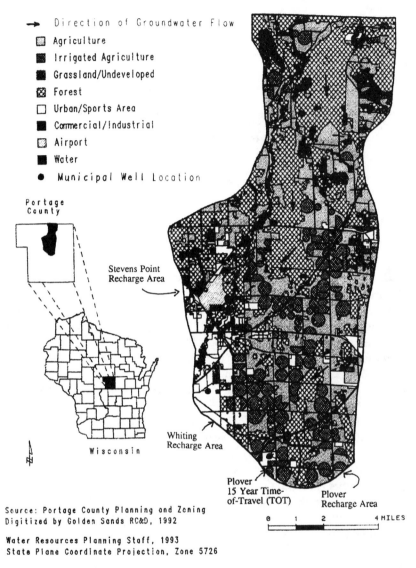

Figure 1. Stevens Point, Whiting, Plover wellhead protection project area.

charge areas are mostly irrigated and dryland cropland. Ground water moves south and southwest toward the urban areas. During the past five years, nitrate levels have increased in four municipal wells (Table 2) shown in Figure 1. Because their wells have exceeded the 10 mg/L NO_3-N standard, Plover and Whiting have installed nitrate removal systems totalling $2.7 million.

Table 1. Stevens Point, Whiting, Plover Wellhead Area Land Use Survey, 1991-1992.

Land Use	Acres	% of Acres
Dryland Alfalfa	9,200	13
Irrigated Alfalfa	1,000	1
Dryland Field Corn	4,600	7
Irrigated Field Corn	1,600	2
Irrigated Potato	3,600	5
Dryland Small Grain	2,100	3
Irrigated Small Grain	300	1
Irrigated Snap Bean	1,500	2
Irrigated Sweet Corn	1,500	2
Irrigated Peas	500	1
Agricultural reserve	600	1
Total Cropland	26,600	38
Woodland & wetland	25,000	36
Urban & Roads	9,500	14
Pasture & Idle Land	7,200	11
Surface Water	800	1
Totals	69,100	100

Table 2. Range of nitrate[†] levels of four municipal wells in the
Stevens Point, White, Plover wellhead area, 1989-1994.

Year	SP4[‡]	SP5[‡]	Whiting	Plover
	\---------------------------mg NO_3-N/L---------------------------			
1989	0-3	5-7	9-13	4-7
1990	0-4	3-6	9-13	6-8
1991	0-3	2-6	10-14	6-8
1992	1-3	2-7	11-15	7-11
1993	2-3	6-7	13-17	8-11
1994[#]	2-3	5-6	14-20	11-15

[†]Drinking water standard = 10 mg/l NO_3-N
[‡]SP4 and SP5 are Stevens Point wells No. 4 and 5
[#]Water sampling taken through July

OBJECTIVES OF THE CASE STUDY

The objective of the case study was to determine if implementing BMPs can improve or protect ground water in the SWP area. To evaluate this question, we chose a sub-area consisting of the 15-year time-of-travel (TOT) to the Plover municipal wells (Figure 1), that is, the area in which ground water can travel to the wells within 15 years. The 15-year TOT was chosen instead of the entire recharge area because (1) its land use is representative of the entire Plover recharge area, (2) its size is more manageable for gathering and interpreting data, and (3) it becomes more difficult to delineate wellhead protection areas having greater distances and travel times (Born et al., 1988). The results of this analysis are scalable to the entire zone of contribution. The Plover 15-year TOT is more intensively farmed with 39% irrigated cropland (Table 3)

Table 3. 1994 major land uses in the Plover 15-Year time of travel (TOT) area.

Land Use	Acres	% of Acres
Irrigated cropland	506	39
Non-irrigated cropland	240	19
Forest	340	26
Urban	132	12
Unsewered residences	35	3
Grass/brush	14	1
Roads	25	2
Total:	1292	100

compared to 14% for the entire SWP area (Table 1). Land use intensity may explain why the Plover well has 1.5-3.0 times more nitrate than the two Stevens Point wells (Table 2).

Integrated Crop Management (ICM) is considered the most important agricultural BMP to protect ground water. It includes nutrient, pesticide, irrigation, and manure management. To evaluate the ICM BMP, we will answer the following questions:

1. What is the steady-state nitrate concentration at the Plover well with and without ICM?
2. If our goal is to get Plover well water below the 10 mg/l NO_3-N standard, how much nitrate must be removed from the system?
3. If ICM does not sufficiently reduce nitrate levels, what are other options?

METHODS AND RESULTS

We used a "mixing model" to estimate NO_3-N concentrations at the Plover well. The model estimates NO_3-N concentrations by dividing annual nitrate loading to ground water by annual ground water recharge.

Nitrate loading rates were assigned to each land use in Table 3. Natural systems are very conservative of nitrogen, so zero was assigned to forest and grass/brush. For the low-density urban development that typically occurs in the Plover 15-year TOT, we also assigned zero to the urban/roads component.

We also determined NO_3-N loading rates for unsewered residential areas, irrigated cropland, and non-irrigated cropland. Septic systems are the major nitrate contributor in unsewered residential areas. We counted 20 residences with septic systems in the area, assumed four residents/household, and assigned a loading rate of 10 lb NO_3-N/person/year (Fetter, 1993). The 20 residences account for an estimated 800 lb NO_3-N/year to ground water.

We used the Nitrogen Leaching and Economic Analysis Package (NLEAP) model, developed by the USDA, Agricultural Research Service (ARS), Fort Collins, Colorado (Shaffer et al., 1991) to estimate NO_3-N leaching under cropland. NLEAP accounts for crop N uptake, nitrate leaching, ammonia volatilization, and N denitrification losses. Since loamy sands and sands make up

80% of the soils in the Plover 15-year TOT, a representative soil, Plainfield loamy sand, was selected for modeling scenarios. We used the model to estimate leaching under typical conventional (CON) management and ICM practices.

Forty acres of 240 acres of non-irrigated (dryland) cropland were actively farmed in 1994. These 40 acres were in a seven-year rotation (one year corn, one year oat/alfalfa seeding, five years alfalfa). Nitrate leached (NL) predicted by NLEAP for the rotation is 56 lb NO_3-N/acre under CON and 43 lb NO_3-N/ acre under ICM (Table 4). The average annual NL for the CON system is 8 (56/ 7) lb NO_3-N/acre/year. The average NL for ICM is 6.1 (43/7) lb NO_3-N/acre/ year. For 40 acres, NL under CON for the rotation is 320 lb NO_3-N/year (8 x 40). For 40 acres under ICM, NL is 244 lb NO_3-N/year (6.1 x 40).

Estimating NL for irrigated cropping is difficult since each operator has his own rotation. Table 5, col. 2, shows the 1994 average acres of irrigated crops in the Plover 15-year TOT area. Table 5, col. 3 and 4, show the amount of NL per acre below the root zone for irrigated crops under CON and ICM systems. NL values came from Table 4. Table 5, col. 5 and 6, give the total predicted NL for crops under CON and ICM. These values are the products of the

Table 4. NLEAP-predicted annual nitrate leached (NL) to ground water under eight cropping systems, with integrated crop management (ICM) and conventional (CON) systems and average rainfall grown on a Plainfield sand in Portage Co., Wisconsin.

Cropping System[t]	N Fertilizer Applied		NO_3-N Leached	
	ICM[t]	CON	ICM[t]	CON
yield/acre	----------------lb N/acre----------------			
3 ton dryland alfalfa	0	0	0	0
5 ton irrigated alfalfa	0	0	0	0
60 bu dryland and irr. oat/alfalfa seeding	30	40	12	15
100 bu dryland 1st year corn after alfalfa	125	170	31	41
140 bu irrigated corn after corn	210	250	61	86
6.5 ton irrigated sweet corn	150	175	55	87
4.5 ton irrigated snap bean	70	110	37	48
20 ton irrigated late potato	220	265	120	138

[t]Includes University of Wisconsin Extension N fertilizer and irrigation scheduling recommendations.

Table 5. 1994 irrigated crops and NLEAP-predicted nitrate leached (NL) under conventional (CON) and integrated crop management (ICM) in the Plover 15-year time-of-travel area.

Crops	Acre	CON	ICM	CON x acre	ICM x acre
		--lb NO_3-N/acre--		total lb NO_3-N Leached	
Sweet corn	205	87	55	17,800	11,300
Late Potato	128	138	120	17,600	15,400
Snap bean	91	48	37	4,400	3,400
Alfalfa	41	0	0	0	0
Field corn	27	86	61	2,300	1,600
Oat/alfalfa seeding	14	15	12	200	200
Totals	506			42,100	31,700

predicted NL/acre for each crop under CON (col. 3) and ICM (col. 4) times the acres of each crop (col. 2).

Table 5 data show that potato, with its long growing season and high N nutrient requirements, has the most NL/acre. ICM reduced NL by 13% to 37% per crop acre. Although potatoes and sweet corn comprise 66% of the irrigated acres, they contribute 84% of the NL to ground water under both CON and ICM.

Next, in Table 6, we totaled the average annual NL by land use under CON and ICM (col. 3 and 4). These data were also summarized on a per-acre basis (col. 5 and 6) and on a percent of NL basis (col. 7 and 8). The predicted data show that irrigated cropland contributed 97% of the total NL within the Plover 15-year TOT.

We calculated the average annual recharge from irrigated cropland and all other land uses within the Plover 15-year TOT (Table 7) based on estimates by Weeks et al., (1965).

Equation 1 below estimates the steady-state (SS) nitrate-N (in mg/L) at the Plover well by dividing the average annual or total NL (in lb NO_3-N) by the average annual recharge (in million lb water) within the Plover 15-year TOT.

SS NITRATE-N $_{PLOVER WELL}$ = (ave. annual N/L)/(ave. annual Recharge)   (1)

Table 6. Summary of predicted nitrate leached (NL in lb NO_3-N) under conventional (CON) and integrated crop management (ICM) by land use in the Plover 15-year time-of-travel (TOT) area.

Land Use	Acres	CON	ICM	CON	ICM	CON	ICM
		------total NL------		-----NL/acre----		---% NL---	
Irrigated Cropland	506	42,100	31,700	83	63	97	97
Non-irrigated cropland	240	320	240	1	1	1	1
Unsewered residences	35	800	800	23	23	2	2
Forest	340	0	0	0	0	0	0
Urban	132	0	0	0	0	0	0
Grass/brush	14	0	0	0	0	0	0
Roads	25	0	0	0	0	0	0
TOTAL	1,292	43,220	32,740	33	25	100	100

Table 7. Average annual ground water recharge in the Plover 15-year time-of-travel (TOT) area.

Land use	Acre	Recharge Rate	Recharge	Recharge
		ft/year	acre-ft	million lb
Irrigated cropland	506	0.50	253	688
All other land	786	0.83	652	1,772
TOTAL	1,292	----	905	2,460

For CON, SS NO_3-N   = (43,220 lb NO_3-N)/(2,460,000,000 lb H_2O)
                        = 18 ppm NO_3-N = 18 mg/L NO_3-N

For ICM, SS NO_3-N   = (32,740 lb NO_3-N)/(2,460,000,000 lb H_2O)
                        = 13 ppm NO_3-N = 13 mg/L NO_3-N

Both steady state predictions are above the 10 mg/L standard. The ICM value of 13 mg/L NO_3-N was similar to the actual 1994 Plover well monitoring results of 11-15 mg/L NO_3-N (Table 2). The CON SS value of 18 mg/L NO_3-N was higher.

DISCUSSION

ICM did reduce NO_3-N nearly 28%, but the objective of meeting the 10 mg/L NO_3-N standard was not met. About 32,740 lb NO_3-N still leached to ground water.

In addition to ICM, we studied changing agricultural land use in order to meet the drinking water standard at the wells. We calculated that 24,600,000,000 lb of annual recharge (from Table 7) that averages 10 mg/L NO_3-N contains 24,600 lb NO_3-N/year. To meet the standard, NO_3-N must be reduced 8,140 lb/year (32,740 - 24,600 = 8,140) or 25% beyond the use of ICM alone.

Using land use and crop NL values in Tables 5 and 6, we developed three land use and management changes to reduce nitrate loading by 8,140 lb/year within the Plover 15-year TOT: (1) convert 25% of irrigated land (129 acre) to forest or grassland; (2) add two years of alfalfa to the typical three-year irrigated vegetable rotation of sweet corn, potato, and snap bean; or (3) convert 40% of irrigated land (203 acre) to 2-acre unsewered residential lots.

Under CON, we need to reduce nitrate-N to 18,620 lb (43,220 - 24,600 = 18,620). Here are three ways that may be accomplished without ICM by using information in Tables 5 and 6: (1) convert 48% of irrigated land (224 acres) to forest or grassland; (2) change irrigated rotation to one year sweet corn, one year potato, one year snap bean and four years hay; or (3) convert 55% of irrigated land (277 acres) to 2-acre unsewered residential lots.

We made several assumptions that affect the study results. Generally, voluntary adoption of ICM is significantly less than 100%. Also, other land use changes, not considered here, may reduce NO_3-N to meet the standard. We assumed the model will reasonably simulate the system. In using the model, we based the NO_3-N loading predictions on an average climate year, including average precipitation.

CONCLUSIONS

This modeling exercise showed that ICM significantly reduced nitrate leached to ground water, but not sufficiently to protect Plover wells. Both ICM and agricultural land use changes would be necessary to meet the 10 mg/L NO_3-N standard. In comparison to ICM, CON systems required more drastic reductions in agricultural land use or intensity to meet the standard. Land owners may be very reluctant to convert irrigated vegetable fields to dryland cropping because such changes probably will reduce profitability.

REFERENCES

Born, S.M., Yanggen, D.A., Czecholinski, A.R., Teirney, B.J., and Hennings, R.G., *Wellhead Protection Districts in Wisconsin: An Analysis and Test Applications*, Special Report 10, Wisconsin Geol. & Natur. Hist. Surv., Madison, Wisconsin, 1988.

Central Wisconsin Groundwater Center, Unpublished data, University of Wisconsin-Stevens Point, Stevens Point, Wisconsin, 1994.

Ebert, W., Moberg, D., and Karim, A., *Stevens Point-Whiting-Plover Wellhead Protection Project, 1993 Annual Report*, Stevens Point, Wisconsin, 1993.

Ebert, W., and Moberg, D., *Stevens Point-Whiting-Plover Wellhead Protection Project 1992 Annual Report*, Stevens Point, Wisconsin, 1992.

Fetter, C.W., *Contaminant Hydrogeology*, New York, New York: Macmillan Publ. Co., 1993.

Hagan, T., Brochure on Nitrate Removal System for Village of Whiting, May, 1992.

Portage County Planning and Zoning Department, Unpublished data, 1992.

Shaffer, M.J., Halvorson, A.D., and Pierce, F.D., Nitrate leaching and economic analysis package (NLEAP): Model description and application, in *Managing Nitrogen for Groundwater Quality & Farm Profitability*, Follett et al., eds., SSSA, Madison, Wisconsin, 1991.

WDATCP, *Nutrient and Pesticide Best Management Practices for Wisconsin Farms*, Wisconsin, Dept. Ag., Trade & Cons. Protection Tech. Bull. 1 - Univ. Coop. Extn. Serv. Bull. No. A3466, 1989.

Weeks, E.P., Erickson, D.W., and Holt, C.L., *Hydrology of the Little Plover River Basin, Portage County, Wisconsin*, USGS Water Supply Paper 1811, U.S. Gov't. Printing Office, Washington, D.C., 1965.

DECISION SUPPORT SYSTEM FOR TOTAL WATERSHED MANAGEMENT

Tony Prato, Chris Fulcher, and Feng Xu

INTRODUCTION

I t is often difficult to identify a single social preference criterion relating to agricultural and environmental policies because the decision making process is influenced by multiple, competing objectives. Agricultural producers are mainly concerned about farm profitability while conservationists and environmentalists emphasize resource conservation. Economic and environmental objectives are oftentimes competitive and non-commensurable, especially in the short term when technology is fixed (Haimes et al., 1975, 1990; Kim and Mapp, 1993; Munda et al., 1994; Xu et al., 1994). In order to improve the choice of farming systems, explicit recognition should be given to economic and environmental tradeoffs for different farming systems. Agricultural policy analysis should consider multiple objectives including farm income, soil erosion, nitrate leaching and any other surface and ground water quality indicators because all of these measures are interconnected.

Both economic and environmental effects of agricultural production systems and their relationships need to be understood in order to prescribe recommendations for economically and environmentally sustainable agricultural production systems. Watershed-level analysis is necessary to understand such economic-environmental relationships at various aggregation levels. The integration of natural resources into the framework of economic theory is a prerequisite for environmental policy (Opschoor and van der Straaten, 1993). Xu et al. (1995) evaluated alternative farming systems using an integrated environmental-economic model. Economic and environmental tradeoffs are estimated using data generated from a multi-objective programming (MOP) model. Their results indicate that tradeoffs exist between economic and environmental objectives and between two selected environmental objectives. Few previous studies have explicitly accounted for economic and environmental tradeoffs in environmental-economic policy analysis.

The objective of this chapter is to introduce the methodology for creating the Watershed Management Decision Support System (WAMADSS). This methodology is one component of on-going dissertation research. Specifically, Fulcher

333

(1993) adopts a landscape perspective, a way to view interactive parts of a watershed, rather than focusing on their isolated components.

The research involves the use of multiple-objective programming to allocate land use activities within a watershed in order to optimize selected planning alternatives consisting of combinations of competing environmental and economic objectives. WAMADSS consists of three components: (1) a distributed-parameter simulation model, (2) a geographic information system (GIS), and (3) an economic model (Figure 1).

The study area is located in the Lower Grand River basin in Missouri. This basin is one of 20 watersheds selected by the Missouri Wetland Watershed

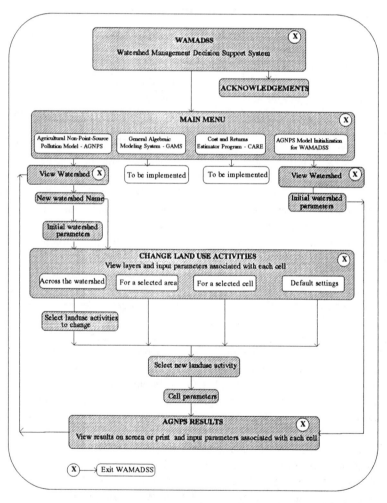

Figure 1. Watershed Management Decision Support System (WAMADSS) Schematic.

Identification Committee as a watershed containing priority wetlands for non-point source pollution protection. Specifically, Bear Creek watershed, located in Linn and Sullivan Counties, is selected as the study area. This watershed has also been designated as a Special Area Land Treatment (SALT) project by the Missouri Department of Natural Resources.

WAMADSS COMPONENTS

The three components are accessed through a graphical user interface (GUI) in order to facilitate analyses. Specifically, AGNPS and the economic models are linked to ARC/INFO via the ARC Macro Language (AML). The interface enables the decision maker to manipulate land use alternatives, run AGNPS, execute the economic models, and view results within ARC/INFO.

The AGNPS model was developed by Young et al. (1987) to analyze the water quality effects of alternative agricultural management practices. It is a distributed parameter model that simulates sediment, runoff, and nutrient (nitrogen and phosphorus) transport from agricultural watersheds. AGNPS is an event-based model that operates on a cell basis in which each cell expresses watershed characteristics and input parameters. The cells are uniformly square areas that subdivide the watershed and form the basis for analyses at any point in the watershed.

Three disadvantages of using AGNPS are (1) the time involved to manually enter or manipulate the large amount of input data required to describe the spatial detail of a watershed; (2) human error in interpreting landscape characteristics such as land slope, slope shape factor, field slope length, or channel slope from topographic maps; and (3) inconsistency in discerning these characteristics across the entire watershed. A GIS can be used to minimize these disadvantages by (1) significantly reducing the time and labor needed to process and manipulate the required input parameters; and (2) generating a surface model from the watershed hypsography layer to address the second and third disadvantages stated above.

A GIS is often defined as the complete sequence of components for obtaining, processing, storing, managing, and manipulating spatial data. ARC/INFO is used as a method to monitor and significantly improve the decision maker's ability to manipulate spatial and non-spatial data in order to evaluate alternative management practices. This approach enhances the "best judgment" decisions offered by conventional simulation models such as AGNPS.

AGNPS parameters are generated in ARC/INFO using AGNPS Source Code, GRID Module, Triangulated Irregular Network (TIN) Surface Analysis and Display Package, INFO Database Manager, and AML. AGNPS output and results generated from executing the program are viewed and printed in the ARC/INFO session using INFO, AML, and ARCPLOT Module.

AGNPS INPUT PARAMETER GENERATION

AGNPS Source Code. AGNPS 4.02 requires 22 input cell parameters to run the model. ARC/INFO generates these parameters using relational tables and five layers or overlays - soils, land use, hydrology, hypsography, and grid cells. Each layer contains attributes that are used to generate the required input parameters (Table 1).

In a previous study, the input parameters were ported to a personal computer (PC) where AGNPS resided. An AGNPS-ARC/INFO input-output linkage had been established; however, it was disjointed as the programs resided

Table 1. Input parameters required for AGNPS data file.

Number	Data	Source
	Watershed Input:	
1	Watershed identification	manual input
2	Description	manual input
3	Area of each cell (acres)	Grid Cells/manual input
4	Number of cells	Grid Cells/manual input
5	Precipitation (in.)	manual input
6	Nitrogen concentration in rainfall (ppm)	manual input
7	Energy-intensity value	manual input
8	Duration (hours)	manual input
9	Storm type (I, IA, II, III)	manual input
10	Peak flow calculations (SCS-TR55/AGNPS)	manual input
11	Geomorphic calculations (yes/no)	manual input
12	Hydrograph shape factor (K coef. / % runoff)	manual input
13	K coefficient	manual input
	Cell Parameter:	
1	Cell number	Grid Cells
2	Cell division	N/A
3	Receiving cell number	Grid Cells, Hypsography
4	Receiving cell division	N/A
5	Aspect (flow direction)	Hypsography
6	SCS curve number	Soils, Land Use
7	Land slope (%)	Hypsography
8	Slope shape factor	Hypsography
9	Slope length (ft)	Hypsography
10	Overland Manning's roughness coefficient	Land Use
11	Soil erodibility factor (K)	Soils
12	Cropping factor (C)	Land Use
13	Practice factor (P)	manual input
14	Surface condition constant	Land Use
15	Chemical oxygen demand factor	Land Use
16	Soil texture number	Soils
17	Fertilizer indicator	Land Use
18	Pesticide indicator	Land Use
19	Point source indicator	Land Use
20	Additional erosion	Land Use
21	Impoundment indicator	Land Use
22	Channel indicator	Hydrology, Hypsography

on different platforms. The output from AGNPS could not be readily viewed in ARC/INFO as the results needed to be ported back to the UNIX workstation and imported as INFO files. To address this problem, a seamless interface between AGNPS and ARC/INFO was developed by establishing both programs on the same platform.

GRID Module. The Grid Module, a raster-based geoprocessing application, is used to establish a cell size for the watershed and to convert the layers in Figure 2 to grid cells. The gridding process can divide the watershed into 1-acre grid cells as readily as 40-acre grid cells. The smaller the grid cells, the greater the resolution and accuracy; however, there is a higher cost in terms of processing speed and storage requirements. The watershed boundary layer is an appropriate layer to initially convert to a grid-based watershed. This layer will then serve as a "cookie-cutter" for the remaining layers that need to be rasterized. Each cell has attributes associated with it that will serve as input parameters for AGNPS.

In order to make WAMADSS a reality, it was necessary to obtain the source code for AGNPS and compile the program for a UNIX-based platform. No equations were modified from the original source code; only input and output format modifications were made for the interface.

ARC Macro Language (AML). AML is the programming language used to interface AGNPS and the economic models in a seamless decision support system framework. This programming language handles all AGNPS-related activities, from generating input files, executing AGNPS, to viewing results in ARCPLOT. In terms of input parameter generation, AML programs are used to (1) create the graphical user interface for entering AGNPS input parameters and (2) transform input parameters from columns in INFO files to an AGNPS compatible input file format (AGNPS.DAT).

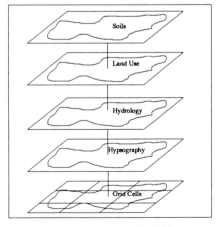

Figure 2. GIS layers for generating AGNPS input parameters.

The user interface is comprised of menus created in ARC/INFO's FormEdit. WAMADSS guides the decision maker through a series of menus depending on the choices made. These menus are positioned and threaded, when appropriate, in order to provide a user-friendly appearance.

WAMADSS permits the decision maker to modify land use activities by (1) selecting a single cell (i.e., change cropland to a feedlot); (2) selecting a group of cells (i.e., change pasture to cropland) or (3) changing one land use activity to another across the entire watershed. Menus guide the user through a series of prompts in order to update all the AGNPS parameters needed to reflect the new land use activity selected. AGNPS is then executed for the altered watershed whereby the results are used as inputs to the economic component of WAMADSS. Competing environmental and economic objectives are then evaluated for a number of possible scenarios. One scenario may involve changing some or all agricultural activities that take place on hydric soils to riparian wetlands.

TIN Surface Analysis and Display Package. The TIN package is used to generate all surface-related input parameters for AGNPS. Specifically, TIN stores the topological relationship between non-overlapping triangles and their adjacent neighbors (ESRI, 1992). This data structure, where triangles are computed from irregularly spaced points with x and y coordinates and z values, allows for the generation of surface models for the analyses and display of terrain surfaces (refer to Table 1 for those input parameters that rely on the hypsography layer).

INFO Database Manager. INFO is a fourth-generation programming language that is used for storing, maintaining, and manipulating all input and output variables for AGNPS. The variables were generated from the gridding procedure mentioned above. INFO stores AGNPS data in a header file and nine relational files for generating an AGNPS input file - AGNPS.DAT (see Table 2). Note that there is a total of 168 variables or columns in these files; 149 variables are AGNPS input parameters (required and optional) for each cell; 10 variables are reserved for AGNPS development testing; the remaining

Table 2. INFO files for generating AGNPS input and output files.

AGNPS input file		AGNPS output file	
Files	variables or columns	Files	variables or columns
Header	23	Watershed	18
Cell	22	Sediment	43
Soil	9	Soil Loss	40
Fertilizer	4	Nutrient	17
Pesticide	17	Pesticide	17
Non Feedlot	5	Feedlot	11
Feedlot	58		
Gully	6		
Impoundment	3		
Channel	21		
Total	168	Total	146

nine variables represent a cell number column in each of the nine files. This column serves as the key field that is used to relate the files.

AGNPS OUTPUT AND RESULTS

INFO Database Manager. After executing WAMADSS, AGNPS generates an ASCII output file - AGNPS.NPS. A number of INFO files need to be established so that the output can be viewed in ARCPLOT. INFO stores information from AGNPS.NPS in six relational files (Table 2). Note that there is a total of 146 variables or columns in these files: 144 variables are AGNPS output parameters while the remaining two variables represent a relational key field (cell number column) for the pesticide and feedlot files.

ARC Macro Language. As stated above, this programming language handles all AGNPS-related activities, from generating input files, executing AGNPS, to viewing results in ARCPLOT. In terms of output parameter generation, AML programs are used to (1) transform output parameters from an AGNPS output file (AGNPS.NPS) to columns in INFO files and (2) create the GUI for viewing AGNPS output parameters in ARCPLOT.

ARCPLOT Module. AGNPS results are viewed in ARCPLOT via WAMADSS. Sixteen AMLs were written in order for the user to interactively view or print AGNPS results. WAMADSS guides the decision maker through a series of menus, depending on the choices selected. AGNPS results are organized at three levels: (1) watershed-level results, (2) cell-level results, and (3) cell-grouping results.

Watershed-level results report findings for the entire watershed while AMLs were written to allow the user to select any cell in the watershed and display the findings for that cell. Similarly for a cell grouping, the user may draw a polygon around any number of cells and display the results for those cells that lie within the polygon. Results at the various levels include runoff, sediment, and nutrient values at the watershed outlet; sediment, feedlot, and pesticide analysis at the watershed, cell, and area levels; and hydrology and soil loss at the cell and area levels.

THE ECONOMIC MODEL

Total watershed management, rather than individual farm firm management, is the focus of this research. Total watershed management suggests a concern for long-run economic returns and related externalities (sedimentation, flooding, decrease in water quality) associated with agricultural production activities.

The economic model includes two components: (1) an agricultural budget generator and (2) a MOP model. The Cost and Returns Estimator (CARE) program serves as the budget generator while the General Algebraic Modeling System (GAMS) is used to execute the MOP model. Both programs reside on a UNIX platform in order to establish a seamless interface with ARC/INFO and AGNPS.

The budget generator calculates costs and returns for each cropping system, conservation management practice, and tillage practice. The MOP model integrates the environmental and economic information in a framework that permits analyses of a range of alternatives. Given this range of alternatives, environmental and economic relationships, impacts, and tradeoffs are identified.

MOP Model. Assume three objectives for total watershed management are expressed as: maximizing $\pi = \pi(x)$, minimizing $\rho = \rho(x)$, minimizing $\gamma = \gamma(x)$, where π is economic profit, ρ is soil erosion, γ is chemical leaching, and x is a set of land use management activities and practices (LUMAPs). There are two major views regarding the current LUMAPs. One view is that the actual x used in the watershed is not most efficient and that a Pareto-superior position can be obtained by changing from x to x' so that $(\pi' \ \rho' \ \gamma') \geq (\pi \ \rho \ \gamma)$, i.e., there exists an alternative LUMAP(s) that provides a preferred set of economic and environmental outputs. In other words, the watershed can achieve higher economic profit without increasing soil erosion and chemical leaching, or soil erosion can be reduced without decreasing profit and increasing chemical leaching. This view is inconsistent with the neo-classical microeconomic theory for a rational decision maker under usual assumptions of perfect competition. An alternative view is that the LUMAPs in the watershed achieve an optimum, i.e., $(\pi \ \rho \ \gamma)$, that is the best output mix possible. There does not exist any alternative LUMAP that results in a more preferred set of economic and environmental outputs relative to the current set. That is, $\nexists x':(\pi' \ \rho' \ \gamma') \geq (\pi \ \rho \ \gamma)$. Therefore, changes from x to x' would entail tradeoffs. Under current technology, higher economic profit cannot be achieved without increasing soil erosion and/or chemical leaching, nor could soil erosion be reduced without decreasing profit and increasing chemical leaching.

However, a new (or modified) Pareto-superior LUMAP x" can be found through technological progress as affected by research and development. Change from x to x" is superior because $\exists x":(\pi" \ \rho" \ \gamma") \geq (\pi \ \rho \ \gamma)$. This implies that a new and more preferred set of economic and environmental outputs is achieved with a new technology. The problem still arises: society may regard both $(\pi \ \rho \ \gamma)$ under the current technology and $(\pi" \ \rho" \ \gamma")$ under the new technology as unacceptable (or undesirable) because environmental impacts are too high and/ or economic profit is too low. Therefore, society's preferred set of outcomes may be different from the efficient agricultural outcomes under either current technology or new technology.

The MOP model as in WAMADSS assumes no *a priori* information about decision maker preferences, and it is primarily used to generate information about non-inferior solutions. Non-inferior solutions are a "set of feasible solutions such that no other feasible solutions can achieve the same or better performance for all objectives and strictly better for at least one objective." (Romero et al., 1987, p.78).

Without losing generality, let x be a vector of decision variables; $f_j(x)$, $j = 1,\cdots,n$ be the jth objective function; and $g_i(x)$, $i = 1,\cdots,m$ be the ith constraint. The feasible region is defined as $x = \{x|g_i(x) \leq 0, i = 1,\cdots n\}$. An optimization problem is then formulated as $\max_{x \in x}[f_1(x),...,f_n(x)]$. Point x* is a non-inferior solution if there exist no $x \in X$ such that $f_j(x) \geq f_j(x^*)$ ∀j, and $f_j(x) > f_j(x^*)$ for at least one j. Non-inferior solutions can be obtained by choosing one of the objective functions as the primary function to optimize and using the other functions as constraints. Haimes et al. (1971, 1990 pp. 72-73) proved that the efficient set of non-inferior solutions to the MOP is unique no matter which of the objective functions is optimized. That is, the efficient surface for the objective functions is identical regardless of which function is optimized.

There are numerous non-inferior solutions for an MOP model (Ringuest, 1992). Obtaining all non-inferior solutions can be computationally expensive and even infeasible. If a subset of non-inferior solutions can represent a "true" set of non-inferior solutions reasonably well, this subset can be used to approximate (or predict) other non-inferior solutions within a relevant range of objective values. Regression analysis can be used for prediction. Determining the number of non-inferior solutions then becomes an empirical question. Only a subset of non-inferior solutions is relevant for a particular decision maker. For example, a plausible "ideal" solution (a particular non-inferior solution) is chosen as the starting point for a particular decision maker. Then other non-inferior solutions close to the ideal solution are identified as the compromise solutions (Zeleny, 1973; Romero, 1987). The drawback of this approach is obvious. An "ideal" solution is difficult to identify *a priori*.

Estimates of effects of economic profitability from CARE and environmental effects from AGNPS are used as inputs required by the MOP model, which determines spatially optimal LUMAPs.

SUMMARY AND CONCLUSIONS

The objective of this chapter is to introduce the methodology for creating the Watershed Management Decision Support System (WAMADSS). The decision support system adopts a landscape perspective, a way to view interactive parts of a watershed, rather than focusing on their isolated components. The three components that comprise WAMADSS are accessed from one common interface in order to facilitate analyses. The interface enables the decision maker to manipulate land use alternatives, run AGNPS, and execute the economic models within an ARC/INFO session.

Procedures presented in this chapter represent an efficient way to obtain tradeoff information among economic and environmental objectives. Advantages of the proposed methods lie in its simultaneously considering multi-objectives from common data bases in WAMADSS. Further efforts should focus on estimation accuracy of economic and environmental impacts of alternative policy options. The methods and procedures described in this study substantially improve the understanding of watershed-level economic and environmental tradeoffs of interest to farmers and policy makers. Policies designed to balance economic and environmental goals should consider these tradeoffs.

REFERENCES

Environmental Systems Research Institute, Inc. (ESRI), *Surface Modeling with TIN - Surface Analysis and Display,* Redlands, 1992.

Fulcher, C., The Role of Wetlands in Improving Aquatic Ecosystems: An Ecological-Economic Assessment, Ph.D. dissertation proposal, Department of Agricultural Economics, University of Missouri-Columbia, 1993.

Haimes, Y.Y., Larson, L., and Wismer, D., On a bicriteria formulation of the problems of integrated systems identification and system optimization, *IEEE Transactions on Systems, Man, and Cybernetics* SMC-1, 296, 1971.

Haimes, Y.Y., Hall, W.A., and Freedman, H.T., *Multiobjective Optimization in Water Resources Systems,* New York: Elsevier Scientific Publishing Co., 1975.

Haimes, Y.Y., Tarvainen, K., Shima, T., and Thadathil, J., *Hierarchical Multiobjective Analysis of Large-scale Systems,* New York: Hemisphere Publishing Co, 1990.

Kim, S.H., and Mapp, H.P., *A farm-Level Economic Analysis of Agricultural Pollution Control,* Selected paper, Amer. Agr. Econ. Association annual meeting, Orlando, Florida, 1993.

Munda, G., Nijkamp, P., and Rietveld, P., Qualitative multicriteria evaluation for environmental management, *Ecological Economics,* 10, 97, 1994.

Opschoor, H., and van der Straaten, J., Sustainable development: An institutional approach, *Ecological Economics,* 7, 202, 1993.

Ringuest, J.L., *Multiobjective Optimization: Computational Considerations,* Boston: Kluwer Academic Publishers, 1992.

Romero, C., Amador, F., and Barco, A., Multiple objectives in agricultural planning: A compromising programming application, *Amer. J. of Agr. Econ.,* 35, 78, 1987.

Soil Conservation Service - USDA, *User Manual - Cost and Returns Estimator (CARE),* Lino Lakes: Midwest Agricultural Research Associates, Inc, 1988.

Xu, F., Prato, T., and Ma, J.C., A Farm-Level Case Study of Sustainable Agricultural Production System, *J. of Soil and Water Conservation,* 50(1), 39, 1995

Young, R.A., Onstad, C.A., Bosch, D.D., and Anderson, W.P., *AGNPS: Agricultural Nonpoint Source Pollution Model: A Watershed Analysis Tool,* Washington, D.C.: Agr. Res. Serv., U.S. Dept. of Agr., Conserv. Res. Rpt. 35, 1987.

Zeleny, M., Compromising programming, in *Multiple Criteria Decision Making,* Cochrane, J.L., and Zeleny, M. , eds., Columbia: University of South Carolina Press, 262, 1973.

NUTRIENT VALUES OF DAIRY MANURE AND POULTRY LITTER AS AFFECTED BY STORAGE AND HANDLING

E.R. Collins, Jr., J.D. Jordan, and T.A. Dillaha

INTRODUCTION

Environmental interests and regulations make it essential that agricultural producers carefully examine operations that place pollutants in air and water resources. Therefore, livestock and poultry operators must carefully evaluate and plan to maximize utilization of manure from their operations. On a typical farm, a high percentage of manure nutrients can be recycled through proper planning and use. In the past, if farmers assigned nutrient credits for manure, they were usually based on tabular values found in Midwest Plan Service (1985) or other similar references. However, questions arise about reliability of available nutrients in manure. Seasonality, management, storage method, and other factors may influence nutrient content. Knowledge about applied organic fertilizers is essential to the future management of livestock, water, and land resources. Farm manure testing experience in Virginia has shown that using tabular estimates is not as reliable as regular manure testing in supporting farm nutrient planning.

FACTORS THAT INFLUENCE MANURE QUALITY

Storage method may influence the fate of nitrogen contained in livestock and poultry manure. Manure nitrogen is typically composed of two fractions: (1) organic nitrogen and (2) ammonia nitrogen. Nitrate nitrogen is generally a negligible part of the total N, especially for the types of manure discussed herein, which are often held in oxygen-limited storage conditions[1]. When manure is permitted to remain on lot surfaces for long periods of time, especially when drying conditions prevail, ammonia nitrogen will likely be volatilized. If precipitation occurs, nitrogen, phosphorus, potassium, and other nutrients will likely be washed from the collection area and will no longer be concentrated in the collection and storage system. Likewise, precipitation collected on lot sur-

---

[1] For this reason, TKN was assumed to represent total nitrogen in this sutdy, and the relatively small amounts of nitrate and nitrite were ignored.

faces and in storage systems may dilute manure and change apparent nutrient levels.

METHODS

Manure and litter test data were compared to determine effects of type of storage/handling and season of year on nutrient levels contained in manure and litter. Manure samples were classified into one of two seasons: spring or fall. Spring samples included all of those received between January 1 and June 30, with most samples being received in April or May, when most storages are emptied prior to spring planting. Fall samples included those received from July 1 through December 31, with most samples being collected following the fall harvest and when storages were emptied in preparation for winter.

Manure and litter samples were analyzed according to Greenberg et al. (1992). Data were statistically analyzed using Minitab (1989), a general purpose data analysis system. The model used was the general linear model (GLM).

RESULTS AND DISCUSSION

DAIRY MANURE

The database used in this study included 675 samples from liquid tank or earthen basin storage systems and 97 samples from semi-solid storage or handling arrangements.

Nitrogen. Liquid dairy manure samples were taken from either concrete or steel tanks (silo-type or in-ground tanks were considered to be the same) or earthen basin storage systems. Total nitrogen (TKN) in stored liquid manure was first examined on a wet basis and was found to be significantly different ($P < 0.001$) between the two methods of liquid manure storage. TKN levels observed from the tank-type storages were generally higher than from the earthen basin storages. Reasons for this difference were thought to be related to the larger exposed liquid surface of the earthen storages, which may afford greater opportunity for nitrogen volatilization and for greater dilution by intercepted precipitation. The observed differences were also thought possibly to indicate difficulty in achieving optimum agitation prior to clean-out and sampling, which might result in sample inaccuracies. Mean values (wet basis) for the factors considered are shown in Table 1. When raw data were converted to a dry basis and statistically analyzed, no significant difference was observed in TKN levels for the two methods of liquid manure storage. Therefore, it can be surmised that there is typically more dilution of manure in the earthen storages, but nitrogen conservation is about equal between the two methods of storage. However, farmers must haul more manure from the earthen storages to distribute equal amounts of TKN. Seasonal effects on TKN and storage x season interactions were not significant.

Ammonium values in liquid manure were significantly different when examined on a wet basis for both manure storage method ($P < 0.007$) and

Table 1. Nitrogen, phosphorus, and potassium in dairy manure.

Storage Type	Total N (TKN) kg/m³ (lb/1000 gal)	Total N (TKN) percent drybasis	Ammonium NH₄ kg/m³ (lb/1000 gal)	Ammonium NH₄ percent	P₂O₅ kg/m³ (lb/1000 gal)	P₂O₅ percent	K₂O kg/m³ (lb/1000gal)	K₂O percent
Liquid manure tank	2.78 (23.21)[a]†	7.10	1.20 (9.98)[aa]	2.96	1.44 (11.99)[aaa]	3.16	2.28 (18.99)[aaaa]	6.12
Earthen manure basin	2.46 (20.49)[b]	7.09	1.10 (9.15)[bb]	3.10	1.31 (10.93)[bbb]	3.41	2.02 (16.84)[bbbb]	5.91

Storage Type	Total N, kg/t (lb/ton)	Total N, percent	NH4, kg/t (lb/ton)	NH4, percent	P₂O₅, kg/t (lb/ton)	P₂O₅, percent	K₂O, kg/t (lb/ton)	K₂O, percent
Semi-solid manure, overall mean	5.17 (10.34)	2.87	1.81 (3.62)	1.04			3.79 (7.58)	2.08
Semi-solid manure, spring					2.56 (5.12)[a]	1.34[aa]		
Semi-solid manure, fall					3.53 (7.06)[b]	1.89[bb]		

†Means in same colum with different subscripts are significantly different.

season ($P < 0.065$). As in the case of TKN, the larger exposed liquid surface of the earthen storages was thought to afford greater opportunity for NH_4 volatilization and for dilution by intercepted precipitation. When sample values were adjusted to a dry basis, there were no significant differences in main effects or interactions for NH_4. Storage x season effects were non-significant when examined on either a wet or dry basis. Mean values are shown in Table 1.

Samples of semi-solid manure were classified as being either from manure stacks or from daily scrape-and-haul operations. Data were analyzed on both a wet and dry basis. No significant differences in effects for either TKN or NH_4 were found between the two methods of semi-solid manure handling or for the two seasons. Values for total nitrogen and NH_4 in semi-solid dairy manure are shown in Table 1.

Phosphorus. Sample values for phosphorus (P_2O_5) were significantly different ($P < 0.026$) between the two types of liquid storage methods and significantly different between seasonal values ($P < 0.009$) when examined on a wet basis. However, when converted to a dry basis, the differences were non-significant. No significant differences were observed for the storage x season interactions on either a wet or dry basis. Mean values of P_2O_5 in liquid dairy manure are shown in Table 1.

An apparent seasonal difference ($P < 0.025$) was observed for phosphorus (wet basis) in the semi-solid manure samples, but the effect of method of manure storage/handling and the storage x season interactions were not significant. When values were converted to a dry basis, the effect of season was highly significant ($P < 0.005$). The higher fall values of P_2O_5 may indicate less loss from exposed lot surfaces during the summer months and more loss during winter months when drying is slower and exposed manure is more easily lost in runoff. Seasonal values of P_2O_5 for semi-solid manure are shown in Table 1.

Potassium. Potassium is not generally considered to be a key nutrient in development of nutrient management plans. Potassium values, as K_2O, were found to be significantly different ($P < 0.001$) when examined on a wet basis for both liquid manure storage types and season. Lower observed values for K_2O for the earthen structures were probably due to greater dilution effects (see Nitrogen section above, and Moisture section below) as compared to tank systems. When data were corrected to a dry basis, significant differences were not observed for any of the main effects or interactions for K_2O in liquid manure (Table 1). No significant differences in K_2O levels were observed between values for storage/handling, season, or interactions for semi-solid manure when compared on either a wet or dry basis (Table 1).

Calcium and Magnesium. No significant differences in liquid or semi-solid manure calcium and magnesium levels on either a wet or dry basis were observed for either storage types, or for seasons. Overall mean values are shown in Tables 2 and 3.

Moisture Content. All sample results were reported to farmers on a wet basis since they are accustomed to handling wet manure and can better esti-

Table 2. Mean levels of calcium (Ca) and Magnesium (Mg) observed in liquid dairy manure (wet basis), kg/m³ (lb/1000 gal), and percent (dry basis).

Element	kg/m³ (lb/1000 gal)	Percent
Ca, Overall Mean	1.20 (10.03)	3.09
Mg, Overall Mean	0.50 (4.15)	1.29

Table 3. Mean levels of calcium (Ca) and Magnesium (Mg) observed in semi-solid dairy manure (wet basis), kg/t (lb/ton), and percent (dry basis).

Element	kg/t (lb/ton)	Percent
Ca, Overall Mean	3.56 (7.12)	1.92
Mg, Overall Mean	1.27 (2.53)	0.69

mate application rates on this basis. The results reported above on both wet and dry basis show how apparent effects are changed through moisture dilution.

Moisture content for liquid dairy manure is shown in Table 4. Significant differences ($P < 0.047$) were observed for the two storage types but not for seasonal effects. This effect seems reasonable since, for a given volume of storage, earthen basins normally include a larger surface area than tank storages. Hence, more rainfall might be intercepted, and greater dilution may occur. The effect would probably not be as dramatic for seasonal effects, especially in this case since "spring" and "fall" months both include roughly equal wet and dry months in Virginia.

No significant differences in moisture content in semi-solid dairy manure were observed for either method of storage/handling or season of the year. The mean moisture content for semi-solid manure is shown in Table 5.

POULTRY LITTER

The database presented here included 451 samples from broiler litter systems and 176 samples from turkey litter systems. Broiler and turkey data sets were compared separately for main effects of storage/handling system, season, and storage x season interaction.

Table 4. Mean moisture content observed in liquid dairy manure, percent.

Type of Storage	Moisture Content, Percent
Tank Manure Storage	94.53[a†]
Earthen Basin Storage	95.02[b]
Overall Mean	94.86

†Means with different superscripts are different, $P < 0.025$ level.

Table 5. Mean moisture content observed in semi-solid dairy manure, percent.

Type of Storage	Moisture Content, Percent
Overall Mean	80.60

Nitrogen. Litter samples were taken from 1) the production house floor, 2) uncovered stacks outside, 3) plastic- (or tarp-) covered stacks, or 4) stacks within roofed storage structures. Total nitrogen (TKN) contained in both types of litter (wet basis) was found to be significantly different ($P < 0.001$, broiler; $P < 0.003$, turkey) among the four methods of litter handling (Table 6). TKN levels (wet basis) from the roofed structure storages were apparently highest for the broiler litter, followed by house litter, plastic-covered stacks, and uncovered stacks. TKN levels (wet basis) for turkey storage systems were highest when taken directly from houses, followed by roofed structure storages, uncovered stacks, and plastic-covered stacks. However, when sample values were corrected to a dry basis, effects of storage and main effect interactions on total N levels were not significant.

A significant effect of season on broiler litter (wet basis) TKN was also noted ($P < 0.01$), and, although not statistically significant, a similar trend was noted for turkey litter. However, when sample values were converted to a dry weight basis, differences were not significant.

After correcting sample values to dry weight basis, total nitrogen and NH_4 values were not statistically different among storage methods and seasons. However, the data suggest that total nitrogen may be better conserved when litter is taken directly from the production houses and spread, or when stored in roofed structures built for that purpose. In either condition, litter will be protected from exposure to rain and other weather factors that dilute nitrogen and other nutrients. Moisture may also cause wet, anaerobic pockets to develop in the stack, hastening the conversion and release of ammonia gas. The data on moisture levels (Tables 7 and 8) support this conclusion. This suggests a "saturation level" of NH_4 in the litter stack with volatilization of excess. A storage x season interaction was noted for broiler litter, suggesting that not all systems performed consistently over both seasons. However, as in previous examples, when data were converted to a dry weight basis, differences were not detected among storage/handling methods. However, seasonal differences were noted for NH_4, suggesting that higher summer temperatures may have negatively influenced the retention of NH_4 in broiler and turkey litter stored through summer and removed from storage in the fall (Table 6).

Moisture Content. As noted previously, all sample results were reported on a wet basis since farmers are accustomed to handling wet litter and can better estimate application rates on this basis. Moisture content for litter is shown in Tables 7 and 8. Not surprisingly, significant effects on moisture levels were found for both storage and season. Moisture levels were lowest for broiler litter in plastic-covered stacks and lowest for turkey litter in roofed structures. Second lowest moisture levels were in roofed sheds for broiler litter and direct house litter for turkeys. Third lowest levels were direct house litter for broilers, and open stacked turkey litter. Highest moisture levels were found in open stacked broiler litter and plastic-covered turkey litter. Moisture pat-

Table 6. Nitrogen in broiler and turkey litter (wet basis) kg/t (lb/ton), and percent (dry basis).

Litter storage method	Spring TKN, kg/t (lb/ton)	Fall TKN, kg/t(lb/ton)	TKN, percent	Spring NH$_4$, kg/t (lb/ton)	Fall NH$_4$, kg/t(lb/ton)	Spring NH$_4$, percent	Fall NH$_4$, percent
Broiler litter							
direct from house	30.97 (61.94)	30.73 (61.45)	4.31	7.69 (15.38)	6.75 (13.50)	1.12[a]	0.94[b]
stacked, uncovered	26.95 (53.90)	29.84 (59.67)	4.30	6.47 (12.94)	7.41 (14.82)	1.04[a]	1.09[b]
stacked/plastic covered	28.51 (57.01)	34.50 (68.99)	4.34	7.17 (14.33)	6.45 (12.89)	1.07[a]	0.86[b]
stacked/roofed shed	33.18 (66.35)	33.48 (66.95)	4.57	9.21 (18.41)	6.70 (13.39)	1.31[a]	0.90[b]
Turkey litter							
direct from house	32.55 (65.10)	31.75 (63.49)	4.77	10.47 (20.93)	10.30 (20.59)	1.67[c]	1.45[d]
stacked/uncovered	26.71 (53.41)	27.48 (54.95)	4.75	11.13 (22.93)	7.11 (14.21)	2.08[c]	1.19[d]
stacked/plastic covered	19.77 (39.53)	24.73 (49.46)	4.33	10.87 (21.73)	8.79 (17.58)	2.37[c]	1.58[d]
stacked, roofed shed	29.44 (58.87)	31.58 (63.15)	4.36	9.52 (19.04)	9.72 (19.43)	1.39[c]	1.38[d]

Storage method and season (TKN and NH$_4$) significant (wet basis). Main effects and season non-significant for TKN when corrected to dry basis. Season significant for NH$_4$ on dry basis.

Table 7. Mean moisture levels observed in broiler litter.

Litter Storage Method	Spring	Fall
	----------------------%----------------------	
Litter, direct from house	29.3	27.6
Stacked Litter, Uncovered	37.1	32.1
Stacked Litter, Plastic Covered	30.4	23.9
Stacked Litter, Roofed Shed	27.6	24.3

NOTES: Storage method significant at $P < 0.001$
Season significant at $P < 0.001$; storage x season non-significant.

Table 8. Mean moisture levels observed in turkey litter.

Litter Storage Method	Spring	Fall
	----------------------%----------------------	
Litter, direct from house	35.2	29.5
Stacked Litter, Uncovered	46.6	38.3
Stacked Litter, Plastic Covered	52.8	45.0
Stacked Litter, Roofed Shed	30.8	28.3

NOTES: Storage method significant at $P < 0.001$.
Season significant at $P < 0.03$
Storage x season non-significant.

terns with respect to storage methods do not seem to follow a consistent pattern between broiler and turkey systems. Covered systems provide better protection from precipitation than outside stacks. The exception for turkeys may be related to the generally higher moisture levels in turkey litter; the higher moisture levels for plastic-covered turkey litter as compared to open-stacked turkey litter may be due to "trapped" moisture under the plastic being greater than net gain from precipitation on the open stacks. Seasonal effects always produced litter with relatively lower moisture content in fall than in spring, likely due to the improved drying conditions during summer and early fall months before fall litter spreading.

Phosphorus. Laboratory sample values for phosphorus (P_2O_5) were not statistically different for either storage method or season when examined on a wet basis. Phosphorous is immobile and does not move appreciably from the litter. However, when sample values were converted to a dry weight basis, a significant difference in the effects of storage on P_2O_5 levels was apparent for both broiler and turkey litter. Higher values were obtained from litter that was stored in systems that are not well protected from weather and other environmental effects. It may be possible that deterioration of the uncovered and plastic-covered stacks, which typically are stored outdoors, contributed to pile shrinkage. Since the P_2O_5 is immobile, it may be concentrated in the pile, with the result of higher mean levels on a dry weight basis than litter taken directly from poultry houses or from roofed storage sheds. Mean values of P_2O_5 in both types of poultry litter are shown in Tables 9 and 10. Effects of season and

Table 9. Mean phosphorus (P_2O_5) values (wet basis)
for broiler litter, kg/t (lb/ton), percent (dry basis).

Litter Storage Method	P_2O_5, kg/t (lb/ton)	P_2O_5, Percent
Litter, direct from house	30.99(61.97)	4.34
Litter stacked, uncovered	31.03(62.06)	4.88
Litter stacked, covered with plastic	29.28(58.55)	4.07
Litter stacked, roofed shed	31.56(63.12)	4.32

NOTES: Effects of storage method, season, and storage x season all non-significant (wet basis).
Effects of storage method significant at $P < 0.001$ (dry basis).

Table 10. Mean phosphorus (P_2O_5) values (wet basis)
for turkey litter, kg/t (lb/ton), and percent (dry basis).

Litter Storage Method	P_2O_5, kg/t (lb/ton)	P_2O_5, Percent
Litter, direct from house	32.26(64.52)	4.78
Litter stacked, uncovered	29.32(58.63)	5.28
Litter stacked, covered with plastic	31.62(63.23)	6.26
Litter stacked, roofed shed	31.56(63.12)	4.52

NOTES: Effects of storage method, season, and storage x season all non-significant on wet basis.
Effects of storage method significant at $P < 0.086$ on dry basis.

storage x season interaction were not significant on either a wet or dry weight basis.

Potassium. As previously noted, potassium is not generally considered to be a key nutrient in development of nutrient management plans. Potassium values, as K_2O, were not generally found to be different in either of the two types of litter (Tables 11 and 12) for the effects of storage or season, and there were no interaction effects. An exception was found for turkey litter where, on a dry weight basis, outdoor stacks of litter were found to have a significantly higher content of K_2O than the other two storage/handling methods. A satisfactory explanation for this difference cannot be given except, perhaps, the one offered above for similar effects for P_2O_5.

Calcium. Seasonal differences ($P < 0.04$) were observed for levels of calcium (Ca) contained in broiler litter on a wet weight basis. It can be speculated that, since Ca is immobile and does not break down in storage, the warmer conditions of summer and early fall hasten organic material breakdown and shrinkage of storage piles. This might cause concentration of Ca in the stored litter.

No significant differences in litter calcium levels (wet basis) were observed in litter samples examined for effect of storage type. However, when samples were converted to dry weight basis, significant differences in broiler litter Ca levels were noted for storage method. The higher overall mean values for fall (Table 13) may be partially accounted for by lower moisture levels found in the fall season. No significant effects were observed for either storage or season, or their interactions in turkey litter. Values for Ca in turkey litter are shown in Table 14.

Table 11. Mean potassium (K_2O) values (wet basis) for broiler litter, kg/t (lb/ton), and percent (dry basis).

Litter Storage Method	K_2O, kg/t (lbs/ton)	K_2O, Percent
Litter, direct from house	14.95(29.90)	2.09
Litter stacked, uncovered	13.60(27.19)	2.08
Litter stacked, covered with plastic	14.48(28.95)	2.02
Litter stacked, roofed shed	15.78(30.00)	2.05

NOTE: Effects of storage method, season, and storage x season all non-significant (wet and dry basis).

Table 12. Mean potassium (K_2O) values (wet basis) for turkey litter, kg/t (lb/ton), and percent (dry basis).

Litter Storage Method	K_2O, kg/t (lb/ton)	K_2O, Percent
Litter, direct from house	11.88(23.76)	1.77
Litter stacked, uncovered	12.67(25.34)	2.23
Litter stacked, covered with plastic	10.87(21.74)	2.11
Litter stacked, roofed shed	13.37(26.74)	1.91

NOTES: Effects of storage method, season, and storage x season all non-significant on wet basis.
Effects of storage method significant at $P < 0.023$ (dry basis).

Table 13. Mean calcium (Ca) levels (wet basis) in broiler litter, kg/t (lb/ton), and percent (dry basis).

Litter Storage Method	Spring Ca, kg/t (lb/ton)	Fall Ca, kg/t (lb/ton)	Spring Ca, %	Fall Ca, %
Litter, direct from house	18.96 (37.92)	19.97 (39.94)	2.67	2.78
Stacked Litter, Uncovered	19.11 (38.21)	20.87 (41.74)	3.15	3.14
Stacked Litter, Plastic Covered	18.50 (37.00)	20.60 (41.19)	2.70	2.71
Stacked Litter, Roofed Shed	19.73 (39.45)	21.38 (42.76)	2.78	2.82

NOTES: Season is significant at $P < 0.04$ (wet basis).
Storage method and storage x season are non-significant (wet basis).
Storage is significant at $P < 0.004$ (dry basis). Season and storage x season are non-significant (dry basis).

Table 14. Mean calcium (Ca) values (wet basis) for turkey litter, kg/t (lb/ton), and percent (dry basis).

Litter Storage Method	Ca, kg/t (lb/ton)	Ca, Percent
Litter, direct from house	22.17(44.34)	3.28
Litter stacked, uncovered	19.36(38.71)	3.51
Litter stacked, covered with plastic	18.70(37.40)	3.72
Litter stacked, roofed shed	20.45(40.90)	2.92

NOTE: Effects of storage method, season, and storage x season all non-significant (wet and dry basis).

Magnesium. Significant differences were observed for the effect of storage on Mg found in litter. This effect may be related to volume reduction of the outdoor stacked litter due to the tendency for the organic material to break down over time more so than in the more protected storage methods. In addition, outdoor exposed stacks may be subject to a greater degree of erosion and transport of material away from the stack due to runoff, thereby concentrating Mg in the remaining stack. Higher values were also observed for the roofed shed stacking systems where composting conditions are often the best. In a similar way, pile shrinkage may occur, concentrating Mg in the stack. Effects of season and storage x season interactions were non-significant. Results for litter are shown in Tables 15 and 16.

Table 15. Mean magnesium (Mg) values (wet basis) for broiler litter, kg/t (lb/ton), and percent (dry basis).

Litter Storage Method	Mg, kg/t (lb/ton)	Mg, Percent
Litter, direct from house	4.07(8.14)	5.71
Litter stacked, uncovered	4.15(8.30)	6.50
Litter stacked, covered with plastic	4.14(8.28)	5.73
Litter stacked, roofed shed	4.39(8.78)	5.94

NOTES: Effects of storage method are significant at $P < 0.04$ (wet basis), and $P < 0.0001$ (dry basis).
Effects of season, and storage x season all non-significant.

Table 16. Mean magnesium (Mg) levels (wet basis) in turkey litter, kg/t (lb/ton), and percent (dry basis).

Litter Storage Method	Spring Mg, kg/t (lb/ton)	Fall Mg, kg/t (lb/ton)	Spring Mg, percent	Fall Mg, percent
Litter, direct from house	3.45(6.82)	3.37(6.75)	0.54	0.48
Stacked Litter, Uncovered	2.86(5.73)	3.68(7.35)	0.55	0.63
Stacked Litter, Plastic Covered	4.37(8.73)	3.15(6.30)	0.92	0.57
Stacked Litter, Roofed Shed	3.10(6.22)	3.98(7.95)	0.45	0.56

NOTES: Storage method and season are non-significant, but storage x season is significant at $P < 0.01$. (wet basis).
Storage method ($P < 0.005$) and storage x season ($P < 0.006$) are significant, and season is non-significant (dry basis).

REFERENCES

Greenberg, A.E., Clesceri, L.S., Eaton, A.D., and Franson, M.A.H., *Standard Methods for the Examination of Water and Wastewater,* American Public Health Association, Washington DC 20005, 1992.

Midwest Plan Service, *Livestock Waste Facilities Handbook,* 2nd Edition, Midwest Plan Service, Iowa State University, Ames, Iowa 50011, 1985.

Minitab, Inc., *Minitab Reference Manual,* Minitab, Inc., 3081 Enterprise Drive, State College, Pennsylvania, 16801, 1989.

EXCESS DAIRY MANURE MANAGEMENT: IMPACTS ON GROUNDWATER QUALITY

Jayaram Daliparthy, Stephen J. Herbert, and Peter L.M. Veneman

INTRODUCTION

Dairy manure, an excellent source of organic fertilizer, provides plant nutrients and contributes to improved soil fertility when properly managed and utilized in crop production systems. Land application of manure has been the most practical method of manure disposal and utilization. However, when manure is applied in amounts exceeding crop nutrient requirements, it becomes a potential source of surface- and ground-water pollution. Nitrate is one of the principal pollutants of the water resource when livestock waste is managed incorrectly, as in improperly maintained manure storage facilities, excessive application rates, high animal-to-land ratios, and use of sensitive hydrogeological sites. Large-scale dairy operations have greatly magnified the problem of handling manure. According to estimates prepared for the Environmental Protection Agency using the 1987 Agriculture Census, Connelly (1991) reported that 64 counties in the United States (US) have manure-to-cropland ratio equivalents in excess of 270 kg N/ha, assuming the entire quantity of manure produced in the county is spread on cropland. Many dairy farms in Massachusetts and the northeastern US have high livestock-to-cropland ratios. High density of animals may offer economic benefits but often results in environmental problems because of spreading manure in close proximity to the areas of confinement.

Madison and Brunett (1985) showed the northeastern US as a problematic area for nitrate contamination because intensive livestock operations coexist with relatively dense rural populations. Nitrogen availability on many dairy farms exceeds the amount required for corn production. In Lancaster County, Pennsylvania, annual manure applications average over 90 Mg/ha (Young et al., 1985). Similar figures have been reported for many dairy farms in southern New England. Corn fields closest to the manure storage area often receive manure applications at rates far above crop requirements. In addition, many farmers apply commercial N fertilizer, further adding to the problem of nutrient excess and potential contamination of ground and surface waters.

Increasing the total land area available for spreading is one of the most practical ways to reduce excess applications of manure. Manure application to perennial forages such as alfalfa is one potential alternative. On dairy farms this alternative practice could increase the land area available for manure application and decrease the rate spread on any field, thus reducing the potential of nitrate leaching. Various researchers have shown alfalfa can be an effective NO_3 scavenger (Muir et al., 1976; Stewart et al., 1968). Levin and Leshem (1974) reported that on a drained peat soil where alfalfa was grown, no nitrate was found in the lower soil profile, whereas under cotton, nitrate accumulated in high concentrations. Alfalfa, a deep-rooted perennial, can reduce nitrate-N concentrations at depths below the root zone of annual crops (Viets and Hageman, 1971; Schuman and Elliott, 1978). Schuman and Elliott (1978) reported N uptake two and one-half to three times more for alfalfa than corn.

Soils rich in nitrate inhibit N_2 fixation by legumes (Phillips and Dejong, 1984), thereby reducing the total N input to the cropping cycle and limiting the leaching of nitrates. Owens (1990) reported NO_3-N concentrations ranging from 15 to 40 mg/L under corn while the concentrations under alfalfa in leachate often were less than 5 mg/L. Nitrates move with the soil water in the profile either by downward or upward fluxes, depending on fluctuations in water potential by depth (Nielsen et al., 1980). Thus, deep-rooted crops such as alfalfa may reduce NO_3 leaching not only by taking up water from deep in the profile, but also by uptake of N.

Alfalfa-corn is the predominant cropping rotation on most dairy farms in the Northeast. Hesterman et al. (1986) reported that this rotation with alfalfa, subjected to a three-cut system, was economically advantageous compared to continuous corn or soybean-corn systems. Alternative strategies, however, are necessary for efficient manure management to prevent excessive applications on corn fields and to minimize nitrate contamination of groundwater. In 1990, long-term field trials were initiated in Massachusetts to study the impact of manure application to alfalfa and corn in alfalfa-corn rotations, particularly with respect to nitrate pollution of ground water. This chapter summarizes three years of field trials evaluating the impact of potentially excess quantities of dairy manure to alfalfa-corn crop rotations.

RESEARCH METHODS

Research trials began in June 1990 on a one-year-old stand of alfalfa in Deerfield and a two-year-old stand of alfalfa in Sunderland, both located in western Massachusetts. The soil at the Deerfield site was an Occum fine sandy loam variant (coarse-loamy, mixed, mesic Fluventic Dystrochrept), which was relatively low in organic matter and has a very fine sandy loam texture. The soil at the Sunderland site was an Agawam fine sandy loam (coarse-loamy over sandy, mixed, mesic Typic Dystrochrept) consisting of a fine sandy loam mantle over coarser sands. Treatments employed for alfalfa were unfertilized (no N), low manure-N (112 kg N/ha from approximately 46,750 L liquid manure/ha), high manure-N (336 kg N/ha from approximately 140,250 L liquid manure/

ha), and low and high fertilizer N (112 and 336 kg N/ha as NH_4NO_3). The experimental design was a randomized complete block with four replications. Individual plot sizes were 3 by 6 m, with a 0.9-m border. Dairy manure (approximately 0.33% total nitrogen, 0.145% ammonia nitrogen) was applied to alfalfa by surface spreading after the first cutting in spring. Water samples were collected through porous cup suction samplers placed at 30-, 60-, 90- and 120-cm depths in each treatment and replicated three times. Water samples were analyzed for NO_3-N by a cadmium reduction method (Technicon, 1977).

Silage corn was planted in 1992, after manuring alfalfa for two years, at the Sunderland site. Alfalfa was plowed under prior to planting corn. Treatments for corn following alfalfa consisted of application of low manure-N (112 kg N/ha) and low fertilizer N (112 kg N/ha from NH_4NO_3); and the residual effects of previously imposed treatments to alfalfa with either no nitrogen (control), low manure-N, and high manure-N. Liquid manure was applied on the surface and disked in immediately prior to planting corn. At the Deerfield site, manuring experiments with alfalfa continued for a third consecutive year to study long-term impact.

RESULTS AND DISCUSSION

EFFECTS OF TWO CONSECUTIVE YEARS OF MANURE APPLICATION TO ALFALFA

Our previous research indicated that application of dairy manure to alfalfa at low rates of 112 kg N/ha had no significant effect on dry matter yield, N accumulation, NO_3-N in the soil water, or on soil NO_3-N levels (Daliparthy et al., 1994). Manure applications at the higher rate (336 kg N/ha) showed a reduction of alfalfa dry matter (DM) yield during the second year in 1991, mainly because of the physical accumulation of manure residue, resulting in delayed regrowth. Mean NO_3-N concentrations in the soil water under alfalfa at the low manure rate averaged at 2.5 mg/L at the Deerfield site. We detected an increase in soil water NO_3-N at the higher manure application rate during the second year of the experiment. Mean NO_3-N concentrations in water samples collected in the spring were significantly lower than those collected in the summer or fall (Daliparthy et al., 1994).

Our research showed that NO_3-N concentrations in soils did not vary much after two years of manure or N fertilizer applications to alfalfa (Daliparthy et al., 1994). Soil NO_3-N levels in the profile generally were low during the spring. We concluded that the reasons for the low soil NO_3-N concentrations were the large amount of N taken up by alfalfa, denitrification, and nitrate leaching. In addition, most of the NH_4^+ fraction in the manure might be lost due to volatilization, which might account for the lack of NO_3-N build-up in soils even at high manure rates. Our experimental results were similar to the findings of Schmitt et al. (1994), who reported that applications of manure or N fertilizer to alfalfa had no significant effect on soil NO_3-N after two years.

We reported that the low rates of manure application to alfalfa did not significantly affect mean annual dry matter yields (Daliparthy et al., 1994). Reduction in alfalfa stand density due to the smothering by manure residues is of concern to dairy farmers spreading manure on hay fields. Research trials in Massachusetts showed no loss in forage yield with low manure application rates (46,750 L liquid manure/ha). However, forage yields were restricted with the application of manure at the high rate (140,250 L liquid manure/ha). Schmitt et al. (1993) from Minnesota showed that preplant manure applications increased alfalfa yields and were not detrimental to alfalfa production in regions in which traditional inorganic fertilizers are recommended for establishment of alfalfa. They also observed reductions in alfalfa yields due to the combined effects of high manure rates and soil compaction from application equipment (Schmitt et al., 1993). Similarly, in Massachusetts, we observed a temporary smothering of alfalfa with the high manure rates, which delayed regrowth.

EFFECTS OF THREE CONSECUTIVE YEARS OF MANURE APPLICATION

To understand the long-term implications and impacts of these alternative manure management practices, manure was applied for the third consecutive year to alfalfa at the Deerfield site in the spring of 1992 after the first cutting. Manure application to alfalfa at the higher rate for three consecutive years showed significantly higher concentrations of NO_3-N in soil water than no nitrogen (control) or low manure plots (Figure 1).

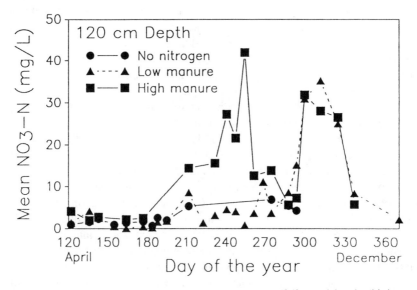

Figure 1. Mean NO_3-N concentrations in soil water under alfalfa receiving the third year of manuring in 1992.

Mean NO_3-N concentrations in soil water, averaged across the depths, under the plots applied with the low manure rates (7.8 mg N/L) did not show any significant increase compared to those under unfertilized plots (4.4 mg N/L). At a few sampling dates, NO_3-N concentrations at the 60-cm depth, especially in the fall, were higher with the low manure rate than those recorded under unfertilized plots. Application of high rates (336 kg N/ha) of manure for three consecutive years showed significantly higher NO_3-N concentrations in the soil water, especially at the 60-cm depth, than under low manure rates or unfertilized plots (Daliparthy et al., 1993). These high NO_3-N concentrations were detected mostly at depths shallower than 120 cm.

Application of manure to alfalfa for the third consecutive year in 1992, either at the low or high rate, had no significant effect on forage yield (Table 1). However, in the second year (1991) of manure application, a significant restriction in alfalfa yields occurred with the higher rate (Daliparthy et al., 1994). This reduction was due in part to low precipitation during the summer of 1991. Manure applications to alfalfa at the high rate followed by dry conditions delayed alfalfa regrowth, thereby affecting forage yields.

NITRATES IN SOIL WATER UNDER CORN FOLLOWING TWO YEARS OF MANURED ALFALFA

It is important to study the residual and added effects of manuring on the crop. At one of the sites, after two years of manure applications to alfalfa, corn was planted in May 1992 immediately upon plowing in of the alfalfa. Residual effects on nitrate leaching and corn yields from the manure applications were recorded with the succeeding corn crop. Water samples collected under corn following two years of alfalfa in the rotation had a high frequency of samples exceeding the 10 mg NO_3-N/L US Environmental Protection Agency drinking water standard, especially at 30-, 60-, and 90-cm depths. Additional manure or inorganic fertilizer applications to corn resulted in increased NO_3-N concentrations in soil water (Daliparthy et al., 1993). Nitrate-N concentrations were higher in all the treatments during June at the shallow depths. Apparently, this increase was due to the mineralization of organic residue. Seasonal variations

Table 1. Effects of three consecutive years of manure application.[†] on alfalfa forage yield in 1992.

Treatments	2nd cut	3rd cut
	--------------Mg/ha--------------	
No Nitrogen	1.6	1.1
Low Manure (112 kg N/ha)	1.8	1.2
High Manure (336 kg N/ha)	1.5	1.2
LSD $_{0.05}$	NS	NS

†Manure was applied after the first cutting in spring.

in NO₃-N concentrations were observed. Since growing corn plants utilize large amounts of water and dry out the soil, there was a limited potential for nitrates to move below the root zone. In concurrence with our interpretation, nitrate-N concentrations in soil water collected at 120-cm depths were lower (Figure 2) than those recorded at shallow depths.

The first year corn following alfalfa in the rotation showed no significant difference in silage yields among the treatments (Daliparthy et al., 1993). Application of manure or fertilizer N at 112 kg N/ha did not show any significant effect on corn silage yield compared to that of corn grown on residual fertility. These results indicated that manure or inorganic fertilizer application to corn following alfalfa in the rotation had no significant effect on silage yield but increased potential for nitrate leaching.

SUMMARY AND CONCLUSIONS

Environmentally and economically sustainable manure management practices are essential for profitability and survival of dairy farms. The results from field trials conducted in Massachusetts show that dairy manure can be applied to alfalfa at the low rates of 112 kg N/ha (approximately 45,000 L liquid manure/ha). Application of low rates of manure for three consecutive years to alfalfa had no significant effect on either forage yield or nitrate concentrations

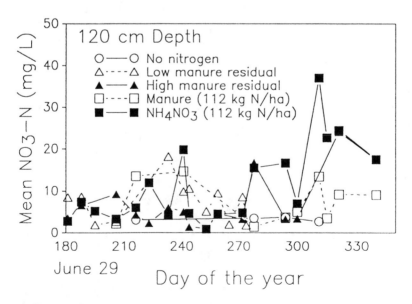

Figure 2. Mean NO₃-N concentrations in soil water under corn succeeding two years of manured alfalfa in the rotation in 1992.

in soil water compared to results obtained with unmanured alfalfa. However, manure applications at a higher rate (336 kg N/ha) significantly increased NO_3-N concentrations in soil water after three years of application and also delayed alfalfa regrowth in the second and third years.

Residual effects of manure applications to alfalfa on the succeeding corn crop occurred. In the alfalfa-corn rotation, corn planted after incorporation of the alfalfa did not require any additional fertilizer N or manure. In general, mean NO_3-N concentrations in the soil water collected immediately after corn planting were high in all treatments at shallow depths, and application of any additional N fertilizer or manure increased NO_3-N concentrations. First-year corn silage yields among the treatments, following alfalfa, showed no significant differences.

Thus, alternative strategies in manure management, like applying manure to alfalfa, will increase the total land area available for spreading and, thereby, will decrease the actual rate of manure on any particular field. Farmers can apply manure to their alfalfa fields at the low rates either for two or three years, depending upon manure availability, stand quality, and total available land area for application. Application of manure to alfalfa instead of to corn in a rotation limits nitrate pollution because of greater accumulation of N by alfalfa and provides a different management option that can improve the efficiency and economy of crop production at dairy farms. Best management practices for manure management such as manure applications to alfalfa in the rotation are continuing to evolve. These practices will promote the environmental and economic sustainability of dairy farming in the northeastern US.

REFERENCES

Connelly, C., Unpublished data prepared for the offices of Policy, Planning, and Evaluation US EPA. 1991.

Daliparthy, J., Herbert, S.J., and Veneman, P.L.M., Dairy manure applications to alfalfa: Crop response, soil nitrate, and nitrates in soil water, *Agron. J.* 86, 927, 1994.

Daliparthy, J., Herbert, S.J., Veneman, P.L., Litchfield, G.V., Akin, T., and Moffitt, J., Nitrate leaching in manured alfalfa-corn rotation, *Agronomy Abstracts*, p. 28-29. American Society of Agronomy 1993 Annual Meetings, Cincinnati, Ohio, 1993.

Hesterman, O.B., Sheaffer, C.C., and Fuller, E.I., Economic comparisons of crop rotations including alfalfa, soybean, and corn, *Agron. J.*, 78, 24, 1986.

Levin, I., and Leshem, Y., Using forage crops to reduce nitrate accumulation in Hula peat soils, *Proc. 5th Scientific Conf. of the Israel Ecolog. Soc.*, Technion-Israel Inst. of Technol., Haifa, Israel. 1974.

Madison, R.J., and Brunett, J.O. Overview of the occurrence of nitrate in ground water of the United States, in *U.S.G.S. National Water Summary*, US Geol. Surv. Water-supply Paper 2275, 1985.

Muir, J., Boyce, J.S., Seim, F.C., Moscher, P.N., Deibert, E.J., and Olson, R.A., Influence of crop management practices on nutrient management below the root zone in Nebraska soils, *J. Environ. Qual.*, 5, 255, 1976.

Nielsen, D.R., Biggar, J.W., MacIntyre, J., and Tanji, K.K., Field investigation of water and nitrate-nitrogen movement in Yolo soil, in *Soil Nitrogen as Fertilizer or Pollutant*, International Atomic Energy Agency, Vienna. 1980.

Owens, L.B., Nitrate-nitrogen concentrations in percolate from lysimeters planted to a legume-grass mixture, *J. Environ. Qual.*, 19, 131, 1990.

Phillips, D.A., and DeJong, T.M., Dinitrogen fixation in leguminous crop plants, in *Nitrogen in Crop Production*, R.D. Hauck, ed., ASA, Madison, Wisconsin, 1984.

Schuman, G.E., and Elliott, L.F., Cropping an abandoned feedlot to prevent deep percolation of nitrate-nitrogen, *Soil Sci.*, 126, 237, 1978.

Schmitt, M.A., Sheaffer, C.C., and Randall, G.W., Preplant manure and commercial P and K fertilizer effects on alfalfa production, *J. Prod. Agric.*, 6, 385, 1993.

Schmitt, M.A., Sheaffer, C.C., and Randall, G.W., Manure and fertilizer effects on alfalfa plant nitrogen and soil nitrogen, *J. Prod. Agric.*, 7, 104, 1994.

Stewart, B.A., Viets, F.G., Jr., and Hutchinson, G.L., Agriculture's effect on nitrate pollution of groundwater, *J. Soil Water Cons.*, 23, 13, 1968.

Technicon Instruments Corp, Technicon AutoAnalyzer II Industrial Method no. 334-74W/B. Technicon Instruments Corp., Tarrytown, New York, 1977.

Viets, F.G., Jr., and Hageman, R.H., *Factors Affecting the Accumulation of Nitrate in Soil, Water, and Plants*, Agric. Handbook 413, USDA-ARS, Washington, DC, 1971.

Young, C.E., Crowder, B.M., Shortle, J.S., and Alwang, J.R., Nutrient management on dairy farms in southeastern Pennsylvania, *J. Soil Water Cons.*, 40, 443, 1985.

BUFFER STRIPS TO IMPROVE QUALITY
OF RUNOFF FROM LAND AREAS TREATED
WITH ANIMAL MANURES

I. Chaubey, D.R. Edwards, T.C. Daniel, and P.A. Moore, Jr.

INTRODUCTION

Buffer strips have been shown to reduce runoff transport of pollutants from cropland and livestock activity areas. Buffer strips are bands of planted or indigenous vegetation used to remove pollutants from incoming runoff. The major pollutant removal mechanisms associated with buffer strips include infiltration of runoff and pollutants, deposition of sediment, filtration of suspended sediment by vegetation, adsorption on soil and plant surfaces, and absorption of soluble pollutants by plants (Dillaha et al., 1989).

Buffer strip effectiveness has been investigated by several scientists (e.g., Doyle et al., 1977; Dillaha et al., 1986, 1988, 1989). These studies indicate that, depending on parameters such as type of vegetation, nature of pollutant source, buffer length, and time since buffer installation, over 90% of incoming runoff nutrients may be retained by buffer strips. Dillaha et al. (1986), however, noted that effectiveness is significantly decreased when concentrated flow patterns develop within buffer strip areas.

Few simulation models have been developed to describe buffer strip behavior in general fashion. Muñoz-Carpena et al. (1992) developed an event-based model with physically based components for evaluating buffer strip performance with respect to sediment. Lee et al. (1989) developed a comprehensive model of phosphorus (P) and sediment transport through buffer strips. Several researchers (William and Nicks, 1988; Flanagan et al., 1989) have predicted the effectiveness of buffer strips for erosion control using the CREAMS (Chemical, Runoff, and Erosion from Agricultural Management Systems) model (Knisel, 1980). Overcash et al. (1981) developed a generalized, event-based model to describe buffer strip performance based on infiltration as the primary mechanism of pollutant filtration.

The objective of this study was to develop and test a model to predict performance of grass buffer strips installed downslope of areas treated with poultry litter based on infiltration as the primary mechanism of pollutant removal. This work complements the efforts of Overcash et al. (1981) by using

363

their model to describe reductions in pollutant concentrations and mass transport but adding physically-based infiltration and overland flow routing components. As will be described later, rainfall and runoff data for the buffer strips are key inputs to the pollutant concentration and mass transport model developed by Overcash et al. (1981). Only infiltration was considered as a mechanism for reducing pollutant transport through buffer strips, because this process is considered to be one of the most significant removal mechanisms of buffer strips performance (Dillaha et al., 1988).

MODEL DEVELOPMENT

The overall model for predicting buffer strip performance consists of the following three linked components: (1) infiltration, (2) overland flow routing, and (3) pollutant concentrations and mass transport prediction.

INFILTRATION COMPONENT

The infiltration component is based on the Mein and Larson (1971) version of the modified Green-Ampt equation as modified by Chu (1978) to accommodate unsteady rainfall. The primary equations are:

$$f_p = K_s \, (1+MS_{av}/F_p) \tag{1}$$

$$K_s(t - t_p + t_s) = F - MS_{av} \, ln[1 + f/(MS_{av})] \tag{2}$$

where f_p is infiltration rate (m/s), K_s is saturated hydraulic conductivity (m/s), M is the difference in average soil moisture before and after wetting (m³/m³), S_{av} is average suction across the wetting front (m), F_p is cumulative infiltration (m), t is the time after the beginning of rainfall (s), t_p is the time to ponding (s), and t_s is the shift of the time scale to produce the effect of having cumulative infiltration at the ponding time, or pseudotime (s). The major output of this component is rainfall excess.

OVERLAND FLOW ROUTING COMPONENT

The overland flood routing component is based on kinematic wave approximations to the hydrodynamic equations:

Continuity:
$$\partial y/\partial t + \partial Q/\partial x = q_e = R - f \tag{3}$$

Momentum:
$$Q = \alpha y^m = (S_o^{1/2} y^{5/3})/n \tag{4}$$

where x is distance in direction of flow (m), t is time (s), y(x,t) is flow depth (m), Q(x,t) is discharge per unit width (m²/s), q_e is rainfall excess (m/s), R is rainfall intensity (m/s), f is infiltration rate (m/s), S_o is slope (m/m), α and m are parameters for equation (4), and n is Manning's roughness coefficient.

POLLUTANT CONCENTRATION AND MASS TRANSPORT COMPONENT

The pollutant reduction model given by Overcash et al. (1981) was used to calculate poultry litter constituent concentrations and mass transport.

$$C_x = C_B + (C_o - C_B) \, e^{[R/(R-f)]ln[W/(W+x)]} \qquad (5)$$

$$M_x = C_x Q_x = C_x \, (W+x)(R-f) \qquad (6)$$

where C_x is concentration of the constituent in runoff at buffer strip length x (mg/L), C_o is initial concentration of the constituent entering the buffer strip (mg/L), C_B is concentration due to rainfall and background pollutant levels (mg/L), R is rainfall rate (mm/h), f is infiltration rate (mm/h), Q_x is rainfall excess (mm/h), W is length of the litter application area (m), and M_x is pollutant mass flux per unit width at buffer strip length x (mg/m).

Using this model requires the assumptions that mass loss of pollutants is only by infiltration and that rainfall and infiltration rates are constant (Overcash et al., 1981). Average rainfall and infiltration rates can be calculated from cumulative rainfall and cumulative infiltration (Overcash et al., 1981). The first two model components provide the rainfall excess, q_e, for each time step. This rainfall excess was integrated over the entire rainfall event and divided by rainfall duration to compute average depth of rainfall excess (R-f). Similar computations were performed to obtain rainfall depth.

APPLICATION AND TESTING OF MODEL

The model was tested using data from a field study on buffer strip effectiveness described by Chaubey et al. (1993). In that study, three fescue plots with dimensions of 1.5 m by 24.4 m were constructed on Captina silt loam soil at the Main Agricultural Experiment Station in Fayetteville, Arkansas. Gutters were installed across each plot at 3.1, 6.1, 9.2, 12.2, 18.3, and 24.4 m down slope to enable runoff collection at those lengths. Poultry litter was applied to the upper 3.1 m of each plot, and the remainder of each plot acted as a buffer. Simulated rainfall was applied to each plot at 50 mm/h until runoff had occurred for 1 h. Runoff samples were collected at buffer strip distances from 0 to 21.4 m and analyzed for ortho P (PO_4-P), total P (TP), ammonia N (NH_3-N), and total Kjeldahl N (TKN). Other parameters were measured but are not included here because they either did not respond to VFS length or are not suitable for use with the Overcash et al. (1981) model. A full description of the procedures, methods of runoff sample analyses, and results of the study was given by Chaubey et al. (1993).

Model outputs were evaluated by comparing observations from the field study to predictions obtained from executing the model in uncalibrated and calibrated modes. Calibrated and uncalibrated executions were performed to demonstrate the value of calibration with respect to buffer strip performance prediction. In addition, observed measures of buffer strip performance were compared to predictions resulting from use of observed runoff values in order to separate the performance of the equations (5) and (6) developed by Overcash et al. (1981) from the performance of the runoff component of the model. Each method of model testing required poultry litter constituent concentrations entering the buffer strip (C_0) as an input. Values of background concentration C_B required in Eq. (5) were based on data reported by Edwards and Daniel (1993, 1994) on runoff concentrations of various parameters from untreated plots similar to those used in the study of Chaubey et al. (1993).

During uncalibrated model execution, average values of 3.33 x 10^{-6} m/s for K_s and 0.024 m for S_{av} were used based on data reported by Thiesse (1984) for the same soil near the location of the plots. The R value was taken as the total rainfall (m) applied during the event, and the average value of (R-f) (m) was calculated for each buffer strip length using the overland flow routing and infiltration components.

The model was calibrated using the observed runoff values measured at the buffer strip length of 21.4 m. In the model calibration, values of K_s, (MS_{av}), and n were identified that minimized the sum of squared differences between observed and predicted runoff rates at the bottom of the plots.

Simple linear regression analysis was performed between predicted and observed concentrations and mass transport of different poultry litter constituents past various buffer strip lengths to assess model accuracy. Concentration and mass transport values at 0 m buffer strip length were not included in the regression analyses. The significance of regression line slopes and coefficients of determination were determined by t-tests.

RESULTS AND DISCUSSION

PREDICTION OF CONSTITUENT CONCENTRATIONS

Table 1 shows the observed and predicted runoff and concentrations of poultry litter constituents past different buffer strip lengths obtained using the three different methods of prediction. As noted earlier, concentrations at 0 m buffer strip were inputs for all these prediction methods in order to calculate the concentrations at other buffer strip lengths.

Poultry litter constituent concentrations predicted using the uncalibrated runoff component were generally less than observed for buffer strip lengths from 3.1 to 15.2 m. Equation (5) indicates that the reduction in incoming constituent concentration depends only on the ratio of rainfall to runoff for fixed

Table 1. Observed and predicted poultry litter constituent concentrations

Variable		\multicolumn{6}{c}{Buffer Strip Length (m)}					
		0	3.1	6.1	9.2	15.2	21.4
		\multicolumn{6}{c}{----------------------------------m^3----------------------------------}					
Runoff	O[†]	0.27	0.54	0.75	0.81	0.93	0.80
	P1	0.10	0.19	0.27	0.35	0.49	0.62
	P2	0.06	0.12	0.18	0.23	0.34	0.44
		\multicolumn{6}{c}{----------------------------------mg/L----------------------------------}					
NH_3-N	O	7.15	2.02	0.90	0.64	0.12	0.05
	P1	7.15	1.01	0.31	0.15	0.07	0.05
	P2	7.15	0.25	0.06	0.05	0.05	0.05
	P3	7.15	2.63	1.35	0.65	0.25	0.07
TKN	O	26.50	6.88	4.68	3.03	1.85	1.67
	P1	26.50	5.05	2.61	2.02	1.74	1.69
	P2	26.50	3.52	2.22	1.84	1.75	1.67
	P3	26.50	10.63	6.21	3.75	2.35	1.76
PO_4-P	O	4.29	1.38	0.75	0.44	0.20	0.17
	P1	4.29	0.73	0.32	0.22	0.18	0.17
	P2	4.29	0.29	0.18	0.17	0.17	0.17
	P3	4.29	1.67	0.93	0.52	0.29	0.18
TP	O	6.72	2.22	1.04	0.59	0.28	0.22
	P1	6.72	1.11	0.47	0.31	0.24	0.23
	P2	6.72	0.42	0.25	0.23	0.22	0.22
	P3	6.72	2.65	1.47	0.82	0.43	0.25

[†]O = Observed (mean of three replications), P1 = Predicted with runoff estimated from the uncalibrated model, P2 = Predicted with runoff estimated from the calibrated model, and P3 = predicted using observed values of runoff.

values of W and x. Table 1 indicates that the uncalibrated runoff component underpredicted runoff for all buffer strip lengths. These runoff predictions could have caused predicted constituent concentrations to reach background concentration levels more quickly than observed. Hypothesis testing indicated that the coefficients of determination were significantly ($P < 0.05$) greater than zero (range 0.95 to 0.98), but the slopes of regression lines for all concentrations predicted using the uncalibrated runoff component were significantly different from one (range 0.44 to 0.50).

Concentrations predicted using the calibrated runoff component reached background levels after about 6.1 m for all constituents (Table 1). As was the case when using the uncalibrated runoff component, all concentrations were underpredicted for most buffer strip lengths. The poor performance of the calibrated runoff component is attributed to variability in soil hydraulic properties. The runoff component was calibrated using data from the 21.4-m buffer strip length, and the calibrated parameters were assumed not to vary within the plot. Table 1, however, indicates that runoff was underpredicted at shorter buffer strip lengths, indicating within-plot variability in the calibrated parameters. As discussed earlier, the predicted runoff values could have contributed to underpredicted concentrations. Hypothesis testing indicated that the slopes of the regression lines were significantly ($P < 0.05$) different from one (range 0.10 to 0.27), but all coefficients of determination were significantly ($P < 0.05$) greater than zero (range 0.93 to 0.97).

When observed values of runoff were used to predict the parameter concentrations by equation (5), the predictions were very similar to the observed values (Table 1). The results obtained from this method of model evaluation tend to validate the basis of equation (5), since this procedure essentially tested equation (5) apart from the other model components. Hypothesis testing indicated that the coefficients of determination were significantly ($P < 0.05$) greater than zero (range 0.98 to 1.0) for all the poultry litter constituents, and slopes of regression lines, except for TKN and PO_4-P, were not significantly ($P < 0.05$) different from one (range 1.18 to 1.3) for all poultry litter constituents.

PREDICTION OF CONSTITUENT MASS TRANSPORT

Observed and predicted mass transport past various buffer strip lengths for each method of prediction are shown in Table 2. Table 2 indicates that mass transport of poultry litter constituents past all buffer strip lengths was underpredicted when using the uncalibrated runoff component. As discussed earlier, prediction of mass transport depends directly on predicted runoff (equation (6)). Predicted mass transport will thus be less than observed if runoff is underpredicted. Underpredicted runoff probably led to underpredictions of mass transport of poultry litter constituents (Table 2).

Predicted mass transport increased after a buffer strip length of 6.1 m for TKN, and 9.2 m for PO_4-P and TP (Table 2). This result was expected since predicted concentrations of these constituents reached the background levels very quickly (Table 1). Therefore, although predicted concentrations did not increase after reaching background levels, mass transport increased due to the continued contribution from background sources. Hypothesis testing indicated that except for NH_3-N, the coefficients of determination were insignificant (range 0.03 to 0.89) for all the poultry litter constituents. The slopes of regression lines for all the parameters predicted were significantly ($P < 0.05$) different from one (range -0.03 to 0.16).

Masses of litter constituents transported past buffer strip lengths were also underpredicted using the calibrated runoff component (Table 2). As noted earlier, this is attributed to variability of soil properties that led to runoff underpredictions at intermediate buffer strip lengths. Similar to results discussed earlier, predicted mass transport of poultry litter constituents increased after a buffer strip length of 9.2 m for NH_3-N and 6.1 m for TKN, PO_4-P, and TP. In all cases, regression line slopes were significantly ($P < 0.05$) different from one, and coefficients of determination were insignificant.

Table 2 indicates that when actual values of runoff were used in equation (6), the predicted mass transport of poultry litter constituents was very similar to the observed values. Hypothesis indicated that the slopes of regression lines

Table 2. Observed and predicted poultry litter constituent mass transport

Variable	0	Buffer Strip Length (m)					
		3.1	6.1	9.2	15.2	21.4	
		-----------------------m³----------------------------					
Runoff	O†	0.27	0.54	0.75	0.81	0.93	0.80
	P1	0.10	0.19	0.27	0.35	0.49	0.62
	P2	0.06	0.12	0.18	0.23	0.34	0.44
		----------------------------kg/ha----------------------------					
NH3-N	O	4.20	2.20	1.30	1.00	0.30	0.10
	P1	1.49	0.41	0.18	0.11	0.07	0.07
	P2	1.00	0.08	0.03	0.01	0.04	0.05
	P3	4.20	3.30	2.50	1.40	0.70	0.20
TKN	O	15.50	9.20	7.10	5.10	4.20	3.40
	P1	5.54	2.02	1.51	1.51	1.84	2.25
	P2	3.64	1.07	0.98	1.04	1.43	1.76
	P3	15.50	13.20	11.00	7.80	5.90	3.70
PO4-P	O	2.50	1.50	1.10	0.70	0.40	0.30
	P1	0.90	0.29	0.19	0.17	0.19	0.23
	P2	0.60	0.08	0.07	0.09	0.13	0.16
	P3	2.50	2.10	1.70	1.10	0.70	0.30
TP	O	4.00	2.30	1.50	0.90	0.60	0.40
	P1	1.35	0.44	0.27	0.23	0.26	0.30
	P2	0.95	0.12	0.09	0.12	0.17	0.21
	P3	4.00	3.40	2.70	1.80	1.10	0.50

†O = Observed (mean of three replications), P1 = Predicted with runoff estimated from the uncalibrated model, P2 = Predicted with runoff estimated from the calibrated model, and P3 = predicted using observed values of runoff.

were not significantly different from one (range 1.41 to 1.62) except for TKN and PO_4-P, in which cases the regression line slopes were significantly ($P < 0.05$) greater than one. Coefficients of determination were significant ($P < 0.05$) for all poultry litter constituents (range 0.95 to 0.98).

SUMMARY AND CONCLUSIONS

Runoff and infiltration components were added to the model of Overcash et al. (1981) to improve assessment of buffer strip performance. The combined model was tested using data on fescue buffer strip performance reported by Chaubey et al. (1993). The model was tested using values of poultry litter constituent concentrations and runoff that were predicted by (a) executing the model in an uncalibrated mode and (b) executing the model with the runoff component calibrated. Observed runoff concentrations and mass transport were also compared to predictions obtained using observed runoff values to test only the equations developed by Overcash et al. (1981).

The ability of the model to accurately predict concentrations and mass transport of poultry litter constituents depended on the accurate prediction of runoff and infiltration. Using either the uncalibrated or calibrated runoff components resulted in underpredicted runoff values at various buffer strip lengths; hence, mass transport at these buffer strip lengths was underpredicted. When observed runoff values were used, the predicted mass transport and concentration values were very similar to observed values. This finding highlights the fact that accurate prediction of runoff is essential in assessing buffer strip performance through a modeling approach.

REFERENCES

Chaubey, I., Edwards, D.R., Daniel, T.C., and Nichols, D.J., *Effectiveness of Vegetative Filter Strips in Controlling Losses of Surface-Applied Poultry Litter Constituents*, Paper No. 932011, ASAE, St. Joseph, Michigan, 1993.

Chu, S.T., Infiltration during unsteady rain, *Water Resour. Res.*, 14(3), 461, 1978.

Dillaha, T.A., Sherrard, J.H., and Lee, D., Long-term effectiveness and maintenance of vegetative filter strips, *VPI-VWRRC-BULL*, 153, 1, 1986.

Dillaha, T.A., Sherrard, J.H., Lee, D., Mostaghimi, S., and Shanholtz, V.O., Evaluation of vegetative filter strips as a best management practice for feedlots, *J. Water Pollut. Control Fed.*, 60(7), 1231, 1988.

Dillaha, T.A., Reneau, R.B., Mostaghimi, S., and Lee, D., Vegetative filter strips for agricultural nonpoint source pollution control, *Trans. ASAE*, 32(2), 513, 1989.

Doyle, R.C., Stanton, G.C., and Wolf, D.C., Effectiveness of forest and grass buffer filters in improving the water quality of manure polluted runoff, Paper No. 77-2501. ASAE, St. Joseph, Michigan, 1977.

Edwards, D.R., and Daniel, T.C., Effects of poultry litter application rate and rainfall intensity on quality of runoff from fescue plots, *J. Environ. Qual.*, 22, 361, 1993.

Edwards, D.R., and Daniel, T.C., Quality of runoff from fescuegrass plots treated with poultry litter and inorganic fertilizer, *J. Environ. Qual.*, 23, 579, 1994.

Flanagan, D.C., Foster, G.R., Neibling, W.H., and Burt, J.P., Simplified equations for filter strip design, *Trans. ASAE*, 32(6), 2001, 1989.

Knisel, W.G., ed., *CREAMS: A Field Scale Model for Chemicals, Runoff, and Erosion from Agricultural Management Systems*, Conservation Research Report # 26, U.S. Dept. Ag., Washington, D.C., 1980.

Lee, D., Dillaha, T.A., and Sherrard, J.H., Modeling phosphorus transport in grass buffer strips, *ASCE J. Environ. Engr. Div.*, 115(2), 409, 1989.

Overcash, M.R., Bingham, S.C., and Westerman, P.W., Predicting runoff pollutant reduction in buffer zones adjacent to land treatment sites, *Trans. ASAE*, 24(2), 430, 1981.

Muñoz-Carpena, R., Parsons, J.E., and Gilliam, J.W., *Vegetative Filter Strips: Modeling Hydrology and Sediment Movement*, Paper No. 92-2625, ASAE, St. Joseph, Michigan, 1992.

Mein, R.G., and Larson, C.L., *Modelling the Infiltration Component of the Rainfall-Runoff Process*, Bull. 43, Water Resour. Res. Center, University of Minnesota, St. Paul, 1971.

Thiesse, B.R., Variability of the physical properties of Captina silt soils, Unpublished M.S. Thesis, Department of Agronomy, University of Arkansas, Fayetteville, 1984.

Williams, R.D. and Nicks, A.D., Using CREAMS to simulate filter strip effectiveness in erosion control, *J. Soil Water Conser.*, 43(1), 108, 1988.

BEST MANAGEMENT PRACTICES FOR ANIMAL WASTE MANAGEMENT IN THE UKRAINE

V.Z. Kolpak and E.V. Skryl'nik

STATEMENT OF THE PROBLEM

U nder the transitory conditions of the Ukrainian move to a market economy, significant destructive processes occurred in the national chemical industry and, subsequently, in agricultural chemical use patterns. In the former Soviet Union, each ruble invested in agrochemicals generated 2.5 rubles as additional returns, and 42% of crop production was correlated with chemical use. These were prerequisites for stable increments in agricultural output.

But an abrupt decrease in production of mineral fertilizers and a correspondingly sharp increase in prices of chemicals, fertilizer-spreading equipment, and energy made them unaffordable for agricultural producers, and especially so for small farmers. As a result, in the Ukraine there has been a declining use of agrochemicals, bringing about a negative balance in nutrients, humus, calcium, and fertility in agricultural soils.

STATE OF THE ART

Under present conditions, a greater use of various organic wastes is a must for the agricultural sector. At the same time, it is necessary to provide for ecological safety of such application. At the present time in the Ukraine there are 148 cattle-breeding and 145 swine-producing complexes including five industrial-type complexes for annual breeding of 108,000 animals.

Efficiency of specialized industrial-type animal breeding complexes has been well demonstrated in practice. But construction and operation of these complexes have brought about a wide range of environmental problems. These problems are caused, to a great extent, by the large volume of manure wastes they generate, amounting at cattle-breeding complexes to about 0.5 million m^3/year, and at large hog-breeding complexes to 1.7 million m^3/year.

These complex problems are connected, primarily, with protection of air, soil, and water against pollution with livestock wastes. A major contributor to an upset of the ecological balance involves changes in management and input cycles to the livestock production complexes themselves and, to a certain ex-

371

tent, to farms that produce fodder for the larger complexes. Soil mineral balance is affected by the above management cycles but is correctable by application of manufactured mineral fertilizers. These management practices also affect the organic matter balance of field soils.

At present the content of humus in all soil and climatic zones of the Ukraine has decreased due to uncompensated input of organics to the soil and humus mineralization. Integrated data indicate that average annual loss of humus in the Polissia soddy podzolic soils amounts to 0.7-0.8 t/ha, in the forest-steppe chernozem soils to 0.6-0.7 t/ha, and in the steppe chernozem and dark chestnut soils 0.5-0.6 t/ha (Nosko et al., 1992).

EXPERIMENTAL STUDIES

The cleansing capacity of soil is mistakenly overestimated when manure and other organic wastes are applied, and this often leads to imbalances in agroecosystems, causing not theoretical but plainly observable differences. At the Edaphology and Agrochemistry Institute of the Ukrainian Academy of Agrarian Sciences (cy. Kharkov), effects of large hog-breeding complexes have been investigated. It was found that long-term (over 15 years) application of manure slurry for irrigation and fertilization caused a negative shift in physical-chemical properties of soils, especially those of the chernozem type. In particular, application of slurries at rates of 150-160 kg/ha total nitrogen brought about a decrease of exchangeable calcium in the soil complex with a concurrent 3-5 times increase of exchangeable sodium content. Downward movement of effluent transfers calcium and magnesium to lower layers of the soil profile where they accumulate. This process causes deterioration of the soil structure, reduces its permeability, increases its bulking in wet periods, and promotes shrinkage cracks and cloddiness in dry periods. An increase in mobility of the finely dispersed fraction of clay-textured soil and humus, with probable downward migration, is also observed. While morphological features of the soil have not changed, there is a shift in redistribution of agronomically valuable aggregates, a decrease in water-stable aggregates, and compaction of the upper layer (Skryl'nik, 1994).

Long-term application of livestock slurry manure, in addition to quantitative changes, also brings about qualitative changes in soil humus. Changes are pronounced by trends toward decreases in the soluble part of humic substances and an increase in humus content. Among groups of humic acids, the main mass is connected with calcium. There appears to be a trend toward an increase in humic acids strongly tied up with the mineral part of soil and stable sesquioxides. But the most significant increment occurs in the fraction represented by weakly connected and free humic acids, which increase 2.1-3.8 times, as indicated by the appearance in the humus of "young" newly formed humic substances. The relationship of carbon of humic acids to carbon of fulvoacids is evidenced by humate type of humus even after multi-year application of livestock slurries (Batsula and Skryl'nik, 1986; Batsula et al., 1994).

Moreover, there are unavoidable losses of nutrients transported from irrigated areas to surface and ground waters. Nitrogen and phosphorus transport is intensified due to failure to follow prescribed procedures, shortage of vehicles and equipment to transport manure, runoff in melting snow and/or storm flow. Content of nutrients in animal wastes amounts to 0.7-30.0% of the annual volume of wastes. In areas of intensive application of animal wastes, nitrate nitrogen content in waters is 4-10 times above the maximum allowable content (MAC).

Delivery of animal wastes to the field environment without proper treatment and disinfection causes human and animal infestation with diseases and helminths. It has been found that a period of 6 to 7 months is insufficient for complete die-off of pathogenic and tentatively pathogenic microorganisms in soils (Tikhonov et al., 1982).

In recent years the Ukraine Scientific-Research, Designing, and Construction Institutes have described and tested various practical systems for the treatment and removal of animal wastes. At the same time, experience has shown that it is impossible to design a universal system applicable to all animal farms.

Scientists at the A.N. Sokolovsky Edaphology and Agrochemistry Institute (EAI), together with others at the Ukrainian Scientific Research and Design Institute of Chemical-Machine Construction, have been developing a method for drying liquid manure slurry and processing the dried material into granulated organomineral fertilizers. The method involves delivery of the liquid slurry to a vibrofilter for separation to fractions. The solid fraction, with 70% moisture content, is thermally treated in a sterilizer and then delivered to the dosimeter of a blending machine where the dried manure is mixed with mineral fertilizers. The organomineral granules thus produced have 10-15% moisture content. The liquid fraction, which contains 1% solids content, is delivered from a storage reservoir to an evaporation unit where the solids content is concentrated to 20-30%. This mass is processed in the blending machine. This technology provides for complete disinfection of the liquid manure.

At the Ukrainian Scientific Veterinary Research Institute, a method has been developed for manure disinfection. The manure, delivered to a holding reservoir, is covered with a layer of mineral oil. Ammonia, in an amount equal to 1-3% of the manure mass, is introduced under the oil layer. Then the manure is mixed and kept in the reservoir 24-72 hours. A production line has been designed that includes the manure pump, a mixer, a unit for anhydrous ammonia batching, the ammonia tank, the reservoir for treated manure, and the unit for field application of the disinfected manure to soil.

Theory and practice confirmed that at present the most promising method of liquid manure treatment includes the manure fractionation to liquid and solid phases. This allows performance of cost-effective separate disinfection of these fractions. Separated fractions can then be transported and applied to the soil by means of standard agricultural machinery.

A method studied at the Sokolovsky EAI utilized thermobiological conditioning of the solid fraction of pig manure. The approach allowed quality improvement of the organic fertilizer, not only agronomically but also from the sanitary and hygienic point of view. The pilot unit consisted of a special chamber, natural and plenum aeration, and testing and data collection instruments. The solid fraction was laced with ammoniated straw for boosting microbiological processes, and with phosphogypsum to reduce ammonia losses (Batsula and Skryl'nik, 1993). Table 1 shows the change in manure after 42 days of treatment with this method.

Experimental data demonstrated that in the process of composting manure, moisture and organic content was reduced by 4-5%, and quantity of nutrients per unit of the compost mass increased 1.5 times. On that basis, one ton of fertilizer contains up to 60 kg of nutrients.

Presently in the Ukraine, the following methods of treatment and application of animal wastes for soil fertilization are most popular:

- Filling spreader tanks with homogenized manure slurry for transportation to fields, and subsequent incorporation into the soil;
- Piping or tanking livestock waste slurry to intermediate storage units, with subsequent surface spreading on fields;
- Piping and surface field spreading through sprinkler irrigation devices.

Within the context of soil fertilization, it became necessary to develop measures aimed at stabilization and prevention of degradation processes in the soil. Concurrently, such measures had to improve properties of already degraded soils as a result of long term animal waste application. A set of measures has been developed for the solution of these problems. These measures, diverse but interconnected and mutually supportive, are as follows:

Table 1. Changes in the main technical parameters and chemical composition of the solid fraction of pig manure under thermobiological treatment.

Index	Fresh Manure	After 42-day Treatment Period
Temperature, °C	26	60
CO_2, %	0.1	8
Humidity, %	65.6	51.0
Ash Content, %	11.8	16.5
Carbon, %	44.9	40.7
C:N, %	17:1	11:1
Total Nitrogen, %	2.64	3.70
Phosphorus, Total %	0.60	1.92
Phosphorus, Mobile %	---	1.53
Potassium, Total %	0.23	0.44
Potassium, Mobile %	---	0.34

- Rational application of animal wastes (optimization of application rates and application time, degree of dilution, etc);
- Scientifically substantiated techniques (biological amelioration);
- Efficient application of chemical methods for prevention of soils salinization and alkalinization (chemical amelioration).

The above approach can be exemplified by the following outline of soil-protective measures applied in a hog-breeding complex of 108,000 hogs per year. The complex is situated in the forest-steppe zone of Ukraine on a typical deep heavy clay chernozem, which also produces grain for fodder. Hydraulic removal of manure results in over 1.5 million m^3/year of slurry. Calculations indicated that for sound use of this slurry, it is necessary to have 1800 ha available for use, and to observe the following requirements:

- To use 40% of the total volume of wastes for water-charging irrigation and 60% vegetative irrigation;
- To dilute each single part of manure slurry with 2.7 parts of low-concentration effluent; nitrogen content of the resulting mixture will be thus reduced 30%;
- The crop-averaged irrigation rate is 1730 m^3/ha; in a case of storage pond overflow or other emergency situations, the rate in the autumn period can be raised to 3000 m^3/ha.

For protection and improvement of the waste-applied soil, the following practical agrotechnical procedures are recommended:

- *Biological amelioration, including stubble incorporation, as well as straw and hog manure solids.* This would lead to soil humus gain, concurrently with organic nitrogen fixation. Field application of straw jointly with slurry applied at the rate of 350-400 kg/ha of nitrogen results in costs that are 20-40% below that for use of stable manure containing straw bedding for the equivalent fertilization effect. It has been found that manuring with slurry and straw considerably enhances the humification coefficient. This practice increases absorbability and moisture holding capacity and improves soil structure while expanding the heat and air regime of the soil.
- *Chemical amelioration, aimed primarily at prevention of soil alkalization processes, which occur as a result of long-term application of animal waste slurries.* For this purpose it is recommended that wet-ground gypsum or phosphate gypsum be applied. Rate of application should be determined based on the absorbed sodium, by calcium absorption from calcium salts solution, or by the mineral colloids coagulation threshold. Furthermore, the chemical composition of the slurry should be taken into account, i.e., quantities and relationships of sodium and calcium delivered to the soil. The chemical ameliorants should be introduced through deep plowing in late autumn.

REFERENCES

Batsula, A.A. and Skryl'nik, E.V., Vlijanie produktov biologicheskoj ochistki zhidkogo svinogo navoza na gumusnoe sostojanie chernozema tipichnogo moschnogo, *Trudy Vsesojuznogo instituta sel'skohozyajstvennoj mikrobiologii*, 56, 83, 1986.

Batsula, A.A. and Skryl'nik, E.V., Termobiologicheskaya pererabotka tverdoj frakcii svinogo navoza, kak sposob uluchsheniya ego kachestva, *Agrochimiya i pochvovedenie*, 56, 55, 1993.

Batsula, A.A., Kravez, T.F. and Skryl'nik E.V., Svyaz sel'skohozyajstvennogo ispolzovaniya pochv s sostavom i svojstvami gumusovyh veschestv, *Vestnik agrarnoj nauki, Kiev*, 5, 31, 1994.

Nosko, B.S., Batsula, A.A. and Chesnyak, G.J., Gumusnoe sostoyanie pochv Ukrainy i puti ego regulirovaniya, *Pochvovedenie*, 10, 33, 1992.

Skryl'nik, E.V., Izmenenie agrofizicheskogo sostoyaniya chernozema pri dlitelnomoroscenii zhivotnovodcheskimi stokami, paper presented at 4 sezde pochvocedov i agrohimikov Ukrainy, 1994.

Tikhonov, P.M. et al., Sanitarnaya ozenka i metody obezzarazhivania svinogo navoza, *Veterinariya*, 3, 24, 1982.

DAIRY LOAFING LOT ROTATIONAL MANAGEMENT SYSTEM FOR NON-POINT SOURCE POLLUTION CONTROL

T.M. Younos, E.R. Collins, B.B. Ross, J.M. Swisher,
R.F. Shank, and K.G. Wooden

INTRODUCTION

In Virginia, a majority of dairy operations are located within river basins that are designated as high-priority watersheds for reducing non-point source pollution. Most of these dairy systems include a loafing lot or exercise paddock. Loafing lots are usually located outside of the free stall barn, normally off concrete, and have a bare soil surface or some unwanted vegetation such as weeds. A typical herd may spend from four to fourteen hours per day on these loafing lots. In many cases, lots are gradually denuded of all vegetation because of continuous use and heavy cow traffic and become susceptible to extensive erosion. In addition, on a daily basis, large volumes of manure and urine are accumulated on loafing lots. Therefore, there is a potential for significant amounts of sediment, nutrients, and pathogens to enter nearby surface water systems with the occurrence of a runoff-producing rainfall event. Runoff from livestock feedlots and dairy operations is a recognized non-point pollution source in receiving surface water systems (Hollon et al., 1982; Reese et al., 1982; Baxter-Potter and Gilliand, 1988; Beck, 1989; Younos, 1990). A comprehensive water quality monitoring program was initiated in the Owl Run watershed (Fauquier County, Virginia) to evaluate water quality problems associated with runoff from dairy operations (Mostaghimi et al., 1989). Monitoring results from the Owl Run watershed and other studies cited above indicate high levels of nitrogen (N) and phosphorus (P) input from dairy operations into streams with concentrations often exceeding the minimum levels needed for algae growth in surface water bodies.

DAIRY LOAFING LOT ROTATIONAL MANAGEMENT SYSTEM

The Dairy Loafing Lot Rotational Management System (DLLRMS) is an innovative best management practice initiated in 1985 as a demonstration project in Augusta County, Virginia (Swisher et al., 1994). Virginia Cooperative Extension, Virginia Tech Biological Systems Engineering Department, USDA

Natural Conservation Service, Virginia Department of Conservation and Recreation, and Headwaters Soil and Water Conservation District cooperated in development of the system. Basically, the DLLRMS enables a dairy operator to rotate the dairy herd out of a single muddy lot into paddocks that are fully vegetated with an appropriate ground cover. Management of the loafing lot system involves rotating the dairy herd between vegetated paddocks with consideration for level of vegetation and soil moisture condition in each paddock.

DESIGN FEATURES

The schematic design for a typical DLLRMS is shown in Figure 1. The system typically requires adequate acreage (15-20 cows/acre), moderately sloped land (4-10%), and well-drained soil capable of supporting a strong sod. The system should be installed adjacent to the dairy complex and no more than 1/4 mile from the barn. Paddocks for typical-sized Virginia dairy farms are usually developed on existing loafing areas by dividing a 10-acre field into three paddocks (2.5, 3, and 3 acres), and a 1.5-acre "sacrifice lot" using an approved electric fence system. The area designated as a sacrifice lot is used during wet periods, or when grass cover is short and likely to be injured by the cows. The grassed paddocks should be located to act as a buffer between the sacrifice lot and receiving streams.

The DLLRMS requires that a thick, vigorous, hardy sod such as 'Kentucky 31' tall fescue be established on each paddock. To accomplish this, each paddock is cultivated by using a heavy cutting disk and then leveled with a cultimulcher. A broadcast seeder attached to a tractor is used to seed the pad-

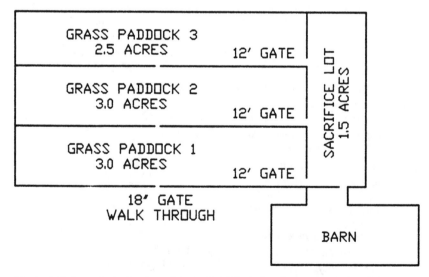

Figure 1. Schematic design for a dairy loafing lot rotational managmenet system.

docks with tall fescue at the rate of 30-50 lb/acre. Following broadcasting of the seed, paddocks are rolled with a cultipacker to obtain proper seed/soil contact for good germination. Usually the soil is tested for nutrient level to determine if additional fertilization is necessary.

If loafing lots are being established on existing sods, the 30-50 lb of Kentucky 31 fescue can be seeded following tilling and preparing the seedbed. No-till seeding can also be accomplished by grazing or mowing the existing cover as closely as possible, allowing 2-3 in. of regrowth, and applying an appropriate herbicide to kill the plants. Successful seedings in either tilled or no-till beds can be made in either late summer (15 August - 1 October) or early spring (1 March - 15 April).

MANAGEMENT FEATURES

A stocking density of 15-20 cows/acre is recommended to maintain a good sod cover on each paddock. Management of the grass paddocks involves rotating the lactating herd from one paddock to another or to the sacrifice lot, depending on condition of the paddocks. During wet periods or times of poor grass cover when the sod is likely to be damaged or destroyed, the cows are restricted to the sacrifice lot and/or free stall barn until paddock conditions improve. Cows on grass paddocks should have ready access to feed and water located in the barn. The cows may show some interest in grazing the lots in spring and late fall. However, grazing is not the purpose of the grass paddocks. If cows are fed a balanced ration, they will not tend to graze on tall fescue. Tall fescue should be mowed periodically to maintain its vegetative growth and sod density. Weed invasion is likely to occur both during and following fescue establishment. Many broadleaf weeds can be controlled by mowing. However, some such as horsenettle, thistles, burdock, and chicory are best controlled by spraying an appropriate herbicide. The sacrifice lot should be scraped periodically to remove accumulated manure.

OTHER DLLRMS ADVANTAGES

The major goal of the DLLRMS is to control non-point sources of pollution from dairy loafing areas. Other advantages of the system include improved cow health and productivity. Weather conditions have a tremendous impact on the loafing lot environment, cow comfort and health. Bare soil loafing lots in winter and rainy seasons are often wet and muddy, conditions that can cause "environmental mastitis" and reproductive infections. At one farm in Augusta County, Virginia, the DLLRMS reduced the time to bring cows to the barn by about 33% because the original 10-acre loafing area was divided into smaller grassed paddocks; therefore, less travel time was required in getting the cows to the barn and milking area (Swisher et al., 1994). Furthermore, cow udders were much cleaner, resulting in less time required to clean udders for milking. Observations indicated a reduction of 15 minutes per milking for 50 lactating cows. Another cited advantage is the difference in temperature between bare lots and grass-covered lots. Bare soils usually average 8°F higher than the grass-covered soils. Heat stress in cows is a serious problem for Virginia milk producers during the summer months. Improved cover condition from a DLLRMS is expected to have a positive effect on milk and fat production as well as reproductive efficiency.

DLLRMS COST-SHARE PROGRAM

The DLLRMS has been approved as a cost-shared "Best Management Practice" under the Virginia Department of Conservation and Recreation through the Division of Soil and Water Conservation (VDCR, 1993). Eligible dairymen may receive 75% cost-sharing for installing the DLLRMS. However, there is a limit of $7,500/applicant/year. This program is funded with state and federal monies through local soil and water conservation districts. County Agricultural Stabilization and Conservation Service (ASCS) money may also be used by utilizing a combination of innovative ASCS practices. The ASCS limit is $3500/applicant/year. From 1992 to 1994, 14 DLLRMS were installed in Virginia, and the requests for six others are pending (Banks, 1994). The total installation cost for the 14 systems was $149,922 of which $76,225 (51%) was provided through the cost-share program.

EVALUATION OF DLLRMS EFFECTIVENESS

Although each year several DLLRMS are installed in Virginia through implementation of a cost-share program, the effectiveness of the DLLRMS on runoff and stream water quality has not been fully evaluated. In 1994, Virginia Tech and Virginia Cooperative Extension initiated a research and demonstration project to evaluate effectiveness of the DLLRMS in reducing non-point source pollution from dairy loafing lots and to develop baseline information that will enable regulatory agencies to make realistic projections with regard to NPS reductions when a DLLRMS is installed on a dairy farm. The project is supported by funds provided through the US EPA 319 Program.

A 60-head dairy operation located in Franklin County, Virginia, was selected to evaluate the DLLRMS effectiveness. The site is within the L74 Hydrologic Unit, a high-priority watershed. A small adjacent stream receives the bulk of drainage from loafing lots (Figure 2). The headwater for the stream is a spring that is not impacted by direct runoff from the loafing lots. The small stream discharges to Mollie Creek, a tributary of Maggodee Creek that flows into the Blackwater River and, finally, into Smith Mountain Lake.

Four sampling stations were established on the study site (Figure 2). A 2-ft H-flume and a Belfort FW-1 stage recorder were installed at Station A (on small stream adjacent to loafing lots) to monitor stream flow. Station A was also equipped with an ISCO automatic water sampler to collect stream water samples from major rainfall events, and a chart recording, weighing-type rain gauge to collect rainfall data. Stations B and C were equipped with a staff

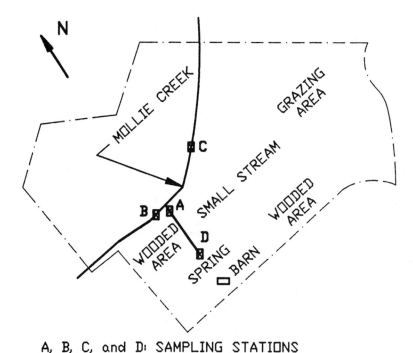

A, B, C, and D: SAMPLING STATIONS

Figure 2. Franklin County, Virginia, dairy loafing lot rotational management system study site.

gauge to monitor the water level in Mollie Creek. Grab samples are being collected from stations A, B and C on a biweekly basis. Station D (the spring) is sampled about every 10 weeks.

Phase I of the monitoring program (Pre-DLLRMS) started on 2 August 1994. On-site measurements of stream water temperature, pH, total dissolved solids, and dissolved oxygen are conducted on a biweekly basis. Both grab and automatic water samples are analyzed in the laboratory for ammonia, nitrate, TKN, orthophosphate, total phosphorus, total suspended solids, volatile suspended solids, COD, fecal and total coliform, and *E. coli* bacteria.

RAINFALL SIMULATION AND PHASE II MONITORING

Work is underway to assess "edge of field" impact of the loafing lot on water quality. A portable rainfall simulator will be used to apply rainfall to temporarily established runoff plots (18 ft x 60 ft) equipped with an H-flume at the outlet. For each simulated rainfall event, the flume discharge will be continuously monitored by a stage recorder. Runoff samples will be collected manually to be analyzed for nutrients, sediment, COD, total suspended solids, volatile suspended solids, fecal and total coliform, and *E. coli*.

Plans are underway to install the DLLRMS at the site in spring of 1995. Phase II of the project (May 1995 - April 1996) will consist of stream monitoring and simulated rainfall studies to evaluate the impact of the DLLRMS on stream water quality and "edge of field" effects.

REFERENCES

Banks, T., Virginia Department of Conservation and Recreation, Division of Soil and Water Conservation, personal communication, 1994.

Baxter-Potter, W.R., and Gilliand, M.W., Bacterial pollution in runoff from agricultural lands, *Jour. Environ. Qual.*, 17(1), 27, 1988.

Beck, L., Review of farm waste pollution, *Jour. Institution of Water and Environmental Management*, 3(5), 467, 1989.

Hollon, B.F., Owen, J.R., and Sewell, J.I., Water quality in streams receiving dairy feedlot effluent, *Jour. Environ. Qual.*, 11(1), 5, 1982.

Mostaghimi, S., *Watershed/Water Quality Monitoring for Evaluating Animal Waste BMP Effectiveness, Owl Run Watershed,* Pre-BMP Evaluation Final Report (Rep. No. O-P1-8906), Department of Agricultural Engineering, Virginia Tech, Blacksburg, Virginia and Virginia Division of Soil and Water Conservation, Richmond, Virginia, 1989.

Reese, L.E., Hegg, R.O., and Gantt, R.E., Runoff water quality from dairy pastures in the Piedmont Region, *Trans. ASAE*, 25(3), 697, 1982.

Swisher, J.M., White, H.E., and Carr, S.B., *Dairy Loafing Lot Rotational Management System*, Virginia Cooperative Extension Pub. 404-252, Virginia Polytechnic Institute and State University, Blacksburg, Virginia, 1994.

VDCR (Virginia Department of Conservation and Recreation), Loafing lot management system, DSWC Specification for WP-4B, Virginia Agricultural BMP Cost-Share Program Manual - 1993, VDCR, Richmond, Virginia, 1993.

Younos, T.M., Manure management and pollution prevention: Challenges for 1990s, *Jour. Environ. and Technology*, 2(3), 54, 1990.

NUTRIENT MANAGEMENT PLANNING

Stuart Klausner

INTRODUCTION

Reducing nitrate contamination of ground water and phosphorus enrichment of surface water has become a regional and national goal. Water quality concerns, in combination with potential or enacted nutrient management legislation in many states, have created renewed awareness about the need for efficient nutrient management in agriculture.

Accounting for residual soil fertility, substituting a major portion of crop nutrient requirements on livestock and poultry farms with animal manure, and maximizing feeding efficiency would serve to ease the quantity of nutrients purchased and improve nutrient recycling. Although this concept appears simple, it is not being achieved in many situations. Most farms do not follow a comprehensive nutrient management plan, and there is little integration between animal and plant nutrition. The lack of nutrient management planning often results in poor nutrient utilization and an increased potential for nutrient loss.

A well-designed nutrient management program helps assure an adequate and sustained supply of high-quality feed and improved nutrient recycling. A successful nutrient management plan must integrate a considerable amount of information about the farm operation and requires a perspective of 1) the movement and quantity of nutrients entering, leaving, and remaining on the farm, 2) the nutrient application schedule to ensure that the rate and timing of manure and fertilizer applications are in concert with crop requirements, 3) a crop selection and rotational sequence to provide quality feed, improve nutrient utilization, and reduce runoff and erosion, and 4) appropriate ration balancing to ensure efficient nutrient recycling from animal to crop and back to animal again.

NUTRIENT PATHWAY

An assessment of the quantity of nutrients entering, leaving, and remaining on a farm is a starting point for understanding nutrient cycling. The outcome of the assessment should be used to determine management options. Figure 1 is a simple illustration of nutrient flows on livestock and poultry farms. Typically, nitrogen (N), phosphorous (P), potassium (K), and other nutrients are brought into the farm in purchased products such as feed, fertilizer, and

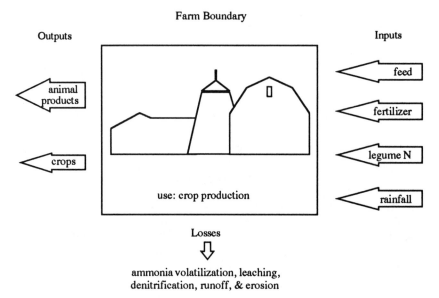

Figure 1. Typical nutrient pathway.

animal replacements. Additionally, nutrients are imported in two other sources: the conversion of atmospheric N into plant proteins by legumes and, to a lesser extent, in precipitation. Nutrients leave the farm in products sold, such as milk, eggs, meat, and crops.

MASS NUTRIENT BALANCE

Dairy farms in New York are used to exemplify the assessment of the nutrient status of farms. A similar assessment can be made for any farming operation. Mass nutrient balances for N, P, and K are shown Tables 1a, 1b, and 1c, respectively, for actual 45-, 85-, 320-, and 500-cow dairy farms. An N balance (P and K data were not available) for an operating 1300-cow dairy is also shown to exemplify a large farm.

Nutrients can concentrate on livestock and poultry farms if more are imported than are exported in products sold. The percentage of the imported N, P, and K that remained on these farms each year ranged from 61-76%, 59-79%, and 67-89%, respectively. The percentage that remained was related to individual feeding and fertilizer practices and not to farm size. However, size dictates the quantity of nutrients that must be managed, and, therefore, an increase in animal numbers requires a corresponding increase in the amount of cropland in order to efficiently utilize nutrients.

A study of mass N balances on 17 farms showed no relationship ($R^2 = 0.07$) between the percentage of the imported N remaining on the farm each year and herd size. The average, range, and standard deviation of the percent N remaining was 69 and 61-79, and 5.0, respectively.

Table 1a. Annual mass nitrogen (N) balance for several dairy farms in New York.

Item	Size of dairy, cows				
	45	85	320	500	1300
	-----------------------------tons/year---------------------------				
INPUT					
purchased feed	3.8	9.7	43.8	78.5	205.0
purchased fertilizer	1.0	2.2	13.5	26.1	9.8
N fixation by legumes	0.8	0.9	14.6	13.9	16.3
purchased animals	-----	-----	0.1	-----	-----
	5.6	12.8	72.0	118.5	231.1
OUTPUT					
Milk	2.0	3.8	18.6	26.4	72.8
Meat	0.1	0.4	1.9	1.9	3.2
Crops sold	0.1	0.5	-----	-----	-----
	2.2	4.7	20.5	28.3	76.0
REMAINDER					
tons	3.4	8.1	51.5	90.2	155.1
%	61	63	71	76	67

Table 1b. Annual mass phosphorus (P) balance for several dairy farms in New York.

Item	Size of dairy, cows			
	45	85	320	500
	-----------------------------tons/year---------------------------			
INPUT				
purchased feed	1.0	1.7	8.4	14.2
purchased fertilizer	1.2	0.9	2.0	10.0
purchased animals	-----	-----	0.03	-----
	2.2	2.6	10.5	24.2
OUTPUT				
Milk	0.4	0.68	3.8	5.5
Meat	0.05	0.10	0.5	0.5
Crops sold	0.02	0.06	-----	-----
	0.47	0.84	4.3	6.0
REMAINDER				
tons	1.7	1.8	6.2	18.2
%	79	68	59	75

Mass nutrient balances are mere estimates of the nutrient status of farms and should be used as a tool to help with management decisions. The mass balances show that purchased feed is usually the primary source and fertilizer the secondary source of imported nutrients (Table 2). When the relative contribution of imported nutrients in fertilizer exceed that in feed, it is usually the result of a low animal-to-land ratio or unnecessary fertilizer purchases. Feed and fertilizer purchases should be carefully matched with *actual* need to reduce nutrient imbalances in the feeding and crop production program and to prevent excessive nutrient loss.

Table 1c. Annual mass potassium (K) balance for several dairy farms in New York.

Item	Size of dairy, cows			
	45	85	320	500
	----------------------------------tons/year----------------------------			
INPUT				
purchased feed	0.9	2.4	12.3	22.8
purchased fertilizer	4.7	1.8	7.3	35.1
purchased animals	-----	-----	0.01	-----
	5.6	4.2	19.6	57.9
OUTPUT				
Milk	0.51	1.0	5.6	8.3
Meat	0.01	0.02	0.1	0.1
Crops sold	0.12	0.4	-----	-----
	0.64	1.4	5.7	8.4
REMAINDER				
tons	5.0	2.8	13.9	49.5
%	89	67	71	85

Table 2. Relative contribution of imported nutrients on dairies
ranging from 45 - 1300 cows.

Input	Nitrogen					Phosphorus				Potassium			
	45	85	320	500	1300	45	85	320	500	45	85	320	500
	--------------------------------------%--------------------------------------												
Purchased feed	68	76	61	66	89	45	65	80	59	16	57	63	39
Purchased fertilizer	18	17	19	22	4	55	35	19	41	84	43	37	61
N fixation, legumes	14	7	20	12	7	--	--	--	--	--	--	--	--

ANIMAL NUTRIENT MANAGEMENT

The feeding program on intensively managed livestock and poultry farms can have a large impact on the mass nutrient balance because most of the nutrient imports are attributed to feed purchases. Because N, P, and K in feed are positively correlated with N, P, and K output in feces and urine (Pell, 1992, 1993; Tamminga,1992), animal nutrient management is a key component of the overall nutrient management program. Proper ration balancing may have a more positive affect on the mass nutrient balance than improved fertilizer management. Animals should not be fed above their nutritional requirements. Feeding management is a means to optimize animal production and reduce nutrient loss. Profitable ration balancing and improved crop cultural practices, to increase feed quality and quantity, are key management practices to avoid excessive nutrient purchases in feed.

CROP NUTRIENT MANAGEMENT

Nutrient accumulations challenge the farmer's ability to manage a soil fertility program in an efficient and environmentally friendly manner. One of the consequences of nutrient accumulation is that manure becomes enriched with essential plant nutrients, making it a valuable replacement for commercial fertilizer. However, on farms with high animal densities, manure may provide more nutrients than can be utilized by the crop rotation.

A further assessment of the nutrient status of a farm can be done by constructing a simple soil-crop-animal nutrient balance comparing the amount of N, P, and K produced in manure with the amount required for crop production. Example soil-crop-animal nutrient balances are given in Table 3 for the previously discussed farms. The crop nutrient requirements on these farms were relatively low owing to the extensive production of alfalfa, which provides N to the following crop, and due to high soil test P and K values as a result of past fertilization and manure application. Because of large nutrient imports, resulting in low crop requirements for supplemental nutrients, the total amount of nutrients produced in manure exceeded crop requirements on these farms.

Unfortunately, the nutrients in manure cannot be substituted for those in fertilizer on a pound-for-pound basis because they are not as readily available, nor can they be as efficiently timed and placed as those in fertilizer. An apparent surplus of nutrients, in terms of total quantity, may not provide a sufficient amount of available nutrients to meet crop requirements. However, the fertilizer replacement value of manure is high, and a decay series can be used to estimate N availability (Pratt et al., 1976; Klausner, 1994a) and soil testing used to determine supplemental P and K requirements. Regional recommendations should be obtained from the respective state Cooperative Extension Service.

Much of the surplus P and K shown in Table 3 remain in the soil, although a relatively small amount is stored temporarily in crops only to be recycled to

Table 3. A comparison of nutrients in manure versus crop requirement, the amount of nutrients per tillable acre, and animal density for several dairy farms.

Number of cows	Nutrient	Produced in in manure	Crop requirement	Surplus or deficit (-)	Amount per tillable acre nutrient	animal units[†]
		----------tons----------			lb	
45	N	4.4	4.0	0.4	112	0.93
	P_2O_5	2.7	0.3	2.4	69	
	K_2O	2.2	0.9	1.3	56	
85	N	17.1	10.0	7.1	127	0.77
	P_2O_5	6.2	4.1	2.1	46	
	K_2O	12.4	2.1	10.3	92	
320	N	53.5	16.8	36.7	178	1.1
	P_2O_5	19.5	6.3	13.2	65	
	K_2O	35.2	7.3	27.9	117	
500	N	75.5	38.4	37.1	151	0.91
	P_2O_5	32.3	9.5	22.8	65	
	K_2O	51.2	9.3	41.9	102	

[†]animal unit= 1000 lb live weight.

the soil again in crop residue and manure, and some may be removed from the farm in runoff and erosion. The accumulation of P and K is reflected in the long-term increase in soil test P and K levels on these farms. The accumulation or depletion of P and K in soil can be monitored by soil testing, which serves as the single-most-important tool for managing P and K. Nitrogen does not accumulate appreciably in soil, and much of the soil N can be lost by leaching and denitrification. The status of available soil N cannot be easily monitored with soil testing; hence, the fertilizer N replacement value of manure has to be estimated.

Estimates of manure N availability are usually based on their ammoniacal and organic N content. About 50% of the total N in fresh manure is present as urea (cattle) or uric acid (poultry), which converts rapidly to ammonia and is easily lost by volatilization. Lauer et al. (1976) showed the average half life of ammonia in manure spread on the soil surface to be about 2.5 days. Therefore, about one-half of manure N may be lost from the cropping system if ammonia is not conserved. Immediate incorporation will prevent ammonia volatilization, but if N is applied too far in advance of crop uptake, or in excess of crop requirement, appreciable N loss will occur in humid regions by leaching and denitrification. The remaining N, namely the more stable organic fraction, must mineralize to an inorganic form before it is plant available. The rate of mineralization is variable and dependent on many soil and climatic factors (Barbarika et al., 1985; Bernal and Kirchmann, 1992; Douglas and Magdoff, 1991; and Klausner et al., 1994). The total amount of N available from manure is equal to the amount mineralized from the more stable organic N fraction (estimated by a decay series) plus the expected N recovery from the readily available urea-ammonium fraction.

Nitrogen availability estimates from manures must be derived for the climatic and physiographic region for which it is suited. By way of example, in New York about 20% of the total N will be available to crops if no ammonia is conserved and there is no residual N contribution from past applications. Nitrogen availability increases to about 75% of the total when there is 100% ammonia conservation and the residual N contribution from past applications is high (Klausner, 1992).

The development of a reliable soil test for N will have a significant impact on improving N management. The recently introduced pre-sidedress N soil test for corn (Magdoff et al., 1984) is useful for identifying fields that are unlikely to respond to additional N and, therefore, helping to avoid the over-application of N.

NUTRIENT MANAGEMENT STRATEGY

Nutrient management involves the development of a planning strategy to ensure that nutrients are used efficiently for the economic production of feed and animal products and for the protection of water quality. Nutrient management must be viewed in an integrated way. Feeds should be analyzed and rations balanced profitably, and soil testing should be used to monitor the nutrient status of soil. Avoiding over fertilization and implementation of good soil and water control practices to reduce nutrient loss are critical aspects of nutrient management.

The following components are important for a successful nutrient management plan. Continued monitoring and periodic adjustments in the plan are a vital part of its success.

Analyze feed and balance rations profitably: Animal nutrient management can have a large impact on the quantity of nutrients that must be cycled through the soil-cropping system each year. Over feeding is economically and environmentally unsound. Use feed analysis to monitor the quality of home-grown and purchased feed and ensure that rations are balanced properly. Monitor animal production efficiency and health to confirm the adequacy of the feeding program.

Animal density: The potential to successfully manage nutrients depends on the amount of cropland available relative to the number of animals and amount of manure produced. An increase in animal numbers requires a corresponding increase in the number of crop acres. Appropriate animal-to-land ratios prevent excessive applications of nutrients and increase the percentage of the feed requirement that is home grown, thereby reducing the import of nutrients in purchased feed. Appropriate animal densities will vary depending on animal type, crop rotation, and physiographic region. By way of example, animal density guidelines for dairy cattle in several northeastern states are given by Lanyon and Beegle (1993) and Klausner (1994).

Determine quantity of manure produced and collected: The quantity of manure produced and the amount collected should be determined to estimate the quantity of manure and nutrients that must be managed in the land application program. Estimates of manure production by animal type are available from various publications (Midwest Plan Service, 1985; American Society of Agricultural Engineers, 1993). The amount collected can be estimated based on the quantity removed from the barn each day or by calculating the volume in storage.

Analyze manure to determine nutrient content: Samples should be analyzed periodically for nutrient content. Analyses should be done frequently at first and periodically once a reasonable estimate of nutrient concentration is determined. The analysis, at minimum, should include total N, ammoniacal N, P, K, and dry matter content. The total quantity of nutrients collected should be calculated.

Estimate nutrient availability in manure: The nutrients in manure cannot be substituted for those in commercial fertilizer on a pound-for-pound basis. The fertilizer replacement value of manure depends on many variables, and estimates of nutrient availability should be derived for the region where they are used. Estimates can usually be obtained from the Cooperative Extension Service.

Use a crop rotation and cultural practices that maximize economic feed production: Determine the dry matter requirements of the feeding program. Develop a crop rotation that provides as much of the dry matter requirement as possible. Verify that soil productivity levels can support the rotation and that good soil conservation practices are followed. Improved cultural practices will increase yield and crop quality and reduce the need for purchased feed and imported nutrients.

Hydrologic evaluation: Evaluate the hydrological sensitivity (runoff, leaching, flooding, and erosion potential) of individual fields and their risk level. Determine the best month(s) to apply manure to reduce the nutrient loss potential.

Soil test to determine crop nutrient requirements: Maintain a good soil testing program to monitor the nutrient status of fields and to determine supplemental nutrient requirements. Determine how much manure is required to meet a given nutrient requirement.

Manure application: Use manure as the primary source of nutrients. Fertilizer should be used only to supplement additional requirements. Base the application rate on the availability and crop requirement of the highest priority nutrient. Apply during periods of the year that maximize the combination of high crop demand and low nutrient loss potential.

Fertilizer management: Fertilizer should be used to supplement additional requirements. Apply at a rate and by a method of placement that are consistent with efficient nutrient recovery.

Storage requirement: Storage may be helpful to improve management. However, a particular manure handling system does not in itself improve nutrient recycling; management does. Storage provides the potential to manage efficiently through proper timing of application. However, if adverse weather conditions prevail when the storage must be emptied, then a storage system may be less advantageous than daily spreading. Daily spreading, on the other hand, may be managed more efficiently if short-term storage is provided during inclement weather.

Soil and water conservation: Work closely with the US Soil Conservation Service to ensure that adequate soil and water control practices are in place.

Do not over feed or over fertilize: These practices are economically and environmentally unfriendly. The development of regional nutrient management planning workbooks, consistent with local recommendations, would be helpful to assist farmers and their advisors construct and implement viable nutrient management plans.

SUMMARY

Economically viable nutrient management plans, which promote efficient nutrient cycling with minimal environmental impact, are an important prerequisite to agricultural environmental planning. An important component of the plan is a critical analysis of the quantity of nutrients imported into the farm each year in comparison to their export in products sold.

Over feeding and over fertilizing are economically and environmental unsound. Present feed and fertilizer purchases should be scrutinized with respect to the actual nutrient requirements of the herd and crop rotation. Soil testing is critical to confirm the nutrient requirements of the crop rotation, and manure and fertilizer application rates should match crop requirements as closely as possible. Applications should be made during periods of the year that maximize the potential for crop uptake and minimize nutrient loss. Feed analysis and profitable ration balancing assure favorable production efficiency and animal health. Good crop and soil management practices can improve yield and quality, thereby increasing nutrient recycling by reducing the need for purchased feed and fertilizer.

REFERENCES

American Society of Agricultural Engineers, *ASAE Standards 1993: Standards Engineering Practices Data.* Amer. Soc. Ag. Eng. 250 Niles Rd. St. Joesph, Michigan, 1993.

Barbarika, A., Jr., Sikora, L.J., and Colacicco, D., Factors affecting the mineralization of N in sewage sludge applied to soils, *Soil Sci. Am. J.,* 49, 1403, 1985.

Bernal, M.P., and Kirchmann, H., Carbon and nitrogen mineralization and ammonia volatilization from fresh, aerobically and anaerobically treated pig manure during incubation with soil, *Biol. Fertil. Soils.*, 13, 135, 1992.

Douglas, B.F., and Magdoff, F.R., An evaluation of nitrogen mineralization indices for organic residues, *J. Environ. Qual.,* 20, 368, 1991.

Klausner, S.D., Nutrient management on livestock farms, Extension Series No. E92-4. Dept. of Soil, Crop, and Atmospheric Sci. Cornell Univ., Ithaca, NewYork, 1992.

Klausner, S.D., Animal density guidelines, in *What's Cropping Up?* Vol 4, No. 1. Dep. Soil, Crop, and Atmospheric Science, Cornell Univ., Ithaca, New York, 1994.

Klausner, S.D., Kanneganti, R., and Bouldin, D.R., An approach for estimating a decay series for organic N in animal manure, *Agron. J.,* 86, 897, 1994.

Lanyon, L.E., and Beegle, D.B., *A Nutrient Management Approach for Pennsylvania: Plant Nutrient Stocks and Flows,* Agronomy Facts 38-B. Pub. Dist. Cntr. Penn State Univ. University Park, Pennsylvania, 1993.

Lauer, D.A., Bouldin, D.R., and Klausner, S.D., Ammonia volatilization from dairy manure spread on the soil surface, *J Environ. Qual.,* 5, 134, 1976.

Magdoff, F.R., Ross, D., and Amadon, J., A soil test for nitrogen availability to corn, *Soil Sci. Soc. Am. J.,* 48, 1301, 1984.

Midwest Plan Service, *Livestock Waste Facilities Handbook,* 2nd Ed., Iowa State Univ. Ames, Iowa, 1985.

Pell, A.N., Does ration balancing affect nutrient management? in *Proc. Cornell Nutrition Confr. for Feed Manuf.*, Rochester, New York Dep. An. Sci., Cornell U., Ithaca, New York, 1992.

Pell, A.N. *Environmentally Friendly Rations,* An. Sci. Mimeo Series no. 168, Dep An. Sci., Cornell U., Ithaca, New York 1993.

Pratt, P.F., Davis, S., and Sharpless, R.G., A four year field trial with manures, *Hilgardia,* 44, 99, 1976.

Tamminga, S., Nutrition management of dairy cows as a contribution to pollution control, *J. Dairy Sci.,* 75, 345, 1992.

NITROGEN RELEASE FROM LAND-APPLIED ANIMAL MANURES

M.L. Cabrera and R.M. Gordillo

INTRODUCTION

Manures generated in confined animal production can be valuable sources of nitrogen (N), phosphorus (P), and potassium (K) for crops. Due to high N requirement of plants and due to the low N-supplying capacity of many soils, the application rate of these animal manures is typically based on the material's capacity to supply N. Consequently, methods for estimating N-supplying capacity of animal manures are essential to calculate correct application rates.

When applied at the rates required to supply adequate N, animal manures often provide more P than crops remove, leading to a buildup of soil P. Once soil-available P reaches a threshold value, runoff water can carry excessive amounts of P to streams and lakes, causing eutrophication. When soil P reaches this threshold value, the rates of application should be based on the P requirement of the crop and on the material's capacity to supply P (Sharpley et al., 1994). When this situation arises, part of the N required by the crop is not met by the applied manure and has to be provided by supplemental fertilizer N. In that case, knowledge of the amount of N supplied by the applied manure is also needed to estimate the amount of fertilizer N to add. Therefore, independently of whether animal manures are applied for N or P, a correct estimation of their N-supplying capacity is needed to ensure adequate crop nutrition and to avoid surface and ground water contamination with N derived from excessive applications.

A MODEL FOR ESTIMATING NITROGEN AVAILABILITY

Sims (1986) proposed the following model for estimating available N from poultry manure: Available $N = A_i N_i + P_m N_o$, where N_i and N_o are inorganic and organic N in the manure, respectively, A_i is the fraction (0-1) of inorganic N that is available, and P_m is the proportion (0-1) of organic N that is mineralizable. A similar model was proposed by Beauchamp (1983) for liquid cattle manure. The relative importance of each term in this equation depends, among other factors, on the relative proportion of organic and inorganic N in the manure.

PROPORTION OF INORGANIC NITROGEN IN ANIMAL MANURES

Fresh and Composted Manures

In general, fresh and composted animal manures have a lower proportion of inorganic than organic N. The proportion of inorganic N in cattle manure typically ranges from 1 to 11% (Table 1), although values as high as 34% have been reported (Paul and Beauchamp, 1993). Similarly, inorganic N in swine manure generally ranges from 4 to 15% of total N (Table 2). On the other hand, values reported for poultry manure have been as high as 50%, possibly suggesting a more mineralizable organic N fraction than that in cattle and swine manures (Table 3). Composted animal manures usually have low levels of inorganic N due to gaseous N losses (Martins and Dewes, 1992; Keener and Hansen, 1992) and to N immobilization during the composting process (Table 4).

Most of the inorganic N in fresh manures is commonly found in NH_4 form, although relatively high NO_3 concentrations have been reported (Sims, 1986; Bitzer and Sims, 1988; Cabrera et al., 1994). The amount of NO_3 in animal manures depends on the presence of nitrifiers, which are likely to be derived

Table 1. Dry matter, total nitrogen (N) contents, and proportion of total N present in inorganic form in cattle manure samples.

No. of samples	Dry matter (g/kg wet manure)	Total N (g/kg dry manure)	Range Inorganic N (% of total)	Mean Inorganic N (% of total)	Reference
1	N.G.[†]	15.3	0.2	0.2	Pomares and Pratt, 1978
1	N.G.	23.2	0.9	0.9	Kirchmann, 1991
1	N.G.	25.6	0.9	0.9	Castellanos & Pratt, 1981
1	170	17.9	4.4	4.4	Douglas & Magdoff, 1991
3	175-198	22.6-25.8	18.7-33.8	26.2	Paul and Beauchamp, 1993
1	167	24.5	5.7	5.7	Reddy et al., 1980
3	235-236	47-56	1.8-28	11.0	Beauchamp, 1986
2	N.G.	19.9-28.7	1.9-2.7	2.3	Castellanos & Pratt, 1981
1	N.G.	22.4	1.6	1.6	Chae & Tabatabai, 1986
3	N.G.	10.5-12.8	0.7-6.1	2.7	Serna & Pomares, 1991

[†]N.G. = Not given

Table 2. Dry matter, total nitrogen (N) contents, and proportion of total N present in inorganic form in swine manure samples.

No. of samples	Dry matter (g/kg wet manure)	Total N (g/kg dry manure)	Range Inorganic N (% of total)	Mean Inorganic N (% of total)	Reference
1	271	39.7	14.9	14.9	Reddy et al., 1980
1	N.G.[†]	38.6	3.7	3.7	Castellanos & Pratt, 1981
1	N.G.	21.2	6.4	6.4	Chae & Tabatabai, 1986
2	N.G.	23.1-28.5	0.4-1.1	0.8	Serna & Pomares, 1991
1	N.G.	30.8	8.3	8.3	Bernal & Kirchmann, 1992

[†]N.G. = Not given.

from soil. Thus, animal manures that have been in contact with soil are likely to have higher NO_3 contents than those that have not. Cabrera et al. (1994) found that two poultry litter samples from houses with "dirt" floors contained 2450 and 2170 mg NO_3-N/kg, whereas a sample from a house with a cement floor did not contain nitrate. The concentration of NO_3 in animal manures in which nitrifiers are active will also depend on the existence of conditions that favor denitrification. Storage of animal manures with high moisture content may result in large losses of NO_3 through denitrification (Cabrera and Chiang, 1994).

Manure Slurries

In contrast to fresh and composted manures, slurries typically contain more inorganic than organic N (Tables 5, 6, and 7). This is apparently the result of mineralization of part of the organic N during storage. Nodar et al. (1992)

Table 3. Dry matter, total nitrogen (N) contents, and proportion of total N present in inorganic form in poultry manure samples.

No. of samples	Dry matter (g/kg wet manure)	Total N (g/kg dry manure)	Range Inorganic N (% of total)	Mean Inorganic N (% of total)	Reference
1	N.G.[†]	51	8	8	Kirchmann, 1991
2	815-869	34-40	10-14	12	Hadas et al., 1983
1	N.G.	45.9	3	3	Castellanos & Pratt, 1981
1	N.G.	22	10	10	Chae & Tabatabai, 1986
1	270	60.3	50	50	Reddy et al., 1980
2	N.G.	21.3-31.2	3.5-4.5	4	Serna & Pomares, 1991
19	N.G.	18.2-81.3	19-55	40	Bitzer & Sims, 1988
3	N.G.	43.9-58.5	6-18	12	Cabrera et al., 1994
3	770-840	27.7-45.9	11-13	12	Cabrera & Chiang, 1994
3	N.G.	40.4-49.3	23-30	28	Sims, 1986
1	661	32.4	16	16	Brinson et al., 1994
15	520-810	26.8-59.6	6.7-18	11	Unpublished
1	N.G.	50.6	11	11	Tyson & Cabrera, 1993
13	N.G.	34.8-46.5	20-25	22	Westerman et al., 1988

[†]N.G. = Not given.

Table 4. Dry matter, total nitrogen (N) contents, and proportion of total N present in inorganic form in composted manure samples.

No. of samples	Dry matter (g/kg wet manure)	Total N (g/kg dry manure)	Range Inorganic N (% of total)	Mean Inorganic N (% of total)	Reference
2	150-809	10.7-14.8	2.6-4.9	3.7	Brinson et al., 1994
2	N.G.[†]	12.5-14.1	2.8-3.4	3.1	Tyson & Cabrera, 1993
1	N.G.	17.0	12.6	12.6	Castellanos & Pratt, 1981
1	N.G.	9.4	1.0	1.0	Tyson & Cabrera, 1993
1	N.G.	19.1-19.7	1.5-1.6	1.5	Castellanos & Pratt, 1981
3	158-216	22.7-27.7	3.1-5.1	4.1	Paul & Beauchamp, 1993

[†]N.G. = Not given.

Table 5. Dry matter, total nitrogen (N) contents, and proportion of total N present in inorganic form in cattle slurry samples.

No. of samples	Dry matter (g/kg wet manure)	Total N (g/kg dry manure)	Range Inorganic N (% of total)	Mean Inorganic N (% of total)	Reference
3	80-93	45.1-59.1	56-60	58	Stevens et al., 1992
6	60.8-85.2	46.9-69.6[†]	60-79	65	Sommer et al., 1992
7	5.6-123.2	29.2-80.1[†]	30-56	43	Chescheir et al., 1985
2	68-71	38.2-53.5	42-58	50	Beauchamp, 1986
1	58.5	67.9	61	61	Husted et al., 1991
1	113	64.6	59	59	Oenema & Velthof, 1993
1	56-62	46.8-57.6	57-76	64	Paul & Beauchamp, 1993
1	113	35.6	26	26	Trehan, 1994

[†]Converted from g N/L to g/kg dry manure with a relationship by Chescheir et al. (1985).

Table 6. Dry matter, total nitrogen (N) contents, and proportion of total N present in inorganic form in swine slurry samples.

No. of samples	Dry matter (g/kg wet manure)	Total N (g/kg dry manure)	Range Inorganic N (% of total)	Mean Inorganic N (% of total)	Reference
9	1.6-132.9	27.7-360.6[†]	18-92	62	Chescheir et al., 1985
8	7.6-68.8	51.1-172.3	45-85	69	Bernal et al., 1992
10	10.5-81.3	66.7-265.9[†]	68-89	78	Sommer et al., 1992
1	131.6	44.9	38	38	Bernal & Roig, 1993
1	101.4	87.2	51	51	Kirchmann & Lundvall, 1993
4	26.4-31.1	96.4-136.4	67-78	75	Rees et al., 1993

[†]Converted from g N/L to g/kg dry manure with a relationship by Chescheir et al. (1985).

Table 7. Dry matter, total nitrogen (N) contents, and proportion of total N present in inorganic form in poultry slurry samples.

No. of samples	Dry matter (g/kg wet manure)	Total N (g/kg dry manure)	Range Inorganic N (% of total)	Mean Inorganic N (% of total)	Reference
1	12.2	167.3[†]	87	87	Chescheir et al., 1985
2	130-147	51-61.5	84-95	89	Beauchamp, 1986
1	83.9	46.9	30	30	Nodar et al., 1992
1	83.3	46.1	37	37	Nodar et al., 1992
1	83.4	42.2	64	64	Nodar et al., 1992
1	82.5	38.0	64	64	Nodar et al., 1992

[†]Converted from g N/L to g/kg dry manure with a relationship by Chescheir et al. (1985)

found that the proportion of inorganic N in poultry slurry increased from 30%, when fresh, to 64% after 14 weeks of storage.

As is the case for fresh manures, most of the inorganic N in slurries is present as NH_4. This is apparently due to the existence of conditions that favor denitrification and discourage nitrification within the slurries (Oenema and Velthof, 1993).

Availability of Inorganic Nitrogen in Animal Manures

According to the model proposed by Sims (1986), availability of inorganic N in manures depends on the value of A_i. Beauchamp (1986) and Bitzer and Sims (1988) used $A_i=0.8$ to estimate the availability of inorganic N from surface applications of liquid cattle manure and poultry litter. The value of 0.8 was selected assuming that 20% of the NH_4 would be lost through NH_3 volatilization. The authors concluded that in addition to NH_3 losses, it is necessary to consider losses through leaching and denitrification and decreases in N availability through immobilization. Thus, the availability factor A_i could be defined as follows:

$$A_i = 1 - F_iNH_3 - F_iLch - F_iDen - F_iImm, \qquad (1)$$

where F_iNH_3 is the fraction (0-1) of inorganic N that is volatilized as NH_3, F_iLch is the fraction (0-1) lost through leaching, F_iDen is the fraction (0-1) lost through denitrification, and F_iImm is the proportion of inorganic N that is immobilized. It should be noted that when estimating the availability of inorganic N in animal manures, it is of interest to calculate the amount of manure inorganic N that would behave as N added with inorganic fertilizer. It is well known that leaching, denitrification, NH_3 volatilization, and immobilization losses also occur with fertilizer N, which is the reason that the efficiency of use is normally less than 100%. The purpose of the factor A_i is to account for leaching, denitrification, NH_3 volatilization, and immobilization losses that occur in addition to those that normally occur with fertilizer N. That is, A_i should account for losses that occur because of the effect of compounds (other than the N) added in the manure. It is the decomposition of these added C compounds that cause extra losses.

Losses of Manure Nitrogen through NH_3 Volatilization

Many studies have shown that surface application of animal manures may lead to losses of N through NH_3 volatilization (Beauchamp et al., 1982; Pain et al., 1989; Thompson et al., 1987; Nathan and Malzer, 1994). These losses are due to the inherent alkalinity of the manure (Husted et al., 1991) and to the increase in alkalinity caused by mineralization of manure N (Tyson and Cabrera, 1993). It is this extra NH_3 loss caused by added alkalinity that should be accounted for by A_i.

Ammonia volatilization losses are commonly high during the first 5 to 10 d after application (Thompson et al., 1990; Cabrera et al., 1993) and show diurnal fluctuations caused by diurnal soil temperature cycles (Nathan and Malzer, 1994). In addition, occurrence of rains immediately after application may reduce NH_3 losses due to dilution of ammoniacal N present in solution and to incorporation of part of the N into soil (Whitehead and Raistrick, 1991).

The magnitude of NH_3 losses can be very significant with certain manures. For example, surface application of swine slurry has resulted in losses ranging from 11% (Hoff et al., 1991) to 78% of the NH_4-N applied (Pain et al., 1989). Similarly, losses of 38 to 70% of the NH_4-N have been reported as a result of surface applications of cattle slurry (Thompson et al., 1987; Pain et al., 1989; Thompson et al., 1990).

Data on NH_3 volatilization from fresh manures are very scarce, but some laboratory studies with poultry manure have shown losses ranging from 37 to 60% of the surface-applied N (Wolf et al., 1988; Cabrera et al., 1993). In a field study, Nathan and Malzer (1994) found that the application of turkey manure on the soil surface caused an NH_3 loss equivalent to 5.7% of applied N. Bernal and Kirchmann (1992) measured NH_3 losses equivalent to 2.3, 14.3, and 4.0% of the total N content of surface-applied fresh, anaerobic, and aerobic swine manure, respectively.

In contrast, NH_3 volatilization from composted manures is relatively low due to their low NH_4 content and to the low rate of mineralization of their organic N (Brinson et al., 1994). It is clear that more field research on NH_3 losses is needed, especially with manures that have shown high NH_3 volatilization in laboratory studies.

Losses of Manure Nitrogen through Denitrification

The application of animal manures or slurries to soil may enhance losses of N through denitrification due to the addition of easily decomposable organic compounds (Paul and Beauchamp, 1989). Aerobic decomposition of these compounds causes a fast depletion of oxygen in the soil atmosphere, which favors development of anoxic microsites adequate for denitrification (Rice et al., 1988). In addition, organic compounds present in manure provide C required for denitrifiers to function. It is this extra denitrification loss, caused by the addition of organic compounds, that should be accounted for by A_i.

Losses through denitrification are usually lower than those through NH_3 volatilization. Egginton and Smith (1986) made several applications of cattle slurry (100 to 200 kg N/ha) to grassland during one year, and measured denitrification losses similar to those of control microplots, which were not fertilized. Comfort et al. (1990) injected dairy slurry into soil and measured denitrification losses that accounted for 2.5 to 3.2% of the slurry's NH_4-N, or 1.0 to 1.3% of total applied N. In a laboratory study, Cabrera et al. (1993) measured denitrification losses from pelletized and non-pelletized poultry litter applied to the soil surface. Losses ranged from 0.2 to 0.6% of the applied N for non-pelletized litter, and from 6.2 to 7.9% of the applied N for pelletized litter. Thompson et al. (1987) estimated denitrification losses that accounted for 7 to 21% of N applied with injected cattle slurry.

Decreases in Nitrogen Availability through Immobilization

Animal manures or slurries may increase N immobilization if they contain easily decomposable C compounds with low N contents (King, 1984). Kirchmann (1991) found that part of the inorganic N present in anaerobically treated manure was immobilized after application to soil. This N immobilization was attributed to rapid decomposition of organic compounds generated during anaerobic treatment. Amounts immobilized, expressed as a percentage of the initial NH_4-N, were 76% for swine manure, 64% for cattle manure, and 21% for poultry manure. It is this extra N immobilization, caused by the decomposition of organic compounds added with the manure, that should be accounted for by A_i.

MINERALIZABLE ORGANIC NITROGEN IN ANIMAL MANURES

Estimation of Mineralizable Nitrogen from Laboratory Incubations

The second part of the N availability model proposed by Sims (1986) refers to the amount of organic N that becomes available through the process of N mineralization. This fraction of organic N is indicated by the factor P_m in the equation. Typically, P_m is estimated with laboratory studies in which animal manures are mixed with soil and incubated at optimum temperature and moisture conditions for periods that commonly range from one to several months. Although limited in scope, data available suggest that in general, P_m is smaller for swine and cattle manures than for poultry manure (Table 8). However, much more additional work is needed to better define P_m values for the different types of manure available.

In addition, limited research results have shown that the mineralization of organic N from manures may differ between soils. Castellanos and Pratt (1981) reported that N mineralization from several fresh and composted animal manures was consistently higher in one of the two soils they studied. Similarly, Chae and Tabatabai (1986) observed that mineralization of N from four animal manures was very low in one of the four soils they used. In some of our recent,

Table 8. Percent organic N mineralized in incubations of animal manures with soils.

Sample Type	No. of Samples	Incub. Length	Temp.	Range Organic N Min.	Mean Organic N Min.	Reference
		days	C	%	%	
swine	2	112	25	8-25	17	Serna & Pomares, 1991
swine	1	182	30	16-52	39	Chae & Tabatabai, 1986
cow	3	112	25	0-13	6	Serna & Pomares, 1991
cow	1	182	30	13-51	35	Chae & Tabatabai, 1986
chicken	1	182	30	21-67	53	Chae & Tabatabai, 1986
broiler	3	150	25	25-40	34	Sims, 1986
broiler	3	150	40	17-64	44	Sims, 1986
broiler	19	140	23	21-110	67	Bitzer & Sims, 1988
broiler	15	112	25	41-85	60	Unpublished
broiler	1	56	25	25-37	31	Tyson & Cabrera, 1993
broiler	1	56	25	43-51	47	Brinson et al., 1994
broiler	1	35	25	60-73	67	Cabrera et al., 1993
compost	2	56	25	3-6	4	Tyson and Cabrera, 1993
compost	2	56	25	0.3- (-4)	-1	Brinson et al., 1994

unpublished work, we observed differences in the kinetics of N mineralization of the same sample of broiler litter decomposing in 15 different soils. Reasons for these differences among soils are not clear. Therefore, additional work is needed to identify soil characteristics that affect mineralization of manure N.

Estimation of Mineralizable Nitrogen from Indices

Estimation of P_m with laboratory incubations is time-consuming and, therefore, impractical to provide farmers with a quick assessment of the fertilizer value of a manure. Consequently, some research efforts have been spent on the search for quick indices of mineralizable N.

Castellanos and Pratt (1981) tested total N, NH_4 released by alkaline and acid $KMnO_4$, N released by pepsin, and NH_4 released by 6N HCl as indices of available N in 10 samples of fresh and composted animal manures. They found that the N released by pepsin could explain slightly more than 80% of the variation in available N. Similarly, Serna and Pomares (1991) reported that N released by pepsin could explain 64% of the variation in available N measured in 10 dry manure samples (including sheep, poultry, swine, and cow manures). In addition, Serna and Pomares (1991) found that N released by autoclaving and by acid $KMnO_4$ showed good correlation with N uptake by maize in a 6-week period, and with N mineralized during 6- and 16-week incubations. In a recent, unpublished study, we found that a fast pool of mineralizable N in 15 poultry litter samples correlated well with uric acid content (r = 0.78; $P < 0.01$) and with soluble organic N in the litter (r = 0.75; $P < 0.01$).

Working with manures, sewage sludges, and soil amendments, Douglas and Magdoff (1991) found a good correlation between the fraction of organic N mineralized and N released into the Walkely-Black acid-dichromate digest (r = 0.91; $P < 0.05$). In a study with municipal and industrial wastes, King (1984) was able to predict potentially available N with a regression equation that included organic N, total N, and C contents as independent variables. These results indicate that certain manure or waste characteristics play a significant role in determining the ease with which organic N will mineralize once in contact with soil. Thus, further work in this area seems warranted.

Availability of Inorganic Nitrogen Released through Mineralization

Not all N released through mineralization is available to plants because of losses through the same mechanisms discussed for the inorganic N initially present in the manure (i.e., leaching, NH_3 volatilization, denitrification, and immobilization). Therefore, a more complete version of the model of N availability may be as follows: Available N = $A_i N_i + A_m P_m N_o$, where A_m is the fraction of mineralized N that is available. The term A_m could be defined as follows: $A_m = 1 - F_m NH_3 - F_m Leach - F_m Den - F_m Imm$, where the different F_m factors represent fractions (0-1) lost through NH_3 volatilization, leaching, denitrification, and immobilization. As indicated previously for inorganic N initially present in manure, A_m should account for extra losses that occur as a result of the addition of manure compounds to the soil.

It would be difficult, if not impossible, to estimate values for all the parameters that make up A_i and A_m from the data available in the current literature. However, consideration of this model in future studies may help to arrive at better estimates of N availability from animal manures.

REFERENCES

Beauchamp, E.G., Response of corn to nitrogen in preplant and sidedress applications of liquid dairy cattle manure, *Can. J. Soil Sci.,* 63, 377, 1983.

Beauchamp, E.G., Availability of nitrogen from three manures to corn in the field, *Can. J. Soil Sci.,* 66, 713, 1986.

Beauchamp, E.G., Kidd, G.E., and Thurtell, G., Ammonia volatilization from liquid dairy cattle manure in the field, *Can. J. Soil Sci.,* 62, 11, 1982.

Bernal, P.M., and Kirchmann, H., Carbon and nitrogen mineralization and ammonia volatilization from fresh, aerobically and anaerobically treated pig manure during incubation with soil, *Biol. Fertil. Soils,* 13, 135, 1992.

Bernal, P.M. and Roig, A., Nitrogen transformations in calcareous soils amended with pig slurry under aerobic incubation, *Agric. Sci.,* 120, 89, 1993.

Bernal, P.M., Roig, A., Lax, A., and Navarro, A., Effect of the application of pig slurry on some physico-chemical and physical properties of calcareous soils, *Bioresource Technol.,* 42, 233, 1992.

Bitzer, C.C., and Sims, J. T., Estimating the availability of nitrogen in poultry manure through laboratory and field studies, *J. Environ. Qual.,* 17, 47, 1988.

Brinson, S.E., Cabrera, M.L., and Tyson, S.C., Ammonia volatilization from surface-applied, fresh and composted poultry litter, *Plant and Soil* (In Press), 1994.

Cabrera, M.L., and Chiang, S.C., Water content effect on denitrification and ammonia volatilization in poultry litter, *Soil Sci. Soc. Am. J.,* 58, 811, 1994.

Cabrera, M.L., Chiang, S.C., Merka, W.C., Thompson, S.A., and Pancorbo, O.C., Nitrogen transformations in surface-applied poultry litter: Effect of litter physical characteristics, *Soil Sci. Soc. Am. J.,* 57, 1519, 1993.

Cabrera, M.L., Tyson, S.C., Kelley, T.R., Pancorbo, O.C., Merka, W.C., and Thompson, S.A., Nitrogen mineralization and ammonia volatilization from fractionated poultry litter, *Soil Sci. Soc. Am. J.,* 58, 367, 1994.

Castellanos, J.Z., and Pratt, P.F., Mineralization of manure nitrogen-correlation with laboratory indexes, *Soil Sci. Soc. Am. J.,* 45, 354, 1981.

Chae, Y.M., and Tabatabai, M.A., Mineralization of nitrogen in soils amended with organic wastes, *J. Environ. Qual.,* 15, 193, 1986.

Chescheir, III, G.M., Westerman, P.W., and Safley, L.M., Jr., Rapid methods for determining nutrients in livestock manures, *Trans. ASAE,* 28, 1817, 1985.

Comfort, S.D., Kelling, K.A., Keeney, D.R., and Converse, J.C., Nitrous oxide production from injected liquid dairy manure, *Soil Sci. Soc. Am. J.,* 54, 421, 1990.

Douglas, B.F., and Magdoff, F.R., An evaluation of nitrogen mineralization indices for organic residues, *J. Environ. Qual.,* 20, 368, 1991.

Egginton, G.M., and Smith, K.A., Losses of nitrogen by denitrification from a grassland soil fertilized with cattle slurry and calcium nitrate, *J. Soil Sci.,* 37, 69, 1986.

Hadas, A., Bar-Yosef, B., Davidov, S., and Sofer, M., Effect of pelleting, temperature, and soil type on mineral nitrogen release from poultry and dairy manures, *Soil Sci. Soc. Am. J.*, 47, 1129, 1983.

Hoff, J.D., Nelson, D.W., and Sutton, A. L., Ammonia volatilization from liquid swine manure applied to cropland, *J. Environ. Qual.*, 10, 90, 1991.

Husted, S., Jensen, L.S., and Jørgensen, S.S., Reducing ammonia loss from cattle slurry by the use of acidifying additives: The role of the buffer system, *J. Sci. Food Agric.*, 57, 335, 1991.

Keener, H.M., and Hansen, R.C., *Practical Implications for the Composting of Poultry Manure*, paper presented at the 1992 National Poultry Waste Management Symposium, Birmingham, Alabama, October 6-8, 1992.

King, L.D., Availability of nitrogen in municipal, industrial, and animal wastes, *J. Environ. Qual.*, 13, 609, 1984.

Kirchmann, H., Carbon and nitrogen mineralization of fresh, aerobic and anaerobic animal manures during incubation with soil, *Swedish, J. Agric. Res.*, 21, 165, 1991.

Kirchmann, H., and Lundvall, A., Relationship between N immobilization and volatile fatty acids in soil after application of pig and cattle slurry, *Biol. Fertil. Soils*, 15, 161, 1993.

Martins, O., and Dewes, T., Loss of nitrogenous compounds during composting of animal wastes, *Bioresource Technology*, 42, 103, 1992.

Nathan, M.V., and Malzer, G.L., Dynamics of ammonia volatilization from turkey manure and urea applied to soil, *Soil Sci. Soc. Am. J.*, 58, 985, 1994.

Nodar, R., Acea, M.J., and Carballas, T., Poultry slurry microbial population: Composition and evolution during storage, *Bioresource Technology*, 40, 29, 1992.

Oenema, O., and Velthof, G.L., Denitrification in nitric-acid-treated cattle slurry during storage, *Neth. J. Agric. Sci.*, 41, 63, 1993.

Pain, B.F., Phillips, V.R., Clarkson, C.R., and Karenbeck, J.V., Loss of nitrogen through ammonia volatilization during and following the application of pig or cattle slurry to grassland, *J. Sci. Food Agric.*, 47, 1, 1989.

Paul, J.W., and Beauchamp, E.G., Effect of carbon constituents in manure on denitrification in soil, *Can. J. Soil Sci.*, 69, 49, 1989.

Paul, J.W., and Beauchamp, E.G., Nitrogen availability for corn in soils amended with urea, cattle slurry and composted animal manures, *Can. J. Soil Sci.*, 73, 253, 1993.

Pomares, F., and Pratt, P.F., Value of manure and sewage sludge as a N fertilizer, *Agron. J.*, 70, 1065, 1978.

Rice, C.W., Sierzega, P.E., Tiedje, J.M., and Jacobs, L.W., Stimulated denitrification in the microenvironment of a biodegradable organic waste injected into soil, *Soil Sci. Soc. Am. J.*, 52, 102, 1988.

Reddy, K.R., Khaleel, R., and Overcash, M.R., Nitrogen, phosphorus and carbon transformations in a Coastal plain soil treated with animal manures, *Agric. Wastes*, 2, 225, 1980.

Rees, Y.J., Pain, B.F., Phillips, V.R., and Misselbrook, T.H., The influence of surface and subsurface application method for pig slurry on herbage yields and nitrogen recovery, *Grass and Forage Sci.*, 48, 38, 1993.

Sharpley, A.N., Chapra, S.C., Wedepohl, R., Sims, J.T., Daniel, T.C., and Reddy, K.R., Managing agricultural phosphorus for protection of surface waters: Issues and options, *J. Environ. Qual.*, 23, 437, 1994.

Serna, M.D., and Pomares, F., Comparison of biological and chemical methods to predict nitrogen mineralization in animal wastes, *Biol. Fertil. Soils*, 12, 89, 1991.

Sims, J.T., Nitrogen transformations in a poultry manure amended soil: Temperature and moisture effects, *J. Environ. Qual.*, 15, 59, 1986.

Stevens, R.J., Laughlin, R.J., Frost, J.P., and Anderson, R., Evaluation of separation plus acidification with nitric acid and separation plus dilution to make cattle slurry a balanced, efficient fertilizer for grass and silage, *J. Agric. Sci.*, 119, 391, 1992.

Sommer, S.G., Kjelleru, V., and Kristjansen, O., Determination of total ammonium nitrogen in pig and cattle slurry: Sample preparation and analysis, *Acta Agric. Scand., Sect. B*, 42, 146, 1992.

Thompson, R.B., Pain, B.F., and Lockyer, D.R., Ammonia volatilization from cattle slurry following surface application to grassland, *Plant and Soil*, 125, 109, 1990.

Thompson, R.B., Ryden, J.C., and Lockyer, D.R., Fate of nitrogen in cattle slurry following surface application or injection to grassland, *J. Soil Sci.*, 38, 189, 1987.

Trehan, S.P., Immobilization of $^{15}NH_4$ by cattle slurry decomposing in soil, *Soil Biol. Biochem.*, 26, 743, 1994.

Tyson, S.C., and Cabrera M.L., Nitrogen mineralization in soils amended with composted and uncomposted poultry litter, *Commun. Soil Sci. Plant Anal.* 24, 2361, 1993.

Westerman, P.W., Safley, Jr., L.M., and Barker, J.C., Available nitrogen in broiler and turkey litter, *Trans. ASAE*, 31, 1070, 1988.

Whitehead, D.C., and Raistrick, N., Effects of some environmental factors on ammonia volatilization from simulated livestock urine applied to soil, *Biol. Fertil. Soils*, 11, 279, 1991.

Wolf, D.C., Gilmour, J.T., and Gale, P.M., *Estimating potential ground and surface water pollution from land application of poultry litter - II.* Publ. no. 137, Arkansas Water Resourc. Res. Center, Fayetteville, Arkansas, 1988.

NITROGEN UPTAKE AND LEACHING IN A NO-TILL FORAGE ROTATION IRRIGATED WITH LIQUID DAIRY MANURE

J.G. Davis, G. Vellidis, R.K. Hubbard, J.C. Johnson,
G.L. Newton, and R.R. Lowrance

PROBLEM DEFINITION AND APPROACH

As the size of dairy herds increases in the southeast, the acreage per head available for land disposal of manure declines. This can lead to excessive manure application in the vicinity of the dairy and can potentially degrade water quality due to nitrate leaching and phosphorus runoff. The objective of this research project is to determine optimum application rates that will minimize nitrate leaching and maximize yield and N uptake.

MINIMIZING NITRATE LEACHING

Excessive dairy slurry application rates can result in shallow ground water nitrate concentrations that exceed drinking water standards (Hubbard et al., 1987). South Georgia receives approximately 50 in. of rainfall annually with about 20 in. falling from December through March. Therefore, nitrate leaching potential is high in the winter due to high rainfall, residual N from the summer crop, and little or no crop growth and N uptake. In addition, soil textures in the southeastern Coastal Plain range from true sands to sandy loams. Many soils have argillic horizons above plinthic layers which can retard water movement, but the plinthite is not continuous, and nitrate can "leak" through it.

MAXIMIZING NITROGEN UPTAKE

In this study, a year-round forage production system was used to minimize winter leaching potential by using N for crop production. The system involved rye planted in the fall into bermudagrass sod and cut twice in winter and early spring, followed by corn planted into the grass sod in March and harvested for silage in July, prior to three bermudagrass hay cuttings in the summer and fall. The triple-crop system allowed dairy lagoon effluent to be applied year-round as it was produced with applications every 10-14 days.

405

NITROGEN REMOVAL AND LEACHING

The liquid manure was applied through a center pivot irrigation system at four rates. The pivot area was divided into four topographically aligned quadrants with each quadrant receiving a different manure application rate: 200 (east), 400 (west), 600 (north), and 800 kg N/ha/year (south). Within each quadrant, there were four sampling sites where manure application, yield, crop removal, soil N, and water NO_3-N levels were measured.

YIELD AND CROP REMOVAL

Yield increased sequentially from 200 to 600 kg N/ha/year, but there were no yield differences between 600 and 800 kg N/ha/year (Figure 1A). Nitrogen concentration increased with increasing application rate for rye and bermudagrass (Table 1). Crop removal of N closely mirrored the yield results with very little increase in N uptake from the 600-kg N/ha/year rate to the 800-kg N/ha/year rate (Figure 1B).

The cumulative N uptake (as a percentage of N applied) ranged from 86-96% at the three lower application rates but declined below 80% at the highest application rate (Table 2). This decline illuminates the potential for N losses to the environment at the highest application rate due to more than 20% of the applied N not being taken up by the crop. Nitrogen use efficiency (NUE) also declined with increasing application rate, but this decline leveled off between the 600- and 800-kg N/ha/year rates since their yields and N uptake amounts

Table 1. Average nitrogen (N) concentration (%) in plants by application rate (from June 1991 through October 1993).

Rate (kg N/ha/year)	Crop			
	Rye	Corn Grain	Corn Stalk	Bermudagrass
200	2.25b[†]	1.57a	0.90a	1.36c
400	2.73a	1.51a	1.10a	1.67b
600	2.70ab	1.80a	1.14a	1.98a
800	3.01a	2.00a	1.20a	2.03a

[†]Rates with a common letter are not significantly different by Least Significant Difference ($P \leq 0.05$).

Table 2. Cumulative nitrogen (N) application, uptake, and use efficiency by application rate (up to day 828; October 1, 1993).

Rate	Cumulative N Application	N Uptake	NUE[†]
kg N/ha/year	kg/ha	% of applied	kg DM/kg N uptake
200	517	86.3	77.2
400	895	95.8	62.2
600	1360	91.0	54.2
800	1703	76.1	51.9

[†]NUE = nitrogen use efficiency; DM = dry matter.

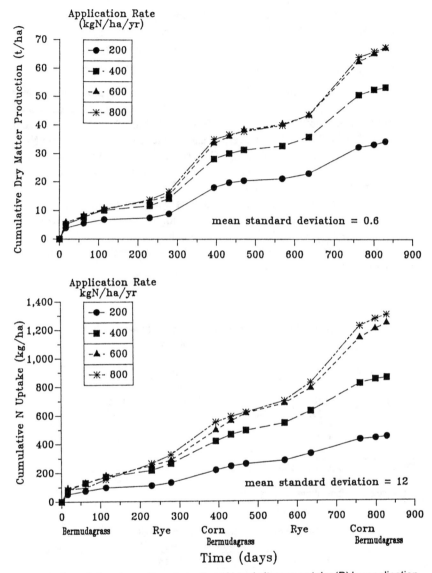

Figure 1. Cumulative dry matter production (A) and nitrogen uptake (B) by application rate.

were very similar in spite of a large difference (343 kg N/ha) in cumulative N application rate.

Denitrification was measured each month in the top 30 cm of soil using the acetylene inhibition technique on intact soil cores (Lowrance and Smittle, 1988). Annual denitrification losses for 1991-92 increased from 32 kg

N/ha/year to 133 kg N/ha/year in the 200- and 800-kg N/ha/year application rates, respectively. Therefore, denitrification accounts for substantial losses of N, thus reducing the amount of nitrate available for leaching.

GROUND WATER QUALITY

Soil samples were taken in duplicate at three of the four sampling sites per quadrant following each crop in the cropping system. Soil NO_3-N levels tended to increase with application rate at the 0- to 6-, 12- to 18-, and 18- to 24-cm sampling depths (Table 3). However, all samples taken below 12 cm had NO_3-N concentrations less than 8 mg/kg. Concentrations generally declined with depth. Soil NO_3-N levels remained relatively low due to the combined effects of crop uptake and denitrification.

Soil NH_4-N levels also tended to decline with soil depth (Table 4), but concentrations increased with application rate in the 0- to 6- and 12- to 18-cm depth increments. Below 12 cm all treatments had soil NH_4-N levels less than 2 mg/kg. Very little NH_4-N built up in the soils, evidently due to the rapid nitrification of manure NH_4-N.

Table 3. Soil NO_3-N concentration (2 N KCl extraction) by rate and depth (day 861).

Rate (kg N/ha/yr)	Soil Depth (cm)				
	0-6	6-12	12-18	18-24	24-30
	---mg/kg---				
200	8.5b[†]	12.7	2.3b	1.3b	2.3b
400	13.3ab	4.8	2.8b	1.5b	0.7b
600	26.3a	6.1	5.4a	6.8a	7.7a
800	25.6a	7.0	5.5a	5.0a	3.2ab
		NS[‡]			

[†]Quadrants with a common letter are not significantly different by Least Significant Difference ($P \leq 0.05$).
[‡]NS = No significant difference ($P \leq 0.05$).

Table 4. Soil NH_4-N concentration (2 N KCl extraction) by rate and depth (day 861).

Rate (kg N/ha/yr)	Soil Depth (cm)				
	0-6	6-12	12-18	18-24	24-30
	---mg/kg---				
200	1.5b[†]	1.1	0.4b	0.6	1.5
400	16.8a	3.0	0.7a	0.5	1.7
600	7.4ab	2.6	0.7a	0.6	0.5
800	9.9a	1.4	0.7a	0.6	0.5
		NS[‡]		NS	NS

[†]Quadrants with a common letter are not significantly different by Least Significant Difference ($P \leq 0.05$).
[‡]NS = No significant difference ($P \leq 0.05$).

Suction lysimeters were installed at 0.5-, 1.0-, 1.5-, and 2.0-m depths in each of the three sampling sites per quadrant, and wells were installed at 3- and 6-m depths. During the course of this project, there was no change in NO_3-N concentration in water from the shallow wells (Vellidis et al., 1995). However, at the 600 and 800 kg N/ha/year application rates, the NO_3-N level increased from year 1 to year 2 at the 0.5-, 1.0-, and 1.5-m depths (Kruskal-Wallis Test $P \leq 0.0001$) . There was no change at the 2.0-m depth for any treatment (Figure 2), and the 200- and 400-kg N/ha/year rates showed no significant change at any sampling depth.

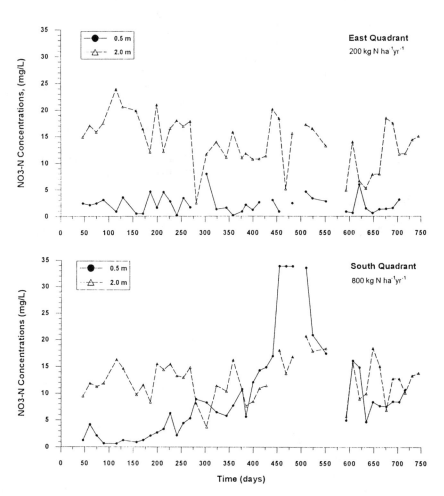

Figure 2. Nitrate concentrations in soil water at the 0.5- and 2.0-m depth in quadrants receiving 200 and 800 kg N/ha/year.

FUTURE NEEDS

In this study, the 600-kg N/ha/year manure application rate resulted in maximum yield and some increased soil and water NO_3-N levels to 1.5-m maximum depth. Long-term research is needed to determine how long and at what rate this practice could be continued before increasing ground water nitrate concentrations. If current trends continue, it appears that an application rate between 400 and 600 kg N/ha/year would result in economic yields without affecting water quality. Additional research on P in runoff is essential to insure that these application rates do not harm surface water quality (Sharpley et al., 1994). Although the no-till system utilized in this study reduces runoff and erosion potential, the soil test P levels have increased with application rate and could pose a surface water hazard.

REFERENCES

Hubbard, R.K., Thomas, D. L., Leonard, R.A., and Butler, J.L., Surface runoff and shallow groundwater quality as affected by center pivot applied dairy cattle wastes, *Trans. of the ASAE*, 30(2), 430, 1987.

Lowrance, R., and Smittle, D.A., Nitrogen cycling in a multiple crop vegetable production system, *Journal of Environmental Quality*, 17, 158, 1988.

Sharpley, A.N., Chapra, S.C., Wedepohl, R., Sims, J.T., Daniel, T.C., and Reddy, K.R., Managing agricultural phosphorus for protection of surface waters: Issues and options, *Journal of Environmental Quality*, 23(3), 437, 1994.

Vellidis, G., Hubbard, R.K., Davis, J.G., Lowrance, R., Williams, R.G., and Johnson, J.C., Jr., Nitrogen concentrations in the subsurface water of a liquid dairy manure land application site, *Trans. of the ASAE*, In Review, 1995.

ESTIMATING PROBABILITIES OF NITROGEN AND PHOSPHORUS LOSS FROM ANIMAL WASTE APPLICATION

A.F. Johnson, D.M. Vietor, F.M. Rouquette, Jr., V.A. Haby, and M.L. Wolfe.

INTRODUCTION

WASTE MANAGEMENT AND WATER QUALITY

Increases in nitrogen (N) and phosphorus (P) imports in feed are associated with growth in number and size of dairies on watersheds in Texas and other dairy states. The unused feed nutrients in solid manure and lagoon effluents can exceed crop requirements on land controlled by large dairies and create a waste disposal problem (Heatwole et al., 1990). This nutrient loading on fields, farms, and watersheds raises concern about N losses to ground water and P contamination of surface waters (Lanyon, 1991; Van Horn, 1991). For example, large effluent loading rates can elevate N in subsoil and P in the upper 15 cm of the soil profile (King et al., 1990).

Federal and state agencies have developed and transferred improved system designs and best management practices (BMP) to help large dairy producers minimize N and P losses and conform to laws enforced by the US Environmental Protection Agency and the Texas Natural Resources Conservation Commission (TNRCC) (Sweeten and Wolfe, 1993). Although new dairies must comply with these laws to receive preliminary TNRCC approval of applications for waste/waste water disposal permits, activist opposition in public hearings of requests is delaying final permit approval for months and even years. This adversarial relationship between producers and their opponents is typical of US approaches to managing resource development and new technology (Covello et al., 1988). Intense loyalty within interest groups, lobbying, competing technical analyses and interpretations, and extensive use of the legal system further typify the situation in the US and on Texas watersheds.

RISK ASSESSMENT

Risk analysis emerged during the 1980s as a methodology for regulatory policy-making in US energy and chemical industries. Under the scrutiny of scientific advisory committees and public hearings, explicit analyses of risks to environment and human health were expected to provide objective and non-arbitrary criteria for developing regulatory policies and mediating policy disputes between industries and their opponents (Covello et al., 1988). Risk criteria were similarly proposed for comparing and choosing management practices. As a part of risk analysis, quantitative risk assessment includes both the identification of hazards to public health and the environment and estimation of the probability and severity of harm resulting from hazards (Rasmussen, 1981; Ex, 1992). Hazards represent material or physical conditions that can potentially cause any type of harm or loss. Risk is defined as the potential for an undesirable consequence resulting from activities related to one or more hazards.

Unfortunately, Probabilistic Risk Assessment (PRA) has not provided the unequivocal criteria for regulatory decisions and for consensus in public-policy debates about new energy technologies that were expected (Linnerooth-Bayer and Whalstrom, 1991). For example, probability estimates for events that contribute to hazards and risk can be scientifically uncertain and vary among analysts and experts. Yet the qualitative exercise of structuring components of technology systems in tree diagrams can help managers understand the relationship among events, whether natural or under their control, that interact to pose hazards to the environment. In addition, the relative probabilities of single events and event sequences serve to identify comparatively large potential hazards (Starr, 1985). Applied to large dairies, PRA can be used to understand relationships among waste system components, including event sequences that result in nutrient release into surface and ground water (Seiler, 1990). Moreover, PRA can be used to evaluate new BMPs in terms of site-specific reductions in the probability and severity of nutrient-release outside the boundary of each dairy.

The methodology of PRA fosters the "systematic, interdisciplinary approach...in planning and decision-making" that was suggested in the National Environmental Policy Act (NEPA). When practiced as a site-specific and interactive process that brings technical advisors from engineering, agronomy, and animal science together with producers and regulatory agents, the value of PRA is represented in the process as much as in estimates of risk (Linnerooth-Bayer and Wahlstrom, 1991). This integrative and participatory process contributes both techniques and language that can bridge the knowledge and site-specific concerns of diverse participants. In addition, the information needs of PRA stimulate empirical research and development of computer simulations for quantifying probabilities of natural and management events.

OBJECTIVE

The major objective of this research is to conduct a site-specific hazards analysis and qualitative risk assessment for a large dairy in eastern Texas. A supporting objective is to develop criteria for estimating the probability of N and P leaching after irrigation of lagoon effluent on perennial crops.

PROCEDURES

RISK ASSESSMENT

The waste management system of the dairy was observed and described in terms of components that could create hazards to water quality. Interviews with the dairy and cropland managers and monitoring of nutrients in lagoon effluent, soil, and runoff from croplands provided information about potential pathways of nutrient flow through system components of the dairy. Starting with waste water discharge from the milking parlor, the potential pathways of nutrient flow are represented as event sequences in an event tree. Event trees systematically link an initiating event through forward logic to subsequent events in sequences that can transform one or more hazards into system failures or accidents (Pate-Cornell, 1984). Producers are asked for feedback about event trees and for estimates of the number of annual occurrences of events that present hazards to water quality.

Fault trees are used to estimate probabilities of those events in the event tree that are difficult to quantify from producer experiences and monitoring of the dairy. Fault trees employ a backward logic to organize and relate relatively simple events that combine to cause a selected failure event in the event tree. The events in fault trees are constructed as binary variables (the event occurs or not) that are related by the logical functions OR and AND (Pate-Cornell, 1984). These logical functions are represented by gates in the fault tree. Outputs immediately above a gate are a function of the input events below the gate. Each input event, in turn, is the logical function or output of yet more input events down to the point at which basic input events cannot be practically analyzed any further. Fault tree construction and probability calculations are aided by computer software (CAFTA) provided by Science Applications International Corporation (Averett, 1988). Plot-scale experiments and stochastic simulations are used to estimate probabilities of basic events that are not readily estimated from site-specific observations of the dairy.

QUANTIFYING LEACHING LOSSES

Plot studies were initiated during the spring of 1992 on a Darco loamy sand in eastern Texas to develop criteria for estimating probabilities of N and P leaching through the root zone of crop and pasturelands on dairies. Four surface applications of dairy effluent per year on 'Coastal' bermudagrass (*Cynodon dactylon*) overseeded with ryegrass (*Lolium multiflorum*) resulted in annual N rates of 0, 250, 500, and 1000 kg/ha with accompanying P. Dairy effluent and poultry litter were N sources (main plots), and N rates were subplots within four replications of this split-plot design. Metal flashing inserted in soil on the downslope border of plots and buffer strips between plots prevented contamination between adjacent treatments. Forage was harvested on 12 dates during a two-year period ending in spring 1994, and soil was sampled at 30-cm increments to a depth of 180 cm in August 1993. Soil water was sampled after large rainfall events at a depth of 180 cm using porous ceramic-cup lysimeters. Total N, NO_3^-, and extractable P of soil; N and P content of forage; and NO_3^- in soil water samples were quantified.

RESULTS AND DISCUSSION

HAZARDS ANALYSIS

Although 400 lactating cows in the dairy were dispersed on pastures, a waste handling system was needed to collect and manage the solid and liquid wastes that were deposited in confinement during two milking and feeding periods that totaled 9 hours/day. The initiating event of waste management was comprised of parlor wastewater combined with solid waste and runoff of rainfall from the paved area occupied by feeding cows. The nutrients in the solid and waste water sources were discharged into a channel that conducted liquid waste to the primary and secondary lagoon (Figure 1).

During this preliminary hazards analysis (Henley and Kumamoto, 1992) of waste management, the producer acknowledged relatively large probabilities for events leading to the hazard of nutrient runoff. For example, events contributing to high soil water content and, as a result, to effluent irrigation rates greater than surface infiltration rate on cropland, could have contributed to runoff into surface waters during 2 days/year (Figure 1). It is noteworthy that the producer had enough confidence at this preliminary stage of analysis to alter his waste management practices. Confidence emerged from the producer's involvement in event tree construction and evaluation and discussions of probability estimates for events. He purchased additional irrigation equipment to

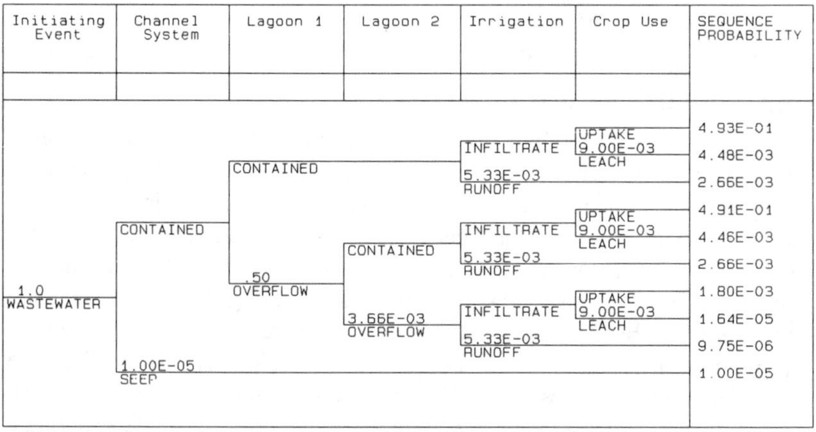

Initiating Event	Channel System	Lagoon 1	Lagoon 2	Irrigation	Crop Use	SEQUENCE PROBABILITY
					UPTAKE	4.93E-01
				INFILTRATE	9.00E-03	4.48E-03
		CONTAINED		5.33E-03	LEACH	
				RUNOFF		2.66E-03
					UPTAKE	4.91E-01
	CONTAINED			INFILTRATE	9.00E-03	4.46E-03
			CONTAINED	5.33E-03	LEACH	
		.50		RUNOFF		2.66E-03
1.0		OVERFLOW			UPTAKE	1.80E-03
WASTEWATER				INFILTRATE	9.00E-03	1.64E-05
			3.66E-03	5.33E-03	LEACH	
			OVERFLOW	RUNOFF		9.75E-06
	1.00E-05					1.00E-05
	SEEP					

Figure 1. Event tree describing sequences of events leading to hazards and risks to water quality.

expand the acreage of cropland that could be irrigated from lagoons and to reduce the probability of runoff.

PROBABILITY OF NITROGEN AND PHOSPHORUS LOSS

The relatively large initial probability estimates for N leaching were developed using knowledge of NO_3^- chemistry and the coarse soil texture, low water holding capacity, and rainfall records for eastern Texas (Figure 1). Yet monitoring of croplands receiving lagoon effluent revealed 18 or less mg/kg of N in the upper 30 cm of soil. Plot studies were initiated to improve estimates of the probability of N and P leaching following irrigation of effluent on land. Annual dry matter yield of plots of bermudagrass overseeded with annual ryegrass increased significantly ($P = 0.001$) as N rates in dairy effluent increased. Annual N removal in forage averaged 66.5, 93.4, 118.7, and 189.9 kg/ha at the four respective N rates in effluent. The mean NO_3^- concentration increased to 8 mg/kg in the uppermost 30 cm of soil (Figure 2) and to 11.5 mg/kg in soil water sampled at the 180-cm depth (Table 1) when 1000 kg/ha of N was applied annually in effluent. Although less than 25% of N applied at the rate of 500 kg/ha was recovered in forage, N concentration in soil and soil water was comparable to control plots (no effluent). Up to 30% of the N in surface applications of effluent could have been volatilized (Christensen, 1986), but additional research is needed to explain the discrepancy between N application rate and N recovery in forage and soil when 250 and 500 kg of N/ha were applied in effluent. The low soil NO_3^- concentrations at depths below 30 cm, even when NO_3^- leached into ground water (N rate = 1000 kg/ha), indicated that NO_3^- was not accumulating in the coarse-textured soil.

Compared to controls, the P applied with N in effluent significantly ($P = 0.001$) increased extractable P in the uppermost 30 cm of soil and P removal in forage at the two highest N rates. Soil P was 54 and 70% greater than controls (40 and 44 mg/kg of soil, respectively) at the two highest N rates (P rates of 95 and 190 kg/ha, respectively). Annual P content of forage increased from 22 to 34 and 50 kg/ha at the same respective P rates. The lack of increase in extractable P at depths below 30 cm indicated that P did not leach through soils even at the highest P rate.

Table 1. Average nitrate concentrations in soil water sampled at the 180-cm depth following large rainfall events.

Nitrogen Rate	Dairy Effluent		
	Mean	Max Obs. Value	Std. Error
	----------PPM NO_3^--N----------		
0 kg/ha	3.8a[†]	8.4	0.7718
250 kg/ha	3.6a	6.9	0.6179
500 kg/ha	1.3a	2.4	0.2377
1000 kg/ha	11.5b	33.4	3.2313

†Means followed by the same letter within columns are not significantly different ($P < 0.05$).

The N rate of 500 kg/ha provides a threshold value for estimating the probability of NO$_3^-$ and P leaching hazards arising from effluent irrigation on crop and pastureland. Low concentrations of P compared to N in dairy effluent and low P recoveries at depths below 30 cm indicate that N represents the greater hazard to ground water quality at N rates up to 1000 kg/ha. Although the annual rates of 250 and 500 kg N/ha in effluent exceeded the N recovered annually in bermudagrass and ryegrass forage, NO$_3^-$ concentrations in ground water were not greater than the controls. Even when maximum NO$_3^-$ concentrations in soil water samples are compared to controls (Table 1), 500 kg/ha of N in effluent is a defendable threshold value for estimating the contribution of effluent applications to risk. Although more than two years of data are desirable for quantifying the upper limits of N and effluent rates on perennial and annual crops on the dairy, the utility of the threshold value in estimating probabilities of leaching is still evident. The plot studies indicate that the hazard to water quality will depend on the probability that the producer applies more than the annual threshold rate (500 kg N/ha) in four or more applications of effluent to perennial forage that is harvested for hay.

QUALITATIVE RISK ASSESSMENT

The producer action of applying "High Nutrient Input from Effluent," i.e., more than 500 kg/ha of N, is included as an event in the fault tree that represents the causal relations among events contributing to leaching of nutrients into soil water beneath roots. The complex event of nutrient leaching through

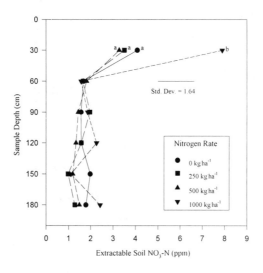

Figure 2. Soil NO$_3^-$ concentration in August 1993 of plots receiving the specified annual N rates through four split applications of dairy effluent.

the root zone is already represented in the event tree (Figure 1) and is the top event of the fault tree. Water percolation through and excess nutrient concentrations in the root zone are two causal or input events that must occur (joined by AND gate) before nutrients leach. Nutrient concentrations in excess of colloidal binding capacity of soil depend on both (notice the AND gate) nutrient excesses over root uptake and a lack of interactions between nutrient ions and soil colloids (Figure 3). Poor monitoring of either (notice OR gate) irrigation volume per acre (High Irrigation Volume) or of nutrient concentration in effluent (High Nutrient Concentration) represent basic events that are dependent on human decisions and skill. Producer behavior is being monitored to quantify the probability that producer error contributes to nutrient inputs greater than 500 kg/ha of N. As was described previously, the producer has already taken action to minimize the probability that excessive volumes of effluent are applied per day or per unit area.

Nutrient runoff and leaching and lagoon overflow are strongly dependent on rainfall and other events that are difficult to quantify as point probabilities. Stochastic simulations are being developed to quantify these events in terms of probability distributions and to complement the empirical estimates of probability described in this study.

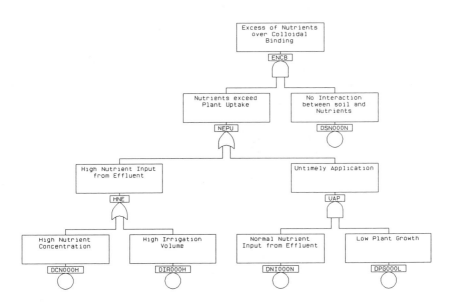

Figure 3. A portion of the fault tree illustrating the causal relationship among events that lead to nutrient concentrations greater than the colloidal binding capability of soil.

ACKNOWLEDGMENTS

This research is supported by the Texas Advanced Technology Program and USDA/CSRS Grant 93-34214-9614.

REFERENCES

Averett, M.W., Fault tree Analysis, *Risk Analysis,* 8, 463, 1988.

Christensen, B.T., Ammonia volatilization loss from surface applied animal manure, in *Efficient Land Use of Sludge and Manure,* A.D. Kofoed, J.H. Williams, and P. L'Hermite, eds., Elsevier Applied Science Publishers, London, 1986.

Covello, V.T., Kawamura, K., Boroush, M., Ikeda, S., Lynes, P.F., and Minor, M.S., Cooperation versus confrontation: A comparison of approaches to environmental risk management in Japan and the United States, *Risk Analysis,* 8, 247, 1988.

Ex, Barbara, Software for assessing hazardous chemical risks, *Pollution Engineering,* January, 70, 1992.

Heatwole, C.D., Diebel, P.L., and Halstead, J.M., Management and policy effects on potential groundwater contamination from dairy waste, *Water Resources Bull.,* 26, 25, 1990.

Henley, E.J., and Kumamoto, H., *Probabilistic Risk Assessment: Reliability Engineering, Design, and Analysis,* Institute of Electrical and Electronics Engineers, Inc., New York, 1992.

King, L.D., Burns, J.C., and Westerman, P.W., Long-term swine lagoon effluent applications on Coastal Bermudagrass: II, Effect on nutrient accumulation in soil, *J. Environ. Quality,* 19, 756, 1990.

Lanyon, L.E., Implications of dairy herd size for farm material transport, plant nutrient management, and water quality, *J. Dairy Sci.,* 75, 334, 1991.

Linnerooth-Bayer, J., and Wahlstrom, B., Applications of probabilistic risk assessments: The selection of appropriate tools, *Risk Analysis,* 11, 239, 1991.

Pate-Cornell, M.E., Fault trees versus event trees in reliability analysis, *Risk Analysis,* 4, 177, 1984.

Rasmussen, N.C., The application of probabilistic risk assessment techniques to energy technologies, *Ann. Rev. Energy,* 6, 123, 1981.

Seiler, F.A., On the use of risk assessment in project management, *Risk Analysis,* 10, 365, 1990.

Starr, Chauency, Risk management, assessment, and acceptability, *Risk Analysis,* 5, 97, 1985.

Sweeten, J.M., and Wolfe, M.L., *The Expanding Dairy Industry: Impact on Ground Water Quality and Quantity with Emphasis on Waste Management System Evaluation for Open Lot Dairies,* Technical Report No. 155, Texas Water Resources Institute, Texas A&M University, College Station, Texas, 1993.

Van Horn, H.H., Achieving environmental balance of nutrient flow through animal productions systems, *The Professional Animal Scientist,* 7, 22, 1991.

IMPACTS OF LONG-TERM MANURE APPLICATIONS ON SOIL CHEMICAL, MICROBIOLOGICAL, AND PHYSICAL PROPERTIES

C.W. Wood and J.A. Hattey

INTRODUCTION

Direct, nutrient-supplying benefits of animal waste applications to crops have been recognized since the beginning of agriculture. Long-term application of these organic materials may also promote plant biomass production indirectly via improvements in soil physical, chemical, and biological properties. Conversely, long-term application of animal waste at heavy rates may have negative impacts on crop production and the environment owing to soil accumulation and/or escape of waste constituents.

As animal wastes are applied to the soil system over time, a shift towards a new equilibrium in soil properties occurs. Attainment of a new steady state in which inputs balance against outputs coupled with little change in soil properties may require many years. In reality, owing to the dynamic nature of soil systems, a steady state may never be realized with soil properties more likely fluctuating within some range (i.e., quasi-equilibrium). Regardless of whether or not true equilibrium is attained, long-term studies are required to determine effects that repeated manure applications have on soil properties and, subsequently, on crop production. The purpose of this chapter is to review recent literature regarding impacts of long-term animal waste applications on soil chemical, microbiological, and physical characteristics.

CHANGES IN SOIL PROPERTIES

Over the past 10 years, a fair number of studies have addressed soil changes owing to long-term applications of animal wastes (Table 1). Long-term, in the context of this chapter, refers to studies in which manure was applied for five or more years. Most studies have focused on alterations in soil chemical properties, and less emphasis has been placed on soil microbiological and physical characteristics.

CHEMICAL PROPERTIES

Since animal manures are carbon (C) bearing materials, with most being applied to supply nitrogen (N), many long-term studies on effects of animal

419

waste application have focused on these elements. Most reports address C and N pool sizes, while some deal with C and N transformations via microbial activity (Table 1).

The majority of studies on organic C and total N indicate that they accumulate in surface soils with repeated application of animal manure (Anderson et al., 1991; Christie and Beattie, 1989; Dormarr and Sommerfeldt, 1986; Johnston et al., 1986; King et al., 1990; Kingery et al., 1994; Mbagwu and Piccolo, 1990; N'dayegamiye, 1990; N'dayegamiye and Côté, 1989; Schjønning et al., 1994; Sharpley et al., 1993; Sommerfeldt and Chang, 1987; Sommerfeldt et al., 1988; Spallacci and Boshi, 1985; Stadelmann and Furrer, 1985; Witter et al., 1993). Perhaps the best example of an "equilibrium" situation is the study by Johnston et al. (1989) at the Rothamsted Experiment Station in England. They found that total N content of unamended soil (3 Mg N/ha) and soil amended with inorganic N (3.5 Mg N/ha) remained relatively unchanged after 140 years, while total N of soil amended with farmyard manure (FYM; 35 Mg/ha/year) was approximately 2.25 fold greater than that fertilized with inorganic N. Johnston et al. (1989) also showed that rate of total N increase with FYM slowed considerably in recent years. A lower rate of increase in soil organic C and total N with time was also shown by Sommerfeldt et al. (1988) in a shorter-term study (11 years) utilizing cattle manure.

Some studies have shown that buildup of organic C and total N in surface soils amended with manures is affected by differences in manure type, manure rate, and soil texture. After 11 years of swine manure application (cumulative = 1109 Mg/ha), percent increases of organic matter on sandy loam and clay loam soils were 15 and 44%, respectively. However, the same cumulative application on a silt loam soil did not increase soil organic matter. After pig and cattle slurries were applied at rates of 50, 100, and 200 m^3/ha/year for 17 years, Christie and Beattie (1989) found that only the highest rate of pig slurry increased surface soil (0- to 5-cm) organic C, while increases with cattle slurry were independent of rate. Spallacci and Boshi (1985) applied cattle slurry at rates to supply 0, 250, 500, and 750 kg N/ha on sandy loam, sandy clay, and clay soils cropped to maize. Following three application years and two residual years (no nutrients added), they found the highest rate of cattle slurry was required to maintain soil (0- to 25-cm) organic C and N at pre-study levels for the sandy loam. Only one-third of the maximum rate was required for maintenance on the clay, and the "maintenance rate" was intermediate for the sandy clay. They concluded that organic matter mineralized at a faster rate in the sandy loam owing to greater aeration. These studies illustrate the control that management and environmental factors exert over soil organic matter dynamics in manure-amended systems.

Alterations in soil organic C and total N amounts owing to long-term manure applications have been accompanied by changes in C:N and fractional distribution of soil C and N. Kingery et al. (1994) observed lower C:N in soils

Table 1. Recent research addressing impacts of long-term manure applications on soil chemical, microbiological and physical properties.

Reference	Waste Type(s)	Years of Manure Application	Soil Variable(s) Studied
Anderson et al., 1991	Swine Manure	11	pH, organic matter, extractable Cu and P
Chang et al., 1990	Cattle Manure	10	pH, EC, SAR, Na, Ca + Mg, SO_4, Cl
Christie, 1987a	Cattle Slurry, Pig Slurry	16	P, K
Christie, 1987b	Cattle Slurry, Pig Slurry	16	P, K, pH
Christie and Kilpatrick, 1992	Cattle Slurry, Pig Slurry	19	pH, extractable P, extractable and total Cu and Zn, mycorrhiza
Christie and Beattie, 1989	Cattle Slurry, Pig Slurry	17	pH, C, N, microbial biomass C and N, extractable Cu and Zn
del Castilho et al., 1993	Cattle Slurry	11	Cd, Cu, and Zn speciation
Dormar and Sommerfeldt, 1986	Cattle Manure	10	C, P, C:N, enzymes, sugars, water stable aggregates, bulk density
Johnston et al., 1989	Farmyard Manure	140+	N
King et al., 1990	Swine Lagoon Effluent	11	pH, extractable P, K, Ca, Mg, Cu, NO_3-N
Kingery et al., 1994	Broiler Litter	15-28	pH, EC, C, N, NO_3-N, extractable P, K, Ca, Mg, Mn, Cu, Zn
Mårtensson and Witter, 1990	Farmyard Manure	31	N-fixing microorganisms
Mbagwu and Piccolo, 1990	Cattle Slurry, Pig Slurry	4-7	aggregate size distribution, C, N, P
N'dayegamiye, 1990	Cattle Manure	8	pH, C, N, P, K, CEC, organic matter
N'dayegamiye and Angers, 1990	Cattle Manure	8	water-stable aggregates, bulk density, soil water, microflora
N'dayegamiye and Côté, 1989	Pig Slurry, Cattle Manure	8	C, N, C:N, CEC, microbial populations, potentially mineralizable N, humic acids
Schjønning et al., 1994	Farmyard Manure	90	pH, CEC, C, particle size, bulk density, plasticity limits, water retention, confined uniaxial compression, drop-cone penetration, annulus shear
Sharpley et al., 1993	Poultry Litter	12-35	pH, C, N, NO_3-N, NH_4-N, P, bulk density
Sommerfeldt and Chang, 1987	Cattle Manure	12	organic matter, bulk density, water retention
Sommerfeldt et al., 1988	Cattle Manure	11	C, N, C:N
Spallacci and Boshi, 1985	Cattle Slurry, Pig Slurry	3-5	C, N, P, K
Stadelmann and Furrer, 1985	Pig Slurry	8	pH, C, N, CEC, microbial populations
Witter et al., 1993	Farmyard Manure	31	C, microbial biomass

(to 30-cm) from farmers' fields amended long-term (15 to 28 years) with poultry litter compared to fields with no history of manure application, signalling potential for higher N mineralization where poultry litter was added. Conversely, Witter et al. (1993) found higher soil C:N where FYM was applied long-term (31 years) than where soil was fallowed or fertilized with inorganic N. Other workers (Dormarr and Sommerfeldt, 1986; N'dayegamiye and Côté, 1989) encountered no changes in surface soil C:N owing to long-term manure additions but observed alterations in fractional distributions of soil C and N. In the Dormarr and Sommerfeldt (1986) study, 10 years of cattle manure application resulted in increased soil chitin-N as a percentage of total N, which presumably would increase formation of water-stable aggregates through its effect on filamentous fungi. They also found that moisture regime played a role in the humification process of manured soils as indicated by increased resin-extractable C in irrigated, manured soils compared to manured, non-irrigated soils. N'dayegamiye and Côté (1989) found increased (over a control) humic acid (HA) concentrations in soils (0- to 15-cm) amended with solid cattle manure for eight years. Long-term application of pig slurry did not alter soil HA concentration. Additionally, they found that E_4/E_6 ratios (an indication of degree of humification and stability of soil organic matter) had increased in cattle manure-amended soil but had decreased in soil on which pig slurry was applied. N'dayegamiye and Côté (1989) concluded that long-term applications of pig slurry could decrease soil organic matter owing to the chemical nature of pig slurry, i.e., low carbohydrate but high N concentrations leading to enhanced decomposition of native soil organic matter. They suggested that if maintenance or accumulation of soil organic matter is a goal where pig slurry is applied long-term, high plant residue producing crops should be grown (e.g., maize as opposed to potatoes).

Although long-term application of animal manures generally improves soil quality owing to increases in soil organic C and N, it can enhance the potential for degradation of water quality via NO_3 leaching. Kingery et al. (1994) found > 40 mg NO_3-N/kg soil at or near bedrock (3-m depth) under fescue pastures fertilized long-term with poultry litter. Conversely, Sharpley et al. (1993) observed no buildup of subsoil NO_3 under poultry litter-amended bermudagrass pastures. However, they sampled to 1.5 m; substantial soil NO_3-N accumulation was observed by Kingery et al. (1994) only between 1 and 3 m. King et al. (1990) applied swine lagoon effluent to bermudagrass for 11 years at rates to supply 335, 670, and 1340 kg N/ha/year, collecting soil samples to 2.1 m at the end of the study. Comparisons of treated soils with a non-treated control revealed that the two lower rates of effluent caused no buildup of subsoil NO_3, while subsoil NO_3-N concentrations were > 30mg/kg under the highest rate where 540 kg NO_3-N/ha had accumulated. Clearly, a need exists for development of reliable manure application recommendations that obviate potential NO_3 ground water contamination while promoting sustained crop production.

Soil reaction is an important soil property owing to the control it exerts over nutrient availability. Several recent studies have addressed the effect of long-term animal waste applications on soil reaction (Table 1), although results among studies are conflicting. Pig slurry had no effect on surface soil pH after 11 (Anderson et al., 1991) or eight (Stadelmann and Furrer, 1985) annual applications. Other workers (Christie, 1987b; Christie and Beattie, 1989; King et al., 1990) observed declining soil pH with long-term pig slurry use, particularly at high application rates. Long-term additions of cattle manure have increased soil pH (Christie and Kilpatrick, 1992; N'dayegamiye, 1990). However, Chang et al. (1990) observed a linear decline in soil pH of -0.004 units/ton cattle manure/year. Kingery et al. (1994) and Sharpley et al. (1993) found increases in soil pH of 0.5 and 0.6 units with applications of poultry litter for periods of 15 to 28 and 12 to 35 years, respectively. The impact of swine and cattle waste applications on soil pH remains unclear, but it appears that long-term poultry litter additions have the potential to raise soil pH.

Cation exchange capacity (CEC) of soils amended long-term with manure remains constant or increases in a linear fashion corresponding to manure type and application rate. Increasing rates of cattle manure increased soil CEC from 10.6 to 15.2 cmol/kg (N'dayegamiye, 1990) and 10.5 to 12.1 cmol/kg (N'dayegamiye and Côté, 1989) after six and eight years of application, respectively. Results regarding effects of pig slurry on soil CEC have been inconclusive. Stadelmann and Furrer (1985) observed that additions of pig slurry increased CEC by 2 and 5 cmol/kg on light and heavy soils, respectively, after eight years of application at 5 tons organic matter/ha/year. N'dayegamiye and Côté (1989), however, found that additions of pig slurry had no effect on soil CEC.

Electrical conductivity (EC, an estimate of soluble salts) levels increase for soils receiving long-term manure applications. Kingery et al. (1994) found significantly higher EC values in soils (to 3-m) under fescue pastures with a long-term history of poultry litter application compared to pastures receiving no poultry litter. Higher EC values were distributed in upper and lower portions of the soil profile, with lower EC values in the middle, suggesting soluble salt movement downward from upper horizons and accumulation in deeper layers. Chang et al. (1990) found increasing EC levels on soils amended with cattle manure for 11 years. Specific rates (dS/m/ton/year) of EC values increased with manure rate at a non-irrigated site but decreased at an irrigated site. Chang et al. (1990) also found that soil sodium adsorption ratio (SAR) had increased. They concluded that specific rate of change for SAR decreased with increasing manure rate, implying that soil CEC was approaching an equilibrium.

The use of feed additives for livestock and long-term manure application has led to significant changes in elemental concentrations in soils. Increased extractable P in the upper 30 cm of soil has been shown for all types of manure applied long term (Christie, 1987a,b; Christie and Kilpatrick, 1992; King et al., 1990; Kingery et al., 1994; Sharpley et al., 1993). Largest increases have been found in the upper 5 cm of soil, indicating potential for P movement via soil transport (Christie, 1987a; Kingery et al., 1994; Sharpley et al., 1993). Sharpley et al. (1993) found that the capacity of surface soil (0 to 30 cm) to adsorb additional P (P sorption index) was diminished by long-term poultry litter applications. A decreased P sorption index suggests a potential for increased soil-P mobility via transport in surface runoff or movement deeper in soil profiles.

Elevated levels of exchangeable cations are also associated with long-term manure applications. Calcium (Ca) increased in the upper 50 cm of manure-amended soil as compared to soil receiving no manure (Chang et al., 1990; King et al., 1990; Kingery et al., 1994). Magnesium (Mg) and potassium (K) also increased in the upper soil profile but were more mobile than Ca (Christie, 1987a; King et al., 1990; Kingery et al., 1994).

Additional cations of importance are the micronutrients copper (Cu) and zinc (Zn). Elevated levels of total and extractable Cu have been found in the upper 50 cm of soil resultant of poultry, swine, and cattle manure applications, with greatest increases in the upper 5 cm (Anderson et al., 1991; Christie and Beattie, 1989; Christie and Kilpatrick, 1992; del Castilho et al., 1993; Kingery et al., 1994). Largest increases have been associated with poultry and swine manures, likely owing to use of Cu as a feed additive. A similar assessment can be made of Zn accumulation in soils having long-term manure addition histories (Christie and Beattie, 1989; Christie and Kilpatrick, 1992; Kingery et al., 1994). It has been proposed that much of the Cu and Zn added in manures is chelated by organic constituents of manure and soil and that movement of these metals is governed by low molecular weight complexes in soil (del Castilho, 1993).

MICROBIOLOGICAL PROPERTIES

Accumulation of C and N, mineralization of N, P, and S, formation of stable aggregates, and many other soil transformations are controlled, in large part, by microorganisms. Long-term addition of organic substrates, such as animal manures, can alter structure and function of soil microbial communities. In a long-term study with pig and cow slurries (both applied at 0, 50, 100, and 200 m^3/ha/year for 16 years), Christie and Beattie (1989) found that soil (0- to 5-cm) microbial biomass C and N decreased with the high rate of pig slurry. The opposite was true for cattle slurry, i.e., soil microbial biomass increased with increasing rates. Amounts of C and N in soil biomass were strongly correlated with soil pH, which decreased and increased as pig and cattle slurry rates increased, respectively. Although microbial biomass C and N were negatively affected by the high rate of pig slurry, nitrifier populations as indicated by potential nitrification rates increased with increasing rates of pig and cattle slurries. In a similar study, N'dayegamiye and Côté (1989) determined the influence of pig slurry and solid cattle manure on microbial populations. They found that bacteria, actinomycete, and fungi numbers and potentially mineralizable N were increased (over a control) by eight years of pig slurry and solid cattle manure applications. Solid cattle manure increased microbial populations and activity more than pig slurry, likely owing to a C deficit in pig slurry compared to solid cattle manure. These studies and others (Stadelmann and Furrer, 1985; Witter et al., 1993) suggest that heterotrophic microbes controlling C and nutrient cycles in soils are influenced by long-term manure additions and that their activity is further modified by manure type and application rate.

Long-term manure applications have also been shown to impact populations of vesicular-arbuscular (VA) mycorrhiza and N-fixing microorganisms in soil. After 19 years of pig and cattle slurry addition to grassland (both applied at 50, 100, and 200 m^3/ha/year), Christie and Kilpatrick (1992) found that increasing application rate of both slurries produced a marked decrease in overall VA mycorrhiza infection of plant roots. Mycorrhizal infection was inhibited more by pig slurry than by cattle slurry. Decline in VA infection was accompanied by increased infection by the fine endophyte, *Glomus tenue*, an indication of environmental stress to VA mycorrhiza. Linear correlations of soil properties and VA mycorrhizal infection indicated that increasing soil P, Zn, and Cu concentrations were responsible for declining VA mycorrhiza populations. Since VA mycorrhizal infection is considered to have greatest impacts on plant nutrition in nutrient-poor soils, declining VA mycorrhiza populations in nutrient-rich, manure-amended systems may have little impact on plant biomass production. Mårtensson and Witter (1990) investigated impacts of long-term (31 years) FYM (4 t ash-free organic matter ha, applied every second year) and other organic and inorganic amendments on N-fixing soil microorganisms under a rotation of cereals, oilseed crops, and root crops; no legumes were grown in the field study. Blue-green algal N-fixation was higher in FYM-amended than unfertilized soils. Relative to the unfertilized control, FYM had no effect on numbers of clover rhizobia (*R. leguminosarium* biovar *trifolii*) or N-fixation by free-living microflora (*Azobacter* spp). However, based on all data gathered in the study, Mårtensson and Witter (1990) observed a positive relationship between soil pH and all N-fixing microorganisms studied. Their study suggests that manure types resulting in soil pH reduction will diminish N-fixing microorganism populations.

PHYSICAL PROPERTIES

Application of manures may alter composition and arrangement of soil solids, thus impacting fluxes of liquids and gases through soil systems. Several studies (Dormarr and Sommerfeldt, 1986; N'dayegamiye and Angers, 1990; Schjønning et al., 1994; Sharpley et al., 1993; Sommerfeldt and Chang, 1987) have shown that long-term addition of animal manure to soil lowers soil bulk density, indicating a positive impact on liquid and gas movement. Percent water-stable aggregates increased with rate of cattle manure, compared to an unfertilized control, after 10 annual applications to barley (Dormarr and Sommerfeldt, 1986). However, the increase was largely due to the presence of undecomposed manure; when undecomposed manure was manually eliminated prior to analysis, percent water-stable aggregates decreased at maximum manure rates. Cattle manure applied (rates of 0, 20, 40, 60, 80, and 100 Mg/ha every two years) for eight years on soil cropped to maize increased, had no effect, and decreased percent water-stable aggregates in the > 2-mm, 0.5- to 2-mm and < 0.5-mm size classes, respectively, proportional to rate (N'dayegamiye and Angers, 1990). Although the N'dayegamiye and Angers (1990) study indicates shifts in water-stable aggregates with increasing rates of cattle manure, Mbagwu and Piccolo (1990) found little effect of pig or cattle slurries on soil dry aggregate distribution after seven annual applications.

Perhaps the most comprehensive study regarding long-term manure application impacts on soil physical properties was conducted by Schjønning et al. (1994). For 90 years at the Askov Experiment Station in Denmark, FYM, inorganic fertilizer (NPK) or no fertilizer (UNF) were applied to a sandy loam cropped to a four-year rotation of winter wheat, beets or turnips, spring barley, and clover/grass mixture. Surface soil (0 to 20 cm) consistency limits followed treatment-induced soil organic C patterns (FYM > NPK > UNF), i.e., FYM soil was more friable at higher water contents than NPK or UNF soils. These data suggest that tillage may be possible under wetter conditions for soils receiving manure applications long-term. Surface soil bulk density was reduced under FYM and NPK compared to UNF, but volume of macropores (> 30 μm) was unaffected by fertilization history. Soil bulk density reductions under FYM and NPK were due to decreased particle density and increased volume of soil pores, particularly smaller pores that affect water availability to plants. The FYM and NPK soils had greater strength than UNF soil at comparable water contents, even though FYM and NPK soils had lower bulk density than UNF soil. Cohesive and frictional strength measurements indicated that UNF soil had a precipitous increase in internal friction and constant cohesion upon drying while FYM and NPK soils increased cohesion with decreasing water potential. Uniaxial, confined compression test data showed that compactability was greater for NPK soils than FYM or UNF soils. The Schjønning et al. (1994) study demonstrates that long-term manuring results in superior physical condition of soil in comparison to mineral-fertilized and unfertilized soils.

SUMMARY

Recent studies indicate that long-term land application of animal manure has a profound impact on soil properties. Positive impacts reported in the literature include soil organic matter buildup; increased soil pH and CEC with application of some manure types; increased soil fertility; greater activity of heterotrophic soil microorganisms with application of some manure types; greater blue-green algal N-fixation; and improvement of soil physical properties. On the other hand, long-term manure application has the potential to pollute ground water with NO_3-N and surface water with P. Long-term application of certain manure types, e.g., poultry litter and pig manure, may also promote unfavorably high concentrations of Cu and Zn in soil. These potential detrimental impacts signal the need for development of manure application guidelines that ensure sustained agricultural productivity coupled with environmental compatibility.

REFERENCES

Anderson, M.A., McKenna, J.R., Martens, D.C., Donohue, S.J., Kornegay, E.T., and Lindemann, M.D., Long-term effects of copper rich swine manure application on continuous corn production, *Comm. Soil Sci. Plant Anal.*, 22, 993, 1991.

Chang, C., Sommerfeldt, T.G., and Entz, T., Rates of soil chemical changes with eleven annual applications of cattle feedlot manure, *Can. J. Soil Sci.*, 70, 673, 1990.

Christie, P., Some long-term effects of slurry on grassland, *J. Agric. Sci. Camb.*, 108, 529, 1987a.

Christie, P., Long term effects of slurry on grassland, in *Animal Manure on Grassland and Fodder Crops. Fertilizer or Waste?* Van Der Meer, H.G., Unwin, R.J., Van Dijk, T.A., and Ennik, G.C., eds., Martinus Nijhoff Pub., Boston, Massachusetts, 1987b.

Christie, P., and Beattie, J.A.M., Grassland soil microbial biomass and accumulation of potentially toxic metals from long-term slurry application, *J. Applied Ecology*, 26, 597, 1989.

Christie, P., and Kilpatrick, D.J., Vesicular-arbuscular mycorrhiza infection in cut grassland following long-term slurry application, *Soil Biol. Biochem.*, 24, 325, 1992.

del Castilho, P., Chardon, W.J., and Salomons, W., Influence of cattle-manure slurry application on the solubility of cadmium, copper, and zinc in a manured, acidic, loamy-sand soil, *J. Environ. Qual.*, 22, 689, 1993.

Dormarr, J.F., and Sommerfeldt, T.G., Effect of excess feedlot manure on chemical constituents of soil under nonirrigated and irrigated management, *Can. J. Soil Sci.*, 66, 303, 1986.

Johnston, A.E., McGrath, S.P., Poulton, P.R., and Lane, P.W., Accumulation and loss of nitrogen from manure, sludge and compost: Long-term experiments at Rothamsted and Woburn, Chap. 10 in *Nitrogen in Organic Wastes Applied to Soils*, Hansen, J.A., and Henriksen, K., eds., Academic Press, New York, 1989.

King, L.D., Burns, J.C., and Westerman, P.W., Long-term swine lagoon effluent applications on 'coastal' bermudagrass: II. Effect on nutrient accumulation in soil, *J. Environ. Qual.*, 19, 756, 1990.

Kingery, W.L., Wood, C.W., Delaney, D.P., Williams, J.C., and Mullins, G.L., Impact of long-term land application of broiler litter on environmentally related soil properties, *J. Environ. Qual.*, 23, 139, 1994.

Mårtensson, A.M., and Witter, E., Influence of various soil amendments on nitrogen-fixing soil microorganisms in a long-term field experiment, with special reference to sewage sludge, *Soil Biol. Biochem.*, 22, 977, 1990.

Mbagwu, J.S.C., and Piccolo, A., Carbon, nitrogen, and phosphorus concentrations in aggregates of organic waste-amended soils, *Biological Wastes*, 31, 97, 1990.

N'dayegamiye, A., Effects of long-term application of solid cattle manure on silage corn production and soil chemical properties, *Can. J. Plant Sci.*, 70, 767, 1990.

N'dayegamiye, A., and Angers, D.A., Effects of long-term cattle manure application on physical and biological properties of a Neubois silty loam cropped to corn, *Can. J. Soil. Sci.*, 70, 259, 1990.

N'dayegamiye, A., and Côté, D., Effect of long-term pig slurry and solid cattle manure application on soil chemical and biological properties, *Can. J. Soil Sci.*, 69, 39, 1989.

Schjønning, P., Christensen, B.T., and Cartensen, B., Physical and chemical properties of a sandy loam receiving animal manure, mineral fertilizer or no fertilizer for 90 years, *European J. Soil Sci.*, 45, 257, 1994.

Sharpley, A.N., Smith, S.J., and Bain, W.R., Nitrogen and phosphorus fate from long-term poultry litter applications to Oklahoma soils, *Soil Sci. Soc. Amer. J.*, 57, 1131, 1993.

Sommerfeldt, T.G., and Chang, C., Soil-water properties as affected by twelve annual applications of cattle feedlot manure, *Soil Sci. Soc. Amer. J.*, 51, 7, 1987.

Sommerfeldt, T.G., Chang, C., and Entz, T., Long-term annual manure applications increase soil organic matter and nitrogen, and decrease carbon to nitrogen ratio, *Soil Sci. Soc. Amer. J.*, 52, 1668, 1988.

Spallacci, P., and Boshi, V., Long-term effects of the landspreading of pig and cattle slurries on the accumulation and availability of nutrients, in *Long-term Effects of Sewage Sludge and Farm Slurries Applications*, Williams, J.H., Guidi, G., and L'hermite, P., eds., Elsevier Applied Sci. Pub., New York, 1985.

Stadelmann, F.X., and Furrer, O.J., Long-term effects of sewage sludge and pig slurry applications on micro-biological and chemical soil properties in field experiments, in *Long-term Effects of Sewage Sludge and Farm Slurries Applications*, Williams, J.H., Guidi, G., and L'hermite, P., eds., Elsevier Applied Sci. Pub., New York, 1985.

Witter, E., Mårtensson, A.M., and Garcia, F.V., Size of the soil microbial biomass in a long-term field experiment as affected by different N-fertilizers and organic manures, *Soil Biol. Biochem.*, 25, 659, 1993.

A PERFORMANCE-BASED NON-POINT SOURCE REGULATORY PROGRAM FOR PHOSPHORUS CONTROL IN FLORIDA

Alan L. Goldstein and Gary J. Ritter

INTRODUCTION

In 1989, the South Florida Water Management District initiated a unique performance-based non-point source phosphorus control program in the Lake Okeechobee basin in south central Florida. The program, its legal basis and implementation strategy, have been described by the authors in previous publication references. Two years of additional implementation experience have occurred since the last publication. This chapter provides a brief summary of the program history and implementation strategy. It then describes interactions among regulators, the regulated, and the scientific support community to develop tools for assessing effectiveness of proposed corrective actions and to provide a process that would minimize the need for or likelihood of litigative enforcement actions. The willingness of the parties to work together cooperatively and the participation of the regulated parties has resulted in a stronger program with greater likelihood of achieving resource protection objectives than with litigative enforcement alone.

BACKGROUND

Lake Okeechobee is a 750-square-mile natural water body located in south central Florida. The lake is recognized as having a premier commercial and sport fishery that generates $28 million annually to the state's economy (Bell, 1987). It also serves as a home and haven for a variety of waterfowl and terrestrial wildlife. In addition, it functions as a water supply reservoir for agricultural irrigation, power generation, and municipal potable water supply for communities adjacent to the lake. It has an essential role in flood control through capability to back pump water into the lake from a number of basins on the southern perimeter. The primary sources of inflow are rainfall, the Kissimmee River, and surface runoff from 30 other tributary basins. The tributary basins (Figure 1) cover a surface area of over 2700 mi^2 (7000 km^2 or 702,500 ha) in portions of six counties. The land uses range from urban residential single- and multiple-family dwellings to large monoculture agriculture, dairies, and

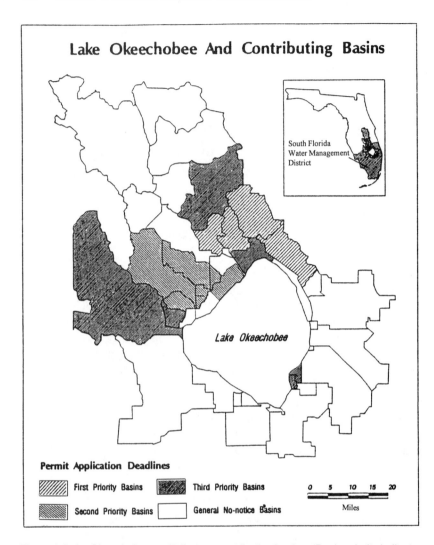

Figure 1. Lake Okeechobee and tributary contributing basins. (Basin priority indicates permitting and monitoring sequence.)

other animal husbandry industries. Large tracts of improved and semi-improved, moderate- to low-intensity-use pasture and native, non-improved areas remaining in a natural state are common. Primary outlets are evapotranspiration and a series of canals that convey water to the lower east and west coasts. With the exception of one inflow tributary (Fisheating Creek), all of the inflows and outlets are controlled by various gated structures or pump stations. These structures and the conveyance canals leading to and from the lake are the heart of the Central and Southern Florida Flood Control Project, a congressionally authorized public works project built by the US Army Corps of Engineers. The Central and Southern Flood Control District was created in 1948 as the local project sponsor. In 1976 that agency became the South Florida Water Management District (SFWMD), a regional water management agency created by the State of Florida and charged with the protection and management of the region's water resources.

The concern over preservation of the state's environmental resources became apparent in the early 1970s concurrent with the completion of some of the project's major channelization works. This concern led to a number of statewide panels and research/demonstration projects (Joyner, 1971; Marshall, et al., 1972; Davis and Marshall, 1975; Central and Southern Flood Control District, 1975; Huber et al., 1976; Federico et al., 1981; Goldstein, 1986) focusing on the health of Lake Okeechobee culminating in 1986 with a major status review by a select panel of scientists and agricultural industry representatives. This Lake Okeechobee Technical Advisory Committee (LOTAC) acknowledged the scientific information that indicated that Lake Okeechobee was being enriched by excess nutrients from agricultural activities in the lake's tributary basins. The committee recommended that steps be taken to reduce or eliminate these nutrient sources, especially from the more intensive animal use areas, particularly dairies (LOTAC, 1986). In response, then-Governor Bob Graham directed the Florida Department of Environmental Regulation (FDER) to develop and implement a "Dairy Rule" with the purpose of controlling and treating animal wastes generated from the 49 large dairy operations located in the lake's drainage basin.

Concurrent with these activities, the state legislature passed in 1986 the Surface Water Improvement and Management (SWIM) Act (Chap. 373.4595 F.S.), which directed the state's five water management districts (WMDs) to develop plans to preserve, protect, and, where necessary, restore all of the priority water bodies for which they had management responsibility. The legislation also directed the districts to implement the plans.

Lake Okeechobee was specifically identified as a priority water body. The SWIM legislation directed the SFWMD to reduce historical total phosphorus loading to the lake by 40% on an average annual basis by July 1992. The specific tonnage was developed using a modified Vollenweider lake eutrophication model developed during the course of scientific studies and evaluations (Federico et al., 1981).

The SFWMD produced the mandated resource management plan for Lake Okeechobee (SFWMD, 1989) that identified the strategy by which the agency intended to fulfill the legislative directive. The plan designated a phosphorus limitation standard for each sub-basin discharging into the lake. Further, each basin's loading limitation was to be achieved by setting phosphorus concentration limitations for off-site discharge from each individually owned land parcel. The limitations were not to be exceeded on an average annual basis. This plan called for a departure from traditional "technology based" regulatory programs that the FDER and the WMDs have historically relied upon to regulate surface water discharges. Instead it proposed a "performance-based" approach, the details of which are unique for a non-point source control program in the United States.

To ensure that the limitations are met, this regulatory program permits all non-exempt land uses in these basins, monitors the parcels for compliance with the target limitations, and forces non-compliers to achieve compliance either voluntarily or through imposing civil penalties and fees. This regulatory program was implemented by Administrative Rule (40E-61 F.A.C.) and is known as the Works of the District (WOD) - Lake Okeechobee Basin SWIM Rule. The key features of the program are as follows:

1. it regulates all uses of land areas greater than 1/2 acre in size;
2. it is retroactive (no one is "Grandfathered" in);
3. only total phosphorus discharges are addressed;
4. off-site concentration limitations are set for each parcel;
5. permitted parcels are monitored for compliance;
6. corrective actions by the owner are required for non-complying parcels.

PROGRAM IMPLEMENTATION

The program implementation strategy was designed to optimize the identification of problem phosphorus sources and thus focus on bringing them into compliance through effective corrective actions. This strategy is accomplished using four elements:

1. parcel identification, land use verification, and permitting;
2. water quality monitoring of permitted parcels;
3. requiring effective corrective actions (BMP implementation) for those cases in which phosphorus discharge limitations are exceeded; and
4. enforcement against parcel owners who do not voluntarily take adequate corrective actions to achieve compliance.

Permitting of all problematic parcels equal to or greater than 5 acres in size was essentially completed by the end of the third year of the program. Administrative maintenance of permit files is an ongoing activity as land uses and ownership changes occur. More than 600 individual permits have been issued under this program.

At least one year of monitoring on all individual permitted properties is required to document compliance status. Compliance is determined based on a twelve-month rolling average of biweekly grab samples taken at locations where surface discharge leaves each parcel. The process of determining the concentration standards that each parcel must meet to achieve compliance is codified in the Rule. Those parcels requiring corrective actions have been identified. As of June 1994, approximately 3% of the permitted properties were found to have one or more parcels with surface discharges not achieving the Rule's performance standards.

The language of the WOD Rule explicitly requires owners of non-complying parcels to take appropriate measures necessary to "ensure" that compliance with limitations is achieved and maintained. The reference to "ensure" is one of the strengths of the Rule. It underscores the seriousness of the intent of the agency to achieve the performance-oriented goals codified in the Rule. On the other hand, the regulations do not state, suggest, or imply what the appropriate measures are to demonstrate that the degree of assurance is provided.

Experience with the implementation of permitting and monitoring elements of the program has been described in previous papers (Goldstein and Ritter, 1992 and 1993). This chapter describes development and features of the corrective action assessment process for the non-complying parcels.

CORRECTIVE ACTIONS

Engineering solutions that rely on extensive works to catch, store, and treat storm water runoff are the only acknowledged methods that are demonstrated to be effective enough to "ensure" that compliance with phosphorus limitations can be achieved. The costs of design, permitting, construction, and long-term operation and maintenance of these systems make these solutions economically impractical for smaller family farms and operators. The size of these facilities would generally need to be quite large given the nature of southern Florida's average annual rainfall of 50 to 55 in. (127 to 140 cm), most of which occurs between the months of May and September.

Due to the prohibitive costs of engineered systems to control non-point source runoff, best management practice (BMP) implementation has been the strategy favored by the agricultural community. The shortcoming of reliance on BMPs is that there is no quantifiable degree of effectiveness to "ensure" that the phosphorus limitation can be met. During the initial permitting phase of the program, generic promises to implement non-verifiable farm management activities (fertilizer control, livestock rotation, etc.) were accepted as adequate actions. Where subsequent failure to achieve the phosphorus limitation was demonstrated, it was difficult to attribute the cause to inadequate practices or failure to implement them properly.

It was apparent that a more sophisticated approach was needed. Site-specific phosphorus audits and nutrient budget balancing appeared to be the best solution. SFWMD staff set out to formulate a process and develop the tools for performing site audits and quantifiable analyses to assess the effectiveness of corrective actions.

As expected, higher-intensity land uses seemed to be the cause of problems on many non-complying parcels. There were several cases, however, in which discharge from parcels with low-intensity land uses had unexpectedly high phosphorus concentrations. In most cases the parcel's historic land use had been intensive. It became apparent that the sources of excessive nutrients in runoff could be from either present-day, ongoing activities or leaching of the phosphorus from relic high-use areas. To adequately "ensure" that corrective actions would be effective, it would be necessary to attack both sources of the problem. To do this, all of the sources of the high concentration discharges had to be identified. Spatially intensive "synoptic" surface water quality monitoring surveys were performed on and around the problem sites. Survey results and known historical land uses were used to identify locations to collect soil sediment information to confirm the location, extent, and magnitude of the suspected hot spots.

Non-structural corrective action strategies applicable to those areas that are identified as relic hot spots through soil phosphorus analyses are those activities that eliminate input of phosphorus and maximize removal through vegetative uptake, harvest, and export from the area. This "common sense" approach encourages the owner to utilize the value of the nutrient in the soils to produce a crop, hay, or forage. This approach is usually economically feasible as a means to continually reduce the amount of phosphorus available for leaching to surface runoff. The time required to reduce the soil phosphorus levels to the point at which concentrations in surface runoff are acceptable becomes the critical issue involved when determining the adequacy of the activity.

Where unacceptably high concentrations are found in surface discharge and there is no evidence of hot spots in the soils, it can logically be assumed that current land use activities on the parcel are too intense, thereby overwhelming natural uptake processes. Effective corrective actions in these instances could be lessening the intensity of the activity, site-specific engineered water management solutions, or some combination thereof.

Non-complying permittees have three choices, each of which requires some degree of sacrifice. The first, conversion to a less-intense land use, implies loss of potential income. The necessity and degree to which land use intensity has to be reduced to meet the phosphorus limitations has been the source of most disagreement and potential for adversarial situations. Construction options, (the second choice) carries high initial costs for engineering, design, permitting, and construction and a perpetual commitment for potentially costly operation and maintenance. Such options are frequently out of the question for low-cash-flow operations typical of family farms and ranches. The third option, to do nothing, incurs the likelihood of regulatory legal actions that incur legal costs and can carry significant fines and penalties. Failure in the legal arena leaves the owner forced with having to implement one of the first two alternatives. Most non-complying permittees choose the management option, which forces a reduction of intensity as the preferable alternative.

There are two pieces of information that are critical to the success of relying on non-structural management options. The first is to know the level of land use intensity that can be expected to have a reasonable chance of successfully achieving the phosphorus concentration targets. The second information needed is an estimate of the amount of time it might require for the surface runoff from the parcel to achieve the phosphorus concentration targets once the new management activities are in place.

Owners of non-complying parcels, predominantly farmers and ranchers, need the former information to make decisions on the most economically feasible combination of corrective actions. Agency policy makers need the latter piece of information to make decisions on how robust corrective actions should be to be considered acceptable.

During the years 1986 through 1994, the District had funded research predominantly through the University of Florida Institute of Food and Agricultural Sciences (U of F IFAS). The purpose of this research was to identify the phosphorus sources and sinks in the Lake Okeechobee tributary basins, the factors that control phosphorus flux in the soils, and the development/enhancement of appropriate field- and basin-scale phosphorus transport models. Many of these studies were being concluded at a time coincident with the developing needs of the regulatory program.

At the request, and with the participation, of the regulated community, the District entered into an additional contract with U of F IFAS to develop and apply land use scenarios using the models to evaluate runoff quality impacts. The results of this effort (Campbell et al., 1993) served as the technical foundation for addressing the two critical questions. Using a 20-year historical hydrology data base, the contractor modelled steady state conditions with an area-specific model, FHANTM (Field Hydrology and Nutrient Transport Model), which is a materials transport component overlay on DRAINMOD, a commonly used field-scale hydrological drainage model.

Inputs were a variety of phosphorus loading scenarios that had been identified as common for various intensities of dairy, beef cattle grazing, and forage production activities in the area. Starting with steady-state phosphorus import and export conditions, the model was used to predict which, if any, of the identified scenarios would be able to achieve the phosphorus concentration targets over the 20-year historical period (Figures 2 and 3).

A second product from the effort was the development of a phosphorus concentration reduction curve with time. The initial conditions assumed that the soil phosphorus loads were characteristic of those found in high-intensity-use areas in herd staging pastures, adjacent to dairy milking barns. Phosphorus concentrations in runoff from these areas were predicted assuming cessation of dairy activity and change of use to another less-intense activity, such as beef cattle grazing, hay, forage production, and fallow land uses (Figure 4).

The model results showed that some activities, by nature of the imported

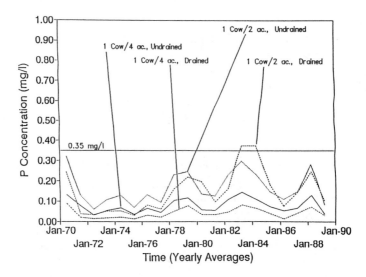

Figure 2. FHANTM simulation of surface runoff phosphorus concentrations for designated land uses. From Campbell et al., 1993.

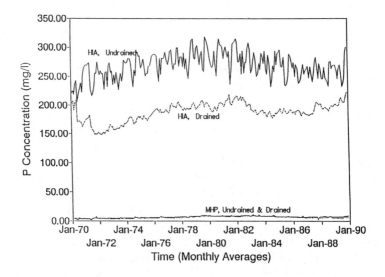

Figure 3. FHANTM simulation of surface runoff phosphorus concentrations for designated land uses. From Campbell et al., 1993.

phosphorus materials to maintain them, were unlikely to achieve target phosphorus concentrations in surface runoff. Other model runs showed that there were land use activities that could most likely achieve the phosphorus concentration targets. The models also showed that under the scenarios in which soils had been heavily loaded with phosphorus from historically intense uses, surface runoff may not achieve target concentrations even after 20 years unless aggressive activity for removing the phosphorus residuals is employed. The most aggressive hay/forage production scenarios modeled indicated that 12 to 14 years might be required to deplete the soil phosphorus loads to the point at which surface runoff can meet the target concentrations in the most highly intense use areas.

Using this information, a mutually agreeable assessment strategy and process were developed. The strategy has eight components. (1) Use FHANTM to evaluate impacts caused by current land uses and intensities. (2) Perform synoptic on-site assessments using event-based grab sampling of surface water to identify areal extent and locations where highest concentrations occur. (3) Use synoptic water quality survey results and historical land use information to identify the location of relic phosphorus "hot spots" caused by historically high intensity land uses or drainage activities. (4) Identify appropriate locations and collect soil samples to confirm the magnitude of the residual in the soils. (5) Where no evidence of relic phosphorus in soils is found, agree to permit land use activities up to the point at which FHANTM predicts target phosphorus concentration levels are likely to be exceeded. (6) Permit activities that are net phosphorus exporters in areas in which high phosphorus residuals are identified in the soils (hay, forage, crop production, etc.). (7) Continue to monitor and evaluate trends in phosphorus concentrations in runoff from the property for a specified period of time (usually one to three years). (8) Re-evaluate effec-

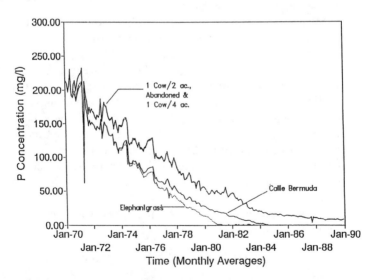

Figure 4. FHANTM simulation of surface runoff phosphorus concentrations following change from dairy high intensity to alternative land use. From Campbell et al., 1993.

tiveness of corrective actions at the end of the specified time. Parcels that still fail to achieve compliance or do not show continued improving trends may be required to implement additional and more robust corrective actions.

The administrative instrument used to formalize these actions is a legally enforceable Consent Agreement in which the permittee commits to implement and utilize the mutually agreed-upon corrective actions. The term of the agreement is limited to the one- to three-year period established for evaluation of effectiveness of the actions.

The regulated community has agreed to this process because it provides many desirable features. (1) A predictive tool is provided for making economic decisions on the best combinations of corrective actions to meet specific needs. (2) The model, which provides the technical basis for assessing corrective actions for adequacy, was developed by experts at the state land grant university. This provides the regulated agricultural community a degree of comfort in its accuracy. (3) The process provides a mechanism to consider time as a factor in the economics of achieving compliance. (4) The process and model constitute a common ground where the regulated community and the regulating agency seek agreement on what are acceptable corrective actions and that they are applied in the right areas. (5) Most of the field data collection efforts are conducted by the regulation agency and/or at owner's request by the Cooperative Extension Service, thereby minimizing consulting services and other expenses to the permittee. (6) The permittee understands that as long as he/she is implementing the corrective actions in accordance with the Consent Agreement, he/she is free from legal enforcement action based on violations of the Rule's phosphorus concentrations limitations.

Advantages to the regulatory agency are all of those for the regulated community. In addition, (1) A mechanism is provided to quantify the problem and the potential effectiveness of the corrective actions by developing site-specific phosphorus budgets with the assistance of the permittee; (2) the model results become the key basis for approving or denying proposed corrective actions. This eliminates the potential for staff to be accused of making arbitrary and capricious decisions in cases in which unacceptable corrective actions are proposed by the permittee; (3) a finite time limit is built into the Consent Agreement process. Once it expires, the agency has the opportunity to re-evaluate the site-specific case and, if necessary, to require the permittee to undertake additional actions. There is always the ability to enforce against an uncooperative permittee.

The process provides a mechanism for all parties to continue work to achieve resource protection in an economically viable, non-litigative manner.

SUMMARY AND CONCLUSION

The performance-based non-point source regulatory program for phosphorus control in the Lake Okeechobee, Florida, basin, initiated in November 1989, has been in operation for over five years. During this time the SFWMD has issued 640 permits covering 832,000 acres (336,840 ha) of land surface area. Field staff have monitored surface runoff at 672 sites for at least 12 months on each of these permitted areas to identify which parcels are not in compliance with the phosphorus concentration limitations established by the Rule. Thirty parcels on 22 permits have been identified as out of compliance. At the request, and with the participation, of the regulated community, the District contracted services with the University of Florida Institute of Food and Agricultural Sciences to utilize existing information and computer models to identify expected phosphorus concentrations in surface runoff from specified land use scenarios common in the basin. The model and the results have been incorporated into a corrective action assessment process used by the district to approve or deny proposed corrective action plans for non-complying parcels. The relationship between the district staff and the regulated community has developed into one of a shared partnership to identify and correct nutrient pollution source problems. To date the program has been judged an institutional success and remains a cornerstone of the SFWMD strategy to preserve and protect Lake Okeechobee's multiple beneficial uses for southern Florida.

REFERENCES

Bell, Frederick W., *The Economic Impact and Valuation of the Recreational and Commercial Fishery Industries of Lake Okeechobee, Florida,* Unpublished study by Department of Economics, Florida State University, Tallahassee, Florida, 1987.

Campbell, K.L., Tremwel, T.K., Bottcher, A.B., and Graetz, D.A., *Performance of selected BMPs for Phosphorus Reduction in High Phosphorus Source Areas,* Agricultural Engineering Research Report No. 93-LS for SFWMD, Institute of Food and Agriculture Sciences, University of Florida, Gainesville, Florida, 1993.

Central and Southern Florida Flood Control District, *Lake Okeechobee-Kissimmee Basin Proposals for Management Actions,* Ft. Pierce, Florida, March 20, 1975.

Davis, F.E., and Marshall, M.L., *Chemical and Biological Investigations of Lake Okeechobee: January 1973-June 1974 Interim Report,* Central and Southern Florida Flood Control District, March, 1975.

Federico, A.C., Dickson, K.G., Kratzer, C.R., and Davis, F.E., *Lake Okeechobee Water Quality Studies and Eutrophication Assessment,* Technical Publication #81-2, South Florida Water Management, West Palm Beach, Florida, 1981.

Goldstein, A.L., *Upland Detention/Retention Demonstration Project Final Report,* Technical Publication #86-2, South Florida Water Management District, West Palm Beach, Florida, 1986.

Goldstein, Alan L., and Ritter, Gary J., A Regulatory Program for Non-point Source Phosphorus Control in Basins Tributary to Lake Okeechobee, Florida, in *Proceedings of the First Annual Southeastern Lakes Management Conference,* March 19-21, NALMS, Florida, 1992.

Goldstein, Alan L., and Ritter, Gary J., A performance based regulatory program for phosphorus control to prevent the accelerated eutrophication of Lake Okeechobee, Florida, *Wat. Sci. Tech.* 28(3-5), 13, 1993.

Huber, W.C., Heaney, J.P., Bedient, P.B., and Bowden, J.P., *Environmental Resources Management Studies in the Kissimmee River Basin,* Dept. Environ. Eng. Sci., University of Florida, Gainesville, Florida, ENV -05-76-2, 1976.

Joyner, B.F., *Chemical and Biological Conditions of Lake Okeechobee, Florida, 1969-70,* US Geological Survey, Open file report 71006, 1971.

LOTAC, *Lake Okeechobee Technical Advisory Committee, Final Report,* Florida Dept. of Environmental Regulation, Tallahassee, Florida, August, 1986, revised November, 1986.

Marshall, A.R., Hartwell, J.H., Anthony, D.S., Betz, J.V., Lugo, A.E., Ver, A.R., and Wilson, S.U., *The Kissimmee/Okeechobee Basin--A Report to the Florida Cabinet, Tallahassee, Florida,* Division of Applied Ecology, Center for Urban and Regional Studies, University of Miami, Miami, Florida, 1972.

South Florida Water Management District (SFWMD), *Interim Surface Water Improvement and Management (SWIM) Plan for Lake Okeechobee* - Part I: Water Quality & Part VII: Public Information. West Palm Beach, Florida, 1989.

LIVESTOCK AND POULTRY PRODUCERS' WASTE MANAGEMENT PRACTICES AND ATTITUDES IN NORTH CAROLINA

Thomas J. Hoban and William B. Clifford

INTRODUCTION

Waste management is now recognized as an important social and political issue. The public is increasingly concerned about the odor and water pollution impacts of livestock and poultry waste. Practices are generally available for minimizing the impacts of waste on the environment. Not all producers, however, are willing and able to use such practices. To address public concerns, North Carolina recently enacted regulations that require some livestock and poultry producers to implement waste management practices within the next few years. These regulations are expected to have a significant impact on many producers.

This chapter examines livestock and poultry producers' attitudes and practices related to waste management. Results are from a statewide telephone survey conducted in early 1994 with over 1000 North Carolina producers. The sample included the following types of operations, selected at random from across the state: poultry producers (n = 415), swine producers (n = 410), dairy producers (n = 130), and beef producers (n = 76). For this chapter, information is presented about their adoption of waste management practices, as well as their attitudes about waste management and government policies. Differences in waste management practices and attitudes based on type of operation are also examined.

For the past 50 years, social scientists have studied the process by which farm operators and others accept and use new practices (Rogers, 1983). During the past 20 years, much attention has focused on farmers' adoption of soil conservation practices (Buttel et al. 1990). A subset of this work has specifically considered farm operators' use of practices to control non-point source water pollution (Hoban and Wimberley, 1992). The research reported here represents the first attempt to focus on farmers' use of poultry and animal waste management practices in the published literature.

Past research has shown that decisions about using new practices represent an ongoing social process for most farm operators (Korsching and Nowak,

1983). Water quality improvements depend on changes in farm operators' attitudes and behavior. Voluntary adoption of waste management practices means that farm operators must initially gain sufficient awareness of and interest in water quality problems and related issues (Napier et al., 1986). Next, farmers must develop favorable attitudes toward the recommended practices. Finally, these attitudes must be translated into adoption behavior. This project builds upon and expands this earlier work by a specific focus on adoption of livestock and poultry waste management practices.

WASTE MANAGEMENT PRACTICES

Livestock and poultry producers were asked about a variety of practices they might use for managing animal waste (Table 1). Land application is an important waste management practice. Over two-thirds of all respondents reported that their waste is applied to land. Dairy producers were most likely, and beef growers were the least likely, to apply their waste. One-third of all respondents said they sold or gave away all or some of their waste for use off the farm. This was particularly likely for poultry producers, but unlikely for the others. Just 40% of those who land applied waste had calibrated their waste application equipment during the previous five years. That practice was most common among poultry and swine producers. Just over half of those who land applied waste reported that they had buffer strips, terraces, or other water quality control practices installed on the fields where they apply waste. This was

Table 1. Adoption of waste management practices and related behaviors.[†]

Practice/Behavior	Type of Producer				Whole Sample	Sign of Difference[‡]
	Poultry	Swine	Dairy	Beef		
Apply waste to land (n=1033)	72%	63%	90%	34%	68%	.000
Sell/ship waste off farm (n=1002)	65%	6%	30%	12%	33%	.000
Calibrated application equipment (n=643)	45%	49%	17%	8%	40%	.000
Install BMP where apply waste (n=691)	55%	52%	70%	50%	56%	.011
Written waste management plan (n=1030)	28%	47%	44%	4%	36%	.000
Soil tested for nutrients (n=1002)	59%	64%	86%	84%	66%	.000
Waste tested for nutrients (n=1000)	42%	41%	37%	10%	38%	.000
Reduced fertilizer use (n=694)	93%	77%	89%	69%	86%	.000

[†]Number of producers answering each question varies depending on their answers to previous questions.
[‡]Significance levels based on chi-square test for statistical differences between groups. Values under .001 indicate that differences are very significant.

highest for dairy producers. Most (86%) of those who land applied said they had reduced their cost of fertilizer by using waste on their land. Poultry and dairy producers were most likely to have reduced fertilizer.

Another important aspect of waste management involves the extent to which waste management practices are carefully planned and monitored. One of the keys to effective waste management is to have a systematic waste management plan prepared by a professional resource manager or agricultural technician. Only about one-third of all the livestock and poultry producers reported having a written waste management plan. Swine and dairy producers were most likely to have such a plan.

Effective waste management planning will also require factual information about the nutrient value of both the soil and waste material. Two-thirds of all producers reported having had their soil tested for nutrients during the previous five years. The number reporting soil tests was higher for dairy and beef cattle producers. Just over a third (38%) of all producers reported having had their waste tested for nutrients during the previous five years. Beef producers were much less likely to report such testing than the other groups. Most of those who applied waste (86%) said they had reduced their use of chemical fertilizers by using waste on their land.

WASTE MANAGEMENT DECISIONS

A number of different factors can influence producers' willingness and ability to successfully implement waste management practices. This is particularly true for those practices that are relatively new or unfamiliar. Respondents were asked, "How important are each of the following in your decisions about which practices you choose to manage or use your animal waste--very important (3), somewhat important (2), or not important(1)?" Results are shown in Table 2, indicating any differences among the four groups of producers.

Table 2. "Very Important" influence on waste management decisions.[†]

Factor	Type of Producer				Whole Sample	Sign. of Difference[‡]
	Poultry	Swine	Dairy	Beef		
Water pollution control	66%	73%	70%	56%	68%	.010
Cost of the practice	54%	55%	63%	44%	55%	.011
Ability to grow profitable crops	53%	50%	72%	52%	54%	.000
Potential to reduce odor	50%	55%	47%	21%	50%	.000
Labor or time required	46%	46%	60%	40%	48%	.008
How easy practice is to use	44%	46%	52%	42%	45%	.004
Availability of land	43%	41%	46%	31%	42%	.221
Experience of other producers	24%	25%	24%	22%	24%	.399
Government cost-sharing	22%	31%	31%	14%	26%	.004

[†]Results show the percentage of producers who said that each factor had a "Very Important" influence on their decisions. The remainder said the factor was either "Somewhat" or "Not" important in their decision.

[‡]Significance level based on chi-square test for statistical differences between groups. Values under .001 indicate that differences are very significant.

The most important stated influence on producers' waste management decisions involves the ability of the practice to control water pollution. Over two-thirds of the whole sample said that was "very important." We find six other influences clustered in the middle, in terms of importance: cost of the practice; potential to reduce odor; ability to grow profitable crops; how easy practice is to use; labor or time required; and availability of land. Two factors appear less important, namely the experience of other producers and government cost-sharing. Dairy producers rated most factors as having the greatest influence, while beef producers tended to see them as less important.

Another important factor that could have an influence on producers' decisions about the use of waste management practices will be the amount of information available from various sources. We asked respondents to rate each of six information sources by asking "How much useful information have you received from the following about ways to use or manage waste--a lot (3), some (2), or none (1)?" Results are shown in Table 3.

Producers in this survey seem to be getting useful information from four of the sources. Farm magazines and newspapers were rated as the most useful source. Three other sources are seen as "useful" by over half the respondents: the Soil Conservation Service, the Cooperative Extension Service, and other producers. Tours and demonstrations provide useful information to only about one-third of the producers. Fertilizer dealers provided relatively little useful information about waste management. Dairy producers tended to report getting more information from each of the sources than did the other types of producers.

To determine the need for more information, producers were asked "Overall, how much more information or technical assistance do you need about how to better use waste?" Almost half (44%) of all producers said they needed "none." Another 20% said they only needed "a little" more information. Over a quarter (27%) indicated that they needed "some." Only 9% said they needed a lot more information. Swine and dairy producers were more likely to see the need for

Table 3. Sources of "Useful" information about waste management.[†]

Source	Type of Producer				Whole Sample	Sign. of Difference[‡]
	Poultry	Swine	Dairy	Beef		
Farm magazines or newspapers	66%	74%	82%	79%	72%	.001
The Soil Conservation Service	59%	70%	78%	49%	65%	.000
The Cooperative Extension Service	61%	66%	72%	67%	65%	.134
Other farmers	52%	56%	66%	50%	55%	.024
Tours and demonstrations	32%	40%	46%	46%	38%	.006
Fertilizer dealers	14%	18%	21%	26%	17%	.026

[†]Results show the percentage of producers who said they had received either "Some" or "A Lot" of information from each source. The remainder said they had gotten "None."
[‡]Significance level based on chi-square test for statistical differences between groups. Values under .001 indicate that differences are very significant.

additional information than were the poultry and beef producers. Most produc-ers seemed to feel they have all the information or assistance they need.

Other results also indicate satisfaction with current practices and general resistance to change. When asked directly, most respondents (82%) said they do *not* plan to make any changes in the way they manage waste. The intention to change was highest among dairy (36%) and swine (23%) producers. Most (79%) said they were "very satisfied" with their current waste management system in terms of water pollution control. Dairy producers tended to be least satisfied. In addition, over two thirds (70%) said they were very satisfied with their current waste management system in terms of odor control. Swine and dairy producers are generally much less satisfied with odor control and more willing to change than the poultry and beef producers. Analysis of the effec-tiveness of producers' waste management practices now underway should help indicate whether, in fact, their assessment is correct.

PUBLIC POLICIES AND ISSUES

Public policies and issues can have an important influence on producers' waste management decisions. Respondents were read the following statement: "The N.C. Environmental Management Commission recently passed new rules about livestock and poultry waste. These require that certain poultry and live-stock producers register with the state by December 31, 1993. These producers will then need to develop and implement a waste management plan." Over a quarter of all respondents had *not* heard of these rules. Awareness varies by type of operation. Almost half of all poultry growers and almost one fifth of all swine producers had not heard of the rules, compared to only five percent of the dairy producers.

Those producers who had heard of the rules (n = 761) were asked some additional questions. Almost all (92%) of the respondents indicated that they believed their operation was already in compliance with the regulations. Dairy producers were least likely to think so. However, less than two-thirds had al-ready registered their operations with the state (in 1994). Poultry and beef producers were least likely to have registered. It is important to note that many of those operations were not, in fact, required to register at the time of the interview.

Respondents' views about the impacts of these rules varied considerably. Over one-third felt the rules would have "a lot" of influence on the way they manage waste. However, one-quarter felt the rules would have "no influence." Swine and dairy producers saw their operations as being most effected. Over a third felt the regulations would have a negative effect on their profits. How-ever, 20% expected a positive effect on profits. The rest (26%) saw no effect on profits. Poultry and beef producers were most likely to believe that the effects of the regulations would be neither positive nor negative.

Several other questions examine producers' attitudes about public issues.

Most (78%) agreed that "Animal agriculture is being unfairly blamed as a cause of water pollution." Most (80%), however, felt that "Public concern over animal waste is really more about odor than about water quality." This was particularly true for swine and poultry producers. Although no research is available, anecdotal information (such as testimony at public hearings) suggests that odor is really the main issue of concern for non-farm rural residents. Earlier research in North Carolina does show that the public is relatively unaware of non-point source pollution (Hoban and Clifford, 1992). Most people associate pollution mainly with point sources and other visible problems (such as litter).

Producers do accept some government intervention in that the majority (68%) disagreed that "Producers should have the right to manage their waste in any way they choose." Swine producers were most likely to see limits on their rights. Just over a third agreed that "Taxpayers should help pay more for water pollution control on farms." Dairy producers were much more likely to feel that way. Under half (40%) agreed that "The new state regulations on animal waste are going to be impossible to enforce." Almost two thirds (61%) agreed that "Environmental laws are getting so strict that many producers will have to quit raising livestock/poultry." This was of particular concern to swine and dairy producers.

CONCLUSIONS AND IMPLICATIONS

This chapter has presented an initial assessment of the results from this major survey. Further analysis remains to be done for the information collected. For example, we will be evaluating the influence of farm size and structure, as well as demographic characteristics, on waste management attitudes and behavior. It is, however, possible to offer some conclusions and implications of this research.

North Carolina livestock and poultry producers seem to be aware of and concerned about animal waste management. However, this awareness and concern may not necessarily be reflected in their behavior. For example, only about one-third of all producers have a written waste management plan. Just over half have BMPs installed on the fields on which they apply waste. Most appear quite satisfied with their current systems. Almost half claim to have no need for additional information or technical assistance.

The groups of producers vary in terms of their attitudes and behaviors. This reflects unique characteristics of each type of farm. Many poultry growers seem to have already settled on the type of waste management system they use. Several reasons for this can be suggested. Almost the entire poultry industry in North Carolina is vertically integrated. Many of the growers' decisions are set by the company with whom they contract. Also, most growers have dry litter management systems that are basically excluded from the regulations. Many have their houses cleaned and waste hauled by contractors.

The swine industry has come under the most public criticism for their waste management. It is not surprising, then, that the swine producers seem to be most concerned about the regulations. Many are also looking for ways to improve their own waste management. Much of the public concern has actually been about odor instead of water quality. This appears to be reflected in their answers. Swine producers seem to have the most to gain from improved waste management practices, especially given the fact that almost all their waste is stored and used on their farm.

In some respects, dairy producers as a group seem to be most advanced and concerned about their waste management efforts. This may be due to the fact that they have a tradition of being more closely regulated, mainly from the standpoint of public health related to their dairy production facilities. The beef cattle industry in North Carolina tends to be relatively small in scale. Most of the beef cattle are kept on pasture rather than in a confined area. As a result, most of these producers are exempt from the regulations. They also are much less likely to have any centralized waste storage and disposal system.

Results of this research provide some guidance for Extension educational programs on waste management. Educational efforts can help build awareness and understanding of waste management technologies (including the importance of having a waste management plan, regular soil tests, and other technical data). Such programs should also try to help producers understand the potential implications of public policies and social issues. Educational programs for citizens and leaders would also be helpful to increase their understanding of the animal waste management challenges that producers face. Results of this survey should also be able to inform the direction of future public policies and minimize the potential adverse impacts on producers of regulations.

ACKNOWLEDGMENTS

Funding for this project was provided by the North Carolina Agricultural Research Service. This project involved an interdisciplinary team of Extension Specialists and others, including Jim Barker, Geoff Bensen, Bob Bottcher, Darwin Braund, Tom Carter, Roger Crickenberger, Frank Humenik, Bob Jones, Dale Miller, Morgan Morrow, Don Wesen, Kelly Zering, and Joe Zublena. The conclusions presented in this chapter are those of the authors and do not necessarily reflect those of others associated with this project.

REFERENCES

Buttel, Frederick H., Larson, Olaf F., and Gillespie, Gilbert W., *The Sociology of Agriculture*, New York, New York, Greenwood Press, 1990.

Hoban, Thomas J. and Clifford, William, *Public Attitudes toward Water Quality and Management Alternatives in the Albemarle-Pamlico Estuarine System (Phase II Report)*, Raleigh, North Carolina: North Carolina Department of Environment, Health, and Natural Resources, 1992.

Hoban, Thomas J., and Wimberley, Ronald C., Farm Operators' Attitudes about Water Quality and the RCWP, in *The National Rural Clean Water Program Symposium*, Washington, DC, U.S. Environmental Protection Agency, 1992.

Korsching, Peter F., and Nowak, Peter J., Social and institutional factors affecting the adoption and maintenance of agricultural BMPs, in *Agricultural Management and Water Quality*, F. Schaller and G. Bailey, eds., Ames, Iowa, ISU Press, 1983.

Napier, Ted L., Camboni, Silvana M., and Thraen, Cameron S., Environmental concern and the adoption of farm technologies, *Journal of Soil and Water Conservation* 41, 109, 1986.

Rogers, Everett M., *Diffusion of Innovations* (Third edition), New York, The Free Press, 1983.

AGRICULTURAL NON-POINT SOURCE WATER POLLUTION CONTROL VOLUNTARY PROGRAM IN MARYLAND

Herbert L. Brodie and Royden N. Powell, III

INTRODUCTION

Programs for water pollution control from agricultural operations in Maryland were initiated in the 1960s. Since that time the concept of volunteer action by farmers has been the foundation of efforts to cause change. Farmers consider themselves stewards of the land. It is important to nurture this feeling of stewardship and pride while providing reason for changing practice because the practice will then be self sustaining. Change takes time. It took more than 40 years for farmers to become fertilizer dependent. One should not expect a hasty break from that dependency. Change on the farm may appear to be too slow to some in this period of rapid development of pollution control objectives, but change is occurring. Recent Maryland programs are aimed at accelerating change on the farm through education, consultation, financial and technical support, and shared participation in watershed planning. For the most part, those farmers adopting change have become allies and some even spokesmen promoting the program objectives.

Maryland is located in the mid-Atlantic region of the United States and is bordered by Delaware and the Atlantic Ocean on the east, Pennsylvania to the north and West Virginia to the west, Virginia and the District of Columbia to the south. Maryland's total land area is approximately 9,800 square miles (25,500 km²). In 1990 the distribution by land use was 32% agricultural, 34% forest, 6% water and 28% urban or developed.

Maryland can be divided into three physiographic provinces that lie approximately parallel to the Atlantic Shore. The provinces are the Coastal Plain, the Piedmont Plateau, and the Appalachian. The Coastal Plain is divided by the Chesapeake Bay into the eastern and western shore--the former a nearly level plain and the latter being more dissected and rolling. The Coastal plain covers nearly 5,000 mi² (13,000 km³), or approximately one half of the area of the state. The Piedmont lies west of the Coastal Plain and consists of roughly 2,500 mi² (6,500 km²) of broad, undulating landscape with some ridges and low knobs with deep narrow valleys at streams. The Appalachian province covers about 2,000 mi^2 (5,000 km^2) of the Blue Ridge Mountains and the Allegheny Plateau to the border with West Virginia.

Surface water from 97% of Maryland flows to the Chesapeake Bay. The exceptions are a small area of the Allegheny Plateau that flows to the Ohio Basin and a portion of the lower eastern Coastal Plain that flows to the Atlantic Ocean. The Chesapeake Bay is the largest estuary in the contiguous United States and receives drainage from approximately 64,000 mi^2 (166,000 km^2) of territory in five states. The Bay proper covers more than 2,200 mi^2 (5,700 km^2), is approximately 200 miles (320 km) long by four to 30 miles (6 to 50 km) wide, and is surrounded by more than 620 mi^2 (1,600 km^2) of marsh.

AGRICULTURE

Agriculture in Maryland is practiced on 15,600 farms covering 2.2 million acres (8,900 km^2). According to 1992 statistics, the value of livestock production was led by broilers at 1.3 billion lb (0.6 million Mg) followed by milk at 1.4 billion lb (0.7 million Mg); cattle and calves at 86 million lb (44,000 Mg); hogs at 74 million lb (38,000 Mg); 855 million eggs; plus other poultry and livestock with total receipts of $804 million. Crops were led by grain corn at 58 million bushels - 470,000 acres (2.04 Mm^3 - 1,900 km^2) followed by soybeans at 18 million bushels - 545,000 acres (0.6 Mm^3 - 2,200 km^2); small grain at 17 million bushels - 290,000 acres (0.6 Mm^3 - 1,200 km^2); hay at 648,000 tons - 220,000 acres (660,000 Mg - 890 km^2); plus vegetables, fruit, greenhouse and nursery, tobacco, forest, and mushrooms with total receipts of $607 million.

The eastern Coastal Plain (Eastern Shore) provides most of the grain, vegetable, and hog production. The lower Eastern Shore provides for the majority of the broiler chicken production. The western Coastal Plain (Western Shore) contains dairy, livestock, and grain farms to the north and vegetable, tobacco, grain, and hogs to the south and is primarily urban in the Baltimore-Annapolis-Washington corridor. The Piedmont and Appalachian regions support most of the dairy and cattle farms plus hogs, eggs, fruit, and hay.

In 1992 agricultural land received an estimated 970,000 dry tons (986,000 Mg) of animal manure containing 23,000 tons (23,400 Mg) of nitrogen (as N) and 12,000 tons (12,200 Mg) of phosphorous (as P). Farms purchased an estimated 56,000 tons (56,900 Mg) of N and 11,000 tons (11,200 Mg) of P as inorganic fertilizer and received 1,500 tons (1,520 Mg) of N and 2,200 tons (2,240 Mg) of P from municipal biosolids. Annual water-eroded soil loss from agricultural land was estimated at 5 tons/acre (11 Mg/ha) in 1990. Erosion is considered a primary threat on 30% of all cropland and pastureland. Nonpoint loss of sediment and nutrients from agriculture is considered a significant contributor to decline of water quality in the Chesapeake Bay.

AGRICULTURAL WATER POLLUTION CONTROL

The Cooperative Extension Service of the University of Maryland (MCES) initiated animal waste management programs in the 1960s. These educational and technical assistance programs were aimed primarily at voluntary adoption of effective and economical manure handling and storage techniques for farms that had outgrown the conventional methods of the time. Water pollution control was limited to keeping manure sediments from entering streams that were very local to the farmstead. During this time the Natural Resources Conservation Service (NRCS) provided technical assistance with the design and construction of water and erosion control structures but had no waste management standards or specifications. The Agricultural Stabilization and Conservation Service (ASCS) provided cost-share funding for NRCS projects and did not include waste management as an approved practice. Maryland regulatory activity on farms was limited to human health issues.

In the early 1970s with the adoption of the National Pollution Discharge Elimination System (NPDES) and the establishment of water quality standards, Maryland water pollution control regulators in the Maryland Water Resources Administration (WRA) recognized that agricultural discharges had to be addressed. Faced with having to show some response, and having neither expertise in agriculture nor funds to support an agricultural program, WRA accepted MCES and NRCS voluntary activities for the solution to agricultural water pollution problems.

A 1971 Memorandum of Understanding (MOU) established that any agricultural waste management facility designed and operated according to the recommendations and specifications of MCES and NRCS was acceptable by the state. A system of state notification prior to animal waste containment or treatment system construction was adopted that provided WRA field inspectors with the location and practice being installed. A letter of approval (not considered a legal permit) was issued in a timely manner prior to construction. If animal waste facilities were to be constructed without MCES and NRCS involvement, the state required detailed plans signed by a professional engineer and a rigorous review.

Although rewritten several times, the intent of the 1971 MOU survived throughout the numerous internal and external reorganizations of Maryland's pollution control agency, which is now the Maryland Department of the Environment (MDE). These reorganizations included the establishment of the Maryland Department of Agriculture (MDA); the planning process for section 208 of the Clean Water Act; the adoption of Chesapeake Bay Initiatives; and other political, economic and environmental forces which shaped the policies and programs that exist today. The basis for adoption of pollution control practices on farms is voluntary in response to established goals.

MDE reserves the right to legal action against specific individuals identified as polluters and issues NPDES permits and Maryland Ground Water Discharge Permits to farms falling within the definitions of those regulations.

CURRENT PROGRAMS

Maryland's non-point program centers around the concept of a 40% reduction of 1985 nutrient input to the Chesapeake Bay by the year 2000 as agreed by the signatories of the Chesapeake Bay Initiatives. The agricultural component of this program relies entirely on voluntary adoption of Best Management Practices by farmers. The combined activities of MDA, MCES, NRCS, ASCS, and MDE with the local Soil Conservation Districts provide farmers with the information for decision making and the technical and financial assistance for practice implementation.

SOIL CONSERVATION AND WATER QUALITY PLANS

MDA and NRCS personnel develop soil conservation and water quality plans for individual farms for erosion and sediment control. Approximately 43% of the state has been planned with a maximum rate of about 200,000 acres/year (800 km^2/year). Planning is conducted through local Soil Conservation District offices. Soil Conservation District committees are farmer driven, giving them a stake in the planning function.

Plans have a 10-year life, and older plans must be reviewed and, possibly, rewritten. The updating process consumes staff time that could be used to expand the program. The current level of agency resources is sufficient to keep up with the review of old plans. The federal government supports this program through the USDA Natural Resources Conservation Service in Maryland with a $4.9 million annual budget. There are 136 NRCS staff persons working in Maryland. MDA supports the agricultural non-point source control program with an annual budget of $4.6 million, providing a total of 92 staff persons and operating support for local programs. Local governments also provide support to programs in local soil conservation districts. There are currently 57 positions in local soil conservation districts funded at the local level.

AGRICULTURAL COST SHARE

The Maryland Agricultural Cost Share (MACS) program, established in 1984 by MDA, provides cost-share money for the construction of agricultural pollution control facilities. Current levels of funding are up to 87.5% of the project cost with a lifetime cap of $50,000 per farm and a cap of $35,000 for animal waste systems. Projects must be designed by NRCS and must be approved by MACS prior to construction. Structures and practices eligible for cost-share range from manure storage to cover crop planting. Since 1984, MACS has paid out $33.5 million on 7,493 projects. During this same time period, $10.2 million has been provided for 1,156 animal waste projects. The costs for MACS program maintenance are not readily retrievable. While we spend $154,000 annually at MDA to run the program, we do not have any way to determine how much is spent locally in terms of staff support to the MACS program by local soil conservation districts.

Participation by farmers in the program varies depending on the agricultural economy as well as on the level of technical design/review staff and the availability of MACS funds. Farmers receive MACS funds with certain caveats. For example, waste storage requires a nutrient management plan, and poultry mortality compost structures require certified attendance at a half-day composting school provided by MCES. MACS funds can be supplemented with funds of the federal Agricultural Cost-Share Program (ACP).

The cost for installation of structures and practices required for a 40% reduction of nutrient release to the Chesapeake Bay was estimated as $336 million in 1990. This estimate did not include the public cost of planning, design, and program management.

NUTRIENT MANAGEMENT PLANNING

A nutrient management program was established by MDA and MCES in 1988 to train farmers how to utilize fertilizer, manure, and biosolid nutrients in an economically and environmentally sound manner. Nutrient management plans match crop yield potential with nutrient inputs on a field-by-field basis. MCES nutrient management consultants contact farmers; learn about the farm; sell the program; assist with soil and manure sampling for free analysis; develop optimum yield goals; prescribe nutrient application amounts, timing, and methods; assist with application equipment calibration; and write plans. Consultants also conduct workshops, maintain demonstration plots, and update previously written plans as well as participate in continuing education activities. The intensity of one-on-one contact required to sell and maintain this program may be considered beyond the scope of public agency involvement, but it is necessary to establish the credibility required to sustain the program.

In the period from 1988 to 1994, MCES consultants produced approximately 4,800 plans covering 360,000 acres (1,456 km²) and updated 2,600 plans covering 209,000 acres (845 km²) at a cumulative program cost of $3.2 million or about $670/plan and $9/acre. Although not all plans provide for nutrient reductions, the overall average change in nutrient use has been estimated as a decrease of 51 lb of nitrogen and 36 lb of phosphorous per acre (57 kg/ha N and 40kg/ha P) enrolled in the program, providing an average farm savings of $21/acre ($52/ha).

The high rate of plan updating interferes with expansion of the client base but may be a response to greater program acceptance by those farmers who allowed only a portion of their fields to be planned initially. It is possible that consultant input will decrease as farmers gain experience with nutrient management techniques.

The 40% nutrient reduction goal is to have 1.5 million acres (6,000 km²) under nutrient management by the year 2000. This cannot be achieved solely with public agency input, which would require almost $14 million. Farmers and others must take a more active role in the program. In 1993, MDA adopted an education and certification program for nutrient management consultants so that certified farmers can write their own plans and private consultants can provide nutrient management services. MCES consultants can then concentrate on education and demonstration and planning for those operators with whom the private consultant is unlikely to work because of size or other factors affecting profitability.

TRIBUTARY STRATEGIES

As a result of the 1991 Re-evaluation of the Chesapeake Bay Program, amendments to the Bay Agreement brought a greater focus to what was happening upstream in the individual watersheds. The State of Maryland was divided into 10 watershed basins for the purpose of planning a strategy that would lead the state to its 40% nutrient reduction goal by the year 2000. The plans or "tributary strategies" were to be comprehensive and unique to the combination of land uses and potential sources of pollution in each watershed. They include specific programs and practices, referred to as options, that address point and non-point sources of nutrients.

In order to provide input during strategy development, a Local Agricultural Tributary Team was formed in each of the 10 watershed basins. Teams were comprised of farmers, local agricultural agency staff, local government staff, agribusiness representatives, environmental interests, and farm and commodity group representatives. These teams have been the "grass roots" element for the agricultural component of the 10 strategies that were presented to the Chesapeake Bay Executive Council for approval in the fall of 1994.

The agricultural community's participation in this process has brought greater credibility to the estimates of what can be done to address non-point source pollution from agriculture. The agricultural community also has a great deal of ownership in the strategies, which should enhance the implementation of the plans. Through three rounds of public meetings, the agricultural perspective has been presented in a positive, proactive, and responsible manner that has served to educate the public about the capacity to address these issues through a cooperative approach.

EDUCATIONAL PROGRAMS

MCES provides educational programs to adults and youths, covering such diverse topics as basic environmental concepts, animal and crop management, farmstead planning, waste management, pesticide and chemical management, and many other related subjects. Youth programs are of particular importance because experience has shown that the opportunity for change occurs when the farming operation is being expanded or otherwise adjusted to accommodate the son or daughter who plans to stay on the farm.

IN DEFENSE OF VOLUNTEER PROGRAMS

Voluntary programs are necessary for several reasons. First, there are insufficient resources to implement mandatory practices on farms within a set time frame. The farms may not have the resources for installation, and agencies may not have the resources to plan, inspect, police, and otherwise administer a program. Second, in order to function, a pollution control practice must be maintained by an interested operator. Mandatory practices, by the very nature of their adoption, produce disinterested operators.

A waste management and nutrient control system is a combination of management, equipment, and structures that allows optimum management of the manure produced on a particular animal enterprise. A single system of set components cannot be developed that will fit the needs of all farms. Each farm is unique and requires its own customized system.

Customized systems are required because each farm has different production activities and goals, topography, existing structures, surface waterway locations, geology and ground water conditions, neighborhood influences, crop capabilities, management abilities, and economic resources. All influence the technology adaptable to that farm for animal waste and nutrient management.

A waste management system may require upland runoff control, a constructed stream crossing, water supply and animal drinking trough development, manure storage, and fencing. Manure storage may be an underground concrete tank, an above-ground concrete or steel tank with or without a roof, an in-ground earthen basin, a wooden-roofed structure, or some variation or combination of construction. The waste management system includes any equipment for manure transport from the animal area to the storage facility, for manure removal from storage, and for field application.

Some waste management systems include technologies for the treatment of manure. These technologies may include anaerobic digestion with the capture of methane for energy production, simple lagooning for nutrient reduction, and compost production. Each of these requires specific design and investment but may be warranted for specific sites.

The nutrient management plan dictates the characteristics of manure storage and other components of a new waste management system. Sometimes the existing waste management system and available land dictate the nutrient management plan.

Component construction must meet certain specifications. The Natural Resources Conservation Service has developed engineering specifications that must be followed for approved construction. Without adhering to such specifications, the potential for failure or catastrophic release of stored waste is increased. Engineers are often accused of over design that increases cost, but the added construction cost is minimal compared to the cost of failure.

With all of these aspects to consider, farmers are often left a little confused. They have to balance economic survival with environmental concern.

How much should be invested in the environment? How much of that investment can benefit economic survival? Because of these conflicts, the farmer must participate in the selection of waste management components for his/her farm. The greater the farmer's input, the greater will be the effectiveness of the system.

Of major importance is education and one-on-one discussion with that farm decision maker. An environmental practice will be adopted and improved when the farmer believes that the right thing was selected. Forcing the adoption of a practice on a farmer or, for that matter, on anyone leaves animosity. If it is believed that some practice will not work, then over a period of time through neglect or purposeful action that practice will fail.

The sales talk, advertising, and education will fail if the farmer perceives that a particular option is being forced. Good sales people listen, observe and develop an understanding of what a customer needs and then package the product to meet those needs.

If we want to sell waste and nutrient management concepts to farmers, we need to make them feel understood on an individual basis. They need to be able to see that we are there to help them survive as well as to help them contribute to the protection of the environment. We must do this through educational programs and person-to-person contact with all agencies working in a comprehensive, coordinated manner. Since 1989, public sector technical assistance has been provided to develop formalized nutrient management plans. With the addition of private sector agriculture professionals through a certification process developed in 1993, 25% of the cropland and pastureland is currently under a nutrient management plan developed by a certified consultant. This demonstrates the willingness of the farmer to participate in such programs, provided adequate technical assistance is available. Farmers must believe that they are part of the solution, not that they are the source of the problem.

It has been said that farmers are gamblers. They borrow money, put seeds in the ground or chicks in the pen and expect them to grow and yield enough to pay off the loan and provide an income. However, they gamble with a great deal of experience, and they use what has been successful in the past. Because they know farming is a gamble, it is difficult to make them change everything at once. Even the dependence they have on inorganic fertilizers took over 40 years to evolve.

Farmers cannot afford to make a big leap, so they make small steps. In nutrient management planning they start with a small field just to see what will happen. They may repeat the test for a couple of years before deciding to expand to other crops or to a greater portion of the farm.

Unfortunately, this kind of educational programming costs money, money that is currently scarce. Mandatory environmental programs that demand that some number of practices be adopted by some deadline fail to provide the dollars necessary to support the technical and educational activities required to make meaningful change. Without investment in this process of social change, we will be unable to meet the goals of the environmental program.

When we educate, we provide a reason to change. Once that reason is accepted, the change will occur by the actions of the educated. We will not have to revisit them to ensure that the change is permanent. However, when we force people to change, they learn how to avoid the changes and make no real progress to better ways. We have to continually revisit them to make the change permanent.

SUMMARY

Programs for water pollution control from agricultural operations in Maryland were initiated in the 1960s. Since that time the concept of volunteer action by farmers has been the foundation of efforts to cause change. Farmers consider themselves stewards of the land. It is important to nurture this feeling of stewardship and pride while providing reason for changing practice because the practice will then be self sustaining. Change takes time. Recent Maryland programs are aimed at accelerating change on the farm through education, consultation, financial and technical support, and shared participation in watershed planning. For the most part, those farmers adopting change have become allies and some even spokesmen promoting the program objectives.

AGRICULTURE AND WATER QUALITY IN CENTRAL NEW YORK'S FINGER LAKES REGION: REGULATORY VERSUS VOLUNTARY PROGRAMS

Judith L. Wright

INTRODUCTION

T he Finger Lakes Region, the result of glacial action as recently as 10,000 years ago, is located in the heart of New York State. Excluding the Great Lakes, the eleven Finger Lakes are the largest of the surface waters in this area and are considered to be the most prominent of New York's 4,000 lakes. This unique region is a system of complicated hydrology creating breathtaking views for visitors and area residents alike.

The seven Finger Lakes that are located in the east are within the Oswego River drainage basin. The predominant land use in this area is agriculture with woodlands following in terms of total area. All eleven of the lakes have varying drainage areas in terms of square miles, which do not relate to the size of the lake (Brower, 1992). Limnologists have divided the lakes into two categories based on lake depth and area. There are six major lakes and five minor lakes.

Water quality problems have slowly emerged as a result of steady growth and changes in development patterns around the lakes and within the drainage area. The need for land-use and lake management brings to focus questions that face society. Increasing demands for use are placed on the lakes, resulting in conflicting management goals and institutions competing for management control. It is interesting to note that the lakes are "sinks" and are affected by every activity occurring in the watershed (Brower, 1992).

Cayuga County is located near the geographical center of New York State as well as the in eastern part of the Finger Lakes Region. Skaneateles Lake forms part of the eastern boundary and Cayuga Lake part of the western boundary with Owasco Lake entirely within the confines of the county. All three lakes are considered major as defined by limnologists. The northern boundary is Lake Ontario. Cayuga County covers 699 miles or 446,360 acres of which 70% is farmland (Soil Survey, 1971). Dairying and cash crops are the principal types of farming, with much of the acreage in crops used to support dairy cattle and livestock. About 30% of the county is forested. Although much of the acreage is small, scattered woodlots, several thousand acres are reforested abandoned farmland.

Most of the soils are glacial deposits that contain various amounts of sandstone, shale, and limestone. For the most part, these soils are deep, gently to moderately sloping, and medium-textured. They are mainly well drained and are medium or high in lime content. They are generally well suited to the type of agriculture common in the county.

THE SITUATION/OVERVIEW

A trend that began over 100 years ago still continues today; there has been a decrease in farm numbers, yet agriculture has been of sustained importance to Cayuga County's economy. While the number of farm owners has decreased, the number of cows and crop acres has remained stable. Major construction on area dairy farms has occurred during the past five to 10 years, resulting in doubling and even tripling of herd sizes.

Recently enlarged farms often draw attention by their size. They do not conform to the public's picture of "Americana." Farms have become specialized; with this specialization comes a business attitude by the farm operator. Yet the desire to be good stewards of the soil and environment remains ingrained in the basic thread of the operator's management philosophy. Good stewardship of the farm's resources is necessary as that farm's livelihood is derived from the soil and surrounding environment. Farmers know that, if they mistreat their basic resources, their profits will decrease.

Farmers share the public's concern for the environment. In fact, they are taking steps to strengthen their environmental stewardship on the farm. The Gallup Organization recently surveyed 1200 farmers (Sandoz, 1993). The survey revealed that three out of five farmers have a heightened awareness regarding the environment compared to five years ago. Nearly one-third of the farmers ranked water quality as their number one concern. Of the farmers surveyed, 80% expected the government's involvement in farm environmental issues to increase, yet they clearly expressed their desire to regulate themselves. An informal survey conducted in Cayuga County showed results consistent with the Gallup results, and the desire to regulate themselves has proven true for Cayuga County farmers as well.

THREE VOLUNTARY PROGRAMS

OWASCO LAKE

Oswasco Lake has been the focus of much attention as a source of drinking water for over 50,000 people and for its recreational value. When a popular swimming area on Owasco Lake was closed during the summer of 1992 due to high coliform counts, agriculture took the brunt of public outrage. The county legislature assigned the Cayuga County Soil and Water Conservation District (SWCD) the development of a voluntary nutrient management program for dairy farms (CCNMP Draft, 1994). A committee representing a cross section of Cayuga County dairy farms met weekly during the winter months to discuss the issue of surface water contamination and how agriculture could address the problem.

Recognizing that there is no one source of non-point source pollution, these farmers developed guidelines for a county-wide voluntary program. This program utilizes best management practices in an effort to reduce nutrient run-off from farm land. Five farms volunteered to be part of the initial phase of the planning process, which evaluates whether the practices recommended 1) are economically feasible and profitable for that farm and 2) will reduce nutrient loading of the area streams and tributaries.

The leadership for the initial phase of the voluntary program was provided by Cayuga County Soil and Water Conservation District and Soil Conservation Service with Cornell Cooperative Extension of Cayuga County providing educational and research expertise to the effort. The five dairy farms that volunteered to be part of the initial process are a cross section of Cayuga County's dairy industry and can be categorized as large (400-500 cows) and small (75-100 cows).

Realizing that a total industry approach is needed, the working committee has been expanded to include agribusiness representation. Also, the consultants who work with farmers for a fee are a part of the committee as the nutrient management planning process is reviewed. A representative of the Owasco Watershed Lake Association (OWL) has also been an integral part of the planning process to date, and area newspaper reporters have sat in on some of the meetings.

SKANEATELES LAKE

Skaneateles Lake is well known for its pristine character and as an excellent source of drinking water for the City of Syracuse, population approximately 165,000. The water is currently not filtered; according to the 1986 Safe Drinking Water Act as established by Federal law, this water must be filtered or qualify for compliance under filtration avoidance criteria. Under the Federal Surface Water Treatment Rule, communities that rely on surface water sources may avoid filtration if they show that their water sources meet Federal and State raw water standards, that adequate disinfection is in place, and that an adequate watershed protection program to reduce risk of waterborne disease can be implemented. If these requirements can be met, the NYS Department of Health issues a letter that sets the criteria to be met to show that the community is implementing a program in the watershed, and this is referred to as avoidance criteria. Agriculture is one of several areas of an overall Watershed Protection Plan.

An Ad Hoc Task Force of farmers developed recommendations for Whole Farm Plans that serves as the agricultural community's recommendations on implementing Whole Farm Planning on its farms (SLAWPP, 1994). This voluntary program differs from the one previously mentioned for Owasco Lake in several aspects. First, this plan looks at the total farm while Owasco Lake evaluates only nutrient management. This is a result of the differences in the two lakes. Owasco Lake has prolific aquatic vegetative growth, but reducing the amount of nutrients reaching the lake may reduce the vegetative growth. While some speculate that recreational fishing on Skaneateles Lake would benefit from some nutrient introduction into the lake, the major focus of concern is turbidity. Although turbidity has approached significant levels, there is some question if it is the result of wave action within the lake or a result of erosion and sedimentation from the watershed. Since agriculture comprises 40% of the land area in the Skaneateles watershed, it is considered the major land use and may be a contributor to the turbidity.

In cooperation with Cornell University, a "Farmer Based Approach to Integrated Information in Management and Decision Making: Manure/Nutrient Balance for New York Dairy Farming" is underway. This project has provided an opportunity for an interdisciplinary approach to solving manure and nutrient management concerns for dairy farms. Research at Cornell has shown that, by adjusting the ration of dairy animals, the nitrogen content of the manure is reduced, and milk production generally increases. The result is a reduction in the nutrient loading on farms, which potentially have nutrient enriched fields. Two Cayuga County dairy farmers were chosen to participate in this extremely intensive study. At the time of this writing, nutrient management plans have been written and presented to the farmers for implementation in the 1995 cropping season. Both farmers are working with consultants; one is working with a crop management association, and the other is working with a consultant. In both cases the agri-service advisor has been made aware of the recommendations and is part of the decision process on that farm.

Although some farmers still feel they will not have to comply with some form of environmental regulation in the future, those taking a realistic look at the situation find voluntary programs developed by their peers a desirable alternative to government intervention and regulation.

REFERENCES

Brower, Robert Nelson, Water Resources Board Helps to Preserve Natural Beauty of Fingerlakes Region, *CLEARWATERS*, Summer, 13, 1992.

Cayuga County Nutrient Management Program (CCNMP), DRAFT, June, 1994.

Sandoz National AgPoll, Sandoz Agro, Inc., presented Washington, D.C. January 12, 1993.

Skaneateles Lake Agricultural Watershed Protection Program (SSLAWPP), Ad Hoc Task Force Report, DRAFT, June 2, 1994.

Soil Survey, Cayuga County New York, USDA-SCS, May, 1971.

PERSPECTIVE ON ALTERNATIVE
WASTE UTILIZATION STRATEGIES

Roland D. Hauck

*What is man, when you care to think upon him, but a minutely set, inge-
nious machine, for turning with infinite artfulness, the red wine of Siraz
into urine.* Isak Dinesen *Seven Gothic Tales* (1934)

INTRODUCTION

Of course, much more than wine is processed, and this remarkable machinery that converts one group of substances to another is not unique
to humans; it is possessed in one form or another by all living organisms, large or small, organisms that eat, digest, and excrete. On the micro-
scale, waste is assimilated in cropland soils, pastures, ranges, waters, woodlands, and within ruminants as well as within digester systems built by humans. On the macro-scale, the exponential growth in human population and
the rise in standard of living direct a corresponding increase in populations of
domestic animals grown for food. The consequent production and recycling of
biological wastes make increasingly greater demands on lands and waters as
waste-converting systems. What alternatives to land application can be made
available for relieving these environmental stresses?

Clearly, humans can do more that convert wine into urine. Presumably,
they not only cause and direct the production of wastes, but when challenged,
and *when it becomes imperative to do so*, they can improve old and discover
new ways of recycling wastes through naturally occurring and commercial food
chains and through human industrial economies.

The basis for this optimism lies inbedded in the phrase *when it becomes
imperative to do so*. Necessity *is* the mother of invention, and in the case of
waste recycling and disposal, necessity eventually will determine that some of
the inventions will in some way be made to be economically feasible. Market
forces will increasingly be required to adjust to pressures based on public concern for environmental quality, domestic animal welfare, wildlife habitat, food
safety, and related issues. In turn, public understanding of production economics will be requisite to the development of affordable and practical waste management systems that eventually will be needed to handle the ever-increasing
quantities of animal and other wastes being generated.

Specialization and intensification have been the major changes in animal husbandry and poultry production over the past 30-40 years. The need to improve animal performance and labor efficiency has led to mechanization of feeding, watering, and other care, with a consequent trend toward animal confinement, especially for broiler and layer poultry in housed systems. These developments have exacerbated the waste disposal problems as the increased number of animals is coupled with increased concentration of waste in relatively small, discrete areas.

Although land application (recycling through microbial systems) will remain the main approach to disposal of wastes from confined animals in the foreseeable future, many alternatives to this approach have been proposed, and several have been adopted to a lesser or greater extent. The intent of this chapter is to provide a brief overview of some viable alternatives, not in detail nor documented with numerous references, but presented in a manner permitting certain observations to be made and conclusions to be drawn. Whether correctly or incorrectly, this is to be done not by one experienced in the art of science of waste management but from the viewpoint of one from outside looking within. The publications cited at the end of this chapter are among the many references that have been consulted to accomplish this intent.

ALTERNATIVE STRATEGIES

BIOGAS GENERATION

Many systems ranging in size from extensive lagoons to the individual privy are in use for treating animal, farm, feed and food processing, and municipal wastes by aerobic or anaerobic means or by a combination of approaches. Intensive anaerobic microbial treatment is attractive because (1) the processes involved are confined to a small area and (2) in addition to waste amelioration, fuel [methane (CH_4)] is produced. Such treatment stabilizes the organic matter, reduces odor, has a low nutrient requirement, and reduces the overall volume of waste, i.e., waste biological sludge production is low.

Anaerobic fermentation occurs in three stages: (1) enzymatic hydrolysis of complex molecules by facultative bacteria, (2) formation of organic acids, mainly acetic acid, and (3) gas production by mesophyllic methanogenic bacteria. A consortium of microorganisms participates in numerous processes that convert diverse complex organic substances into simpler forms, including carbon dioxide (CO_2) and CH_4. Depending on the substrate, hydrogen sulfide (H_2S) and considerable ammonia (NH_3) may be formed, both of which may accumulate to concentrations that inhibit methanogenic activity.

Study of anaerobic fermentation in ruminants discloses that the sequential processes occur in different compartments of the stomach and gut. During digestion in a body environment of 37 °C, the food slurry may contain about 15% solids and 15% crude protein (depending on diet). In addition to the conventional single- or two-stage systems, single-chamber digesters have been designed that simulate the compartmentalized sequence of processes occurring in the ruminant digestive system, e.g., sequential processing can be achieved in an elongated digester in which animal waste is fed in at one end, moved along with gentle transverse mixing, and discharged at the opposite end. Control of parameters for optimal performance in each section of the digester is possible. Fresh material can be quickly heated to about 35ºC and maintained at this temperature; the C/N ratio of slurry solids can be adjusted to 20-30; pH can be maintained between 6-8; NH_3 can be kept at concentrations that do not inhibit or stop methanogenic activity (1200 mg/L and 1700-1800 mg/L, respectively); and loading rates can match digester capacity with (1) animal waste properties and (2) quantity of volatile solids added vs. quantity already present.

Biogas can burn directly in air but cannot be used interchangeably with commercially available natural gas or liquid petroleum gases because of its high CO_2 content (about 40%), plus other gas impurities. Raw biogas cannot readily be liquefied on a small scale (the critical pressure is >5000 psi). Because the cost of removing CO_2 (e.g., via absorption) can be as high as 25% of the total product cost, CO_2 recovery and use may give added value to animal waste treatment using anaerobic digesters.

Successfully operating anaerobic treatment systems requires (1) a relatively constant supply of substrate waste conforming to system design criteria, (2) heat and other energy inputs, (3) machinery, and (4), to be cost-effective, considerable operational knowledge and management skill. Despite these requirements, biogas-producing digesters using cattle or swine wastes have been built and are in operation worldwide, mainly in developing countries of the tropics or subtropics. Information on the value of the CH_4 produced often is not reported or, at best, is anecdotal. For competitive farm enterprises, an important consideration in the design of the digester system is whether anaerobic treatment of the animal waste is to be made mainly for pollution control or for energy production. If the latter, then the biogas energy should be converted conveniently and economically to a form that the farm can use, and a continual supply of this energy should be needed. It can readily be seen that the cost-effectiveness of a digester as a fuel source very much depends on the cost of other available fuels.

Considering the need for process control and the expanded opportunity for using process by-products, economic generation of CH_4 from animal wastes appears (at least for US producers) to be more suited to large-scale integrated operations than to individual farm enterprises. This observation will be revisited later in this paper.

ANIMAL WASTE FEED PRODUCTS

Recycling animal waste as feedstuffs makes direct use of protein, amino acids, and fiber that otherwise would be lost were the waste to recycle through soil. Poultry excreta (high in protein), waste from poultry and finishing cattle (high in digestible nutrients), and swine wastes have characteristics acceptable for refeeding. Most work has been directed toward making recycled feedstuffs for ruminants because microbial activity in their digestive systems efficiently converts non-protein nitrogen and fiber. Feeding processed animal excreta to cattle is not universally acceptable. For example, such practice is considered unnatural in Europe and is banned in the U.K. However, in the US, more than 20 states have regulations that permit the marketing of animal waste as a feed ingredient.

Broiler litter has been the preferred waste for use as a feed ingredient. Not all broiler litter is acceptable for use. Litter should be free of dirt and stones; contain no toxic amounts of chemical residues (medicines, growth stimulants); be free of pathogens; contain >3% total nitrogen and > 8% crude protein, of which <25% is insoluble (bound to acid detergent fiber); and contain <30% potential ash. Dried caged layer waste and ensiled beef cattle excreta and poultry litter also can be used as a feed ingredient. Substances that have been used for ensiling animal excreta include corn forage and grain, hay, molasses, blood, and rumen contents (semi-digested feed substances in animal stomachs after slaughter). Waste management practices such as handling and storage can markedly affect the value of the waste as a potential feed ingredient. For example, heat generated during stacking of poultry litter may contribute to nitrogen loss via volatilization and to increase in the amount of nitrogen associated with the acid detergent fiber fraction, i.e., increase in percentage of insoluble nitrogen.

Although interest is growing in the processing of cattle, poultry, and swine wastes as feed, commercialization has been limited by processing cost and because the value of the final product offered for sale is reduced as a result of its lower nutrient content, reasons that reduce product competitiveness with other feed.

COMPOST

Composting is an aerobic treatment of wastes, mainly by thermophylliic bacteria (temperature range 45-70°C). The waste material usually is a solid of low water content placed in piles, covered containers, or windrows in such a manner that air can diffuse into the mass, be blown or drawn through it, or can be incorporated by mixing the composting materials. The process is exothermic and requires no additional heat input, but temperatures should be maintained at >55°C to kill pathogens and <75°C to prevent destruction of beneficial microorganisms. Water content of the compost mixture should be kept at about 50%. The end result ideally is a stable, dark, humic-like substance, friable, of lower density and volume than the original waste, and free of noxious odors, toxins, and viable weed seeds. Nitrogen loss during formation may be negligible to substantial (e.g., as much as 50%), depending upon composting method used and management.

Composting methods range from simple passive systems that require little or no attention to closed reactors with precise control of aeration, water content, and mixing. Simple stacking in unsheltered piles is not recommended because both aerobic and anaerobic fermentation can occur, resulting in a slow rate of decomposition, nitrogen loss through NH_3 volatilization and/or nitrification followed by denitrification, odor formation, and other undesirable effects. Choice of method and scale of operation depend on a variety of factors, including volume and type of animal waste, space available, cost of aeration or turning equipment, building costs, labor requirement, and end use of the compost. External factors such as federal and state legislation also may be important in this regard, as discussed later.

Where composted wastes are to be land-applied, compost quality centers upon ease of application and nutrient content. Although the benefits of compost applied to soil can readily be demonstrated (improved aeration, drainage, and water-holding capacity; reduced compaction; and improved plant growth), these benefits currently offer little incentive for large-scale, farm-site composting of most animal wastes. On the other hand, composting is attractive for solving special problems, such as the disposal of litter and dead birds from poultry operations. Alternate layering of carcasses with about 15 cm each of straw and/or poultry (or other) litter to a height usually not exceeding 2 m produces an acceptable (but not bone-free) product, provided that suitable control of aeration, temperature, and water content is maintained in the composting mass. Despite the initial cost of the sheltered composting facility and front-end loader or other turning or mixing device, this method of dead bird disposal is cost-effective in time; moreover, the process is relatively odor-free, does not attract pests, and, generally, is more environmentally sound than alternate disposal methods. The compost can be sold or applied to the producer's land, either in lieu of fertilizer or as a fertilizer supplement. Where a market for clean poultry litter exists for use as a feed ingredient, other waste carbon sources can be used.

A large-scale, off-farm market for compost has not yet developed. Potential high value markets are for compost use in nurseries, greenhouses, and home gardens, and for mulching, landscaping, soil conditioning, and for transplanting shrubs and trees. Requisite to developing such potential is to improve the art and science of composting and to develop composting procedures and standards that will result in products of known quality and characteristics designed to meet various market needs. Legislative attempts thus far have failed on federal and state levels to mandate use of composted materials on road banks and for land reclamation, but such use represents a large potential outlet.

AQUACULTURE

Along both banks of many rivers in Southeast Asia can be seen large cages with confined fish. Stabled near these cages are cattle, goats, sheep, and swine, their excreta feeding the aquatic life of water coursing through the cages. This ancient practice of aquaculture (variations of which have been used by the Chinese for several thousand years) is the basis for developing aquatic reclamation systems using livestock wastes to feed bacterial, phytoplankton, and zooplankton communities as the first links of a food chain. Growing fish in ponds fertilized with liquid animal manures is common worldwide. The Chinese have developed polyculture systems in which combinations of several fish species, all filter-feeders differing in food and feeding habits, more efficiently use fish food (including detritus) of different partical sizes.

Excreta from a variety of livestock have been used to improve the productivity of fresh water fish ponds. Manures from buffalo, cattle, chickens, goats, horses, sheep, and swine have been used in brackish water shrimp ponds. However, some animal wastes are more suitable for use in aquaculture than others. For example, dairy waste contains straw, pesticides, disinfectants, and large volumes of water. Poultry broiler waste is not fluid, being removed mainly as solid by mechanical means. However, swine waste flushed from piggeries is particularly suited for use in an integrated system of which aquaculture is a component.

An increasing interest in regulating aquaculture pond effluents and concern for conserving water should stimulate development of systems in which effluents can profitably be recycled. The need for such systems is even more apparent with effluent carrying the residues of animal waste added to enhance aquaculture productivity. Where effluent re-use is desirable, where effluent treatment becomes necessary, or where the accumulation of fish plus added excreta limits the productivity of ponds, these problems can be alleviated using a sequential treatment recycling system. For example, fish waste water can be led into a constructed wetland containing a sand bed planted to water chestnut (*Eleocharis dulcis*). Waste water circulation can be continuous or intermittent during the water chestnut growing season. Fresh manure waste can be added after fish harvest and the sand beds reconstituted after harvest of chestnut corms and hay. Numerous variations of such integrated aquaculture production systems can be devised. All are greatly affected by economy of scale, larger systems generally being more profitable than smaller ones. Valid comparisons of reported cost analyses cannot justifiably be made because the analyses were made at different times in different countries for different systems. For example, various cost estimates considered by the author as reasonable suggest that, for optimum economic operation, pond area sizes can be as small as 1 ha or, from other studies, should be as large as 8 ha for 100 pigs. Such discrepancies are not surprising considering the wide differences in construction design and costs, operational costs, and market value of end products.

BIOMASS PRODUCTION

Algae. Although animal wastes can be used for such purpose, growing bacterial biomass as a source of protein is not attractive because of the highly indigestible structure of bacterial cells. For this reason, much of the protein of swine manure, which may comprise 30% of the manure dry weight, is not readily available to animals and many microorganisms such as algae because the manure protein is locked within bacterial cells, protected by tough walls containing techoic acid and other structural materials difficult to digest. In addition, swine manure has an undesirably high lignin content.

Algal cell dry matter consists of about 50% protein, which is readily digestible. Much of the interest in growing algae stems from its potential as a source of protein or other biochemical. Productivity can be high, up to 25 times the protein yield from an equivalent area of corn (*Zea maize*). Algae are cultured not in slurries of raw manure but in diluted liquid fractions or acid hydrolysates of animal wastes, including swine wastes. Of the thousands of microalgal species available, many are suited for use as food and feed. However, a typical production process would involve acid hydrolysis of stored animal wastes, solids separation (lignins and other insoluble substances), neutralization, heavy (toxic) metals precipitation and removal, algal culture, and harvest, followed by final processing, as needed.

Obviously, the algal by-product, to be economically competitive must have a value substantially greater than the value of alternative uses for the animal wastes, e.g., for use as a nutrient source for land application or as a feedstuff ingredient. Development of a high-quality feedstuff from algae would be welcome, i.e., a product having value commensurate with the cost and degree of sophistication needed for acceptably pure algal production, harvest (there is a current lack of convenient, economic harvesting methods), and processing. *Spirulina platensis*, used as food in Africa for thousands of years, shows promise, based on its currently expanding market in Japan as both feed and food. Alternatively, specific genera or species of algae can be grown for extraction of high value substances. For example, blue-green algae of the genera *Anabaena*, *Anabaenopsis*, and *Nostoc* are particularly high in protein; some *Chlorella* species have high lipid contents; species of *Scenedesnus* and *Chlamydomonus* are high in polysaccharides, as are the marine algal species of *Nitzschia* and *Porphyridum*; and *Phaeodactylum* has a unique fatty acid content and balanced protein composition. The industrial chemicals that can be derived from these cultures are useful in such materials as paints, drying oils, emulsifiers, and thickeners, among others. Other integrated schemes have been and can be suggested, limited in theory only by the imagination, e.g., processes to produce algae, yeast, biogas, and sludge. Assuming biological and mechanical problems can be overcome, such schemes prove to be economical (in the absence of subsidies or other cost-mitigating factors) only when the end products are unique, meet special market needs, or can be produced more cheaply than alternative products.

Duckweed. An invasion of duckweed (*Lemma minor*) can rapidly cover the surface of ponds enriched with natural marsh effluent or nutrient-bearing waste water. Duckweed is fast-growing and prolific, with species such as *L. gibba* being capable of producing up to 10 metric tons/ha of high-quality protein (high content of lysine and methionine). Fish ponds can be inoculated with duckweed as feed for herbivorous tilapia (e.g., *Oreochromis niloticus*; *O. aureus*), or can be used as a livestock feed ingredient. However, sufficient information is not yet available to evaluate the potential of different species of duckweed as scavengers of nutrients from animal wastes and waste waters or as feed or high protein feed amendments.

Insects. The author has observed night watchmen in Africa sweeping up insects that fell to the ground below lights that burned through the night. The yield was low but sufficient to supplement their diet. Elsewhere in Africa and Asia, yields of insects harvested from wastes have been reported sufficiently high to feed domestic animals. An objective of some current research is to grow soldier fly (*Stratiomyia* spp.) or house fly (*Musca domestica*) larvae on chicken or cattle manure, respectively. The edible larvae contain >40% protein and, after drying and processing, are palatable to cattle, chickens, pigs, and fish as a feed supplement. The marketing goal is to substitute insect larvae for corn and soybean as feedstuff ingredients.

DISCUSSION

PRIVATE AND PUBLIC SECTOR OBJECTIVES

The foregoing synopsis of alternatives to direct land application of animal wastes should suffice to illustrate the wide range and intensity of effort in progress to accomplish either or both of two broad objectives: (1) to dispose of animal wastes in an economical, environmentally sound, and socially acceptable manner and (2) to treat the waste as a recyclable resource. The two objectives are not mutually exclusive; meeting them will require developing a complex mix of interrelationships among private and public interests. In the long term, as the trend toward producing more livestock on less land continues, the overarching objective must be waste volume dilution by dispersing it more uniformly than now throughout the earth's atmosphere, hydrosphere, and terrasphere. For example, a given volume of animal waste fed to livestock generates additional waste, perhaps about 80% of the original mass that was fed. Such recycling through the animal slows down the rate of solid animal waste accumulation but, in the long term, results in mass reduction only if the rate of growth in use of recycled feed is faster than the rate in growth of animal numbers.

When animal waste management no longer addresses only materials handling problems but also problems of waste reduction to sustain acceptable environmental quality, costs rise accordingly. New questions now are asked: Who pays? Who is involved? Who decides? What is the role of government? To what extent can apparently conflicting objectives among individuals or groups with diverse interests be met while maintaining a viable animal production industry? These questions, of course, are not unique for animal production; they are being asked about all waste-producing activities. They will continue to generate policies and actions that will seek to balance environmental quality expectations with need and costs.

INTERNALIZING EXTERNALITIES

The animal production industry is no different from other economic enterprises, large or small, in that producers allocate scarce resources to maximize personal benefits against costs according to market signals. Third-party costs and benefits (costs and benefits external to the enterprise and market action) usually are not considered by the producer in making cost/ benefit decisions. Because some adverse effects of production occur outside of the producer decision-making process, the producer receives personal benefits from external costs (e.g., cost of pollution abatement) imposed on someone else (society). Some of the cost of pollution control initially is absorbed by the environment without usually being apparent to society. Eventually, when the effects of pollution become of *societal concern* and the costs of control are not borne by the producer, pollution may increase through failure to internalize the external costs (i.e., permitting external costs to substitute for internal costs may lead to pollution).

On the other hand, society receives benefits from some imposed production costs for which society is unwilling to pay its share (e.g., society may support the imposition of product or environmental quality control measures but not support product price increases or other cost/share measures). Thus, society can internalize costs external to itself. The process of bringing internal costs into the decision-making process is inherent in making sound policy. Such policy requires compromise and fosters production at optimum levels and with what is considered at any given time to be a permissible level of waste discharge.

Several strategies can be adopted to stimulate the internalization of external costs. Producers can be encouraged to change their behavior through education, being made aware not only of all production costs but also of alternative approaches to reducing costs and optimizing production. Society can be made aware of its responsibility in internalizing the costs of benefits received by supporting market incentives, such as cost/share programs, subsidies, permits, and taxes. Finally, regulation forces internalization of costs.

ROLE OF INTEGRATOR

Clearly, many of the alternatives to land application of animal wastes have greater opportunity to be economically viable when pursued on a large scale, not only because of the economies of scale inherent in a particular process, but also because, on a large scale, several processes can be linked together to make maximum recovery of all by-products of the combined processes. Thus, heat from an alcohol production facility can be used to heat water in fish ponds that receive waste water from biogas production that has been purified in a hydroponic system. Biogas CO_2 can be separated and re-circulated to enhance greenhouse production or be solidified as dry ice. Volatilized NH_3 can be absorbed and used as a nitrogen source. Animal waste not used in the biogas digester can be converted into feed along with by-products of the ethanol production.

On a less complex scale, an integrator located centrally in a livestock production area can accept some responsibilty for the waste produced by animals under contract [the integrator would be internalizing (into his/her cost of doing business) some or all of the producer's cost of waste disposal)]. By so doing, the integrator demonstrates an economic interest in preserving the base for his/her business (the producer). On a more complex scale, the integrator could (1) help transport the animal waste; (2) provide large, central storage and composting facilities; (3) provide the money to establish a waste recycling facility; (4) provide the labor to operate the facility; (5) provide the means for sophisticated monitoring and improving the operation of thefacility, i.e., increase the intensity of in-house management; (6) make existing markets available and establish new markets; and (7) help educate producers about technical, economic, and regulatory issues.

The integrator would benefit not only as a processor and distributer of poultry or livestock but also through receipt of low-cost waste that could be converted into higher-value products and, by being sensitive and responsive to growing environmental pressures, gain leverage in negotiating practical solutions to environmental problems. Although the producer would lose some independence, being part of such a multi-use operation would alleviate cash-flow and waste disposal problems. However, the integrator probably would need to work on minimum margin and be faced with the risk of losing production when any single component of the production complex fails. Adequate insurance against economic loss and measures to mitigate the costs of compliance with federal and state regulations probably will be needed to stimulate the establishment of such large-scale waste disposal and re-use operations. Such measures would include an infusion of public monies (e.g., through subsidies, tax credits) and promulgation of regulations that make the best compromise among environmental quality and production objectives. Important in this regard is increased cooperation between integrators and their trade associations and among related trade organizations.

ART, SCIENCE, AND ENGINEERING

The scientific bases for several of the ancient arts of animal waste management and use have progressed to the point at which current practices are limited more by mechanical problems than lack of knowledge. This observation does not minimize the need for continued identification of new microbial systems and for improved understanding of microbial biochemistry and energy conversions in the cycles of nature. Research benefits both small- and large-scale operations. But the above observation does emphasize the need for increased association of biologists, chemists, and engineers in planning and designing efficient, large-scale waste-conversion systems. If an ultimate objective of society is to prevent the undesirable accumulation of waste by reducing its volume, then existing technologies need to be improved and new ones developed, technologies such as hydrogenation, hydrogasification, or other thermochemical processes involving high temperatures and pressures. The cost-effectiveness of pyrolyzing animal and other wastes improves as environmental costs are internalized by society, *when it becomes imperative to do so.*

PUBLIC RELATIONS

Environmetal quality and food safety issues breed misconceptions, misinformation, and consternation in both private and public sectors of society. Education and increased dialogue among groups with diverse interests and objectives are medications for this societal disorder. Every interested person should be aware of the trade-offs required to sustain a relatively cheap and plentiful variety of animal and other food products. All should be aware that for waste disposal and other environmental problems, there are no solutions, only alternatives. The alternatives relate to how and in what concentration and form wastes are dispersed on land, in waters, and into the atmosphere.

Because the response of public opinion to emergencies and perceived personal threat often creates public policies, the producer has interest in lessening the threat before such policies adversely affect his/her enterprise. One approach to lessening this threat is to inform the consuming public in interesting and persuasive ways of the many remarkable means by which animal waste problems are being addressed, from alleviating the bio-security problem with dead-bird composting to establishing complex multi-process facilities. Of interest, if packaged effectively, might be the disposal of waste petroleum or trinitrotoluene by composting these substances with chicken litter, or the search for and potential use of psychrophillic bacteria.

The public image of animal waste as a resource to be used is vague, if perceived at all. In any educational program, a little bit of humor is well received. Yet the humor that has evolved about manure and its various other names reflects rather negative thoughts about a product that almost everyone manages on a daily basis. Thus, worthy of note are Crappy Critters, Dung Bunnies, and Poopets, molded out of dried, composted manure into forms of cats, ducks, rabbits, swans, turtles, and the like, for use in flower pots and gardens, gradually eroding away and degrading, releasing their nutrients. Their impact on reducing the enormous volume of the nation's animal waste is negligible, but the creatures *are* profitable and they *do* present a positive image.

PARABLE

A cock has great influence on his own dunghill.
Pubilus Syrus (ca. 1 BC)

Every cock is proud on his own dunghill.
John Haywood *Proverbs* (1546)

Yes, humans can take pride in the quality and quantity of their food animals, and even more pride when all, humans and their animals, are well fed. As humans are masters over the dunghills they are instrumental in creating, so they can influence the manner of the final disposition of these dunghills.

REFERENCES

Agricultural and Food Processing Waste, Proc. Sixth Intl. Sym. on Agricultural and Food Processing Wastes, Chicago, Illinois, American Society of Agricultural Engineers, St. Joseph, Michigan, 1990.

Agricultural Waste Management and Environmental Protection, Proc. 4th Intl. Sym. of CIEC, Vol. I, Braunschweig, FRG, International Centre of Ferilizers (CIEC), Vienna, Austria, 1987.

Animal Waste Treatment and Utilization, Proc. Intl. Sym. on Biogas, Microalgae & Livestock Wastes, Chung, P., ed., Council for Agricultural Planning and Development, Taipei, Taiwan, 1980.

Agricultural Waste Utilization and Management, Proc. Fifth Intl. Sym. on Agricultural Wastes, Chicago, Illinois, American Society of Agricultural Engineers, St. Joseph, Michigan, 1985.

Foster, T. H., personal notes from seminar on concept of externality, 1994.

Managing Livestock Wastes, Proc. 3rd Intl. Sym. on Livestock Wastes, Urbana-Champaign, Illinois, American Society of Agricultural Engineers, St. Joseph, Michigan, 1975.

National Livestock, Poultry and Agriculture Waste Management, Proc. of the National Workshop, Kansas, Missouri, American Society of Agricultural Engineers, St. Joseph, Michigan, 1991.

Overcash, M.R., Humenick, F.J., and Miner, J.R., Utilization, in *Livestock Waste Management,* Vol. II, CRC Press, Inc., Boca Raton, Florida, 1983, chap. 2.

ENERGY PRODUCTION FROM ANIMAL WASTES

Phillip C. Badger, Janice K. Lindsey, and John D. Veitch

INTRODUCTION

L ike other biomass resources, fuel from animal wastes (manure) can be obtained directly by direct combustion or can be converted into a gaseous fuel by either biological or thermochemical processes. Animal manures may also be converted into liquid fuels such as ethanol; however, there is very little literature on the topic. Therefore, liquid fuels from animal manures will not be covered in this paper.

THERMOCHEMICAL CONVERSION

Manure has been used for centuries around the world for fuel. Today, many foreign countries still use dried manure for cooking, and anaerobic digesters are commonly used to generate biogas for cooking in India and China. In the United States, the primary use of directly combusted or gasified manures is for relatively large-scale, independent power production. Typically, a grate combustion system is used in the United States to burn manures.

Table 1 provides information on manure production and characteristics. Devices for the thermochemical conversion of manure must take into account its higher ash content, higher moisture content, and greater slagging potential relative to other biomass materials. It is recommended that the fuel and ash characterization analyses listed in Table 2 be conducted as the basis for thermochemical conversion system design.

Higher volumes of ash inherent in manure should not be a problem if accounted for properly in the initial system design. Problems occur primarily when conventional combustion units are used without modification for handling manure.

In comparison to green wood at 50% moisture content and dry wood at 10% moisture content, the moisture content of *fresh* manure can range from 68 to 92%. Frequently, depending on livestock production practices, manure is mixed with bedding materials, thus changing its moisture content and physical and chemical characteristics. Due to the energy required to evaporate the water present before combustion can occur, materials over 65% moisture content will not sustain flame on their own. Thus manure over 65% will require either

Table 1. Fresh manure production and characteristics per 1,000 kg live animal mass per day (standard ASAE D384.1, 1993).

Parameter units*			Dairy 640 kg†	Beef 360 kg	Veal 91 kg	Swine 61 kg	Sheep 27 kg	Goat 64 kg	Horse 450 kg	Layer 1.8 kg	Broiler 0.9 kg	Turkey 6.8 kg	Duck 1.4 kg
						Typical live animal masses							
Total manure‡	kg	mean§	86	58	62	84	40	41	51	64	85	47	110
		std. dev.	17	17	24	24	11	8.6	7.2	19	13	13	**
Urine	kg	mean	26	18	**	39	15	**	10	**	**	**	**
		std. dev.	4.3	4.2	**	4.8	3.6	**	0.74	**	**	**	**
Density	kg/m³	mean	990	1000	1000	990	1000	1000	1000	970	1000	1000	**
		std. dev.	63	75	**	24	64	**	93	39	**	**	**
Total solids	kg	mean	12	8.5	5.2	11	11	13	15	16	22	12	31
		std. dev.	2.7	2.6	2.1	6.3	3.5	1.0	4.4	4.3	1.4	3.4	15
Volatile solids	kg	mean	10	7.2	2.3	8.5	9.2	**	10	12	17	9.1	19
		std. dev.	0.79	0.57	**	0.66	0.31	**	3.7	0.84	1.2	1.3	**
Biochem. oxygen demand, 5-day	kg	mean	1.6	1.6	1.7	3.1	1.2	**	1.7	3.3	**	2.1	4.5
		std. dev.	0.48	0.75	**	0.72	0.47	**	0.23	0.91	**	0.46	**
Chem oxygen demand	kg	mean	11	7.8	5.3	8.4	11	**	**	11	16	9.3	27
		std. sev.	2.4	2.7	**	3.7	2.5	**	**	2.7	1.8	1.2	**
pH		mean	7.0	7.0	8.1	7.5	**	**	7.2	6.9	**	**	**
		std. dev.	0.45	0.34	**	0.57	**	**	**	0.56	**	**	**

*All values wet basis.

†Typical live animal masses for which manure values represent. Differences within species according to usage exist, but sufficient fresh manure data to list these differences were not found.

‡Feces and urine as voided.

§Parameter means within each animal species are comprised of varying populations of data. Maximum numbers of data points or each species are: dairy, 85; beef, 50; veal, 5; swine, 58; sheep, 39; goat, 3; horse, 31; layer, 74; broiler, 14; turkey, 18; and duck, 6.

@All nutrients and metals values are given in elemental form

#Mean bacteria colonies per 1000 kg animal mass multiplied by 10¹⁰. Colonies per 1000 kg animal mass divided by kg total manure per 1000 kg animal mass multiplied by density (kg/m³) equals colonies per m³ of manure.

** Data not found.

Table 2. Methane potential and fuel and ash analysis (Miles, 1994; Owens, 1994)

Fuel and Ash Analysis

• Source, animal
• Moisture Content, % wet basis
• Type of Bedding Material
• Estimated % bedding material, dry basis
• Viscous liquid_____ or Solid_____
• Bulk density, lb/cf

The following ASTM procedures will be used to determine fuel composition. These ASTM methods have been used for previous fuel analysis by most biomass plants. (Methods underlined are preferred.)

	Biomass	Coal
Calorific Value	D2015[†] E711	D2015
Proximate composition		D3172
Moisture	E871	D2013, D3173
Ash	D1102	D3174
Volatiles	E872/E897	D3175
Fixed Carbon	By difference	By difference
Ultimate analysis		D3176
C,H	E777	D3178
N	E778	D3179
S	E775	D4239, D3177
Cl	E776, AOAC 969.10	D2361

Water soluble alkali (K, Na, Ca)
    Soak overnight in water @ 90° C. Analyze by AA

Residual or Ash Elemental Composition:
    Special precautions must be taken to prepare ash fuel samples for elemental analysis so that some of the constituents will not be volatized. Microwave digestion or wet ashing methods are preferred. ASTM D4278, AOAC or U.S. Bureau of Mines.

Ash preparation (600° C)	D1102
Ash elemental	D3682, D2795
(Si, Al, Ti, Fe, Ca, Mg, Na, K, P, Cl, CO2)	

Optional:
    Ash sinter test (observed sintering or fusion during ashing)
    Ash fusion temperatures   D1857

Methods of Measurement:
    The following procedures are recommended for the analysis of biomass fuel and its composition:

continued

Table 2. continued

ASTM E870	Standard Test Methods for Analysis of Wood Fuels
ASTM D1102	Standard Test Method for Ash in Wood
ASTM E711	Standard by the Bomb Calorimeter Test Method for Gross Calorific Value of Refuse-Derived Fuel
ASTM E775	Standard Test Method for Total Sulfur in the Analysis Sample of Refuse-Derived Fuel
ASTM E777	Standard Test Method for Carbon and Hydrogen in the Analysis Sample of Refuse-Derived Fuel
ASTM E778	Standard Method for Nitrogen in the Analysis Sample of Refuse-Derived Fuel
ASTM E871	Standard Test Method for Volatile Matter in the Analysis of Particulate Wood Fuels
ASTM D3178	Standard Test Method for Carbon and Hydrogen in the Analysis Sample of Coal and Coke
ASTM D1756-04	Standard Test Method for Carbon Dioxide in Coal

Methane Potential Analysis

ASTM E 1196-87	Standard Test Method for Determining Anaerobic Biodegradation Potential of Organic Chemicals (Biochemical Methane Potential Assay)
APHA, 1989*	Standard Methods for the Examination of Water and Wastewater for the following:

Total Solids (TS)
Volatile Solids (VS)
pH
Electrical Conductivity (mmhos/cm)
TKN
NH3-N
Total P
Total K
Total C
Total S
*Or a more recent version, if available.

†ASTM methods underlined are preferred.

flame stabilization for combustion, dewatering, or use in biological conversion systems that are better suited to these high moisture content materials. If flame stabilization is necessary, the device generates no net energy and becomes an incinerator.

Drying devices using heat for manure are typically not cost effective while efficient mechanical dewatering devices are expensive to purchase and typically cannot achieve moisture contents below 50%. Most manure used for energy in the United States comes from confined, high-intensity livestock production systems. In these systems, bedding is usually not used for cattle and swine, and the manure naturally dries to some degree before collection and use --unless hydraulic flush manure removal systems are used.

Broilers, turkeys, horses, sheep, and goats, if kept confined, will all use bedding. Animal bedding materials commonly used are straw, rice and peanut hulls, and sawdust or wood shavings. The bedding materials used depend on local customs, cost, and material availability. Since these bedding materials are relatively dry (< 25% moisture) and their proportion relative to manure usually is large (> 50%), their use reduces the overall mixture moisture content and serves as a form of dewatering.

The biggest problem with thermochemical conversion of manure with or without bedding materials is slagging. Slagging occurs when the fusion temperature of an ash is lower than the combustion temperature. This condition causes the ash to melt during the combustion process and to deposit on firebox and heat transfer surfaces. Slagging is primarily due to the presence of two alkali metals, potassium and sodium, and silica, all elements commonly found in animal feeds (Miles et al., 1993). Potassium and sodium metals, whether in the form of oxides, hydroxides, or metallo-organic compounds, tend to lower the melting point of ash mixtures containing various other minerals such as silica (SiO_2). The high alkali content (up to 35%) in the ash from burning crop residues such as straw lowers the fusion or "sticky temperature" of these ashes to as low as 1300° F (in comparison to wood ash at 2200° F). Research has shown that even small percentages (10%) of some of these high-alkali residues burned in conventional boilers will cause serious slagging and fouling in a day or two, necessitating a system shutdown.

Research by Miles et al. (1993) has led to a method used in the coal industry to roughly classify various materials relative to slagging and deposit formation. The method involves calculating the weight in pounds of alkali (K_2O + Na_2O) per million BTU in the fuel as follows:

$$\text{Slagging Index} = \frac{\text{lb Alkali}}{\text{BTU/lb}} = \frac{1 \times 10^6}{\text{MM BTU}} \times \% \text{ Ash} \times \% \text{ Alkali of the Ash}$$

$$\text{where Alkali} = K_2O + Na_2O$$

This method combines all the pertinent data into one Slagging Index number. A Slagging Index number below 0.4 lb/MM BTU is considered a fairly low slagging risk. Values between 0.4 and 0.8 lb/MM BTU will probably slag, with increasing certainty of slagging as 0.8 lb/MM BTU is approached. Above 0.8 lb/MM BTU, the fuel is virtually certain to slag and foul. Slagging Indexes for various manures are not known; however, research by the Tennessee Valley Authority (TVA) on broiler litter found a Slagging Index above 6 and ash fusion temperatures above 2100° F (Lindsey, 1994). Thus, the poultry litter tested should not slag if combusted below 2100° F; however, if the ash is exposed to temperatures above 2100° F then slagging would occur very quickly.

Direct combustion of biomass and thermochemical gasification of biomass are similar processes. Gasification differs by limiting the amount of oxygen present during combustion. Typically about 33% of the air required for combustion is used for gasification (Rajvanshi, 1986). Sub-stoichiometric conditions prevent complete combustion of the volatiles driven off by the partial combustion process. The typical gas composition for producer gas from wood is carbon monoxide (18-25%), hydrogen (13-15%), methane (3-5%), carbon dioxide (5-10%), water vapor (10-15%), and nitrogen (45-54%) (National Academy Press, 1983). The gas has an energy content of about 150 lb/cf and can be used as a boiler fuel, fuel for internal combustion engines, or in fuel cells. Since gasification occurs at temperatures below direct combustion, gasification systems are better able to handle materials that have greater slagging potential.

Updraft, downdraft, and cross-draft gasifiers are fixed-grate systems classified according to air flow direction through the fuel bed in the gasifier. Fluidized bed gasifiers also exist. Each system has its advantages. Fluidized bed gasifiers have the most flexibility to deal with high-ash fuels; however, fluidized bed systems are the most complex and expensive.

There are a few commercial manure burning projects in place. *Waste Age* (1989) reported that the first US cattle manure-fueled power plant was built in the Imperial Valley of California by National Energy Associates. The plant burned roughly 1000 tons/day of manure to produce 15 MWe of electricity, which was sold to Southern California Edison. The manure reportedly had a high salt content and contained weedseeds, thus making it unfit for fertilizer applications and imposing a cost for disposal.

More recently, research in the United Kingdom has focused on the use of broiler litter for on-farm space heating of broiler houses and large-scale, off-site electricity generation. The on-farm research was based on a modified commercial Bioflamm™ unit that uses two separate stages--gasification and combustion.

Economic analysis showed that the most cost-effective method of heating broiler houses was a central system serving multiple houses on a farm. A firetube coal boiler was retrofitted with the Bioflamm™ unit mounted on the front of the boiler. Initially, the project was intended to cogenerate electricity with a turbine; however, subsequent research found that a small steam engine was more cost-effective.

> The project has demonstrated an ability to dispose of poultry litter reliably, safely, efficiently and without the aid of supplementary fuels. As the fuel is free, it has proved possible to increase ventilation rates through the poultry house which has led to: a drier litter, improved bird welfare; a reduction in odor problems from the poultry houses; improved bird quality resulting in higher value stock (Dagnell, 1992).

On a large scale, there are presently two 12-MWe central power generating stations in the United Kingdom operating on broiler litter. Each plant consumes about 140,000 tons/year of litter from within a 25-mi. (40 km) radius of the plant (Dagnell, 1992). The combustion unit uses four reciprocating grates, each split into four zones. The air supply to each zone can be varied independently. A three-pass, water tube boiler is used to produce 943 psi (65 bar) stream at 840° F (450° C).Turbines rated at 14 MWe coupled with air-cooled condensers generate 12.6 MWe of net power. Stack emissions are cleaned by a three-stage electrostatic precipitator before discharge into an 82-ft (25 m) steel stack. The ash is used for fertilizer.

The internal rates of return were over 42% for the on-farm system and over 23% for the large-scale, off-site power plant. Unfortunately, the high capital costs have made farmers cautious about investing in this technology. Large-scale use is also most sensitive to capital costs; however, utilities have easier access to capital. Dagnells (1992) states that further projects are already being planned and that the market will grow to use about 80% of the poultry waste (in the UK) over the next 5-10 years.

The TVA has initiated similar projects, although lower energy costs in the United States make such systems more difficult to justify economically. Initial efforts have focused on developing cost-effective systems for broiler litter on-farm space heating. Economic analysis indicates that the *total* energy system cost will have to be in the range of $10,000 to $12,000/house (40 ft x 500 ft with 24,000 birds) to be cost effective. This cost assumes an energy use equivalent to 3,750 gal of propane/year at $0.79/gal, a 15-year system life, and a $10/ton opportunity cost for the litter. About 1,250 gal of propane are still required for periods of high and low heat demand.

Preliminary information indicates that commercial systems could be developed that could be cost effective under these conditions. In the southern US, adequate litter is produced in each broiler house each year to heat the house. Although relatively high initial costs may hamper adoption of the technology, increasing environmental regulations stemming from land application of litter may force wide-spread adoption in the United States.

BIOLOGICAL CONVERSION (ANAEROBIC DIGESTION)

Biological conversion via anaerobic digestion is typically used and is most cost effective for organic materials containing high moisture contents (> 90%). The TVA and others have researched anaerobic digestion for livestock wastes over a number of years. Historically, anaerobic digestion for energy production has been difficult to justify in the US on an economic basis. However, the opportunities for anaerobic digestion of livestock wastes have improved due to new technological developments that improve the cost effectiveness of biogas recovery and use, growing use of lagoon type livestock waste treatment systems, growing concern about odor control from livestock wastes, and the rapid increase of large-scale livestock production facilities--which work to improve the economy of scale. Additionally, there is increased government interest in methane as a global climate change gas; hence government programs are being established that should assist with the further development of more cost-effective technologies and assist with the implementation of digestion technology.

Materials derived from green plants form the basis for animal feeds and hence animal wastes. Green plants consist of various complex organic polymers composed primarily of carbohydrates, with some lipid, protein, and inorganic material. The hemicellulose and lignin components of carbohydrates are undigestible and are hence found in animal wastes.

Anaerobic digestion of organic polymers occurs through a complex interaction of a number of microorganisms. A broad spectrum of fat-, cellulose-, and protein-decomposing facultative microorganisms initiates the breakdown into soluble compounds; acetogenic organisms convert the soluble compounds into organic acids (primarily acetic). Methanogenic organisms produce methane by fermenting the acetic acid to methane and carbon dioxide or by reducing carbon dioxide to methane using hydrogen gas or formate produced by other organisms. Methanogenic organisms are anaerobic and are inhibited by even small amounts of oxygen. They are also sensitive to changes in pH and to highly oxidized materials such as nitrates or nitrites.

Additionally, these organisms are sensitive to temperature, and their productivity decreases with declining temperatures. Traditionally digesters were operated in the mesophilic (25-40° C or 87-104°F) and thermophilic (40-65° C or 104-149°F) temperature ranges. As temperatures decline and net microbial growth rate decreases, the minimum solids-retention time for process stability increases. One benefit of slower reaction times associated with declining temperatures is more opportunity for the operator to detect and correct any system upsets.

Previous anaerobic digester work in the US focused on tank digesters such as the continuous stirred tank reactors (CSTR) or plug flow digesters. These digesters were costly to build, required significant management, and, because of the cost of the vessel and the exponential decline of gas production with time, could cost effectively recover only about 50% of the potential methane.

Anaerobic lagoon digesters operate under low temperature (psychrophilic) conditions (< 20° C or 68°F) and essentially overcome these problems. Lagoons are required by the newer, large-scale livestock production operations to contain the volume of liquid inherent in the hydraulic flush manure removal systems they prefer to use. Under normal conditions, these lagoons are naturally anaerobic. Due to their size, the lagoons are relatively stable in temperature and, to help maintain temperatures, can be designed to have greater depth to minimize surface-to-volume ratios.

Research by Safley and Lusk (1990) with anaerobic lagoon digesters for cow manure found that, due to the relatively simple design, the initial cost of lagoon digesters was significantly lower than that of a CSTR and virtually no maintenance was required. Additionally, the large volume of the lagoon provided a hydraulic retention time (HRT) roughly three times that of a CSTR. This relatively long HRT allows for most of the potential methane to be captured. As an extra benefit, the biogas is 10 to 15% richer in methane and produces relatively little H_2S. A lagoon digester on a commercial swine operation in Virginia produces about 6-7 ppm H_2S in comparison to a CSTR that would produce around 50 ppm (Moser, 1994). This level of H_2S allows the gas to be used without scrubbing in internal combustion engines (Moser, 1994).

It is recommended that the methane potential analyses in Table 2 be used as the basis for anaerobic digestion system design. Safley and Lusk (1990) used digester loading rates of 0.010 lb of volatile solids/ft^3 of digester volume to be added to the digester each day. He suggests that loading rates up to 0.012 lb of volatile solids/ft^3 may be used without overloading the digester. Operating above these loading ratesincreases the potential for digester upsets and the production of offensive odors.

The lagoon is covered with a flexible membrane such as XR-5 or Hypalon™. Newer covering techniques use foam floats to channel the gas to a header for collection and removal with a vacuum pump. To avoid the need for expensive foundations and seals, the floating cover is anchored in place with rope guylines to stakes on the banks. Skirts roughly 2 ft in length hang down into the lagoon to form a gas seal. Cover costs are currently high and range from approximately $1/ft^2, uninstalled, to $4/ft^2 installed.

The methane content of the biogas from a lagoon digester ranges from 60 to 80% with the balance primarily CO_2. This methane content is equivalent to roughly 600 to 800 BTU/ft^3. Biogas may be substituted for fuel in virtually any natural gas application, including internal combustion engines, boilers, absorption chillers, and hot water heaters.

A comparative study of the economics of psychrophilic anaerobic lagoon digesters versus mesophilic CSTRs by Safley and Lusk (1990) for dairy cows showed the lagoon system to have almost double the IRR (15.5 vs. 8.3). This analysis discounted the indirect effects and the economic value of recoverable high-nutrient co-products useful as fertilizers or other goods.

SUMMARY COMPARISONS

Both thermochemical and biological conversion systems offer potential for disposal of livestock wastes. Both methods have advantages, and both need further research--the thermochemical more for technical reasons and the biological more for economic reasons. Thermochemical systems require less land area, have lower cost, and are best suited for low-moisture-content materials. Additionally, more flexibility on energy use is provided since it is easier to store the dry manure until energy is needed.

Biological systems are best suited for high-moisture-content materials; however, due to their high moisture content, these materials are more difficult to store economically, and their storage can produce environmental concerns such as odors. Thus, energy use from biological systems must parallel manure production fairly closely. Biological systems, however, keep the organic benefits of the manure which can provide significant byproduct value as fertilizer and soil conditioners.

REFERENCES

American Society of Agricultural Engineers (ASAE) Standard ASAE D384.1, Manure Production and Characteristics, *ASAE Standards 1993*, ASAE, St. Joseph, Michigan, 1993.

Dagnell, S.P., Poultry litter as a fuel in the UK--A Review, *Proceedings of the 1992 Incineration Conference*, Albuquerque, New Mexico, 1992.

Lindsey, Janice K., personal communication, TVA Biotechnology Research Department, Muscle Shoals, Alabama, April 1994.

Miles, Thomas R., Consulting Design Engineers, Portland, Oregon, personal communication, 1994.

Miles, Thomas R., Miles, Thomas R. Jr., Baxter, Larry L., Jenkins, Bryan M., and Oden, Laurance L., Alkali slagging Pproblems with biomass fuels, *Proceedings of the First Biomass Conference of the Americas: Energy, Environment, Agriculture, and Industry*, August 30-September 2, Burlington, Vermont (Published by the National Renewable Energy Laboratory, Golden, Colorado), 1993.

Moser, Mark, personal communication, RCM Digesters, Berkeley, California, June, 1994.

National Academy Press, *Producer Gas: Another Fuel for Motor Transport*, Washington, DC, 1983.

Owens, John M., Full Circle Solutions, Inc., Gainesville, Florida, personal communication, 1994.

Rajvanshi, A.K., Biomass Gasification, Chapter 4, in *Alternate Energy in Agriculture*, Volume II, CRC Press, Boca Raton, Florida, 1986.

Safley, L.M., and Lusk, Philip D., *Low Temperature Anaerobic Digestion*, North Carolina Department of Economic and Community Development, Division of Energy, Raleigh, Report under SERBEP contract TV-74816A, 1990.

Waste Age, Cow Manure Burned in Place of Oil, May, p. 16, 1989.

COMMERCIAL AND ON-FARM PRODUCTION AND MARKETING OF ANIMAL WASTE COMPOST PRODUCTS

Lewis Carr, Ray Grover, "Bo" Smith, Tom Richard, and Tom Halbach

INTRODUCTION

Composting is the biological controlled process of decomposing organic materials. Microorganisms are responsible for this process. Given the proper environment, these microbes will stabilize the organic materials and produce compost with time.

Compost has benefits in addition to stabilizing organic materials. Some of the potential benefits (Rynk et al., 1992), are as follows:

* Improved manure handling.
* Enhanced soil tilth and fertility.
* Reduced environmental risk.
* Destruction of pathogens and viruses.
* Freedom from unpleasant odors.
* Destruction of weed seeds.

This chapter will address the principles of composting, types of composting systems, compost quality, commercial versus on-farm production of compost, and marketing of compost. Composting is one alternative to excess manure application on farm land.

PRINCIPLES OF COMPOSTING

There are nine key elements associated with composting. They are as follows:

1. **Organic Materials** for composting will either be high in nitrogen (3-6%) or high in carbon. Manures, sewage sludges, food processing sludges, and grass clippings are examples or organic materials high in nitrogen. Sawdust, wood chips, leaves, corn stover, straw, peanut hulls, and paper products are examples of materials high in carbon. Some of the nitrogen-rich materials may also be high in moisture. Sludges may have

a high moisture content, some as high as 97% unless dewatered. Some sludges may be unsuitable for quality compost because of high proportion of heavy metals and other undesirable materials such as PCBs or other trace organics.

2.   **Microbes** are the "work horses" of the decomposition process. The microorganisms may be classified as bacteria, fungi, and actinomycetes. Each participating group has mesophilic species that operate best at temperatures less than 43°C and thermophilic species that operate best at temperatures ranging from 43°C to 66°C.

3.   **Mass and free air space** are important because the process is aerobic. Oxygen infusion is necessary for the process to be as odorless as possible. Microbial decomposition occurs on the particle surface. A smaller particle has more surface-to-volume ratio than larger particles and, thus, a greater rate of bio-oxidation. Rubin and Shelton (1993) suggest optimum particle size depends on many things, but particles 0.65 cm to 2.54 cm in size may be the optimum range. Porosity of an initial compost mix should be about 30% for proper aeration. Bulking agents with large, stiff particles will assist in providing the porosity needed in a compost mix. These particles can be recycled by screening from the composted product.

4.   **Carbon (C), Nitrogen (N), and Phosphorus (P)** ratios are important elements in developing a composting recipe. Rubin and Sheldon (1993) suggest a C:N ratio of 20:1 to 30:1 and a C:P ratio of 100:1 to 150:1. As the C:N ratio increases, there is less nitrogen available for metabolism and thus a slower composting process. If more nitrogen is available, such as a C:N ratio of 15:1, there is the possibility of ammonia being released in the composting process and the creation of undesirable odors. It can be concluded that C:N and C:P ratios are very important in proportioning nutrient-rich and nutrient-poor substrates. Brodie (1994) developed a computer program to determine least-cost compost mixes based on feedstock inputs.

5.   **Oxygen** is essential for producing aerobic conditions in a compost mix. Air is about 21% oxygen. Oxygen should be monitored in composting operations. As oxygen falls below 5% in a compost mix, anaerobic, potentially toxic conditions begin to exist, which discourages the multiplication of aerobic microorganisms. Oxygen should be added immediately.

6.   **Moisture** is essential for providing a proper environment for the microorganisms to bio-oxidize the organic materials in the compost mix. The moisture should range from 50 to 60%. If the moisture content is too low, microbial activity can be affected; if the moisture contest is too high, the transfer of oxygen can be inhibited, resulting in anaerobic conditions. When the mix is less than 40% moisture, nutrients are also less available, which can also contribute to the cooling of the mixture.

7. **Temperature** in a compost mix increase as the microbes metabolize the organic materials. Composting occurs essentially in two temperature ranges; mesophilic (10-43°C) and thermophilic (43-66°C). The thermophilic range is more desirable because pathogens, virus, insect larvae and weed seeds can be destroyed in the composting process. It will take temperatures as high as 63°C to kill weed seed.

8. **pH** of feedstocks for composting can vary from 5.5 - 8.5. However, the extreme ends may not compost as well as those close to neutral. Ammonia release becomes a problem in high nitrogenous mixes above pH 8. Alkaline feedstocks such as lime, wood ash, etc., should not be added where the pH is 8 or greater. This will cause the conversion of ammonium (a product of nitrogen metabolism) to volatile ammonia, thus reducing the nitrogen content of the final compost product. Sweeten (1988) suggested that the initial compost mix should have a pH in the range of 6.5 - 7.2. As the compost mix decomposes, pH change occurs in the mix. Organic acids formed may shift the pH to slightly acid. This shift may not be as great if broiler litter is used in the process because of the high calcium content in the litter (Carr et al., 1990). At times, mixes containing broiler litter may have a pH of 8 or greater. If this occurs, Carr and Brodie (1992) suggest the use of ferrous sulfate to lower the pH.

9. **Time** for composting is dependent upon the recipe (C:N ratio), moisture, temperature, frequency of aeration, feedstocks used, particle size and end use. The degree of stability required depends on the final use of the compost. Complete curing, the slow maturation of the compost after active composting, may not be required if the compost is applied to cropland. However, the C:N ratio of the compost should not exceed 30:1 even when it is applied to cropland. Municipal solid waste (MSW) compost with a C:N ratio of approximately 60:1 has caused problems in Maryland when applied to farm land. The compost used the available soil nitrogen to continue the composting process, thus "depriving" the plant system of its nitrogen needs. Table 1 shows typical composting times for selected methods and materials.

COMPOSTING TECHNIQUES

Four composting techniques will be discussed. These techniques, differing widely in degree of sophistication, cost, and production rates, will "speed up" the composting process over natural composting. The techniques are static pile, aerated static pile, windrow, and in-channel.

Static pile is where the compost mix is piled and not disturbed for a long period of time. It may be turned, but not frequently. To assist in natural aeration, the initial compost mix should have a porosity of approximately 30% or a bulk density of approximately 540 kg/m³.

Aerated static piles can be active or passive in mode of operation. The

Table 1. Typical composting times for selected combinations of methods and materials.

Method	Materials	Range	Active composting time Typical	Curing Time
Passive composting	Leaves	2-3 years	2 years	-----
	Well-bedded manure	6 months to 2 years	1 year	-----
Windrow-infrequent turning†	Leaves	6 months to 1 yr.	9 months	4 months
	Manure + amendments	4-8 months	6 months	1-2 months
Windrow-frequent turning‡	Manure+amendments	1-4 months	2 months	1-2 months
Passively aerated windrow	Manure-bedding	10-12 weeks	-----	1-2 months
	Fish waste + peat moss	8-10 weeks	-----	1-2 months
Aerated static pile	Sludge + wood chips	3-5 weeks	4 weeks	1-2 months
Rectangular agitated bed	Sludge+yard waste or Manure + sawdust	2-4 weeks	3 weeks	1-2 months
Rotating drums	Sludge and/or solid wastes	3-8 days	-----	2 months#
Vertical silos	Sludge and/or solid wastes	1-2 weeks	-----	2 months#

†For example, with bucket loader.
‡For example, with special windrow turner.
#Often involves a second composting stage (for example, windrows, or aerated piles).

Source: *On-Farm Composting Handbook*, NRAES-54, Northeast Regional Agricultural Engineering Service, 152 Riley-Robb Hall, Cooperative Extension, Ithaca, NY 14853, (607) 255-7654, with permission.

active piles normally draw air through the compost mix by using pipes or ple-nums placed in the compost mix and fans attached to a duct system. Monitor-ing systems can be used to insure proper temperature and oxygen control. Air discharge from the fan system can be filtered through a biofilter for odor con-trol. Another aerated pile system is passive in operation. The passive system uses a series of perforated 10.1- or 12.7-cm diameter plastic pipes underneath the compost pile. The pipe ends are left open, and a natural convective process provides oxygen to the compost mix. A porosity of approximately 30% or a bulk density of approximately 540 kg/m³ is also desirable for the aerated pile system.

Windrow composting can be accomplished outside or in a large, covered structure. Windrows are normally turned with some type of turning equipment. The equipment can be as simple as a front end loader or self propelled equip-ment that straddles the windrow and turns it in one pass. However, good mix-ing may not be as effective when a front end loader is the turning device. A porosity of approximately 30% or use a bulk density of approximately 540 kg/m³ is desirable.

In-channel techniques primarily use a turning device that runs down a rail of some type. (Sometimes referred to as a rectangular agitated bed.) It is possible to have parallel bays with common walls so the turning device can be moved from bay to bay. This type of system is expensive but may be an appro-priate system for short composting times. The in-channel system may also be used in conjunction with an aerated system. Fans and air ducts are placed throughout the system and will speed up the composting process by continu-ously providing oxygen to the compost mix. This may be of great benefit if the compost mix is highly volatile. Air from the fans can be discharged into a biofilter for odor control. A 30% porosity or a bulk density of approximately 540 kg/m³ will also assist in this process.

COMPOST QUALITY

Thought must be given to the compost product use before developing the initial compost mix. Final use can also influence the composting technique used and the acceptable cost of production. The end product will be no better than the feedstock used to make the initial mix. Therefore, it is very important to have a reasonably current nutrient analysis of each feedstock used in "recipe making." Tables 2 and 3 illustrate nutrient parameters associated with hatch-ery and dissolved air flotation (DAF) compost mixes (Carr and Brodie, 1992).

A decision has to be made concerning end use and compost quality. If the compost is going to be used as a field manure source, the refinement or quality of feedstocks does not have to be as great as that used in home landscaping. Figure 1 illustrates compost refinement based on end use. Feedstock selection should consider "keeping out" undesirable metals and organics such as PCBs from the compost mix.

Table 2. Hatchery compost composition (Wet Basis).

Item	TKN	Moisture	Total Solids	Carbon	pH	Bulk Density
	%	%	%	%(DB)†		kg/m³
Wood shavings	0.01	18.62	81.38	55.29	5.5	136
Centrifuged hatchery waste	1.61	24.51	75.49	21.96	8.6	780
35-day compost	1.19	25.01	75.99	31.34	8.4	360

†Dry Basis

Table 3. Selected ingredient and compost mix parameters
for dissolved air flotation (DAF) compost (Wet Basis).

Item	TKN	Moisture	Total Solids	Carbon	pH	Bulk Density
	%	%	%	% (DB)†		kg/m³
DAF Skimmings	0.75	80.0	20.0	55.10	5.42	919
Broiler Litter	2.44	30.0	70.0	48.27	8.29	490
Sawdust	0.47	48.6	51.4	54.36	5.5	423
Initial Compost Mix	1.53	56.9	43.1	50.78	7.85	618
35-day compost	1.90	43.6	56.4	50.50	7.60	415

†Dry Basis

To assist in determining if compost is matured, respiration rates of the compost can be determined by laboratory procedures. A field determination can be made by collecting a compost sample, saturating it with water (but not soaking, dripping wet), placing it in a sealed plastic bag, and storing it in a warm place (21-30°C) for one week. After one week, the bag is opened. If there are no bad odors, the compost has stabilized.

Quality compost will have a C:N ratio of about 15:1. The time to achieve quality compost will depend on the technique used to compost. It may take one year or more to achieve a quality compost using static piles whereas a quality compost may be achieved in 2-3 months using mechanical systems (Table 1).

COMMERCIAL VS. ON-FARM PRODUCTION OF COMPOST

The principles of composting are the same, regardless of where the compost is made. In a commercial setting, one thinks of a central location where feedstocks are pooled and mixed according to a recipe with an end product specified. Normally the feedstocks are taken in for a fee, and the compost produced is sold. This setting is normally not located on a farm.

On-farm composting occurs in conjunction with a farm operation. The operation normally produces part of the feedstocks (manure) and may purchase or charge a fee for other materials. In many locations, if the farm operation uses the compost in their operation, the composting operation will be considered farming. However, an on-farm compost operation may be classified commercial, in some locations, if the compost is marketed and sold.

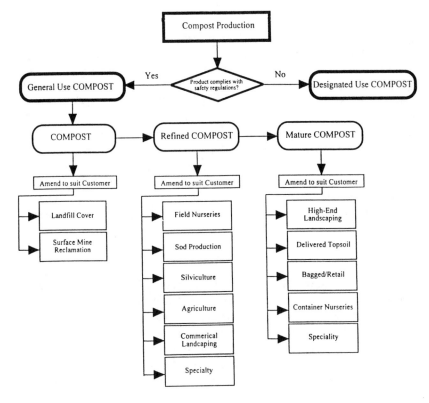

Figure 1. Compost refinement illustration. Source: The Composting Council, 114 South Pitt Street, Alexandria, VA 22314, with permission.

COMPOST MARKETING

The aspect of composting most often overlooked and underestimated is the marketing of the products to the public and the agricultural community. It is easy to give compost to farmers. It is much more difficult to recoup your operating expenses from the sale of the product.

There is a wide variety of potential markets for animal (poultry, horse and cattle) manure compost. Examples include:

1. Agriculture, both high dollar crops and row crops.
2. Land reclamation from mining or construction activities.
3. Turf and nursery operations.
4. Professional turf and landscape care, including golf course construction.
5. Retail lawn and garden sales.

It is a universal fact that good soil containing organic matter is essential to growing any plant. But not just any organic source will do. Raw manure can be composted into a quality compost. Quality compost is the key to marketing.

Raw or poorly composted materials have little, if any, value due to their odor, high moisture content, volatile and leachable nutrients, and pathogens. A well-composted product eliminates these problems by stabilizing the raw material into a true organic complex. Once stabilized, the manure compost has value and can be sold in the market place through the same channels of distribution conventional materials are marketed and sold. Compost can be marketed in bulk or bag and refined based on the customer's need.

Once a quality product is in place, each of the sales opportunities above can be reached by direct mailing information to them or calling on the customer, trade shows, "word of mouth," and using different advertising methods to move the product. A good product will bring repeat sales and a growing customer base. The bottom line is having a good, consistent product. When the markets above find a quality consistent source, demand can outweigh the supply.

SUMMARY

A brief overview of why and how compost works has been presented in this chapter. The final compost will be no better than the initial mix of feedstocks and the practices utilized during the process. Current nutrient analyses of the feedstocks are necessary in formulating the initial mix recipe. Refinement or feedstock quality of a compost mixture will be determined by its end use. A stable, consistent animal waste compost can be marketed. The market will dictate the form of the compost (e.g., pelletitized, bagged, bulk, odor, degree of maturity, and particle size).

REFERENCES

Brodie, H.L., *Multiple Component Compost Recipe Maker,* ASAE Paper 94-3037, International ASAE Meeting, Kansas City, Missouri, June 19-22, 1994.

Carr, L.E. and Brodie, H.L., *Composting Hatchery By-Products and DAF Skimmings,* 1992 Midwest Poultry Federation Convention, Minneapolis Convention Center, Minneapolis, Minnesota, 1992.

Carr, L.E., Wheaton, F.W., and Douglass, L.W., Empirical models for determining ammonia concetrations from broiler chicken litter, *Trans ASAE,* 33(4), 1337, 1990.

Rynk, R. ed., *On-Farm Composting Handbook,* NRAES-54 Publication, Northeast Regional Agricultural Engineering Service, Ithaca, New York, 1992.

Rubin, A.R. and Shelton, J., *Basic Principles of Composting,* Proceedings of the National Extension Compost Utilization Conference, Minnesota Extension Service, University of Minnesota, St. Paul, Minnesota, 1993

Sweeten, J.M., Composting Manure and Sludge, in *Proceedings of the National Poultry Waste Management Symposium,* Ohio State University, Columbus, Ohio, 1988.

FEEDING BROILER POULTRY LITTER AS AN ALTERNATIVE WASTE MANAGEMENT STRATEGY

T.A. McCaskey

INTRODUCTION

Broiler poultry litter is a major by-product of the three leading poultry producing states in the US: Arkansas, Alabama, and Georgia. The quantity of litter produced in these states is estimated to be over 6 million tons (wet basis) annually. In addition, significant quantities of caged layer manure and turkey litter are produced throughout the US. Because by-products of poultry and turkey production have significant potential to affect the quality of the environment, federal and state regulatory agencies have focused attention on measures to reduce environmental pollution due to these by-products. If improperly managed, poultry wastes can pollute the environment, principally by contaminating surface and ground water. However, poultry excreta and litter have economic value associated with their nitrogenous and mineral components. Poultry wastes have long been recognized for their beneficial effects on soil fertility and for their content of fertilizer nutrients. Also the value of the wastes as a source of low-cost dietary nutrients for ruminant animals has been recognized for over 40 years (Noland et al., 1955). The purpose of this chapter is to review alternative waste management strategies that have promoted the use and value of poultry excreta and litter as low-cost feed sources.

HISTORY OF FEEDING OF BROILER POULTRY LITTER

The first documented attempts of feeding animal manures to livestock is believed to have occurred in the early 1940s (Hammond, 1942; Bohstedt et al., 1943). Reports by Noland et al. (1955), Southwell et al. (1958), and Fontenot et al. (1963) demonstrated that cattle will consume rations containing poultry excreta and that the material has nutritive value for ruminant animals. Following these early reports, the practice of feeding poultry production wastes, as well as other types of animal wastes, became more widespread. During the 1960s several feeding trials conducted by researchers in Virginia documented the benefits of feeding poultry litter. Currently the practice is well known in the US and in many countries. As the practice of feeding poultry wastes became more widespread in the US, the Food and Drug Administration (FDA) questioned the impact of the feeding practice on animal and human health (Kirk, 1967). In 1967, the FDA published a statement that the practice was not condoned because animal manures and related wastes might contain infectious diseases and chemical residues that when fed to food-producing animals, might be a health hazard to the animals and also to the persons who consume the animal products. The potential health risks questioned by the FDA are listed in Table 1.

Much has been discussed about the safety of feeding animal wastes, particularly manures, to food-producing animals. There are many ways one can justify the safety of the feeding practice, and there are ways also to justify its condemnation. A major disadvantage of recycling excreta from one animal species back to the same species is the potential risk of transmitting specific viral, protozoan, and bacterial diseases through the waste. Furthermore, some diseases common to animals also can be acquired by humans. Although the potential health risks have not been fully resolved, and might again become a hotly contested issue, much has been learned through research conducted during the past 20 years that today plays a role in how animal wastes are managed, processed, and used as feedstuffs. Several reviews have been published on the safety of feeding animal wastes, and these address many of the potential health risks questioned by the FDA (Fontenot and Webb, 1975; Webb and Fontenot, 1975; McCaskey and Anthony, 1979; McCaskey et al., 1985).

Excreta from beef cattle (Anthony, 1966) and swine (Kornegay et al., 1977) have been explored in addition to poultry litter as potential feedstuffs for animals, but poultry wastes, particularly broiler poultry litter, are more amenable as a feedstuff compared to other types of animal wastes. Characteristics of broiler poultry litter that make it valuable as a feed ingredient are shown in Table 2. Smith and Wheeler (1979) compared the relative nutrient contents of wastes from four classes of food-producing animals (Table 3). Poultry wastes, including dehydrated layer excreta and broiler poultry litter, contain greater quantities of nitrogen and total digestible nutrients that contribute to their value as alternative feed sources for animals. Layer poultry excreta is high in nitrogen, but it has a high moisture content, which makes it difficult to manage and preserve, and the ash content is 54% higher than broiler poultry litter, which reduces its feed value (Fontenot et al., 1983). For these reasons broiler poultry

Table I. Potential health risks of feeding broiler poultry litter to food-producing animals.

Pathogenic Microorganisms	Antibiotics and Drugs
Microbial Toxins	Hormones
Mycotoxins	Pesticides
Parasites	Coccidiostats
Viruses	Heavy Metals
Arsenicals	Trace elements

From: McCaskey and Anthony, 1979.

Table 2. Desirable characteristics of broiler poultry litter
for use as a feedstuff compared to other animal wastes

Higher nutrient content
Relatively dry
Easily stored and preserved
Amenable to handling with conventional feed equipment
Essentially all collectable
Readily available at low cost in poultry production areas

Table 3. Nutrient composition of animal excreta.

Nutrients	Poultry Litter	Dehydrated Poultry Excreta	Cattle Excreta	Swine Excreta
	------------------------percent of dry matter------------------------			
TDN	73	52	48	48
CP	31	28	20	24
CF	17	13	20	15
Calcium	2.4	8.8	0.9	2.7
Phosphorus	1.8	2.5	1.6	2.1
Magnesium	0.4	0.7	0.4	0.9
Potassium	1.8	2.3	0.5	1.3

From: Smith and Wheeler, 1979. Total digestible nutrients (TDN), crude protein
(CP), crude fiber (CF).

litter has been extensively evaluated as a potential feedstuff for ruminant animals.

PERFORMANCE OF RUMINANTS FED POULTRY LITTER

Ruminant animals have played a major role in the utilization of many by-products, including poultry litter, because of their unique ability to use non-protein nitrogen (NPN) and plant fibers such as cellulose and hemicellulose. Broiler poultry litter is a rich source of NPN, minerals, and also fiber, which is derived from poultry bedding materials such as wood chips. Many cattle feeding trials have been conducted over the years, and the studies have demonstrated a positive value for litter as a source of dietary nitrogen for cattle. Smith and Wheeler (1979) summarized cattle performance data from eight broiler litter feeding trials. They concluded that the mean daily gain of cattle fed litter was 0.1 lb less than the gain for cattle fed traditional feed supplements. Daily feed consumption was increased 0.66 lb when 24% of the diet was poultry litter. The feed/gain ratio was 10% higher for cattle fed litter compared to traditional feed ingredients. Although cattle fed litter do not perform as well as cattle fed conventional feedstuffs, the economics of cattle production favor the use of litter because of its low cost.

Cullison et al. (1976) demonstrated that dried poultry excreta (DPE), with no litter bedding, when fed at 5.8% of the ration, could supply half of the dietary protein supplement for steers. Average daily gain (ADG 2.6 lb) for steers fed the ration containing DPE was the same as for steers receiving all their dietary protein from soybean meal. Cullison et al. (1976) also demonstrated that steers fed 20% of their diet as wood shaving broiler litter (ADG 2.6 lb) performed better than steers fed a 20% diet of peanut hull litter (ADG 2.4 lb) or steers fed 13% of their diet as dried hen manure (ADG 2.0 lb). The value of the poultry manure was calculated to be $52/ton for the wood shaving litter, $34 for the peanut hull litter, and the dried hen manure had a negative feed value. Some feeding trials have shown improved animal performance when broiler litter was incorporated into diets of steers. Cross et al. (1978) reported that ADG was increased 25% from 1.58 to 1.98 lb when 10% broiler litter was incorporated into the diets of steers, and increased 31% when 30% litter was added to the diet. The feed/gain ratio was improved incrementally up to 30% incorporation of litter into the diet.

Stocker cattle production is a common beef cattle production practice in the southeastern US and is dependent upon animal grazing and dietary supplementation to meet the protein requirements of the growing animals. Research has shown that the carrying capacity (animal density) of winter ryegrass pastures could be increased by supplementing the diet of grazing steers with a ration containing 41% broiler litter (Duffy et al., 1994). The highest ADG (2.84 lbs) was achieved by steers stocked at 1.25 animals/acre and fed the supplement *ad libitum*. The litter supplement increased ADG 0.15 lb/day. The highest return/acre was at the two animal/acre stocking rate with *ad libitum* litter supplement. The return/acre was $55 compared to $28 for grazing without litter supplement. A study with beef calves on winter pastures showed improved animal performance with corn/litter supplementation over corn supplementation alone (Davis et al., 1994). ADG for mixed lots of both heifers and steers fed a corn supplement was increased 0.97 lb/day, and animals fed a corn/litter supplement gained 1.41 lb/day more than the control animals fed no supplement. The net increase in value of the calves fed the corn/litter supplement was $17.58/head for the heifers and $49.18/head for the steers. Other economic benefits of feeding broiler poultry litter have been demonstrated with stocker cattle fed in a drylot (McCaskey et al., 1994). Mixed-breed, beef heifer calves were purchased from local stock sales and assigned to two feeding groups. One group was fed a conventional ration with corn grain, cottonseed hulls, and soybean meal as the principal feed ingredients, and the other group was fed a 50/50 mix of broiler poultry litter and ground corn grain. The animals were fed *ad libitum* in confinement for 112 days receiving only the diets described above. Heifers fed the corn/litter diet had an ADG of 2.12 lb and a feed/gain ratio of 10.8:1; and heifers fed the conventional diet had an ADG of 2.53 lb and feed/gain of 8.7:1. Although the animals fed the conventional diet performed better than the animals fed the corn/litter diet, the low cost of the litter diet and the cost/gain favored the corn/litter diet. The cost/gain was $0.66 for the conventional diet and $0.46 for the corn/litter diet, or a feed cost savings of 30% for the corn/litter diet. Cross et al. (1978) reported a feed cost saving of 32% for a steer diet containing 30% litter.

Reports on the value of animal manures as sources of dietary nutrients vary widely from one study to another. There are many opportunities to vary feed formulations containing animal manures, and variations in manure management practices also might contribute to differences in feed values of animal manures. Smith and Wheeler (1979) estimated the value of manures from four classes of animals as sources of protein. The values of the manures were determined by least-cost feed formulations and by estimating the dollar value of the manure protein relative to the cost of soybean meal (44% protein). The dollar values reported by the scientists have been adjusted relative to current soybean meal prices (Table 4). Poultry litter and dehydrated poultry excreta were reported to have the highest values followed by swine excreta and lowest for cattle excreta. On a wet basis the poultry litter had an approximate value of $123/ton, poultry excreta $82, swine excreta $71, and cattle excreta $60. Based on the studies described above, animal manures, and particularly poultry litter, have value as feedstuffs for ruminant animals. The value of litter is not fixed like traditional feed ingredients because the dietary benefit of litter varies widely depending on such factors as chemical composition and feed quality of the litter, and how the litter is used in feeding regimes. Therefore, it is difficult to assign a feed value to broiler poultry litter, but some scientists report that the feed value of litter ranges from $52 to $123/wet ton (Table 5). Most agree its feed value is about $100/ton. The economic value of poultry litter is directly related to its substitution value for more costly resources. The major uses of litter are for fertilizer and for feed. Reports indicate that the highest economic value that can be assigned to litter is for its feed value, which exceeds its fertilizer value by 3 to 10 times (Smith and Wheeler, 1979; Stephenson et al., 1990; Fontenot et al., 1983). Broiler litter will continue to be used as a feedstuff, provided that its cost is low relative to traditional feedstuffs. When litter can be purchased for $10 to $15/ton, beef cattle producers have a strong economic

Table 4. Estimated value[†] of animal excreta as protein sources.

Beef Cattle	Poultry Litter	Dehydrated Poultry Excreta	Cattle Excreta	Swine Excreta
	-----------------------Dollar value/ton (dry basis)----------------------			
Growing Steer	148	148	107	127
Finishing Steer	149	134	110	121
Dry Cow	148	145	107	126
Lactating Cow	148	147	108	126
Average	148	144	108	125
	(123)[‡]	(82)	(60)	(71)

From: Smith and Wheeler, 1979.
[†]Values adjusted to current price of soybean meal (44% protein).
[‡]Values in parentheses expressed on wet basis of excreta.

Table 5. Reported values of broiler litter for ruminant animal feed.

$ Value/ton (wet basis)	Source
114[†]	Free, 1977
123[†,#]	Smith and Wheeler, 1979
52[‡]	Cullison et al., 1976
114[†]	Stephenson et al., 1990
106[‡]	McCaskey et al., 1994

[†]Based on nutrient content of litter relative to the nutrient value of conventional feed ingredients.
[‡]Compared to performance of animals fed conventional feed.
[#]Value adjusted to current price of soybean meal (44% protein).

incentive to feed litter. If the cost were to approach $50/ton, many producers might limit the feeding of litter. This scenario has become apparent for processors who pellet litter and litter/feed combinations for sale to cattle producers. Pelleted litter has been marketed for about $90/ton and pelleted litter feed mixtures for $135 to $165/ton. Pelleting adds about $30/ton to the cost of the feed. When corn costs $110/ton and a 50:50 combination of pelleted corn and litter sells for $135/ton, litter therefore costs $100/ton. Many broiler litter/feed processing operations have failed because of the high market price of the pelleted litter feed. When the performance of beef stocker cattle fed a 50:50 corn/litter diet was compared to the performance of animals fed a traditional diet formulated to provide about the same level of nutrients, the breakeven price of the 50:50 corn/litter diet relative to the cost of the traditional diet was $123/ton (McCaskey et al., 1994). This indicates that a beef producer could pay up to $123/ton for a 50:50 corn/litter feed and feed costs relative to animal performance would favor the corn/litter diet. But as the cost of the corn/litter feed increases above $123/ton, the economic advantage of feeding litter diminishes.

MANAGEMENT OF POULTRY LITTER AS A FEEDSTUFF

The composition and feed quality of broiler poultry litter are highly variable. The type of bedding material used in poultry houses, the number of broods of birds reared on the litter, and soil contamination of the litter during removal of the litter from broiler poultry houses are some factors that affect the chemical composition and thus the feed quality of the litter (McCaskey et al., 1990). A survey of the proximate and mineral analyses of 106 litter samples collected throughout Alabama showed that the feed value of litter is extremely variable (Stephenson et al., 1990). Crude protein (CP) is one of the more valuable components of litter when it is considered as a feed ingredient. The CP of the 106 samples averaged 24.9% and ranged from 14.4% to 37.5%. Ash content of the litter was variable, and values above 15 to 18% indicate soil contamination of the litter as it was removed from dirt floor poultry houses. Ash content of samples collected after litter was removed from the poultry houses was higher

(28.6%) than before the litter was removed (20.2%). The average ash content of the 106 litter samples was 24.7% and ranged from 8.9 to 54.4%. Moisture content of the litter ranged from 4.7 to 39% and averaged 19.5%. The effect of chemical composition of litter on its feed value is obvious when litter contains, for example, 40% ash and 30% moisture. On a wet basis, 70% of such litter would have essentially no feed value. Litter that is acceptable for feeding should have less than 25% moisture, less than 30% ash, and more than 19% crude protein (McCaskey et al., 1990). Of the 106 litter samples surveyed in Alabama, 62% would have met these criteria. Seventy nine percent of the litter samples collected prior to removal from broiler houses would have met the criteria, but only 52% of the samples collected after the litter was removed from the houses would have been acceptable for feeding. Increased ash content associated with litter removed from dirt floor broiler houses was the principal obstacle of meeting the feed quality criteria.

After litter is removed from the poultry house, how it is managed affects its feed value. Because enteric pathogenic bacteria, such as *Salmonella*, might be in the litter, all litter intended to be used as a feed ingredient must be processed prior to feeding. The most common and the most economical method of storing and processing litter is by the method called deep stacking or dry stacking, which is essentially piling the litter into a conical pile or stack as it is removed from the broiler house. Heat is spontaneously generated in the stack and should not exceed 140°F. Litter with more than 25% moisture, if not covered to exclude air, will generate temperatures in excess of 140°F, which damages the feed value of the litter (McCaskey et al., 1990). The damage is due to insolubilization of the litter nitrogen, which decreases the digestibility of the litter nitrogen by ruminants. Some researchers are of the opinion that insoluble or bound nitrogen of litter might serve as by-pass or escape protein for ruminants. This concept has not been widely accepted because composted broiler litter, which is heated to high temperatures and has high levels of bound nitrogen, is a poor crude protein source for ruminants. Spontaneous heating of litter stacks is due to microbial activity. The organisms use the most readily available nutrients in the litter, and when their activity subsides, as evidenced by a decrease in the litter temperature, the readily available dietary protein and energy contents of the litter have been depleted of dietary value for ruminants. Litter that is very dry and black and looks like large particles of coffee grounds is typical of heat-damaged broiler litter. As a general guideline, litter with more than 25% of its nitrogen bound or insoluble should not be used for feeding. One of the more important improvements that has been made in recent years relative to the use of poultry litter as a cattle feed has been the education of cattle producers that all broiler litter is not acceptable quality for feeding. Cattle producers have become more aware of good quality litter, and their satisfaction with the feeding of litter has greatly improved also.

Other methods of processing broiler litter to preserve its feed value and to render it safe from pathogens include 1) mechanical heating such as pelleting of litter and 2) acidification of litter either by direct addition of acid or by lactic acid fermentation of the litter. A description of these processing methods and how they eliminate enteric bacteria from litter has been published earlier (McCaskey and Anthony, 1979). All these methods, if properly used, are effective in preserving the nutrient value of litter and also ensuring that the litter is safe from enteric pathogenic bacteria.

One way to promote marketing of by-products is to add value or to increase the value of the by-products. The feed value of broiler poultry litter might be improved by increasing the crude protein content and, simultaneously, decreasing the ash content of the litter. The crude protein is the most valuable component, and ash and fiber are low-valued components. Sieving broiler litter has been evaluated to separate the higher and lower valued components of litter. A small amount of crude fiber, especially from litters that have large quantities of visible wood chips, can be removed with 4-mm screens (Koon et al., 1992). Sieving litter through progressively smaller screens increases the crude protein content of the smaller litter particles. However, the benefit is nullified because the ash content associated with the smaller litter particles also increases. Screening litter does have some benefits, but not for increasing the crude protein content of litter. Large fragments of wood and hardware are sometimes found in litter, and screening to remove these materials is essential to protect the health of animals fed the litter.

Pelleting is another method of adding feed value to poultry litter. Although the process does not add nutritive value, it increases the bulk density of litter and changes the physical form of litter, making it amenable for marketing as a commercial feedstuff. Pelleting also has the advantage of preventing stratification of litter components and other feed supplements that might be mixed with litter. Because many commercial feeds are pelleted, pelleting of broiler litter would appear to be a logical process for marketing litter as a feedstuff. A major disadvantage has been the high cost of pelleting litter. Pelleting adds about $30/ton to the cost of the litter, and most of these operations have had limited success in marketing pelleted litter as feed.

The acceptance of broiler poultry litter as a feedstuff is widely recognized in the southeastern US where poultry litter is plentiful and inexpensive. The acceptance was not immediate but was gradually acquired over several years. Many cattle producers began feeding litter when they learned of the economic benefits of feeding litter from neighboring cattle producers. The feeding of poultry litter to beef cattle has become an accepted feeding practice in many of the broiler poultry production areas of the southern US. Provided that feed-quality litter can be purchased at low cost relative to traditional feeds, the economic incentive to use litter as a feedstuff is a strong motivation to manage litter as a valuable resource, which greatly diminishes its impact on the environment.

REFERENCES

Anthony, W.B., Utilization of animal waste as feed for ruminants, *Proc. Nat'l. Symp. on Animal Waste Management*, ASAE Pub. SP-0366, 109, 1966.

Bohstedt, G., Grummer, R.H., and Ross, O.B., Cattle manure and other carriers of B-complex vitamins in rations for pigs, *J. Anim. Sci.*, 2, 373, 1943.

Cross, D.L., Skelley, G.C., Thompson, C.S., and Jenny, B.F., Efficacy of broiler litter silage for beef steers, *J. Anim. Sci.*, 47, 544, 1978.

Cullison, A.E., McCampbell, H.C., Cunningham, A.C., Lowery, R.S., Warren, E.P., McLendon, B.D., and Sherwood, D.H., Use of poultry manures in steer finishing rations, *J. Anim. Sci.*, 42, 219, 1976.

Davis, G.V., Troxel, T.R., Burke, G.L., and Duncan, R.L., Supplemental corn and broiler litter for stocker cattle on legume-grass pastures and subsequent feedlot performance, University of Arkansas Cooperative Extension Service Report, September 26, 1994.

Duffy, P.A., Nawrocki, G., McKee, D.D., and Novak, J.L., The economics of using a broiler-litter/grain supplement for winter stockers on rye cover in peanuts, in Alabama, *Proc. 1994 Nat'l Poultry Waste Management Symposium*, University of Georgia, 292, 1994.

Fontenot, J.P., McClure, W.H., Kelly, R.F., and Litton, G.W., *The value of poultry litters as feedstuffs for fattening beef steers*, Va. Agr. Exp. Station Livestock Res. Prog. Report, 9, 1963.

Fontenot, J.P., Smith, L.W., and Sutton, A.L., Alternative utilization of animal wastes, *J. Anim. Sci.*, 57(Suppl. 2), 221, 1983.

Fontenot, J.P. and Webb, K.E., Jr., Health aspects of recycling animal wastes by feeding, *J. Anim. Sci.*, 40, 1267, 1975.

Free, J.W., Economic aspects of feeding waste to ruminants, Proc. Alternative Nitrogen Sources for Ruminants, Atlanta, Georgia, 86, 1977.

Hammond, J.C., Cow manure as a source of certain vitamins for growing chickens, *Poulty Sci.*, 21, 554, 1942.

Kirk, J.K., Statements of general policy or interpretation on use of poultry litter as animal feed, *Fed. Register*, 32(171), 12714, 1967.

Koon, J.L., Flood, C.A., Turnbull, R.D., McCaskey, T.A. and Brewer, R.N., Physical and chemical characteristics of pine shavings poultry litter, *Trans. ASAE*, 35, 1653, 1992.

Kornegay, E.T., Holland, M.R., Webb, K.E., Jr., Bovard, K.P., and Hedges, J.D., Nutrient characterization of swine fecal waste and utilization of these nutrients by swine, *J. Anim. Sci.*, 44, 608, 1977.

McCaskey, T.A., and Anthony, W.B., Human and animal health aspects of feeding livestock excreta, J. Anim. Sci., 48, 163, 1979.

McCaskey, T.A., Ruffin, B.G., Eason, J.T. and Strickland, R.C., Value of broiler poultry litter as feed for beef cattle, Proc. 1994 National Poultry Waste Management Symposium, University of Georgia, 267, 1994.

McCaskey, T.A., Stephenson, A.H., and Ruffin, B.G., Factors that influence the marketability and use of broiler litter as an alternative feed ingredient, *Sixth Internat'l Symp. on Agr. and Food Processing Wastes*, ASAE Publication 05-90, 197, 1990.

McCaskey, T.A., Sutton, A.L., Lincoln, E.P., Dobson, D.C., and Fontenot, J.P., Safety aspects of feeding animal wastes, *Proc. Fifth Internat'l. Symp. on Agr. Wastes*, ASAE SP13-85, 275, 1985.

Noland, P.R., Ford, B.F. and Ray, M.L., The use of ground chicken litter as a source of nitrogen for gestating-lactating ewes and fattening steers, *J. Anim. Sci.*, 14, 860, 1955.

Smith, L.W., and Wheeler, W.E., Nutritional and economic value of animal excreta, *J. Anim. Sci.,* 48, 144, 1979.

Southwell, B.L., Hale, O.M., and McCormick, W.C., *Poultry House Litter as a Protein Supplement in Steer Fattening Rations,* Ga. Agr. Exp. Station Memo, Series N.S. 55, 1958.

Stephenson, A.H., McCaskey, T.A. and Ruffin, B.G., A survey of broiler poultry litter composition and potential value as a nutrient resource, *Biol. Wastes,* 34, 1, 1990.

Webb, K.E., Jr., and Fontenot, J.P., Medicinal drug residues in broiler litter and tissues from cattle fed litter, *J. Anim. Sci.,* 41, 1212, 1975.

MULTIPLE POLICY INSTRUMENTS:
AN EVOLUTIONARY APPROACH
TO ANIMAL WASTE MANAGEMENT

W.G. Boggess and M.J. Cochran

INTRODUCTION

One-third of all agricultural non-point source pollution in the US has been attributed to animal operations, particularly those relating to dairy and poultry operations (US EPA, 1994). Public concern over the contamination of water from livestock, dairy, and poultry wastes is renewing interest in producer regulation. In addition, there is an on-going effort at the federal level to set overall water quality goals and provide general guidance for management of non-point source pollution. Hence, animal production industries are faced with increasing prospects for federal, state and local environmental policies.

There are many economic dimensions to the development of public policy to manage animal wastes. This chapter will focus on the contributions that economics can provide to assist policy development. The selection of preferred policy instruments is fraught with conflicts about what group action is appropriate. In some situations, the well-being of some groups can be improved without imposing costs on other groups (i.e., Pareto improvement). However, it is much more common that trade-offs exist so that to enhance the well-being of one group it is necessary that another group be asked to sacrifice. Economics provides a framework from which the trade-offs can be evaluated, and under certain circumstances policies that maximize the well-being of society can be identified. The policies that maximize social welfare are sometimes referred to as "first best" solutions and are found where the marginal benefit is equal to the marginal cost. At this point, alterations to the policy will result in an overall loss to society because adjustments will produce incremental losses that exceed incremental gains.

Economics has more to contribute than just "first best" solutions. Often the information necessary to identify "first best" policies that maximize social well-being is not readily available. In these cases, economics still can contribute with a focus on cost effectiveness by identifying which policies will achieve standards specified by the political process at the lowest possible cost (Baumol and Oates, 1988). Both the distribution of the benefits and costs and the impact

503

on the efficiency of the economic system are important criteria in policy design. These are all areas in which economics has unique contributions to make to policy design.

The purpose of this chapter is to 1) discuss the key economic dimensions of water quality problems associated with animal waste; 2) describe alternative policy instruments that could be used to manage these problems; and 3) discuss how animal waste policies may evolve.

ECONOMIC DIMENSIONS OF ANIMAL WASTE AND WATER QUALITY

EXTERNALITIES AND DECISION-MAKING

To understand the economic dimensions of conflicts arising from animal waste management, it is necessary to briefly relate some key economic concepts on decision-making at the firm level. Neo-classical theory of the firm suggests that businesses behave by adopting the management options that maximize profits. However, there are some characteristics of the animal waste problem that may not be directly reflected in either the costs of production or in the revenue received by growers. These characteristics are "external" to the calculus of profit maximization and hence in an unregulated market may not influence management decisions. Externalities are not unique to animal production, but, as the First Law of Thermodynamics implies, they are examples of residuals that are inherent aspects of any production process. The management challenge is to change the magnitude, form, location, or time of distribution of the residuals to minimize their undesirable consequences.

Externalities can cause conflicts about animal waste management since many of the associated benefits and costs are not reflected in market signals. These non-market benefits and costs can disrupt the market efficiency and generate conflicts about whose preferences will count in how animal wastes are managed. The policy dilemma then becomes a question of how best to correct for these impacts.

ECONOMIC/POLICY PERFORMANCE MEASURES

A number of performance measures are relevant to the evaluation of animal waste management policies. The most common performance measures include (1) social benefit/cost analysis, (2) cost effectiveness analysis, (3) distribution of costs and benefits, and (4) philosophical or ethical concerns.

Social Benefit/Cost Analysis

Maximizing social well-being requires the economically efficient production of goods and services. Efficient production ensures that resources are not wasted and that the cost of producing (including externalities) any given level of output is minimized. Operationally, social benefit/cost analysis procedures are commonly used to evaluate social well-being implications of alternative policies. Social benefit/cost analysis rests on the assumption that, given the existing distribution of income and resources, competitive market prices for goods and service provide socially optimal weights for aggregating the output of goods and services into a single benefits value. Social benefit-cost analysis becomes problematic when market prices are distorted by government policies (e.g., milk price supports) or market structure (e.g., concentration in the broiler and hog industries) or when externalities (e.g., nutrient runoff) are present since important cost or benefit items escape competitive market valuation processes. In these cases, the economic analyst uses non-market valuation and shadow price techniques to estimate the appropriate prices.

An even bigger conceptual problem confronting the use of social benefit/ cost analysis is the absence of a unique social well-being maximum, since the current set of market prices is determined by the existing distribution of resources and property entitlements (Bromley, 1990). Thus, unless the current distribution of income and resources is deemed socially optimal, the resulting market prices will not reflect society's optimal weighting of various goods and services.

Recognition of the limitations of social benefit/cost analysis as a means of identifying "optimal or first best" policies often results in the political specification of alternatives for consideration. Economic analysts are then often called upon for cost-effectiveness analyses to evaluate the relative costs of the alternative policies. Cost-effectiveness analysis in the pure sense is designed to determine which of a set of alternatives meets the specified policy goal at the lowest cost to society.

Distributional Considerations

Distributional measures refer to indicators of who benefits from and who pays for a particular policy. If a particular policy maximizes social well-being, then theoretically everyone could be made better off since gainers from the policy could compensate losers. However, compensation is difficult and generally not an integral part of a policy (Boadway and Bruce, 1984). If compensation is not paid, then potential losers will likely prefer their bigger piece of the current pie to a smaller piece of the potentially bigger pie. In addition, the identity of gainers and losers depends upon the particular set of property rights in effect. Thus, philosophical considerations often dominate discussions of distributional implications of alternative policies.

At times economists have restricted discussions of distributional impacts of policies to issues of Pareto optimality. However, for most policy alternatives, distributional implications will be a key component of political discussions, and notions of justice become relevant. Three notions of justice are particularly germane to the evaluation of environmental policies. One is the polluter pays principle, which implies that the entity responsible for environmental damages should bear the associated costs (Young, 1992). One problem with a full application of the polluter pays principle is that it reduces incentives for potential victims to avert damages. The second notion of justice of import to environmental policy analysis is intergenerational equity (Page, 1991). Intergenerational equity refers to the appropriate weighting between current and future benefits and costs. The third important notion of justice relates to private property regimes and the associated issues of "takings" and "just compensation" if these regimes are affected by a policy. As distributional concerns are likely to dominate political discussions of alternative environmental policies, policy analysts need to be prepared to identify gains and losses from alternative policies under alternative specifications of property entitlements and distributions of income.

Philosophical Considerations

Philosophical differences among key interest groups may also present difficulties in evaluating the performance of alternative policies (Castle, 1993). The prevailing economic theory (i.e., neoclassical theory) is philosophically anthrocentic, utilitarian, and consequentialist. Anthrocentric implies that human values provide the weights for evaluating alternative policy impacts. Utilitarianism implies that the weights for evaluating alternative policy impacts arise from individuals' assessments of what yields the highest level of net satisfaction or utility. Finally, consequentialist implies that consequences matter, in particular economic consequences of alternative policies matter.

There appear to be two alternative philosophical perspectives in the environmental community. One group has much in common with neoclassical economists. They also are anthrocentric, utilitarian, and consequentialist. The differences revolve around which consequences are important, how the consequences are measured, and how the anthrocentric interests are represented. This group of environmentalists tend to focus on long-run carrying capacity and the environment's capability to support mankind. They also recognize that prices are a function of the existing set of property entitlements, which reflect the prevailing economic view of important consequences and relative values. They believe that these entitlements need to be changed and, thus, that a different set of consequences and weights are relevant for policy analyses.

The second set of environmentalists are biocentric rather than anthrocentric. Biocentric environmentalism reflects a belief in the intrinsic beauty and moral goodness of the natural world. Taken literally, biocentrism would accord no more importance to humans than to any other part of nature. This is a fundamentally different philosophical starting point that provides much less opportunity for compromise.

REGIONAL ECONOMIC IMPACT ASSESSMENT

Social benefit/cost analysis attempts to determine socially optimal policy alternatives. As such, distributional considerations are secondary and wealth transfers from one segment of society to another are ignored in the calculations. Regional economic impact assessments, on the other hand, are commonly done to determine how a particular policy will affect the economy of a specific region. Thus, regional economic impact assessments focus only on benefits and costs that accrue within the specified region, including transfers into or out of the region.

Many times environmental policies affect the profitability of agricultural activities. If these policies are only implemented in specific regions or if the impacts differ across regions, regional costs of production and, thus, comparative advantages will often be affected. In these cases, the impact on the regional economy associated with changes in agricultural production may be of interest. For example, concerns that the state would be placed at an economic disadvantage played a role in eroding legislative support for a proposed per-bird tax in Arkansas to manage environmental loadings from poultry litter.

It is important to recognize that existing patterns of regional production arose in response to historical comparative advantages. If important externalities exist in the production of particular agricultural products (e.g., erosion, water pollution, etc.), these costs have historically not been reflected in costs of production underlying comparative advantages. Thus, policies that address the costs of externalities may be merely correcting historically biased comparative advantages.

TECHNOLOGICAL CHANGE AND ADOPTION

For the past century, new information reflected in new technologies has allowed multi-factor productivity in agriculture to increase at an average annual rate of 1.53% (Huffman and Evensen, 1993). Technological change has been the driving force behind this steady increase. Technological change can arise via two mechanisms. First, basic research may result in new information that provides spinoff applications in the form of new products or new production techniques. Second, economic conditions may change relatively abruptly (e.g., new environmental regulations), inducing producers to adapt in order to remain competitive (i.e., induced technical change).

Induced technical change is an important response to policy changes. Producers react to changes in incentives brought about by environmental policy changes. The result is that the direct impacts of policies are generally smaller than would be predicted based on current technologies and costs. For example, a series of environmental regulations have been imposed on dairies in the Lake Okeechobee watershed in southern Florida (Boggess, 1994). Compliance with these regulations stimulated significant innovation in dairy technologies and management practices. During the three-year period prior to implementation of the regulations (1986-1988), average milk production per cow in the state as well as in the two counties most affected by the regulations increased by approximately 3%. During the three-year period following compliance, average milk production per cow in the affected counties increased by 19%, compared to an increase of only 2% in the rest of Florida. (Calculated from Florida Agricultural Statistics Service data.)

Technological change commonly has the unintended impact of altering economies of size. The tendency toward greater concentration in livestock operations may be accelerated by environmental regulations. In many cases, there are significant economies of size associated with complying with environmental regulations.

INFORMATION AND UNCERTAINTY

Information to guide the policy process is often lacking. Externalities by their very nature are external to existing market accounts and difficult to measure. In cases of non-point source pollution, the matter is further complicated because of geographic dispersal and stochastic processes so that measurement costs on a per-unit basis are exceedingly high. The stochasticity of externalities is important because it introduces variability in the consequences of waste management practices and disguises cause and effect. The same practice implemented under different weather, soil, and/or hydrologic conditions can produce externalities of varying magnitudes. However, policy instruments may have a more consistent impact on costs of production. Since the costs of a policy instrument are often much easier to document than the benefits, the impact on the policy debate of uncertainty may be asymmetrical. Policy makers are concerned about the certainty of effect, and those alternatives with less certainty may be viewed less favorably.

Policy making under uncertainty is further challenged by the fact that cleanup of an environmental problem is often more expensive than prevention. Many will advocate a safe minimum standard to manage risks so that present actions do not unduly restrict future options foraction. However, evidence is accumulating that damage avoidance or treatment of select, target populations may be more cost effective than prevention in some instances (Stipp, 1994).

POLICY MECHANISMS FOR WATER QUALITY IMPROVEMENT

GOALS OF POLICIES

The goals of public policies can be quite diverse and at times inconsistent with one another. They can be categorized into three broad sets. One major set of policy goals is to reduce or resolve conflicts by establishing rules under which exchanges can be conducted in an efficient and economical manner. A second set of goals focuses on the incentive structure to alter the magnitude, form, location, or time of distribution of residuals to control their undesirable consequences. Another design issue is the relative attention paid to controlling the probability of damage occurring or the magnitude of damage. Incentives can be different depending upon whether the focus is on low probability, high outcome events or high probability, low outcome events. The final set of goals worthy of note is the resolution of distributional issues. The public policy process must address questions not only of efficiency but of equity and economic justice as well.

POLICY RESPONSES/MECHANISMS

Public policy mechanisms available to help solve animal waste management problems can be grouped into four categories: (1) public action designed to facilitate private solution; (2) public establishment of ex ante economic incentives; (3) government regulation; and (4) judicial options. (See Bohm and Russell, 1985, for a detailed comparative analysis of alternative policy mechanisms.) This taxonomy reflects four fundamentally different means of providing incentives for private decision makers to modify their behavior. Public action to facilitate private solution and judicial options require the least government involvement. Public establishment of ex ante economic incentives entails a moderate level of government involvement. Finally, direct government regulation requires the most direct government involvement.

A number of public actions have been used to facilitate private solutions to environmental problems. One of the most common is funding research designed to develop new technologies to help provide cost-effective alternatives for producers. A second approach is to provide cost share money to make it economically attractive for producers to adopt specified best management practices. Third, education programs designed to inform producers about environmental problems have been coupled with moral suasion to encourage producers to voluntarily adopt best management practices.

Public establishment of ex ante economic incentives can be achieved via charges, marketable permits, tax credits, or easements. Charge approaches are based on the assumption that an optimal "price" of pollution can be established and then private decision makers will adjust their production practices to reflect this additional cost. Charge approaches set the price of pollution and then let the market determine the resulting level of pollution. Charges are designed to reflect the costs of pollution so that these costs are explicitly accounted for in production and management decisions. Charges may be assessed based on the level of input use, output produced or effluents discharged. Effluent charges are the most direct means of reflecting pollution costs, but non-point effluent discharges are generally too expensive to monitor. Input and output charges represent indirect measures of effluent discharges but have the advantage of being cheaper to monitor and enforce.

All charge programs face the technical difficulty of determining what level of tax is optimal. Theoretically, the charge should reflect the marginal cost of pollution associated with the discharge of one more unit of effluent, the use of one more unit of input, or the production of one more unit of output. In practice, this value is difficult to determine without considerable search via trial and error adjustments. Thus, many charge programs have been designed with relatively low tax rates to encourage some reduction in pollution, but with the primary goal of generating revenue to help fund research and cost share programs.

Marketable permit programs provide an alternative means of establishing economic incentives that avoid the technical difficulty of specifying the optimal charge. In a marketable permit program, the government creates a fixed quantity of pollution permits, which can be bought and sold by firms. Each pollution permit gives the owner the right to discharge a fixed amount of effluent. Firms are free to buy or sell permits based on a comparison of the market price for permits versus their own private costs of reducing effluent discharges. When pollution abatement costs are non-homogeneous, marketable permits allow market forces to be exploited to obtain cost-effective pollution reduction. However, implementation of a marketable permit program is equally as dependent as effluent charges upon the ability to monitor effluent discharges in a cost-effective manner.

Two types of direct government regulation have been used to achieve environmental policy goals. Technology (design) standards specify certain technologies and production practices that must be implemented by producers (e.g., lined lagoons, with land application of dairy waste). This approach is also known as "command and control" since the government dictates and controls production practices. The two biggest drawbacks of command and control approaches are the inflexibility (i.e., all firms may be forced to use the same technology) and the indirect linkage between technology employed and actual emission reductions achieved. Another major drawback of technology standards is that they eliminate incentives for innovation or the development of improved techniques. Porter and van der Linde (1993) argue from a competitiveness perspective that dynamic efficiency considerations swamp static efficiency, and thus it is imperative that environmental policy stimulate innovation.

Performance standards avoid the major drawbacks associated with design standards by specifying maximum levels of allowable effluent emissions but not specifying how firms meet the standard. However, implementation of performance standards requires that it be cost effective to monitor effluent emissions. A combination of design and performance standards has been imposed on dairies in the Lake Okeechobee watershed in an effort to control phosphorus discharges (Boggess, 1994). The "Dairy Rule" design standard requires that dairies implement technologies to capture runoff from a 25-year, 24-hour storm for all of the highly impacted areas of the dairy, and that this runoff be land applied at agronomic rates. Dairies with both high and low phosphorus discharges were subject to the same basic design requirement. Compliance with the design standard required only that the dairy construct an approved plan regardless of the ultimate effect on phosphorus discharges. Subsequent to the "Dairy Rule," a performance standard was imposed limiting the phosphorus concentration in discharge waters from dairies to 1.2 mg/L. No specific technologies or management practices are mandated. The performance standard provides more flexibility for producers to control phosphorus discharges than does the technology standard; however, the monitoring and enforcement costs associated with the performance standard are much greater than for the technology standard.

Judicial options include use of strict liability and negligence-based liability rules. Under strict liability the generator of damages is liable for the damages, regardless of whether appropriate precautions were taken. Under negligence-based liability rules, the generator of the external damages is liable only if he/she failed to exert "due standard of care." Liability instruments have the advantages of dynamically adjusting over time to exogenous changes and providing continuing incentives to seek new technologies to reduce expected damages. Drawbacks of liability systems include their ex post nature (i.e., damages are compensated, rather than prevented), enforcement through the courts is expensive, and the burden of proof can be substantial.

GENERAL ASSESSMENTS OF POLICY INSTRUMENTS

In the previous sections, we discussed a number of alternative policy instruments and performance measures that can be used to compare instruments. Each policy instrument has its strengths and its drawbacks--there are no silver bullets in the environmental policy arsenal. The evaluation of public policy can employ a number of performance measures. These include 1) static efficiency; 2) the costs and value of required information; 3) government failure; 4) litigation costs; 5) dynamic efficiency; and 6) distributional and ethical concerns. Table 1 provides a subjective comparison of policy mechanisms based on alternative performance measures. Many of the key policy mechanisms and important performance measures are included in the table, but neither set is comprehensive. The comparisons are necessarily subjective and attempt to reflect general differences and tradeoffs among alternative mechanisms. In practice there is considerable flexibility in how individual mechanisms or combinations of mechanisms might be applied that could greatly affect performance on certain measures. For example, technology standards might be applied with sufficient flexibility that static efficiency is not seriously impacted.

In general, the appropriate policy instruments will depend upon the particular characteristics of the problem, which may evolve over time. Boggess (1994) provides a case study description of programs designed to control phosphorus runoff from dairies that illustrates how key components (e.g., technologies, monitoring programs, and policy mechanisms) evolved over time. Technologies evolved to effectively convert a primarily non-point source into a point source. Monitoring programs evolved in purpose and design from an initial focus on problem assessment, to measuring efficacy of practices, and, finally, to providing a basis for determining compliance with performance standards. Finally, policy mechanisms evolved from purely voluntary with full cost sharing, to a voluntary easement program, to a mandatory technology based standard with partial cost sharing, to finally a performance based standard with no cost share.

Historically three types of policy instruments have been most commonly used to address environmental problems. The most pervasive approach in common law countries is negligence liability rules. Under common law doctrine,

Table 1. Subjective comparison of policy mechanisms based on alternative performance measures.

| | Policy Mechanisms[z] | | | | | | | | | | | |
| | Group 1 | | | Group 2 | | | | | Group 3 | | Group 4 | |
Performance Measures[y]	Gov't funded research	Gov't cost share	Education/ moral session	Purchase easements	Effluent charge	Input/ output charge	Market- able permits	Tax credits	Techno- logy standards	Perfor- mance standards	Strict liability	Tort liability
Government Expenditures	L[x]	H	L	H	R	R	RorM	M	L	L	L	L
Static Efficiency	N.A.	M	H	L-M	H	M	H	L-M	L-M	H	L	L
Information/Enforcement/ Litigation Costs	N.A.	L	N.A.	L	H	M	M	L	L-M	H	M	H
Dynamic Efficiency	M	L	L	L	H	M	H	L	L	H	M	L
Polluter Pays Principle	I	I	N	I	C	C	C	I	C	C	C	N
Property Rights Concerns	L	L	N.A.	L	M	M	M	L	H	M	L	L

[z]Group 1 - Public action designed to facilitate private solution; Group 2 - Public establishment of *ex ante* economic incentives; Group 3 - Government regulation; Group 4 - Judicial options

[y]Comparative indicators of performance:

[x]L=Low, M=Medium, H=High, R=Generates Revenue, N.A.=Not Applicable, N=Neutral, I=Inconsistent, C=Consistent

affected parties have the right to sue perpetrators for the damages caused. The second most prevalent approach in the agricultural environmental policy arena is use of public action to promote private solutions, including research and development, cost share, and education combined with moral suasion. Since these approaches are voluntary, participation rates are often low if there are no clear economic advantages. Cost share or subsidizing participation has been used to increase participation rates, but these approaches require significant government outlays.

Command and control approaches have been the most common approach to point source environmental problems. Technology standards have the attractions of providing a sense of a "certainty of effect," they tend to be preferred by the regulated firms, environmentalists believe that they stigmatize pollution rather than providing a "license to pollute," as is implied by permit approaches, and they incur lower implementation and monitoring costs. With the exception of monitoring costs, the other perceived advantages often are illusionary. A design standard specifies only that a particular technology be installed; there are no explicit incentives for the technology to be operated efficiently.

Public establishment of ex ante economic incentives is generally the approach favored by economists on efficiency grounds. Economic incentives facilitate decentralized decision making by firm managers in response to environmental price signals. This enhanced flexibility allows firms to address pollution reduction in the least cost fashion for each firm. Short-run efficiency of economic incentives is generally as good as or better than regulations. Long-run efficiency incentives are much stronger using economic instruments. The major limitation to the use of economic incentives to address non-point problems is the high costs of measurement.

Several factors can be instrumental in increasing the efficiency of environmental policies, regardless of the particular policy instrument or combination of instruments employed. First, maintaining as much flexibility for individual firms as well as groups of firms (i.e., trading schemes) to find the least cost option for reducing pollution will reduce the overall costs of the policy. Second, it is important that policies provide incentives for firms to innovate over time to find new, more effective control mechanisms. Finally, sufficient lead time should beprovided to allow firms to make the necessary adjustments in an orderly fashion.

CONCLUSIONS

Programs designed to solve complicated, non-point pollution problems generally will evolve over time. The political process required to address distributional concerns, uncertainty and lack of information about the problem and alternative solutions, and administrative inflexibility all but guarantee a cautious step-by-step approach. Policy design can be improved by explicit recognition of key economic dimensions of non-point pollution problems, including the non-linear nature of control cost functions, the rapidly escalating costs of achieving greater reliability of controls, and the difficulty and cost of monitoring and enforcing non-point source pollution policies. Given the complexity of non-point pollution problems, no single policy instrument is likely to achieve the desired goals of maximizing static and dynamic efficiency, minimizing information and enforcement costs, avoiding government failure, minimizing litigation costs, and equitably addressing distributional and ethical concerns. The appropriate mix of policy instruments will depend on the particular characteristics of the problem and will often evolve over time. However, despite the evolutionary nature of non-point policy, policy makers need to minimize regulatory uncertainty in order to decrease the option value associated with a "wait and see" strategy and increase firms' willingness to undertake costly investments in pollution reducing technologies.

REFERENCES

Baumol, W.J. and Oates, W.E., *The Theory of Environmental Policy*, Second Edition, Cambridge University Press, New York, 1988.

Boadway, R.W., and Bruce, N., *Welfare Economics*, Blackwell, Oxford, 1984.

Bohm, P., and Russell, C., Comparative analysis of alternative policy instruments, Chap. 10, in *Handbook of Natural Resource and Energy Economics*, Vol. 1, Kneese, A.V. and Sweeney, J.L., eds., Elsevier Science Publishers B.V., 1985.

Boggess, W.G., A case study of nutrient management for Florida dairies, in *Economic Issues Associated with Nutrient Management Policy*, SRIEG-10 Workshop Proceedings, Southern Rural Development Center, Mississippi State, Mississippi, March 1994.

Bromley, D.W., The ideology of efficiency: Searching for a theory of policy analysis, *Journal of Environmental Economics and Management*, 19, 86, 1990.

Castle, E.N., A pluralistic, pragmatic and evolutionary approach to natural resource management, *Forest Ecology and Management*, 56, 279, 1993.

Huffman, W.E., and Evenson, R.E., *Science for Agriculture: A Long-Term Perspective*, Iowa State University Press, Ames, Iowa, 1993.

Page, T., Sustainability and the problem of valuation, Chap. 5, in *Ecological Economics: The Science and Management of Sustainability*, R. Constanza, ed. Columbia University Press, 1991.

Porter, Michael E., and van der Linde, Class, *Towards a New Conception of the Environment-Competitiveness Relationship*, Working Paper, Harvard Business School, November 1993.

Stipp, D., Prevention may be costlier than a cure, *Wall Street Journal*, July 6, 1994.

US EPA, *Office of Water, National Water Quality Inventory: 1992 Report to Congress*, EPA 841R94001, March, 1994.

Young, M.D., *Sustainable Investment and Resource Use*, The Parthenon Publishing Group, 1992.

LIVESTOCK POLLUTION:
LESSONS FROM THE EUROPEAN UNION

Philip W. Gassman and Aziz Bouzaher

INTRODUCTION

Major structural changes are occurring in the US livestock industry, driven by economic, institutional, and technological factors, resulting in consolidation of production with a greater concentration of animals per livestock farm. Simultaneously, there has been a rise in environmental problems and public conflicts associated with intensive livestock production. Over the past few decades, there has been a similar intensive build-up of animal production in several European regions. Concurrently, an increase in environmental pollution has occurred from the disposal of excess animal manure. As a result, an increasing number of regulations and directives are being implemented at the national and European Union (EU) levels to mitigate these pollution problems.

The objectives of this chapter are (1) to describe emerging policy and regulatory trends at both the EU and national levels and (2) to discuss potential lessons and implications of the European experience for the US livestock industry. The primary focus of the chapter is on those countries visited by the authors during June 1993 (Belgium, Denmark, England, Germany, and the Netherlands).

BACKGROUND

The severity of the livestock pollution problems in EU-member nations can be linked in part to animal density. Table 1 shows total livestock populations and animal densities in 1987 for the 12 EU member nations (Brouwer and Godeschalk, 1993). The highest animal density calculated was 4.0 livestock units per hectare (LU/ha) for the Netherlands, followed by Belgium, Germany, Denmark, and Luxembourg at 2.9, 1.5, 1.4, and 1.4. In the Netherlands, livestock production dramatically increased between 1950 and 1990 (Dietz and Hoogervorst, 1991), with the total number of cattle, poultry, and pigs increasing by factors of 2, 4, and 7 (to 4.9, 92.8, and 13.9 million animals, respectively). One unintended result of this production increase was a nutrient imbalance with a massive manure surplus (Table 2).

Table 1. 1987 livestock populations and animal densities
in the 12 European Union (EU) member nations.[†]

Country	Cattle	Other grazing animals[‡]	Pigs	Poultry	Total livestock	Animal density
	------------------1000 animals------------------				1000 LU	LU/ha
Germany, F.R.	15231	1329	23989	68696	17552	1.5
France	21856	11650	11777	224908	23204	0.8
Italy	8907	9437	8795	144350	11300	0.7
Belgium	3071	218	5844	23243	3915	2.9
Luxembourg	217	8	75	99	175	1.4
Netherlands	4895	1083	14349	98669	8016	4.0
Denmark	2351	134	9266	15540	3967	1.4
Ireland	6765	5033	911	7844	5773	1.2
United King.	12087	38529	7899	138918	15830	0.9
Greece	678	12730	909	29752	2416	0.6
Spain	5358	23191	12744	125338	11406	0.5
Portugal	1387	3260	2362	31499	2301	0.7
EU-12	82801	106602	98921	908856	105855	0.9

[†]Adapted from Brouwer and Godeschalk (1993); a LU is one livestock unit, representing the nutrient value of feed. See Brouwer and Godeschalk for calculations of LU and LU/ha.
[‡]Other grazing animals include mainly sheep and goats.

Table 2. 1992 manure production and surplus by animal type in the Netherlands
(manure amounts in million tonnes; nitrogen (N), phosphorus (P),
and potassium (K) amounts in million kg).[†]

Animal Type	Production				Surplus[‡]			
	Manure	N	P	K	Manure	N	P	K
Cattle	66.5	380.6	47.0	344.2	2.0	8.1	1.3	6.9
Pigs	16.7	150.3	30.3	86.7	10.4	96.6	19.2	55.0
Poultry	2.6	75.1	16.1	33.5	2.3	66.3	14.4	29.5
Total	85.9	606.0	93.4	464.4	14.7	171.0	34.9	91.4

[†]Adapted from Poppe et al. (1994).
[‡]Surplus refers to excess manure at the individual farm level.

Table 2 also shows the total excess nitrogen (N), phosphorus (P), and potassium (K) produced in the Dutch manure surplus. In 1990, it was estimated that 43 and 82% of Dutch grassland and maizeland was saturated with P (Breeuwsma and Silva, 1992). This P saturation has resulted in contamination of Dutch surface waters via discharge in sub-surface flow. Other environmental problems connected with the manure surplus in the Netherlands include nitrate leaching to ground water, odor, and acidification resulting from ammonia volatilization. These problems are particularly acute in the most vulnerable, sandy soil areas, located in the south and east of the country.

Similar problems have developed in other parts of Europe. In Denmark, the average concentration of nitrate measured in 10,920 wells across the country has increased from 4.3 mg/L to 13.3 mg/L over the past 30-40 years (Jensen,

1993), which is attributed to both manure and inorganic fertilizer applications. The greatest problems occur in the western part of the country, including the southwest where dairy production is concentrated on shallow, sandy soils. Eutrophication of marine waters along the Danish coast from N loadings is also considered a major problem.

In Germany, 50% of N emissions to the atmosphere are attributed to ammonia volatilization from agriculture, resulting in damage to forests (Rustemeyer, 1993). Although estimates vary, it has been calculated that there is a general surplus of 100 kg/ha of N applied across the country. Some of the most acute environmental problems have occurred in Lower Saxony, especially in the Landkreis (County) Vechta, where poultry and pig production has greatly expanded over the past 20 years (Witte and Kramer, 1991). As a result, nitrate levels almost twice the allowable standard were observed in two-thirds of the private wells in Vechta.

Concern about intensive livestock production-related pollution exists in other EU nations as well, such as the Flanders region of Belgium and the region of Brittany in France (Brouwer and Godeschalk, 1993).

THE POLICY AND REGULATORY ENVIRONMENT

Responses to increased environmental damage from livestock production by EU member nations are a function of both EU and national goals. At each level, production and environmental policies affect measures designed to mitigate livestock waste pollution.

At the EU level, this can be illustrated by the milk quota policy (production) and nitrate directive (environmental). Milk quotas were instituted by the EU to limit the milk production of each member nation. Ths has resulted in reductions of both the overall EU and member nation dairy herds (Jördans, 1993). Member nations in turn set individual farm limits. Although not part of the original intent, these restrictions also affect the total amount of dairy cow manure production for specific regions in each country as well as for the member nations as a whole. The EU nitrate directive is designed to control manure applications by limiting applications of N to 170 kg/ha in all "vulnerable areas" by the year 2000. Designation of vulnerable areas was to be determined by the end of 1993 by each individual country (Dietz, 1993). Ultimately, producers in these vulnerable areas who have excess manure surpluses will be required to pay a high tax.

At the national level, an initial attempt was made by the Dutch government in 1984 to address the manure surplus problem by introducing restrictions that prevented pig and poultry farmers from increasing their herd sizes (Dietz and Hoogervorst, 1991). Since that time several countries have adopted regulations to mitigate livestock waste pollution problems, which are briefly summarized here.

Manure application standards based on P_2O_5 and/or N (Table 3) have been

Table 3. Initial manure application limits in terms of P and N in Belgium,
Denmark, England, and the Netherlands[†]

Country	P_2O_5 (kg/ha)			N (kg/ha)			
	Grass-land	Maize land	Other cropland	Grass-land	Crop-land	Pig land	Cattle land
Belgium	200	150	150	400	400	--	--
Denmark[‡]	--	--	--	--	180	180	240
England	--	--	--	250	250	--	--
Netherlands[‡]	250	350	125	--	--	--	--

[†]Data sources: Belgium and England (Hacker and Du, 1993); Denmark (Dubgaard,1992);
TheNetherlands (Dietz and Hoogervorst, 1991).
[‡]Pig and cattle land is all land (grass and crop) on Denmark pig and cattle farms.

adopted in the Netherlands (in 1987), Denmark (in 1988), Belgium (in 1990), and England (in 1991), to reduce the risk of nutrient pollution in ground and surface water. In Belgium, P_2O_5 application limits will be further tightened to 125 kg/ha on all land by 2001 (Hacker and Du, 1993). Similarly, P_2O_5 application limits in the Netherlands will be reduced to 175 and 125 kg/ha on grassland and cropland (including maize) by 1995 (Dietz and Hoogervorst, 1991). Restrictions of the total animal units per hectare (usually 2 to 3 units) have been implemented in four German states (Rustemeyer, 1993) that limit the total amount of N applied in manure.

Beginning in the mid-1980s, livestock producers in Denmark were required to store liquid manure in impermeable tanks and solid manure on impermeable material where runoff can drain to a storage tank (Dubgaard, 1992). By 1995, all producers will be required to have 7- to 9-months storage capacity. Storage requirements also exist for several other EU nations (Brouwer and Godeschalk, 1993).

Manure applications are currently banned during certain time periods in several EU countries. Examples include: (1) between September 1/October 1 and December 31/January 31 in the Netherlands, depending on vegetation (grassland or cropland) and soil vulnerability to nitrate leaching (Brouwer and Godeschalk, 1993) (2) between November 2 and February 15 in the Flanders region of Belgium (Brouwer and Godeschalk, 1993), (3) liquid manure between fall harvest and February 1 in Denmark (Dubgaard, 1992), (4) solid manure in the fall before October 20 in Denmark (unless a green crop is established), and (5) between October 15 and February 1, and October 15 and February 1, on grassland and cropland in the state of Lower Saxony in Germany (Brouwer and Godeschalk, 1993).

Regulations in Germany vary widely among different states and are closely tied to the administration of water protection areas that are designed to protect drinking water sources (Manale, 1991). For example, 5-m filter strips are required along all significant waterways in the German state of Hesson. Application of manure within 10 m of any water course, and within 50 m of a spring, well, or borehole that supplies water for human consumption, is forbidden in

England (Hacker and Du, 1993). In Denmark, uncultivated filter strips of at least 2 m are required along all water courses and lakes (Dubgaard, 1992).

In 1990, the System of Registration of Minerals (SRIM) was introduced in the Netherlands to track nutrient inputs via feed to livestock and the resulting nutrient output in manure (Hacker and Du, 1993). A mandatory nutrient book-keeping system was also introduced in Denmark in 1993 that requires producers to submit nutrient accounts to a central authority (Dubgaard, 1992).

A National Manure Bank was established by law in 1986 in the Netherlands that facilitates efficient redistribution of cattle manure from manure surplus regions to areas that can effectively use it for cropping purposes (Voorburg, 1993). "Surplus manure farmers" must pay a fee for the bank's services. Taxes are imposed in Belgium for excess N and P that can not be applied to a producer's land (Hacker and Du, 1993). In addition, if excess manure is generated that cannot be applied to a producer's land or neighboring land, it must be taken to a manure bank. Further taxes are assessed on excess manure hauled to these banks.

Processing of pig manure has also been initiated by the Dutch government to produce fertilizer pellets for export and domestic use (Voorburg, 1993). The goal is to process 6 million tonnes of manure by 1995. Composting is being attempted in several European countries, but lack of markets has greatly hindered efforts to date.

Measures are being implemented by the Dutch government to reduce ammonia emissions by 70% by 2000 (Hacker and Du, 1993). Beginning in 1994, all surface applications of manure were banned in the Netherlands to reduce ammonia emissions. Likewise, all manure storage facilities must now be capped to prevent ammonia releases to the atmosphere. Belgium regulatory officials are considering installing sensors in livestock facilities to monitor ammonia emissions, with the ultimate purpose of taxing facilities that exceed a standard for ammonia emissions (Berckmanns, 1993).

LESSONS FOR THE LIVESTOCK INDUSTRY
IN THE UNITED STATES

European policy responses to increasing livestock pollution over the past decade have been characterized by three major elements: (1) a stricter regulatory approach, including production quotas, manure manifests, application standards (rates, mode, and timing of application of manure), and, in some cases, charges or levies on inputs; (2) public and private investment in technological research for production automation and waste processing; and (3) environmental education, emphasizing recycling, extension, and local involvement in policy making.

The effectiveness of the European approach is yet to be determined. Already, a debate has emerged around three key points. The first is that livestock manure is not properly utilized because of the lack of economic incentives. For instance, Dubgaard (1992) proposed that a tax or tradeable quota be applied on nitrogen fertilizer in Denmark to encourage the effective use of the nutrient value of manure. He believes this is necessary to overcome an excess utilization of plant nutrients by Danish producers, especially N, in spite of the application standards that have been implemented.

The second issue is that some of the current regulations may not be delivering the desired environmental benefits. For example, the P_2O_5 application standards adopted in the Netherlands may need to be broadened to protect ground water resources from N leaching. Dietz and Hoogervorst (1991) proposed an alternative policy that incorporates both P_2O_5 and N application standards. They calculated that while the alternative policy would cost the Dutch government an additional $55-60 million dollars/year, it would result in negligible costs for the agricultural sector and would provide greatly improved environmental benefits compared to the regulations implemented by the Dutch government.

The third key point is that the economic impact of environmental regulations could be substantial. In a study by Veenendaal and Brouwer (1991), the potential impacts of two policy options designed to reduce ammonia emissions for Dutch agriculture were analyzed. It was estimated that agricultural revenue would decrease by 35% over the period 1985-2010 (annual net revenues were estimated to decrease by roughly 675 million dollars by 2010). However, the total amount of N loadings to the environment were calculated to decrease from 783 kton in 1985 to 131 kton in 2010, indicating clear environmental gains.

The economic impacts of environmental regulations are a major issue in the EU. Protection of producers and markets is a key emphasis in the countries visited, which is reflected in national agricultural and applied research policies targeted to higher levels of efficiency. However, national production goals may increasingly conflict with environmental goals, especially at the EU level. In the Netherlands, for example, it is politically infeasible to reduce animal production at this time, in spite of the environmental problems that have developed. However, environmental improvement may not be realized in the Netherlands without a reduction in stocking rate (Dubgaard 1993). Indeed, EU directives and regulations are likely to play an ever-increasing role in "leveling the playing field" between member countries in terms of "fair" environmental and production policies.

In the United States, the current policy approach to livestock pollution at the federal level is the regulation of Confined Animal Feeding Operations (CAFOs) through the 1972 Clean Water Act (Pagano and Abdalla, 1994) and its amendments. The relevant features of this policy are (1) the requirement of no pollutant discharge from any CAFO with greater than 1000 animal units and (2) the regulation of the disposal of manure (via cropland and other pathways) through a permitting process according to state and local laws, rather than at the federal level. This has resulted in a myriad of state and local regulations (Hacker and Du, 1993) that are applied in an uneven manner.

Current consolidation and integration trends in the US Livestock Industry have led to increased production in relatively small regions, especially for the poultry, hog, and dairy industries (Pagano and Abdalla, 1994). This trend, together with increasing public sensitivity to livestock pollution issues, may portend more stringent environmental regulations at the regional, state, and federal levels, following the pattern that has occurred in the EU. However, a pure regulatory approach may not provide the desired solutions; instead, a judicious mix of regulations and economic incentives combined with stakeholder participation, may prove to be more effective. We recommend that further research be conducted to better quantify (1) the economic and environmental impacts of regulations that have been implemented in the EU and (2) the implications of adopting similar regulations for the US livestock industry.

WILL GATT HAVE AN IMPACT?

Future environmental livestock environmental regulations in the US and EU may also be impacted by the General Tariff and Trade Agreements (GATT) concluded in 1994. Macroeconomic projections averaged for the period 2000-2002 at the sector level indicate impacts on EU and US production levels (which are indicators of herd size levels) are shown in Table 4 (CARD, 1994). With enhanced efficiency in production, income effects are expected to result in an increase in consumption. This is especially true in the pork sector, where the US stands to reap substantial economic gains. The projected production level increases shown in Table 4 could fuel a continuation of the consolidation and integration trends in the swine and poultry industries in the US, leading to a further exacerbation of animal waste management problems.

Table 4. Average production changes projected for European Unit (EU)
and United States (US) livestock sectors for the period 2000-2002[†]

Livestock Type	EU (%)	US (%)
Beef	+3.7	+ 0.6
Pork	- 0.1	+ 3.8
Poultry	- 3.4	+ 1.2
Dairy	+ 0.03	- 0.3

[†]Source: CARD (1994).

ACKNOWLEDGMENTS

This is Journal Paper No. J-16257 of the Iowa Agriculture and Home Economics Experiment Station, Ames, Iowa, Projet No. 2872.

REFERENCES

Berckmanns, D., Personal Communication, Dept. of Agrc. Engr., Catholic Univ., Leuven, Belgium, 1993.

Breeuwsma, A., and Silva, S., *Phosphorus Fertilisation and Environmental Effects in The Netherlands and the Po Region (Italy),* Report 57, DLO The Winand Staring Centre, Wageningen, The Netherlands, 1992.

Brouwer, F.M. and Godeschalk, F.E., *Pig Production in the EC: Environmental Policy and Competitiveness,* Publikatie 1.25, Agricultural Economics Research Institute, The Hague, The Netherlands, 1993.

CARD, The Impacts of the Uruguay Round on U.S. Agriculture, A CARD Symposium, Washington, D.C., Center for Agrc. and Rural Devlp., Iowa State Univ., Ames, Iowa, 1994.

Dietz, F.J., Personnel Communication, Erasmus University, Rotterdam, The Netherlands, 1993.

Dietz, F.J., and Hoogervorst, N.J.P., Towards a sustainable and efficient use of manure in agriculture: The Dutch case, *Environmental and Resource Economics,* 1, 313, 1991.

Dubgaard, A., The Danish Aquatic Environment Programmes - An assessment of policy instruments and results, in *Proceedings of the International Symposium on Soil and Water Conservation: Social, Economic, and Institutional Considerations,* Honolulu, Hawaii, 1992.

Dubgaard, A., Personal Communication, Dept. of Econ. and Natural Res., The Royal Veterinary and Agrc. Univ., Copenhagen, Denmark, 1993.

Hacker, R.R., and Du, Z., Livestock pollution and politics, in *Nitrogen Flow in Pig Production and Environmental Consequences,* Verstegen, M.W.A., den Hartog, L.A., van Kempen, G.J.M., and Metz, J.H.M., eds., Pudoc Scientific Publishers, Wageningen, The Netherlands, 1993.

Jensen, H., Personal Communication, Dept. of Agrc. Sciences, The Royal Veterinary and Agrc. Univ., Copenhagen, Denmark, 1993.

Jördans, R., Personnel Communication, Bundesministerium für Ernährung, Landwirtschaft, und Forsten, Bonn, Germany, 1993.

Manale, A.P., European Community programs to control nitrate emissions from agriculture, *International Environmental Reporter,* 345, 1991.

Pagano, A.P., and Abdalla, C.W., Clustering in animal agriculture: Economic trends and policy, in *Proceedings of the Great Plains Animal Waste Conference on Confined Animal Production and Water Quality, Balancing Animal Production and the Environment,* Storm, D.E. and Casey, K.G., eds., National Cattlemen's Association, Englewood, Colorado, 1994.

Poppe, K.J., Brouwer, F.M., Welten, J.P.P.J., and Wijnands, J.H.M., *Loundbouw, Milie en Economie,* Agricultural Economics Research Institute (LEI-DLO), Periodieke Rapportage PR-68/92, Editie 1994.

Rustemeyer, F.C., Personal Communication, Bundesministerium für Umwelt, Naturschutz, und Reaktorsicherheit, Bonn, Germany, 1993.

Veenendaal, P.J.J., and Brouwer, F.M., Consequences of ammonia emission abatement policies for agricultural practice in The Netherlands, in *Environmental Policy and the Economy,* Dietz, F., van der Ploeg, F., and van der Straaten, J., eds., Elsevier Science Publishers, Amsterdam, The Netherlands, 1991.

Witte, T., and Kramer, M., Ökologisch-Ökonomische Modelle zur Beschreibung und Lösung von Konfliktfällen in Agrarraum, Verhand. der Gesellschaft für Ökologie 19, 467, 1991.

Voorburg, J., Personal Communication, Institute of Agricultural Engineering (IMAG-DLO), Wageningen, The Netherlands, 1993.

TECHNOLOGY TRANSFER IN THE POULTRY INDUSTRY: FACTORS ASSOCIATED WITH INCREASED PRODUCTION

Clare Narrod and Carl Pray

BACKGROUND

Growing population and rising per capita income have lead to an increasing demand for poultry products around the world (Unnevehr et. al., 1992). Countries have attempted to meet this demand by raising domestic poultry production. The number of chickens in the world increased by 53% between 1978-80 and 1988-90, with growth of over 100% in Asia alone (WRI, 1993). Adoption and spread of new technology could be a contributing factor to this increasing production. Technology transfer is crucial for developing countries, where production in the poultry industry is being increasingly commercialized to meet the growing demand. However, the increase in poultry production has resulted in increased manure production, the improper disposal of which may lead to water and air pollution. Concern is, now being expressed, therefore, about the rapid growth of the poultry industry in countries that lack a substantial land base for disposal of the manure in an environmentally sound manner. Apprehensions have been particularly strong as large integrated poultry operations are primarily clustered around locations close to main feed sources or urban markets (de Haen, 1993). These concerns have thus led to policies, aimed at mitigating the problem, which in turn can reduce productivity by slowing the downward shift of the poultry supply curve.[1]

With the above in perspective, this chapter attempts to address three main issues related to the following objectives. The first objective is to estimate the impact of new technology and other supply-side factors on poultry production. The second objective is to substantiate the premise that some countries will face acute problems in absorbing the manure that will be produced by the year 2005. The third objective is to examine the major policies that countries have adopted to address the problems.

---

[1]Aho (1989) notes that a redistribution of poultry growers from 25 miles to 35 miles from a processing plant would double the spatial costs to $2 million/year/plant.

523

TECHNOLOGY CHANGE AND SUBSEQUENT TRANSFER IN THE POULTRY INDUSTRY

Technology change contributes to productivity growth, reflecting changes in the economy's information set over time. However, it is difficult to trace and quantify how results of technology change are carried through from the initial research into production, in view of insufficient data. Thus, various strategies have been developed to measure such change. The first attempt to quantify the benefits of agriculture and extension are detailed in Schultz (1953). Here, an economic surplus approach was used to measure the social benefits of agricultural research, with specific attention to increases in productivity in the US. A series of analyses for various commodities followed. Griliches (1957) study on the economic returns of hybrid corn research in the US is a widely used method to measure technology change.

Technology change can originate from many sources: domestic or foreign, public or private. Technology transfer is necessary for countries that lack an adequate research base, which are unable to develop the needed technologies indigenously. Under these circumstances, countries find it cost effective to transfer technology from abroad with the aid of the public sector. For instance, the Green Revolution was successful in transferring high-yielding varieties of wheat and rice from temperate countries to countries of South and Southeast Asia, the Middle East, and Latin America, though local adaptive research was needed (Pray, 1981). The technology[2] associated with the "Poultry Revolution," differs from the Green Revolution experience in that it can be transferred to anywhere in the world, provided a country has access to these inputs. Consequently, countries may increase poultry productivity using transferred technology, rather than having the public sector invest in basic research and development.

DETERMINANTS OF POULTRY SUPPLY

Technological innovation aims at increasing efficiency in production as it may be more practical for countries intending to expand production, to alter the form of production in use rather than devote more resources to old production technologies. Extending this argument to the poultry industry, the impetus for technology change is greatest in countries in which the price of chicken products, relative to corn, is the highest. Peterson's (1966) work on the poultry industry utilized a supply function to estimate benefits from agricultural research, measured in terms of changes in the consumers' and producers' surpluses in the US. It concluded that the downward shifts in a farm's poultry supply schedule arose from the use of more efficient inputs obtained through research. However, more recent models projecting poultry supply, such as that

---

[2]The technology consists of hybrid chicks imported initially from the United States or Europe, raised in containment facilities and fed a compound feed diet, the principal component of which is corn (Nelson and Unnevehr, 1988).

in FAPRI (1994), do not include variables for technical change. The major exception is Nelson and Unnevehr (1988) study of Asian poultry, which concludes that technology is an important factor for consideration.

This chapter uses a supply function to measure the response of poultry producers to price and technological changes. The implicit form of this function is given as:

$$Q = f(P, \tau, Z) \tag{1}$$

where Q stands for output, P depicts price, τ represents factors related to technology, and Z connotes other exogenous supply shifters of interest.

The relative price of meat to corn is of particular interest in assessing poultry supply because the price of corn constitutes approximately 60% of the variable cost of poultry production (Nelson and Unnevehr, 1988). In addition, at any given price, improvement in technology can shift the supply curve outward. Much of the technological progress that has occurred in the poultry industry is embedded in baby chicks (Nelson and Unnevehr, 1988). Table 1 shows a complete list of the variables of interest used in this supply function.

The following linear function form (2) was then estimated, for 25 large poultry producing countries[3], in the world, using ordinary least squares:

$$Q_{it} = \sum_p \alpha_p Z_{pt} + \sum_\tau Z_{\tau t} + \mu_c \varepsilon_{it}. \tag{2}$$

where Q_{it} stands for output; Z_{pt} is vector input prices; and τ is vector supply shifters related to technology. The component μ_c is the unobserved country specific effects, and ε_{it} is the idiosyncratic error term. In order to eliminate bias in the coefficient estimates, the supply function is estimated in fixed effects. There are country-specific effects that do not change over time but affect the supply of poultry. For instance, in the poultry industry, the suppliers of baby chicks might prefer one genetic line over another. This bias may arise as a result of correlations between the technology variables and the error term. Thus the country-specific effects are unmeasured and contained in the error term. The results of the estimation are presented in Table 2.

The ratio of the variable price of meat to the price of corn was chosen to determine effects on supply due to changes in prices. The number of animal health personnel was used as an indicator of the supply of services within a country. This is expected to increase production as access to veterinary services enables better realization of the benefits from a technology. The research variables were included to measure the impact of US public poultry research, as well as the effect of public research in the country to which the technology is

---

[3]Countries used are Argentina, Australia, Brazil, Canada, France, Germany, Hungary, India, Indonesia, Iran, Italy, Japan, Korea, Malaysia, Mexico, Netherlands, Nigeria, Pakistan, Philippines, Poland, S. Africa, Spain, Thailand, UK, and U.S.

Table 1. Description of variables used in supply function.

Variable	Definition	Source	Dates
Q_{it}	MT of meat+eggs	FAO Production	1960-90
P_{it}	Price of meat/price of corn	FAO Prices Paid to Farmers	1960-87
Z_{it}	Animal Health Services	FAO Animal Health	1960-90
τ_{it}	HYT 1†	FATUS	1960-90
τ_{it}	HYT 2	FATUS	1960-90
τ_{it}	US research	USDA	1960-90
τ_{it}	National research	VET and Beast-cd rom	1973-90

†HYT 1 represents high yielding technology embodied in the number of hybrid chicks exported from the US.; HYT 2 denotes the cumulative effect of this technology.

Table 2. Regression results for explanatory variables.

Variable	Significance†	t-statistic
Price Meat/Corn	1.63*	(7.89)
U.S. Research	.00003	(0)
Imports HYTs	-.007	(-0,589)
Cumulative HYTs Imports	-.005*	(-7.200)
Veterinarians	.06*	(6.717)
Moving avg. res	2.6*	(10.828)
Country and time dummies not reported		
Adjusted R2	.993	

†Significance at 5% level: MT = metric tons; HYTs = high yielding baby chicks exported from US; t-stats. in parentheses.

being transferred. The only uniform data on public sector poultry research in most countries is the number of research publications abstracted in the Veterinary Abstracts and Breeding Abstracts of the CABI data base (Boyce and Evenson, 1975). Since information in these publications continues to be useful over time, a moving average of publications for the past five years was used as an indirect measure of the cumulative effect of the local effort on poultry research. However, the research variable will fail to pick up the private sector's activity in applied research as most of the information is kept in house. It should be recognized that there is considerable research in the private sector, particularly in poultry breeding. Poultry breeding research is concentrated at the headquarters of multinational corporations in the developed countries, and its results are directly transferred to local affiliates or joint ventures in LDCs (Pray and Echeverria, 1991). As these data are not available, it has not been included for measure in this model. To determine the impact of US public poultry research, data on the actual amount of money spent by year (in 1985 dollars) was used. It was assumed that the benefits from this source would be readily available to other countries. The high yielding technology variables measure the direct transfer of technology embodied by importing baby chicks. Here, the model includes the direct and cumulative effect of this transfer since the impact will continue to be felt after the initial transfer of the technology.

Since other countries also export this input, it could be argued that the number of baby chicks imported is not a perfect measure of this input. However, considering that the US is the principal exporter and in the absence of data for other countries, imported baby chicks (HYTs) has been used as an indirect measure.

The statistical analysis of the model provides interesting results. The coefficients of both the ratio of price of meat to price of corn and animal health workers were highly significant on production. It would thus be expected that if the price of corn were increased through changes in the world market, or trade barriers, production would stand reduced. Since the number of animal health workers also has the expected positive impact on production, it is possible that further gains in productivity could accumulate from expansion of this input as opposed to increases in bird number. In addition, the national research variable was significant. It would thus appear that considerable benefits could accrue from adaptive research at the national level, despite the fact that much of the technology is being transferred privately. In contrast, the research variable from the US research centers was not significant, possibly because the gains from public poultry research expenditures occurred in the early years of the time series. It may be interesting to note that the high yielding technology variables (baby chicks) had negative but significant coefficients. This could be attributed to the fact that the variables are in aggregate form.

IMPLICATIONS: EXTERNALITIES OF PRODUCTION

There are three major externalities associated with the production and processing of poultry products. Externality is a term used to describe a situation in which an individual in the course of rendering some service also renders services or disservices to other persons. The first externality relates to manure. When spread on cropland, manure can be a positive by-product of the production process and a valuable inorganic fertilizer source. Application beyond what is absorbed by plants however, may cause problems if excess nitrogen and/or phosphorous leaches into the ground water or contaminates surface water. A second externality arises when the by-products from processing pollute surface waters or leach into ground water. The third externality is caused by the release of ammonia into the air, leading to air pollution, acid rain, and odor problems. This chapter will focus exclusively on the first externality--the application of manure to cropland.

In order to determine the extent of water pollution from increased production in the poultry industry, as well as growth of other livestock industries, the future supply to the year 2005 for the poultry, cattle, and swine industries was projected, assuming policies remained unaltered. For projecting the effect of potential changes in poultry supply, the average growth for the period 1980-1986 was used as a base, and coefficients of the variables were estimated from the supply model. The period 1980-1986 is expected to capture the average growth rate since major technological changes have occurred. However, it should be noted that in the case of Indonesia, the estimated growth rate for the number of chickens is high, which is perhaps resulting from rapid growth due to changes in laws that have until recently limited the size of operations. These results were analyzed for the year 2005 using structural equations from the supply model. For the cattle and swine industries, projections involved interpolating future growth trends in the number of animals. From the estimated manure production the quantities of organic N and P_2O_5 that would be produced at current growth rates[4] was calculated. Refer to Table 3 for the projection of number of animals for selected DCs and LDCs, currently experiencing rapid growth in the livestock industry.

These figures were used to obtain a rough estimate of a country's potential to utilize the nutrients (refer to Table 4). This computation is indicative of a country's assimilative capacity of nutrients based on future growth of the livestock industry. For instance, in the absence of regulatory policies in the Netherlands, an estimated 88,970 MT of organic N and 35,795 MT of P_2O_5 would be produced from these four livestock sectors by the year 2005. If it is then assumed that corn is grown on all the permanent cropland (930,000 ha) in the Netherlands and only one yield is harvested per year, the result would be an excess of 351,000 MT of nitrogen and 58,000 MT of P_2O_5 that could affect the water table and lead to other consequences. This is a simplified calculation, but it clearly corroborates the concern of the Dutch about reducing livestock numbers, developing policies to control wastes, and exploring methods to export dried manure to other countries. It can be inferred from Table 4 that neither the Netherlands nor Japan would be able to absorb all the nutrients generated by the year 2005, if applied solely to their permanent cropland. This inference is in consonance with the fact that both of these countries, along with other large livestock-producing countries, are considering methods to mitigate potential negative externalities. Bower and Godeschalk (1993) predict that stricter environmental policies will result in major changes in the competitiveness of livestock production among countries.

TECHNOLOGY CHANGE, POLLUTION PROBLEMS AND IMPLICATIONS FOR ENVIRONMENTAL POLICY AND INTERNATIONAL TRADE

The trends in poultry demand, technology change, and pollution problems are leading to new policies and new patterns of poultry trade. The increasing demand is being met through a combination of increased inputs and new technology. Increase in poultry production is coupled with increased manure production contributing to ground water problems in countries that have highly

---

[4]Due to missing time series data for the Netherlands and Indonesia, interpolated growth trends of number of animals were used.

Table 3: Projected growth in the number of livestock industry
and animal nutrients in 2005.

Country	Poultry	Swine	Cattle	Dairy cow	Organic N[z]	P$_2$O$_5$[z]
Indonesia	721,800	8,503	11,817	572,855	36,345	14,844
Japan	125,700	16,398	5,541	1,500,727	92,935	37,363
Netherlands	133,301	20,212	6,079	1,435,450	88,970	35,795
Philippines	273,350	8,780	2,088	22,091	1,730	807
Thailand	305,300	4,837	6,451	91,000	6,150	2,612
U.S.	808,150	53,107	105,621	8,671,909	540,067	217,757

[z]These figures are in metric tons (MT); Calculation for yearly amount of N and P$_2$O$_5$ are based on the average that 5 pigs = 1 animal unit (au); 150 lb of N and 118 lb of P$_2$O$_5$ per pig au; 1 beef cattle = 1 au; 109 lb of N and 69.5 lb of P$_2$O$_5$ per beef cattle au; 1 dairy cow = 1 au; 135.7 lb of N and 54.3 lb of P$_2$O$_5$ per dairy cow au; 250 chickens = 1 au; 298 lb of N and 209 lb of P$_2$O$_5$ per chicken au.

Table 4: Potential available nutrients (in 1000 Mt of N and P$_2$O$_5$)

| Nutrient/Cropland | DCs | | | LDCs | | |
	Japan	Netherlands	U.S.	Indonesia	Thailand	Philippines
N available[z]	705	481	10,681	1,646	583	402
N removed[y]	643	130	26,588	3,080	3,100	1,116
P$_2$O$_5$available[z]	727	110	3,983	600	321	106
P$_2$O$_5$[z] removed	257	52	1,064	1,232	1,240	446
Crop land(ha)	4,596	930	189,915	22,000	22,140	7,970
Diff. in N	-61	-351	15,907	1,434	2,517	714
Diff. in P$_2$O$_5$	-470	-58	2,919	632	919	341

[z]Denotes mass balance and includes nutrients from manure and commercial fertilizers. It assumes that there is no overlap.
[y]Represents the potential amount of nutrients removed if all cropland is planted with corn. Assuming that the average yield of corn is 125 bu/acre, it is estimated that 125 lb (140MT) of N and 50 lb (56 MT) of P$_2$O$_5$ is removed. There are 2.471 ha in an acre.

concentrated livestock operations, import much of their feedstuffs, and have little agricultural land to absorb the nutrients from the manure. In fact, growers were often highly concentrated around processing facilities to take advantage of economies of scale (Aho, 1989) and to be near points of entry for imported feeds (de Haen, 1993). Thus, countries experiencing rapid expansion of the livestock industry are expected to face serious problems (as elaborated in previous section). These countries, which include Japan (Taha, 1994), Netherlands (Dietz and Hoogervorst, 1991), Singapore and Hong Kong (Nelson and Unnevehr, 1988), are now adopting polices that restrict poultry production and are instead importing from countries that have larger land areas and less stringent environmental policies.

Japan and the Netherlands have adopted environmental policies that restrict poultry production. In 1987, the Netherlands enacted the Manure and Fertilizer Act not only to prevent a further increase in manure production, but to also stimulate solutions for areas with excess manure (Steenvorden, 1989).

A National Manure Board was established to analyze and control for manure quality, provide manure storage, and subsidize manure transport to storage facilities. Currently, the Dutch require that farmers and enterprises transporting processed manure keep records of their activities and are taxed for the surplus they produce. These tax revenues form a Manure Fund that lends support for research in manure-related problems.

The Japanese are in the process of requiring manure management strategies of every poultry producer. Unlike the US and the Netherlands, Japan has provided farmers minimal assistance to aid adoption of environmentally friendly manure strategies. This has caused Japan's poultry production to reduce over the years. With decline in production, imports from land-rich and less-regulated economies have been on the rise. For instance, Japan currently has a low tariff rate (12%) for imported chicken, which encourages imports from Brazil and Indonesia (Taha, 1994). The combination of policies plays an important role in limiting the growth of the Japanese poultry industry.

Singapore and Hong Kong are, however, currently restricting animal production through the use of tax incentives and zoning laws to move commercial poultry and swine operations away from cities and into neighboring countries. Both countries basically allow free trade in poultry (Nelson and Unnevehr, 1988).

In contrast to the above-mentioned countries, the US has less restrictive environmental policies. For instance, the US is promoting the use of voluntary measures such as best management practices, nutrient management programs, and government cost-sharing for the adoption of waste-mitigating technologies (Narrod et al., 1993). These polices have not succeeded in reducing poultry production in the country; production has continued to grow, and exports have increased. Interestingly, the US Environmental Protection Agency has recently experimented with ways to make companies internalize the negative externalities associated with production (Narrod et al., 1993).

Poorer countries such as Indonesia and Brazil possessing adequate land to absorb manure do not have laws to control water pollution or even enforce existing laws effectively. Leng et al. (1993) suggest that only the US and other industrialized countries have significant regulatory measures to control pollution. They also argue that pollution control has not been realized in most developing countries because policy makers lack interest in enforcement. This may be because policy makers desire to expand production, to take advantage of the demand from neighboring countries that have restrictions on growth of their own poultry industry.

Since commercial poultry technology can be transferred relatively easily from industrialized countries, growing demand in industrialized countries creates opportunities to increase income through poultry exports. As a result, poultry production and exports have expanded very rapidly in many LDCs. In addition, many of the LDC countries are trying to increase production to meet growing demand. Thus, it is highly unlikely that they will choose one of the livestock growth limiting policies that have been used in places like the Netherlands. Nevertheless, it is important for policy makers to recognize the potential problems associated with rapid growth of the livestock industry in feed and land-limited areas. Specifically, countries need to be aware of the potential tradeoffs between water quality and productivity as they promote policies that increasepoultry production.

REFERENCES

Aho, P., *Spatial Costs and Economics of Manure Management*, paper presented at Symposium on the Clean Water Act and the Poultry Industry, Washington, DC, 1989.

Bower, F., and Godeschalk, F., *Pig Production in the EC: Environmental Policy and Competitiveness,* Agricultural Economics Research Institute, The Hague, 1993.

Boyce, J., and Evenson, R., *National and International Agricultural Research and Extension Programs*, New York, Agricultural Development Council, 1975.

de Haen, H., Livestock and the environment interactions and policy implications, *Quarterly Journal of International Agriculture*, 32 (1), 122, 1993.

Dietz, F., and Hoogervorst, N., Towards a sustainable and efficient use of manure in agriculture: The Dutch case, *Environmental and Resource Economics* 1, 313, 1991.

FAO, *Data Base for Trade, Production, Fertilizer, and Animal Health*, 1960-1994.

FAO, *Statistics on Prices Paid by Farmers for Means of Production and Prices Received*, 1982, 1988, 1992.

Food and Agricultural Policy Research Institute (FAPRI), *FAPRI 1994 International Agricultural Outlook*, Ames, Iowa State University, FAPRI Staff Report #2-94, May, 1994.

Griliches, Z., Hybrid corn: An exploration in the economics of technological change, *Econometrica*, 25(4), 501, 1957.

Leng, S., Srichai, W., and Ludwig, H., Monitoring and enforcement for management of industrial wastes: Case study for Thailand, *The Environmentalist*, 13(4), 277, 1993.

Narrod, C., Reynnells, R., and Wells, H., *Potential Options for Poultry Waste Utilization: A Case Study of the Delmarva Peninsula,* USDA/EPA White Paper, Washington, D.C., 1993.

Nelson, G., and Unnevehr, L., *The Product Cycle of Foreign Trade and Asian Demand for Poultry Imports*, Working Paper, University of Illinois, 1988.

Peterson, W., *Returns to Poultry Research in the United States*, PhD. Dissertation, University of Chicago, Illinois, 1966.

Pray, C., The green revolution as a case study in transfer of technology, *The Annals of the American Political and Social Science*, November, 1981.

Pray, C., and Echeverria, R., Private sector agricultural research, in *Agricultural Research Policy: International Quantitative Perspective*, Pardey, Rosebom, and Anderson, eds., Cambridge University Press, Cambridge, 1991.

Schultz, T., *The Economic Organization of Agriculture*, New York, McGraw Hill, 1953.

Steenvorden, J., Manure Management and Regulations in the Netherlands, *Proceedings from the Dairy Manure Management Symposium*, Syracuse, New York, 1989.

Taha, F. Animal waste and chemical fertilizers add to Japan's environmental problems, in *Asia and Pacific Rim Situation and Outlook Series*, International Agriculture and Trade Reports, Economic Research Reports, Dec. 1994.

Unnevehr, L., Eales, J., Nelson, G., and Kim, Y., *East Asian Poultry Markets: Technology Transfer and Demand Dynamics*, Working Paper, University of Illinois, 1992.

US Foreign Agricultural Trade Statistical Report (FATUS) for Calendar Years 1960-1990.

World Resource Institute (WRI), *World Resources*, Oxford University Press, New York, 1994.

ANALYZING MANURE MANAGEMENT POLICY: TOWARD IMPROVED COMMUNICATION AND CROSS-DISCIPLINARY RESEARCH

Amy Purvis and Charles W. Abdalla

INTRODUCTION

Structural change in animal agriculture in the US has given rise to a set of critical issues related to the current and future ability of this industry to meet societal goals. Recently, the evolution of agriculture to a more industrialized state has been accelerated by technological innovation, changes in marketing arrangements and practices, and the emergence of new public policies. Concern is growing among stakeholders in the agricultural system, policy makers, and citizens about implications of this structural change for the performance of the system. A key area of concern is how well animal agricultural operations are able to manage manure without degrading land and water resources or air quality.

Knowledge about environmentally acceptable animal agricultural practices is not readily available because much previous research on manure management issues has been fragmented. Most studies have been confined to a single commodity focus, such as dairy or poultry. Many have been site specific and/or technology specific. Disciplinary specialization and demands for greater scientific rigor have resulted in research projects emphasizing (1) production, (2) its environmental effects, (3) policies pertaining to animal agriculture, or (4) the industry's structure and marketing linkages. There has been progress in linking these four subsets of the animal agriculture-environment interface through the collaborative efforts of social and physical scientists as well as through disciplinary research.

Cross-disciplinary efforts have assessed the cost-effectiveness of new manure management technologies and modeled the environmental and economic ramifications of changes in physical flows and residuals from modifying production processes. The interface between animal agriculture and the environment is complex; thus it is necessary to conduct some economic analysis without complete information on the environmental ramifications from changing a particular manure management practice, and vice versa. Comparative analysis across sectors--and even in the same sector across regions--has been undertaken only infrequently. This poses challenges and, at the same time, opportunities. There are information gaps. Perhaps policy options are being overlooked that have potential to bring the performance of today's animal agriculture into line with society's expectations.

Physical and social scientists agree that these issues are urgent. To tackle them, many of us agree that cross-disciplinary collaboration is essential. This chapter is about how we might work together to frame and prioritize research questions on policy options for animal agriculture and how we might modify existing cross-disciplinary research to get more satisfying answers.

BACKGROUND

ANIMAL AGRICULTURE AND ITS ENVIRONMENTAL EFFECTS

Diversity in animal agriculture is tremendous--across sectors, across regions, and across farm sizes and types. From an environmental standpoint, it makes a difference whether nutrients from manure are utilized in crop production or whether residuals accumulate due to lumpiness (spatial and temporal) in manure disposal (Roka et al., 1995). Economies of size in both production technologies and manure management have reinforced trends toward bigness in the production sector of animal agriculture (Schwart et al., 1995). At the same time, these larger production units tend to locate in clusters because of economies of size in processing, specialized infrastructure requirements, or strong vertical marketing linkages (Pagano and Abdalla, 1994). This phenomenon has happened with the poultry industry on the Delmarva peninsula (Narrod et al., 1993), in swine production and processing in North Carolina (Roka et al., 1995), in dairy farming in central Texas (Pagano, 1993) and eastern New Mexico. In some cases in which several large production units have located in close proximity, neighbors have become outraged about air and water pollution.

Dilemmas that occur at the local level due to the industrialization of animal agriculture can be viewed as a mismatch between "micromotives and macrobehavior" (Schelling, 1978), as an outcome of "the logic of collective action" (Olson, 1971), as "the tragedy of the commons" (Hardin, 1968). Dilemmas occur because individual decisions and behaviors made within existing institutional rules can lead to cumulative outcomes that are inconsistent with the goals or preferences of most participants over the long run. Economists often label these dilemmas as market failure, but they are usually better framed as institutional failure (Shaffer, 1980). Favorable economic and environmental conditions attract new animal-agriculture operations to an area and/or cause existing production units to get larger. Similar to the overgrazing of an unfenced "commons," today's livestock and poultry producers--having unrestricted access to the capacity of the local environment to assimilate waste--often exhaust this resource. Initially, given an abundance of assimilative capacity, a single producer's decision to locate or to increase the size of his/her operation may be unimportant. As more producers make decisions to expand or relocate in a particular locale, driven by private benefit motives and ignoring the additional demands they make on the commons, the assimilative capacity of the environment gets used up. The result is elevated nutrient levels in ground or surface water that adversely affect producers as well as other present or future users of the water resource.

Growth often occurs rapidly in animal agriculture; thus constraints to the assimilative capacity of receiving airsheds and watersheds often become binding, and conflict among stakeholders becomes sharp before local institutions can react. When there was plenty of assimilative capacity, conflicts among stakeholders were rare; thus only sketchy guidelines exist on who can do what, when, and how under conditions of scarcity. Once overburdening is a problem, however, negotiation and compromise become increasingly difficult. Facilities for animal production and processing require substantial capital investments, often with a significant irreversible component. Neighbors to a large or growing animal agriculture sector worry about surface water degradation, ground water contamination, and odor from overloaded assimilative capacity. Adverse effects on environmental amenities and on land values are also viewed as irreversible, or costly to reverse. Both stakeholders in animal agriculture and its neighbors argue that they have claims--property rights--to the land, water, and air resources.

A SELECTIVE INVENTORY OF POLICY PRESCRIPTIONS

In agriculture generally and in animal agriculture in particular, environmental policies have emphasized voluntary technology adoption to minimize adverse effects on land and water resources. Emphasis has been on education and technical assistance to promote technologies that enhance productivity. Alternatively, cost-sharing is provided so that, in principle, compliance is, at worst, a break-even proposition. As budget constraints become binding, these policy strategies appear to be incapable of reaching increasingly stringent environmental quality standards. Federal guidelines for controlling surface water pollution from confined animal feeding operations (CAFOs) were spelled out in the 1972 Clean Water Act (Schwart et al., 1995). More detailed technical requirements for handling surface water runoff are the domain of state and local rule-making, as are nuisance odor and ground water protection. For both regulators and producers, having technology-based regulation in place is advantageous: what constitutes compliance is straightforward. The downside is an implicit disincentive to technological innovation. For experimentation with new technologies to be permissible (and permitted) requires overcoming significant inertia (Pagano et al., 1994).

Over the past five years, in response to increasing public concern about the environmental effects of animal agriculture, new federal policies have been developed. Examples include amendments to the National Pollution Discharge Elimination System general permit for CAFOs (Federal Register, 1993), as well as proposals pertaining to CAFOs in the Coastal Zone Act Re-authorization Amendments of 1990. In addition, a patchwork of state and local policies is emerging. As a result, environmental compliance is beginning to place constraints on local production management decisions. Differences in compliance costs across states affect regional comparative advantage in animal agriculture. In support of the design and implementation of existing policies most research has focused on evaluating the physical performance and/or the cost-effectiveness of alternative manure management technologies.

There is a small but growing body of research on regions with clusters of animal agriculture activities where nutrient flows are measured at the farm level and at the watershed level using a materials-balance conceptual framework (Smolen et al., 1994). Such research has fostered an appreciation of the character of non-point source pollution from animal agriculture. It is difficult to link cause and effect, especially at the watershed level or above. Increasingly, there is awareness of obstacles to designing and implementing policies that require an environmental monitoring network to keep track of elevated levels of nutrients in the environment. Policies requiring animal agriculture to meet performance standards (specified discharge limits, such as those existing for point sources) have often been deemed both technically problematic and prohibitively expensive to implement. Nonetheless, economists and others continue to think about opportunities for creating markets to facilitate the flow of nutrients from surplus to deficit regions. Trading of marketable emission allowances has been described in theory (Baumol and Oates, 1988), but little success has been achieved in real-world applications to water quality. A well-functioning market requires water quality data linking off-farm nutrient levels with on-farm nutrient management and/or historical water quality data, which often do not exist (Boggess, 1992).

OPPORTUNITIES FOR POLICY INNOVATION

Technical assistance, cost sharing, and voluntary adoption of pollution control technologies are the tried-and-true policy approaches to agricultural non-point pollution. Yet problems remain, with respect to animal agriculture in particular. The US Environmental Protection Agency (US EPA) reported in 1992 that agriculture is the largest contributor to the nation's non-point pollution, and animal agriculture is named as a main culprit (US EPA, 1994). Especially for citizens concerned about environmental quality but having no direct experience in agriculture, the manure management issue is compelling and graphic. A recent *Wall Street Journal* profile of Erath County, Texas, opened with two juxtaposed points: that annual milk sales were $161 million in 1992 but that 70,000 cows produced 175,000 tons of manure (Gerlin, 1994). In public hearings in Erath County, the local concerned citizens emphasized that a

1000-cow dairy generates a waste water stream with volume and nutrients equal to a city of 17,000. They asked why the government does not enforce the same rules for dairies as their own small city must follow.

For researchers to make positive contributions to effective policy innovation may require fresh approaches to framing research problems. It may require new ways of thinking about how to improve the synergy between research and policy design through better coordination and sequencing. In this chapter, we argue that cross-disciplinary collaboration is the key to creativity. As individuals and as disciplinary professionals, we see things differently. If we can manage to weave our disparate insights together, then our diversity can be a source of creative strategies for both research and policy innovation. Such an integrative effort will require a flexible conceptual framework for organization, communication and analysis.

AN ANALYTICAL FRAMEWORK: THE S-S-C-P MODEL

A framework that appears to be both flexible and robust enough to assist with integrative cross-disciplinary studies of manure management issues and policies is the Situation-Structure-Conduct-Performance (S-S-C-P) model (Shaffer et al., 1987; Thompson et al., 1994). This framework was designed and has been applied in problem-solving research settings, mostly by agricultural economists. Its intellectual roots are in the public choice literature (Schmid, 1987) and in industrial organization (Scherer, 1979). The dual focus of the S-S-C-P model is (1) describing the institutional and physical context for individual and collective decision making and (2) identifying effective policy mechanisms for improving coordination, equity, or production outcomes. The appropriate framing of policy analysis is the aim of S-S-C-P modeling. It offers a conceptual framework for identifying which variables to monitor and for selecting variables with potential for being modified to produce more desirable outcomes. Emphasis is on relationships among varieties of *situation* (that is, sources of interdependence); policies or the institutional *structure*; the actions taken or the *conduct* of stakeholders in response to incentives and disincentives provided by the existing institutional structure; and the cumulative set of outcomes that comprise *performance*.

SITUATION: HUMAN INTERDEPENDENCE

Every natural resource and environmental amenity has a set of characteristics that determine the human interdependence with respect to its use. Understanding the roots of these interdependencies is the starting place for postulating changes in community institutions that might lead to different performance outcomes (Schmid, 1987).

The term *situation* is defined as circumstances of human interdependence that arise from the innate characteristics of resources, environmental amenities, goods, or services. These attributes are inherent to these products regardless of institutional structures that govern how they are produced, purchased, or consumed. Categories of situational characteristics include 1) *incompatible use*: when one person's usage of a product makes it unavailable for others' use; 2) *high exclusion cost*: when use of an existing good cannot be limited to those contributing to the costs of its production and costs of excluding unauthorized users are high; 3) *joint-impact*: cost of allowing additional people to use a good is zero; 4) *contractual costs*: costs of making a decision or reaching an agreement with another party; 5) *information costs*: costs of obtaining information about a product; and 6) *economies of scale*: cost per unit of output for a product declines with increases in output or population and then may rise again (Schmid, 1987; Shaffer et al., 1987).

The *situational* characteristics of products are a joint function of physical, biological, or technological factors. Schmid (1987) notes that "There is both a social and a technological dimension to the categories of human interdependence. This can change over time, but is fixed for the period of impact analysis." In applying the S-S-C-P model, specifying situation defines the scope of the analysis and also frames the problem being studied.

Consider the community of Westminister, Maryland, located 35 miles northeast of Baltimore. "For yet another summer...swarms of flies from two local chicken egg businesses have disturbed the serenity of the valley....At least two people have been hospitalized after fly bites." A retired couple "...has not used their swimming pool in at least two summers....Jars lined with chemicals to kill the flies quickly fill up with thousands of flies--maybe millions" (Janofsky, 1994).

The Maryland egg producers opened their farms 15 years ago; the largest farm currently houses approximately half a million chickens. Migration of urbanites and suburbanites has been more recent. These emigrants have recently organized a concerned citizens' group. The group recently learned that "state laws provide no sanctions against egg farmers who improperly manage their manure." Further, a spokesman for the Maryland Department of Health said that "he doubted the flies could cause a health problem" (Janofsky, 1994).

Presence of the manure and flies generated by the egg production facilities have an effect on the environmental amenities they share with their neighbors (incompatible use). Currently, it appears costly to definitively prove a causal link between the flies and health risks (information costs). Since there are ill-defined rules and procedures for resolving conflicts concerning manure management, egg producers and the concerned citizens are negotiating in an *ad hoc* process (high contractual costs). If a technological solution became available at a higher cost to the producer, it would be difficult to determine who should pay and how much (high exclusion costs). Private costs of production are lower as producers add chickens on their farms (economies of scale). However, if costs due to environmental effects and the associated negotiations are counted, then bigness may become increasingly expensive.

Analysis of situation appears to offer opportunities for physical scientists to make important contributions to analysis of the policy and institutional dimensions of manure management issues. By providing information about the physical, biological, or technological factors giving rise to new situations of human interdependence, insight can be gained into the nature and extent of the policy conflicts and potential solutions. For example, entomological information about the life cycle and geographic dispersion of flies determines the spatial and temporal boundaries of the issue, including who are parties to the dispute and when they will be affected. Also, more complete delineation of a problem's scope and accurate framing can lead to innovative suggestions for technical fixes, such as incorporating manure into the soil, or new policies, such as establishing buffer zones between producers and residences.

STRUCTURE

"Situation is inherent and structure is chosen" (Schmid, 1987). Structure is the composite of the institutional factors that make a difference in the choices of individual community members as well as group behavior. By providing incentives and disincentives for particular actions, it defines and limits the opportunity sets open to interdependent parties. The institutional or policy structure sets up the "rules of the game." Public choice can be the result of a vote or other consensus-oriented decision making. Otherwise, the lack of deliberate public choice creates structure.

Structure is best described in terms of property rights. Broadly defined, property rights are the rights and obligations established by law, custom, and covenant that define the relationships among participants. Property rights originate with the community, since no one can actually have a right unless the relevant community acknowledges it (Shaffer et al., 1987). Property rights are reciprocal in nature: one person's rights represent a restriction on another's freedom to act. For example, the F/R Cattle Company in Erath County, Texas, is a CAFO housing 5000 baby calves. A neighbor filed a lawsuit for damages from nuisance odor. The case was heard in the Texas Supreme Court and was decided in favor of the F/R Cattle Company. This decision turned on the fact that F/R Cattle Company was located in a predominantly agricultural area and that odors from the facility were deemed to be from "natural processes" (Kelton, 1992). Thus, freedom of a CAFO to produce some level of nuisance odor-- given that they are operating with the best management practices prescribed by their operating permit--places restrictions on the rights of their neighbors to breathe odor-free air.

Property rights are dynamic. They define how individuals or groups can have access to control of resources. Property rights are guidelines stating the community's concepts of fairness and equity governing the conduct of community members with regard to use of resources that involve interdependencies.

CONDUCT

Conduct is behavior. Conduct encompasses all the choices, decisions, or strategies adopted by participants in response to the opportunity set provided under the institutional structure. Conduct links structure and performance. Implementation of policies, as well as the strategic behavior and/or motives driving policy design, are in the domain of conduct.

Circumstances characterized by "impacted information" illustrate the link between structure and performance. Impacted information

> ...is mainly attributable to the pairing of uncertainty with opportunism. It exists in circumstances in which one of the parties to an exchange is much better informed than is the other regarding conditions germane to the trade, and the second party cannot achieve information parity except at great cost--because he cannot rely on the first party to disclose the information in a fully candid manner (Williamson, 1975).

Consider circumstances in which a new CAFO has located near outraged neighbors who are complaining about nuisance odor and water pollution. The state regulatory agency holds a public hearing. The CAFO owner presents a plan to install a new manure management technology and presents scientific evidence of its effectiveness. To pay for the new technology, the producer proposes to increase the number of animal units on the farm. The neighbors are unwilling to be convinced that under any circumstances, under any technology, more animals would result in less odor or less water contamination. The regulators have no first-hand experience with the technology. The concerned citizens are unwilling to trust the scientific evidence presented by the producer. Regulators often are forced to break the impasse by falling back on enforcement of rules written into administrative codes, rules that prescribe proven technologies. Even if the producer were willing to bear the risk of investing in a promising new technology, neither the neighbors nor the regulators are willing or able to accommodate that opportunity. Such a scenario--specifically, the case of sluggish adoption of free stall dairy technology in central Texas--was documented by Pagano (1993).

PERFORMANCE

Performance is the flow of outcomes or consequences that result from participants' conduct given the opportunity set provided by a particular institutional structure and the situation. Outcomes can produce costs for some members while at the same time generating benefits to them and/or others. Whether consequences are deemed good or bad depends on the preferences of the parties involved. These judgments are related to individual and community goals, preferences, values, and images of well-being. Discussions of most public issues will involve several different goals that have value to community members.

These multiple dimensions of performance include equity, fairness, productivity, and progressiveness, as well as other services being obtained. Each policy alternative is likely to yield a mix of outcomes that meet some performance criteria and fail others. Due to the multidimensional nature of performance, policy choices often require tradeoffs among competing goals, preferences, and values (Shaffer et al., 1987).

TOWARD IMPROVED COMMUNICATION AND COLLABORATION

Perhaps the best thing to emerge from grappling with S-S-C-P analysis of a complex policy dilemma is awareness of the orientations we each bring to the research process. Our values, as well as our disciplinary training, make a difference in how we frame research topics and how we distill the policy implications from our research. Where we start and what we emphasize makes a difference in what gets done. Our values affect how we define problems and what kinds of solutions we propose. A simple issue like whether scientific and policy discussions are framed as "animal waste" or "manure management" can influence the tone of a cross-disciplinary discussion of "situation."

Thompson (1993) used the S-S-C-P paradigm to demonstrate how our ethical orientation shows up in our research processes. He described two orientations. On one hand, policy inquiry can focus on performance or on the end state. A policy is deemed bad or good according to the outcome it achieves. On the other hand, policy inquiry can focus on structure or on how results are achieved. A policy option is judged according to whether it promotes equity or justice or another "right." We balance the trade-offs between means and ends differently.

In analysis of manure management policy, it has been common to frame manure management problems fundamentally as farm-level materials balance problems. Information was seen as the missing link: farmers were not taking into account the on-farm value and the off-farm externalities associated with manure. Thus research focused on farm-level technology and management practices and on education and technical assistance to encourage voluntary technology adoption. The resulting policy prescription--provision of technology and information, leaving adoption a voluntary choice--is consistent with an ethical belief in the importance of individual sovereignty. Others have framed manure management issues from an aggregate level using a watershed-, basin-, or airshed-level materials-balance approach. From this perspective, the goal is to manage water or air quality in order to achieve optimal performance of the overall system, rather than to optimize at the field or farm level. Coordination was seen as the missing link. Accordingly, policy prescriptions have aimed to improve the overall effectiveness of incentives through targeting, for example, or to change property rights as a means of achieving overall improvements in water or air quality. Implicitly, efficiency may take precedence over individual sovereignty.

Taken separately, neither the micro-level approach nor the macro-level approach is sufficient. We suggest the S-S-C-P model as a starting place for framing and sequencing of collaborative cross-disciplinary research efforts. Working together to arrive at a common understanding of "situation" can set the tone and focus the scope of the analysis. For example, a team made up of faculty from poultry science, entomology, rural sociology, and agricultural economics was recently assembled at Penn State University to address community conflicts over flies from manure at egg production facilities in southeastern Pennsylvania. Much progress has been made in framing the problem, identifying crucial knowledge gaps, and discovering communication barriers among stakeholders. The chemistry of social scientists posing questions about the situational aspects of the policy issue to well-informed physical scientists set the tone and uncovered important hidden areas worthy of research inquiry. Continued collaboration should lead to the research and extension education strategies on technical and policy solutions that will help in resolving conflict among producers and rural residents.

In sum, the best we can hope for in discussions of policy options is to be aware and honest about our value orientations, and our attitudes regarding working together and willingness to compromise. We recommend the S-S-C-P paradigm as providing the rudiments of a vocabulary for cross-disciplinary communication and policy analysis.

REFERENCES

Baumol, William J., and Oates, Wallace E., *The Theory of Environmental Policy*, Cambridge: Cambridge Press, 1988.

Boggess, William G., *On the Use of Marketable Emission Credits to Help Preserve the Everglades: Observations and Suggestions*, University of Florida, Food and Resource Economics Department Staff Paper SP93-13, June, 1992.

Federal Register, National Pollution Discharge Elimination System General Permit and Reporting Requirements for Discharges from Concentrated Animal Feeding Operations, *Notice*, 58, 24, 1993.

Gerlin, Andrea, As the state's milk production increases, so do the screams of environmentalists, *Wall Street Journal*, Texas edition, 25 May 1994.

Hardin, Garrit, The tragedy of the commons, *Science*, 162, 1243, 1968.

Janofsky, Michael, A plague of flies harasses a valley, *New York Times*, 21 July 1994.

Kelton, Steve, Calf raising facility snubbed by high court, closure to come, *Livestock Weekly*, 17 September 1992.

Narrod, C., Reynnells, R., and Wells, H., *Potential Options for Poultry Waste Utilization: A Focus on the Delmarva Peninsula*, USDA/EPA White Paper, Washington, D.C., 1993.

Olson, Mancur, *The Logic of Collective Action: Public Goods and the Theory of Groups*, Cambridge: Harvard University Press, 1971.

Pagano, Amy Purvis, *Ex Ante Forecasting of Uncertain and Irreversible Dairy Investments: Implications for Environmental Compliance,* Ph.D. dissertation, University of Florida, August, 1993.

Pagano, Amy P., and Abdalla, Charles W., Clustering in animal agriculture: Economic trends and policy, in *Balancing Animal Production and the Environment,* Proceedings of the Great Plains Animal Agriculture Task Force, Denver, Colorado: October 19-21, 1994.

Pagano, Amy, Sims, Kimberly, Holt, John, Boggess, William, and Moss, Charles, Environmental Permitting and Technological Innovation, in *Balancing Animal Production and the Environment,* Proceedings of the Great Plains Animal Agriculture Task Force, Denver, Colorado: October 19-21, 1994.

Roka, Fritz, Hoag, Dana, and Zering, Kelly, Making economic sense of why swine effluent is sprayed in North Carolina and hauled in Iowa, *Animal Waste: The Land-Water Interface,* Forthcoming.

Schelling, Thomas, *Micromotives and Macrobehavior,* New York: W.W. Norton and Company, 1978.

Scherer, F.M., *Industrial Market Structure and Economic Performance,* Chicago: Rand McNally, 1979.

Schmid, A.A., *Property, Power and Public Choice: An Inquiry into Law and Economics,* New York, New York: Praeger Publishers, 1978 (first edition) and 1987 (second edition).

Schwart, Robert B., Holt, John, and Outlaw, Joe, Economic factors driving regional shifts in livestock production: Opportunities for policy innovation, *Animal Waste: The Land-Water Interface,* Forthcoming.

Shaffer, J.D., Food system organization and performance: Toward a conceptual framework, *American Journal of Agricultural Economics,* 62(2), 310, 1980.

Shaffer, James D., Schmid, Allen A., and van Raavensway, Eileen, Community Economics: A Framework for Analysis of Community Economics Problems, Michigan State University, Unpublished manuscript, 1987.

Smolen, Michael D., Kenkel, Philip L., Peel, Derrell S., and Storm, Daniel E., Mass balance analysis of nutrient flow through feed and waste in the livestock industry, in *Balancing Animal Production and the Environment,* Proceedings of the Great Plains Animal Agriculture Task Force, Denver, Colorado: October, 1994.

Thompson, Paul B., Ethical issues facing the food industry, *Journal of Food Distribution Research,* February, 1993.

Thompson, Paul B., Matthews, Robert J., and van Ravenswaay, Eileen O., *Ethics, Public Policy, and Agriculture,* New York: Macmillan Publishers, 1994.

US Environmental Protection Agency, Office of Water, *Managing Nonpoint Source Pollution: Final Report to Congress on Section 319 of the Clean Water Act (1989),* EPA-506/9-90, Washington, D.C.: January, 1992.

Williamson, Oliver E., *Markets and Hierarchies: Analysis and Anti-Trust Implications,* New York: The Free Press, 1975.

MAKING ECONOMIC SENSE OF WHY SWINE EFFLUENT IS SPRAYED IN NORTH CAROLINA AND HAULED IN IOWA

Fritz M. Roka, Dana L. Hoag, and Kelly D. Zering

INTRODUCTION

Owners of large confined animal feeding operations have been criticized by environmentalists and sustainable agriculture advocates as being wasteful of their manure nutrients (Magdoff, 1991; Honeyman, 1990). There exists a widespread view that if livestock producers gave full credit to manure nutrients in cropping decisions, farm income and environmental quality would be simultaneously enhanced. This view encourages the adoption of manure management systems that conserve nutrients and incorporate their application in profitable crop enterprises. Environmental reasons alone may justify restricting the loss of manure nutrients, but it is less certain that economic reasons exist for uniformly conserving manure nutrients.

Observing manure management practices across different production regions leads one to conclude that nutrient conservation is not a universal goal among producers. One response to this observation would be that some producers waste nutrients and would benefit from education. An alternative response is that producers are fully aware of the value of manure and choose not to utilize it. This chapter adopts the latter view, which suggests that manure value encompasses more than just nutrient content. The objective of this chapter is to describe a more complete definition of manure value and provide an economic rationalization for different manure management goals. The analysis below focuses on swine manure management and, specifically, compares the general management practices utilized in Iowa with general management adopted in North Carolina.

SWINE MANURE MANAGEMENT IN IOWA AND NORTH CAROLINA

Swine producers in Iowa typically collect manure in deep anaerobic pits situated underneath a feeding floor. The material, or slurry, is approximately 5% solid and is hauled to crop fields in tank wagons. Nutrient concentrations per 1,000 gal of slurry average 12, 16, and 12 lb of nitrogen, phosphate, and potash, respectively (Barker, 1991). In contrast, manure handling systems in North Carolina center around anaerobic lagoons. Lagoons serve a dual function--to store collected effluent and to treat organic compounds. Solids are reduced to less than 1%, making it possible to transport lagoon effluent through irrigation lines. Lagoon treatment reduces the nitrogen concentration in fresh manure by over 90%. In addition, 90% of the potash and 96% of the phosphate settle with the sludge. Consequently, the average nutrient concentration of 1,000 gal of irrigated lagoon effluent is 2.5, 1.4, and 3.4 lb of nitrogen, phosphate, and potash (Barker, 1991).

Iowa swine producers are encouraged to apply their manure to corn. Chase et al. (1991) demonstrated that lower application rates better utilize manure nutrients and increase overall farm profits. While soils in North Carolina are not as productive as the soils in Iowa, corn still ranks as the most profitable feed grain crop. However, the greatest volume of swine effluent is irrigated to coastal bermudagrass hay, which has a zero or very low output price. Bermudagrass hay demands twice the nitrogen quantity that is applied to corn, thereby reducing the required acreage for effluent disposal.

North Carolina and Iowa hog producers are separated by important socioeconomic differences, most important of which is that Iowa producers generally operate integrated crop and livestock farms. Many North Carolina producers are specialized hog producers. Feed supplies are imported on to the farm site, and crops serve only as a means of nutrient disposal. This chapter assumes that producers seek to maximize net whole-farm income. Therefore, the relationship between livestock and cropping enterprises is irrelevant. Crops that maximize whole-farm income will be grown regardless of whether they are consumed directly as a livestock feed. While North Carolina producers utilize fewer manure nutrients than Iowa producers, it is the premise of this chapter that both Iowa and North Carolina producers choose management practices that *maximize* manure value.

MANURE VALUE

A common approach to assess manure value has been to sum the monetary values of individual plant available nutrients (Badger, 1980; Honeyman, 1990). This approach, however, assumes that all manure nutrients contribute positively to crop yield. If nutrients applied in excess of crop requirements do not contribute to crop yield, they should not be counted. More importantly, a monetary summation is a measure only of manure benefit, not of net value. A measure of value must consider not only benefits but also handling costs.

Manure management is an interrelated set of decisions involving three components--biological treatment, effluent transport, and crop selection. Manure value incorporates the effects of all decisions on total net farm income.

BIOLOGICAL TREATMENT

Biological treatment utilizes anaerobic and aerobic bacteria to stabilize organic compounds and reduce nutrient concentrations. Slurry from anaerobic pits loses between 15-30% of the nitrogen produced in raw manure (MWPS-18, 1985). In North Carolina, anaerobic lagoons volatilize over 90% of the excreted nitrogen (Barker, 1991).

Per-head storage coefficient (K), per-head treatment coefficient ($T(\delta,$ *climate*)), and number of animals (h) determine the total volume of a facility. Treatment costs can be summarized as:

$$C(treatment) - P_T \cdot [K + T(\delta, climate)] \cdot h . \qquad (1)$$

The discounted unit cost of construction, P_T, converts total treatment volume into an annualized value. A lagoon's storage capacity ($K \cdot h$) is assumed to be independent of treatment efficiency.

Treatment capacity, $(T(\cdot) \cdot h)$, depends on level of treatment, δ, and climate. Treatment level, δ, is a variable between zero and one and represents the percentage of desired nutrient reduction. Nutrient concentration of effluent after treatment is represented as γ^T and equals:

$$\gamma^H \cdot (1-\delta),$$

where γ^H is the concentration of excreted manure. Increasing treatment level, δ, decreases nutrient concentration, γ^T. North Carolina data have shown that treatment efficiency is correlated with lagoon size. Greater treatment levels are achieved with bigger lagoons (Safley, 1992). Climate, in particular winter air temperatures, has an effect on treatment efficiency (MWPS-18, 1985). Warmer air temperatures encourage increased bacterial activity. Therefore, a lagoon built in a warmer climate achieves greater nutrient reduction than a lagoon of equal size built in colder climates.

TRANSPORT DECISION

Transport involves two parts--hauling and spreading. Hauling moves the effluent from a treatment facility to a field site. Spreading applies the effluent within a field. Transport costs are the sum of hauling and spreading costs, or:

$$C(transport) = s(L) \cdot A + t(A) \qquad (2)$$

The per-acre spreading cost, $s(\cdot)$, is a function of nutrient loading rate, L. A per-acre loading rate is calculated by dividing the nutrient demand for the *ith* crop (Q^i_N) by the concentration of treated effluent (γ^T) or:

$$L = \frac{Q^i_N}{\gamma^i}$$

Increasing treatment level, δ, and/or selecting a crop with greater nutrient demands, Q_N, increases loading rates and thereby increases the per-acre spreading cost. The predicted change in total spreading cost ($s(L) \cdot A$) is ambiguous because an increase in per-acre spreading cost is offset by fewer required acres over which manure is spread.

Hauling costs are a function of the total acres required for manure disposal. Required acres increase with manure produced by the livestock enterprise. Assuming that effluent is spread first on fields nearest a treatment facility, hauling costs, $t(\bullet)$, increase with the distance between a treatment facility and a field site. The acres required to satisfy the disposal constraint is determined by dividing total volume of stock (Q_M) by the effluent loading rate (L):

$$A = \frac{Q_M}{L}.$$

Increasing treatment level, δ, or crop nutrient demand, Q^i_N, increases per-acre loading rates, which in turn, decreases acres, A, and hauling costs.

CROP SELECTION

A producer considers two attributes when selecting a crop to receive manure effluent--profitability and capacity to absorb nutrients. In the absence of manure, the crop with the highest per-acre net return would be selected. When manure disposal is required, crops with lower net returns but greater per-acre nutrient demand may become economically feasible because they lower treatment and transport costs. If these cost savings offset lower crop returns, overall farm profit increases.

Returns from utilizing manure on crop production are expressed as:

$$R(crop) = [(R_M^i - R_F^j) + V^i(Q_N)] \cdot A. \tag{3}$$

Per-acre net returns, R_M^i and R_F^j, are functions of their respective crop prices and input costs. The *ith* crop receives manure effluent, and the *jth* crop receives commercial fertilizer. In the absence of manure, the *jth* crop is the most profitable crop that can be grown. Since manure nutrients are not purchased, the monetary benefit of replaced commercial fertilizer for the *ith* crop, $V(Q^i_N)$, is added. Nutrient demand, Q^i_N, is associated with the *ith* crop and its expected yield. Expected yield depends on soil productivity or quality. Differences in soil quality are reflected by yield potentials.

An alternative crop that increases nutrient demand implies:

$$\frac{\partial (R_M^{\ i} - R_F^{\ j})}{\partial Q_N} \leq 0$$

This is offset by the change in value of replaced fertilizer. As the level of Q_N increases, the value of replaced commercial fertilizer, $V(Q^i_N)$, increases, or:

$$\frac{\partial V}{\partial Q_N} \geq 0.$$

The overall change in returns to crop production from manure utilization is ambiguous.

Combining equations (1), (2), and (3), manure value is expressed as:

$$\Pi(manure) = R(crop) - C(transport) - C(treatment) . \qquad (4)$$

A producer's manure disposal problem is a constrained maximization problem where treatment level (δ), land receiving effluent (**A**), and rate of nutrient application (Q_N) are chosen to maximize equation (4) subject to two constraints. First, manure produced by the swine operation must be removed from the house and disposed either through treatment or crop harvest. Second, per-acre nutrient application rates can not exceed predetermined limits that minimize the risk of nutrient runoff and leaching. Optimal levels of the decision variables-- treatment level (δ), land receiving effluent (**A**), and rate of nutrient application (Q_N)--depend on the following exogenous variables: crop prices, fertilizer prices, other input prices, climate, soil productivity, environmental limits, and the total annual quantity of manure produced.

EMPIRICAL ANALYSIS

The objective function (equation 4) was solved by linear programming techniques. Activities in the basic LP model included four treatment options (high, moderate, and low treatment lagoons and slurry pit), irrigation capacity up to 120 acres, and two crops (corn and coastal bermudagrass hay). For a given stock of manure, net returns were computed for the optimal combination of treatment level, crop, and necessary acreage that satisfied the disposal and nutrient application constraints. This value was compared to the net return from planting an equivalent number of acres in the most profitable crop with commercial fertilizer. Manure value was computed as the difference in net crop returns with and without manure.

An inverse relationship between lagoon size and nutrient concentration was estimated from North Carolina data (Safley, 1992). Nitrogen concentration from a fourth treatment option, anaerobic pit slurry, was assumed to be a 25% reduction from excreted amount (MWPS-18, 1985). Amortized costs of lagoons varied with treatment capacity and herd size. Annual lagoon costs ranged from $304 for 600-head capacity and low treatment to $5829 for 4800-head capacity and high treatment (Roka, 1993). Irrigation, slurry hauling, and crop production costs are explained in Roka (1993). Assuming a storage period of 180 days (six months) and an average steady state animal weight of 135 lb, annual irrigated volume is estimated to be 0.035 acre-in./head (Barker, 1991). Land value enters in the programming model as a rental rate equal to the value of lost income from not planting the best alternative crop.

Nutrient demands vary with yield expectations. A yield-nitrogen response function was assumed to be a linear-plateau function. Response slopes for corn and bermudagrass are 1.2 and 45 lb N/unit of yield, respectively. A slope of 1.2 lb N/bu corn corresponds to nitrogen fertilization guidelines in both Iowa and North Carolina.

Two versions of the linear programming model were specified to compare optimal manure management systems under soil and climate conditions that distinguish Iowa from North Carolina. Soil quality in North Carolina is poor relative to soil quality in Iowa. Based on USDA Agricultural Statistics between 1984 and 1991, state average corn yields in North Carolina and Iowa were 80 and 125 bu/acre.

Winters are milder in North Carolina than they are in Iowa. Mean daily air temperatures in North Carolina during December, January, and February average 41.1°F (Ruffner and Bair, 1987). In Iowa, the corresponding mean daily air temperature averages 22.9° F. Lagoon loading rates are based on volatile solids and are correlated with average temperature conditions (ASAE, 1990). The recommended lagoon loading rate in North Carolina is 72 g of volatile solids (VS)/m^3/day. Alternatively, in Iowa, where winter air temperatures average under 23°F, the corresponding recommended loading rates are only 56 g of volatile solids/m^3/day. Lower loading rates require Iowa lagoons to be 28% bigger to achieve equal treatment efficiency.

Results for each set of soil and temperature conditions are presented in Table 1. Manure management systems were simulated over a range of manure quantities. Negative manure values (profits) indicate that, for the soil and climatic conditions considered, manure disposal imposes a cost on Iowa and North Carolina hog producers. In other words, the benefit of replaced commercial fertilizer is offset by the costs of delivering manure nutrients. However, the differences in soil and climate that distinguish Iowa from North Carolina are important in determining how manure is handled. Under Iowa conditions more acres received manure. A low treatment system and corn were chosen. Under North Carolina conditions, fewer acres received manure effluent by selecting a high treatment system and irrigating remaining nutrients to coastal bermudagrass hay. The net effect of manure management systems indicated that manure value was higher under Iowa conditions than under North Carolina conditions, implying that Iowa producers should conserve nutrients more intensely than North Carolina producers.

Table 1. Manure profits and manure disposal systems at various herd sizes under Iowa (IA) and North Carolina (NC) conditions.

Herd	Manure Value[†]		Treatment System[‡]		Crop[#]		Acres	
	IA	NC	IA	NC	IA	NC	IA	NC
600	-1676	-2235	L	H	c	b	9.3	5.3
1800	-4473	-6512	L	H	c	b	28.0	15.8
3000	-6718	-10565	L	H	c	b	46.7	26.3
4200	-8642	-14436	L	H	c	b	65.3	36.8
5400	-10480	-18169	L	H	c	b	84.0	47.3

[†]Manure value defined as the difference in crop net returns without manure and crop net returns utilizing manure.

[‡]Treatment system:
   L = low treatment lagoon system, 80 lb plant-avail. N per acre-in.
   M = moderate treatment system, 62 lb plant-avail. N per acre-in.
   H = high treatment lagoon system, 45 lb plant-avail. N per acre-in. (Safley, 1992)

[#]Cropping activities:
   c = corn
   b = coastal bermudagrass.

A per-head cost of manure disposal is derived by dividing manure values by herd sizes. For 600-head finishing floor capacity, per-head costs were $3.73 in North Carolina and $2.79 in Iowa. These costs decreased with herd size. Economies of scale associated with irrigation were the primary reasons for decreasing per head manure disposal costs.

CONCLUSIONS

This analysis has demonstrated that two swine producers can strive for the same object--maximization of manure value--yet implement significantly different manure management systems. The results suggest that manure value is endogenous to a farmer's given set of circumstances. Climate and soil quality are two important variables that affect how a producer utilizes manure. In Iowa, where soils are productive and treatment options are not as efficient, manure value will be greatest when nutrients are conserved. Conversely, in North Carolina soils are not as productive, and warmer winter climates enable greater treatment efficiency. Consequently, manure value is greatest in North Carolina when nutrients are volatilized and crops are chosen for their high nutrient demand characteristics rather then for their income potential.

The structure of the hog production industry is moving toward fewer and more highly concentrated farms. The trend has been particularly strong in North Carolina. The results in this chapter suggest that economies of scale with respect to irrigation encourage larger farm sizes. As Iowa production units increase, their manure management systems will adjust, perhaps utilizing lagoon technology. Nevertheless, regional production differences will affect the opportunity costs for manure nutrients use and consequently affect the design and management of manure handling systems.

Valuing manure solely on the basis of its nutrient content ignores the important effects regional differences can play on how manure is utilized. One implication of this analysis is that policies intended to protect the environment and encourage sustainable agriculture must take into account regional production conditions. A second implication of this analysis is that research and technology development are more likely to achieve policy goals if economic models that identify feasible solutions are incorporated.

REFERENCES

ASAE, *Design of Anaerobic Lagoons for Animal Waste Management*, ASAE Engineering Practice EP403.1, revised March 1990.

Badger, D.D., Economics of manure management, in *Livestock Waste: A Renewable Resource, Proceedings 4th International Symposium on Livestock Wastes*, pages 15-17, Amarillo, Texas, 1980.

Barker, J.C., *Livestock Manure Characteristics,* Biological and Agricultural Engineering, North Carolina Cooperative Extension Service, 1991.

Chase, Craig, Duffy, M. and Lotz, W., Economic impact of varying swine manure application rates on continuous corn, *Journal of Soil and Water Conservation,* 46, 460, 1991.

Honeyman, Mark, Sustainable swine production in the U.S. Corn Belt, *American Journal of Sustainable Agriculture,* 6, 63, 1990.

Magdoff, Fred, Managing nitrogen for sustainable corn systems: Problems and possibilities, *American Journal of Alternative Agriculture*, 6, 3, 1991.

MWPS-18, *Livestock Waste Facilities Handbook,* Midwest Plan Service, Iowa State University, 1985.

Roka, F.M., *Analysis of Joint Production Relationships between Pork and Swine Manure: Should Manure be Flushed or Bottled,* Ph.D. dissertation, North Carolina State University, December 1993.

Ruffner, J.A., and Bair, F.E., eds, *Weather of U.S. Cities,* Gale Research Company, Detroit, Michigan, 1987.

Safley, L.M., Professor, Biological and Agricultural Engineering, North Carolina State University, unpublished data, 1992.

WASTE MANAGEMENT REGULATIONS AND THEIR IMPACT ON REGIONAL COMPARATIVE ADVANTAGE AND DAIRY INDUSTRY STRUCTURE

Robert B. Schwart, Jr., John Holt, and Joe L. Outlaw[1]

INTRODUCTION

The interaction of agriculture and the environment has become one of the most widely discussed issues over the past decade. Developed economies such as the United States have introduced regulations with the intent to lessen the impacts of harmful substances on the physical environment. Environmental regulations are becoming more inclusive as, in the case of animal agriculture, the definition of who is covered widens. These regulations prohibit or require modifications of some production practices. These regulations may include minimum discharge requirements, particle size, and bacteria counts in addition to nitrogen, phosphorus, and other chemical standards. Further, these regulations may include requirements for construction of waste containment structures or processing structures and suggested or required management practices.

Adding structures or changing processes can change the cost structure of the production unit. As the regulations become more inclusive, concerns increase about regulatory impacts on the profitability of animal agriculture, on regional comparative advantage, and on industry structure. This chapter focuses primarily on the potential ramifications of environmental regulations on the dairy industry as a case in point; however, the conclusions are applicable to the swine, beef, and poultry industries.

THE REGULATORY ENVIRONMENT

In 1972 the US Congress enacted the Clean Water Act (CWA), and amended it in 1987, aiming at control of both point and non-point sources of water pollution. The CWA designated confined animal feeding operations (CAFOs)

---

[1] In the 1990s, the authors have developed detailed financial information, including waste management investments and operating costs, from focus-group interviews with dairymen operating both "family" and "large" dairies in California, Washington, New Mexico, Texas, Wisconsin, Missouri, New York, Vermont, Georgia, and Florida. Those interviews support some judgement calls in this chapter, which are made despite lack of documentation available in the literature.

as possible point sources of pollution. The US Environmental Protection Agency approved the National Pollution Discharge Elimination System (NPDES), which detailed the types and sizes of CAFOs required to have permits.

If a CAFO is under 1,000 animal units (700 dairy cows), and will not discharge waste into the waters of the United States except during a 24-hour, 25-year storm event, then no permit is necessary. Location matters: a 24-hour, 25-year storm event is about 3 in. in New Mexico but 9.5 in. in southern Louisiana. However, if a complaint is lodged against a dairy, an investigation is triggered, and a permit may be necessary.

Since state regulations must at least equal the federal regulations, most states followed the federal pattern of regulating larger operations and ignoring the smaller ones. In states where most dairies are smaller than the CAFOs, less regulatory attention has been focused on dairies (Atwood and Hallam, 1990). And in states where animal agriculture is a major industry, state regulators have tended to be lenient in enforcing waste management regulations (Outlaw et al., 1993).

The recent trend, however, is for more regulation of animal agriculture in general, especially in environmentally sensitive areas. This is particularly true for dairies in heavily populated areas where both ground and surface water quality issues may be political issues. The Coastal Zone Act Reauthorization Amendment of 1990 (CZARA) encompasses 35 states, 29 of which have Coastal Zone Management (CZM) programs approved by the National Oceanic and Atmospheric Agency (NOAA). State and/or local regulators are focusing more attention on non-point source pollution, so smaller dairies in more states are coming under regulatory authority.

The regulatory picture for the different states is a patch-work quilt (Outlaw et al., 1993):

- Regulatory activity in some states is complaint driven. There is no regulatory activity until a complaint has been filed.
- Some states grant permits to operate animal agriculture activities, with no recognition of discharge limitations.
- The EPA permits are permits to discharge water into waterways.
- Some states have stricter guidelines than the EPA permit standards for CAFOs.
- There are differences between states as to what agency enforces the regulations.

RECENT EXPERIENCES WITH DAIRY WASTE REGULATIONS

The most dramatic regulatory impact is for violations. Texas, for example, allows fines of $10,000/day/event. In an attention-getting case, fines of $60,000 were levied on a Texas dairy.

The permitting process itself requires both time and money. Costly engineering studies may be necessary to ensure regulatory compliance and obtain permits. In Florida, a dairy being built in 1994 took 8 months and $13,000 to get approval at the local level before applying at the state level. The relatively simple engineering plan (Holloway et al., 1994) for a 1,400-cow dairy using rotational grazing cost $60,000, up from $20,000 for a more complex 2,800-cow confinement dairy four years earlier. Similarly, in Texas, one dairyman went through more than a dozen hearings, and said "I spent two years and about $100,000 trying to get a legal permit" (Stalcup, 1992).

Some states, including Texas and Florida, have water quality regulations that exceed the Federal guidelines for regulated CAFOs. Those regulations matter. When the Department of Environmental Regulation (DER) ruled that dairies in the Lake Okeechobee drainage basin must have an approved system for collecting the waste water and runoff from milking parlors and high-intensive use areas, 19 out of 49 dairies in the Okeechobee, Florida, area closed or relocated, rather than comply. This exodus occurred despite the state paying $355/cow toward the required investment cost (Boggess et al., 1991). Direct milk sales from the dairies that closed were more than $30 million annually, resulting in total annual impacts on the local economy of somewhere around $50 million (Clouser et al., 1994).

THE SIZE FACTOR

Dairy size drew special mention in the legal description of CAFOs, and size continues to be a lightning rod attracting regulatory attention. Larger barns and facilities are more visible. Larger herds generate more waste and may be more odorous and, hence, more odious to some of the public, particularly those who think that large farms are more like factories than way-of-life farms. Other things equal, the more non-farm people there are in the area, and the larger the dairy, the more likely it is that a complaint will be lodged.

Even so, economies of size are a fact in dairying, so large dairies are being built in states in which expansion is occurring. Economies of size hold for investments in waste management facilities, and the newer dairies typically incorporate state-of-the-art waste treatment methods.

Small dairies can be, and some are, as profitable on a per-hundredweight basis as large dairies (Knutson et al., 1994), but 50 cows generate less investable income than 500 cows. The cost-price squeeze continues, and small dairies face serious problems in preserving adequate family incomes. Expansion is an option, but retrofitting old barns is expensive and disrupts operating incomes; land may not be available for more forage production and waste disposal; and the less income cushion, the greater the uncertainty about any expansion.

Uncertainty from any source adds to the investment hurdle rate as indicated in Pagano, 1993. Recent milk price variability has added to economic uncertainty, and environmental concerns add regulatory uncertainty. Especially for small dairies, the growing likelihood of new waste management regulations add both difficulty and immanence to the decades-old, but always agonizing, decision of getting bigger or getting out (Atwood and Hallam, 1990).

REGIONAL COMPARATIVE ADVANTAGE

One problem with generalizing about regional comparative advantage is that there is at least as much difference between producers within a region as across regions. However, on average, producers in certain regions of the United States (e.g., California, Wisconsin, Florida, and New Mexico) have tended to adopt, within those regions, similar technology, management practices, and growth strategies.

These observed patterns are composed of a complex collage of regional milk prices, forage production patterns, land price and availability for expansions, real-estate development patterns, and personal preference. Collectively, these characteristics have tended to impact regional milk production.

CHANGES IN REGIONAL MILK PRODUCTION

Two obvious structural trends in dairying are increases in dairy farm size, and shifts to warmer climates (Forste and Frick, 1979). Forty years of change in the Wisconsin and California dairy industries document the drifting United States dairy industry.

In 1954, Wisconsin had 128,800 producers managing 2.2 million cows, averaging 17 cows/dairy. By 1993, herd size had increased to about 57 cows with 30,000 producers managing 1.7 million cows in the state. California had 34,000 producers in 1954 milking about 791,000 cows for an average herd size of 23 cows. By 1993, there were 2,300 California producers managing 1.15 million cows, and average herd size had increased to 501 cows.

Wisconsin dairy farms decreased about 77%, and state cow numbers dropped 23%. Cows per farm, however, increased over the 1954 to 1993 period by 2.3 times. In California, there was a 93% reduction in dairy farm numbers and a 127% increase in total cows, and herd size increased a whopping 20.8 times. At the same time, milk prices in California are significantly lower than in the rest of the United States.

The California example shows that there are significant economies of size in dairying in California. Recent trends in New Mexico and Texas indicate that there may be comparative advantages in dairying in other warm climates as well as in some colder climates such as New York (Knutson et al., 1994).

IMPACT OF WASTE MANAGEMENT REGULATIONS

Effective waste management is a modern cost of doing business in the US, and so will it be in dairying. But state and regional differences create the need for flexibility in regulatory regimes. In Wisconsin, where "Dairy nonpoint pollution is 'a grave problem'... The obstacles are attitudes and finances..." (Merrill, 1994). There are few generally applicable technical solutions; constructed wetlands may work well in Oregon but cause salt concentrations in drier areas such as California (Merrill, 1994).

Any regulation that increases expenses more than incomes lowers dairy profitability and lowers the ability of the dairy to invest in new waste management methods. This can, of course, affect comparative advantage. In California, recent expansions have been made with more attention to regulations; therefore, future regulatory impacts should be minimal (Outlaw et al., 1993). States such as Wisconsin, stocked to capacity with cows in small herds managed in old barns with old waste-handling mindsets and machinery, might be impacted heavily by increased regulatory attention if, and/or when, it comes.

New regulations may require significant changes in dairying techniques and large investments. In Florida, investments necessary for impounding and managing waste water from high-intensity use areas cost dairies, on average, $923/cow, even without buying or leasing land for effluent disposal (Boggess et al., 1991). Construction took 11 months, and revenue losses during construction averaged about $352 per cow. Even 100% cost share for the Florida DER approved facilities would not have covered the cost of rule compliance. Larger dairies (1,409 cows) were better able to control the revenue losses during construction, and invested more effectively in facilities than did smaller dairies (768 cows). That study concluded: "In sum, it appears that despite almost identical investment costs per cow for the DER components, the dairy rule will ultimately cost small dairies roughly 50% more per cow than it does large dairies" (Boggess et al., 1991).

The upshot is that dairymen who are already in compliance with "modern" waste management regulations will not likely see a change in their relative comparative advantage. Those producers forced to make substantial investments will feel economic pressure to milk more cows in order to meet their income goals and cover the added costs of complying with waste management regulations. Retrofitting old facilities to meet economic and environmental objectives will, in many cases, be prohibitively expensive.

POTENTIAL IMPACTS OF REGULATION TYPE
ON DAIRY STRUCTURE

New and more stringent regulations applied to all dairy farms would likely change dairying cost structure most in regions of the country that are not already in compliance. Cost of production will be affected by the type of regulation introduced.

To be in compliance, technology-based regulations require that specific structures be installed (a specified number of wires, wire type and gauge, and corner-bracing patterns, say, for fencing cows out of waterways in Florida). Or technology-based regulations can specify "best management practices" in order to be in compliance (primary and secondary lagoons, land-spreading of effluent, and removal of forages from that land). In contrast, standards-based regulations would require only that cows be kept out of waterways and that effluent contain specified minimum amounts of nutrients, leaving the means of doing so to the managers while holding them accountable for complying.

Standards-based regulations would allow decision-makers to develop more cost-effective technologies and practices to meet science-based standards. However, in practice, waste management problems arise before there are science-based standards for dealing with them. So there is a tendency to adopt technology-based "prescriptions" in hopes of alleviating the perceived waste management problems.

Technology-based regulations are a stop-gap political palliative, and they can be easily counted by regulators who lack scientific skills or testing procedures. They can also be ineffective and costly for producers (Baumol and Oates, 1988). For example, requirements for lining lagoons with a clay layer that must be 2 ft thick were written into the Texas regulations without knowing the environmental and economic efficacy of the practice. Sinking limited investment capital in the mandated technology or practices thus lowers firms' reserves for investing in environmentally and economically sound technology and/or management practices as they become known.

However, not all environmentally driven investments are economic disasters. Investments in roofed, concrete-paved, fan-cooled, feeding barns in the Okeechobee area appear to be paying for themselves (Boggess et al., 1991). However, it is doubtful that the perceived *ex ante* profitability of these investments would have been enough to attract investment, absent the dairy rule. As indicated above, enforcing new regulations did force some producers out of production.

Ideally, scientists are developing safe environmental standards that can be the basis for regulations specifying tolerance standards for managing waste products that would achieve environmental goals. Firms, and their consultants, would be developing improved means of meeting those standards in the most economically efficient way. Thus the regulatory environment would do minimum damage to firm-level profitability without distorting regional comparative advantage or industry structure.

Some "old" technology offers both environmental and economic promise in some areas. Rotational grazing systems, where they are practical, cheapen forage costs, spread manure more uniformly with minimal investment, improve run-off problems, and appear to reduce odor and flies. This technology appears to have more potential in warmer climates.

FUTURE STRUCTURE OF THE DAIRY INDUSTRY

The answer to the question of the impacts of waste management regulations on the structure of the United States dairy industry is--it depends. It depends on the uniformity of regulations and their enforcement across states--which has been sporadic at best (Outlaw et al., 1993). Texas and Florida made costly mistakes with technology-based standards but are adjusting toward science-based standards and implementing more flexibility in enforcements. If other states develop and implement standards that make both economic and environmental sense for their conditions, the impact on industry structure could be minimal.

Various states must respond to their environmental problems in their own political way. Regardless of how they do that, added regulations add to the pressure for dairies to get larger (Atwood and Hallam, 1990; Schertz, 1979). And, we believe, the shift to warmer climates will continue.

SUMMARY AND CONCLUSIONS

Economic pressures are forcing increases in dairy farm size, measured in cow numbers. Environmental pressures usually increase the acreage needed to manage waste. The net effect is that added regulations add to investments necessary for managing waste.

The best hope for meeting environmental and economic objectives in the dairy sector is for scientists, dairy producers, and regulators to concentrate on crafting systems based on good science, good management, and good regulations that are appropriate for the area.

Cooperation is key; it appears to be happening in the dairy industry. Scientists are striving to understand complex biological systems; regulators are taking more reasonable stances; and dairymen are seeking, and investing in, new systems for sustaining or enhancing environmental quality without sacrificing their economic health.

REFERENCES

Atwood, J.D. and Hallam, A., Farm structure and stewardship of the environment, in *Determinants of Farm Size and Structure*, Arne Hallam, ed., Proceedings of the program sponsored by the NC-181 Committee on Determinants of Farm Size and Structure in North Central Areas of the United States, Albuquerque, New Mexico, January 6,8-9, 1990.

Boggess, William G., Holt, John, and Smithwick, Robert P., *The Economic Impacts of the Dairy Rule on Dairies in the Lake Okeechobee Drainage Basin*, Food and Resource Economics Department SP #91-39, I.F.A.S., University of Florida, November, 1991.

Baumol, W.J., and Oats., W.E., *The Theory of Environmental Policy*, Cambridge University Press, New York, New York, 1988.

Clouser, Rodney L., Mulkey, David, Boggess, Bill, and Holt, John, The economic impact of regulatory decisions in the dairy industry: A case study in Okeechobee County, Florida, *J. Dairy Sci.*, 77, 325, 1994.

Forste, R.H., and Frick, G.E., Livestock production: Dairy, in *Another Revolution in U.S. Farming?* Lyle P. Schertz, ed., Economics, Statistics, and Cooperatives Service, US Department of Agriculture, Washington D.C., Agricultural Economic Report No. 441, December 1979.

Holloway, M.P., Bottcher, A.B., and St. John, Ron, Design of a rotationally grazed dairy in North Florida, in *Proceedings of the Second Conference: Environmentally Sound Agriculture*, University of Florida, American Society of Agricultural Engineers, St. Joseph, Michigan, April 1994.

Knutson, R.D., Outlaw, J.L., Miller, J.W., Richardson, J.W., and Schwart, R.B., Jr., *Status and Prospects for Dairying 1994-1998*, Agricultural and Food Policy Center, Texas Agricultural Experiment Station, Texas A&M University, College Station, Texas, AFPC Policy Working Paper 94-2, February 1994.

Merrill, Lorraine Stewart, What's hot in manure management, *Hoard's Dairyman*, October 25, 1994.

Outlaw, J.L., Schwart, R.B., Jr., Knutson, R.D., Pagano, A.P., Miller, J.W., and Gray, A.W., *Impacts of Dairy Waste Management Regulations*, Agricultural and Food Policy Center, Texas Agricultural Experiment Station, Texas A&M University, College Station, Texas, AFPC Policy Working Paper 93-4, May 1993.

Pagano, A.P., *Ex Ante Forecasting of Uncertain and Irreversible Dairy Investments: Implications for Environmental Compliance*, Ph.D. Dissertation, University of Florida, Department of Food and Resource Economics, 1993.

Schertz, L.P., The major forces, in *Another Revolution in U.S. Farming?* Lyle P. Schertz, ed.. Economics, Statistics, and Cooperatives Service, U.S. Department of Agriculture, Washington D.C., Agricultural Economic Report No. 441, December 1979.

Stalcup, Larry, Big stink in Texas, *Dairy Today*, August, 1992.

CONTRIBUTORS

Abdalla, Charles, Pennsylvania State University, 112 Armsby Building, University Park, PA 16802-5600, (814)865-2562, FAX (814)865-3746.

Ackerman, Eric O., State of Illinois-Environmental Protection Agency, Peoria Regional Office, Division of Water Pollution Control, 5415 North University, Peoria, IL 61614, (309)693-5463, FAX (309)693-5467.

Ahlstedt, Steven, Water Management, Aquatic Biology Laboratory, Tennessee Valley Authority, Norris, TN 37828, (615)632-1424, FAX (615)632-1693.

Alston, George, Texas Agricultural Extension Service, Rt 2 Box 1, Stephenville, TX 76401, (817)968-4144, FAX (817)968-3759.

Andres, Scott, University of Delaware, Delaware Geological Survey, Delaware Geological Survey Building Room 205, Newark, DE 19716-7501, (302)831-2847, FAX (302)831-3579.

Aumen, Nicholas, South Florida Water Management District, Department of Research, P. O. Box 24680, 3301 Gun Club Road, West Palm Beach, FL 33416-4680, (407)686-8800 ext. 6601, FAX (407)687-6442.

Auvermann, Brent, Colorado State University, 416 Pearl Street, Fort Collins, CO 80521-1742, (303)491-1192, FAX (303)491-7369.

Badger, Philip, Tennessee Valley Authority, P. O. Box 1010, CEB 3A, Muscle Shoals, AL 35660-1010, (205)386-3086, FAX (205)386-2963.

Blevins, Robert L., University of Kentucky, Department of Agronomy, N122 Agri Science North Building, Lexington, KY 40546-0091, (606)257-8750, FAX (606)257-2185.

Boggess, W.G., University of Florida, Department of Food and Resource Economics, 1105 McCarty Hall, P. O. Box 110240, Gainesville, FL 32611, (904)392-5081, FAX (904)392-8634.

Bouzaher, Aziz, Iowa State University/CARD, 573 Heady Hall, Ames IA 50011, (515)294-1183, FAX (515)294-6336.

Breeuwsma, A., DLO Winand Staring Centre for Integrated Land, Soil and Water Research, Marijkeweg 11/22, P. O. Box 125, 6700 AC Wageningen, The Netherlands, (31)8370-74200, FAX (31)8370-24812.

Brodahl, Mary, USDA-Agricultural Research Service, Great Plains System Research Unit, Federal Building, 301 South Howes, Fort Collins, CO 80525, (303)490-8338, FAX (303)490-8310.

Brodie, Herbert, University of Maryland, Department of Biological and Agricultural Engineering, College Park, MD 20742, (410)778-7676, FAX (410)778-9075.

Bubenzer, Gary, University of Wisconsin-Madison, Department of Agricultural Engineering, 406 Henry Mall, Madison, WI 54707, (608)262-0096, FAX (608)262-1228.

Bundy, Larry, University of Wisconsin, Department of Soil Science, 1525 Observatory Drive, 263 Soils Building, College of Agriculture and Life Sciences, Madison, WI 53706, (608)263-2889, FAX (608)265-2595.

Burks, S.L., Stover Biometric Labatories, P. O. Box 2056, Stillwater, OK 74076, (405)743-1435, FAX (405)624-0019.

Cabrerra, Miguel, University of Georgia, Department of Crop and Soil Science, 3111 Plant Science Building, Athens, GA 30602-7272, (706)542-1242, FAX (706)542-0914.

Carr, Lewis, Department of Agricultural Engineering, LESREC-PAF, 11990 Strickland Drive, Princess Anne, MD 21853, (410)651-9111, FAX (410)541-9187.

Chakraborty, P.K., Uniformed Services University of the Health Sciences, Building A, Room 3077, 4301 Jones Bridge Road, Bethesda, MD 20815, (301)295-3126, FAX (301)295-6774.

Chasteen, Eric, Texas Natural Resource Conservation Commission, P. O. Box 13087, Austin, TX 78711-3087, (512)239-1413, FAX (512)239-1300.

Chaubey, Indragent, Oklahoma State University, Department of Agriculture and Engineering, 109 Agri Hall, Stillwater, OK 74078-0497, (405)744-5431, FAX (405)744-6059.

Chen, Shulin, Louisiana State University, Department of Civil and Environmental Engineering, Baton Rouge, LA 70803-6405, (504)388-8549, FAX (504)388-8662.

Clifford, William, North Carolina State University, Department of Sociology and Anthropology, Box 8107, 1911 Building Room 301, Raleigh, NC 27695-8107, (919)515-1676, FAX (919)515-2610.

Cochran, Mark, University of Arkansas, Department of Agricultural Economics and Rural Sociology, 221 Agriculture Building, Fayetteville, AR 72701, (501)575-2256, FAX (501)575-5308.

Cocke, Mark A., USDA-Natural Resource Conservation Service, 2121C, 2nd Street, Davis, CA 95616, (916)757-8283, FAX (916)757-8382.

Collins, E.R., Virginia Polytechnic Institute and State University, Department of Biological Systems Engineering, 310 Seitz Hall, Blacksburg, VA 24061-0303, (703)231-6813, FAX (703)231-3199.

Combs, Sherry, University of Wisconsin, Soil and Plant Analysis Lab, Department of Soil Science, 5711 Mineral Point Road, Madison, WI 53705-4453, (608)262-4364, FAX (608)263-3327.

Correll, David, Smithsonian Environmental Research Center, P. O. Box 28, Edgewater, MD 21037-2238, (410)798-4424, FAX (301)261-7546.

Cothren, Gianna, Louisiana State University, Department of Civil and Environmental Engineering, Baton Rouge, LA 70803-6405, (504)388-8549, FAX (504)388-8662.

Coyne, Mark, University of Kentucky, Department of Agronomy, N-122 Agricultural Science Building-North, Lexington, KY 40546-0091, (606)257-4202, FAX (606)258-1952.

Daliparthy, Jayaram, University of Massachusetts, Department of Plant and Soil Sciences, 210 Bowditch Hall, Box 30910, Amherst, MA 01003-0910, (413)545-1843, FAX (413)545-0260.

Daniel, T.C., University of Arkansas, Department of Agronomy, PTSC 115, Fayetteville, AR 72701, (501)575-5720, FAX (501)575-7465.

Davis, Jessica G., University of Georgia, Coastal Plains Experiment Station, Tifton, GA 31793, (912)386-7093, FAX (912)386-3219.

DeRamus, H. Alan, University of Southwestern Louisiana, Department of Renewable Resources, Lafayette, LA 70504, (318)482-6640, FAX (318)482-5395.

Dillaha, Theo, Virginia Polytechnic Institute and State University, Department of Biological Systems Engineering, 310 Seitz Hall, Blacksburg, VA 24061-0303, (703)231-6813, FAX (703)231-3199.

Doll, Billi Jo, USDA-Natural Resources Conservation Service, P. O. Box 189, Chinook, MT 59523, (406)357-2310, FAX (406)357-2087.

Ebert, William, Natural Resources Conservation Service, Stevens Point-Whiting-Plover Wellhead Protection, Gilfrey Building, 817 Whiting Avenue, Stevens Point, WI 54481, (715)345-5977, FAX (715)345-5966.

Edwards, D.R., University of Kentucky, Department of Agriculture and Engineering, 128 Agri Engr Building, Lexington, KY 40546-0276, (606)257-3000, FAX (606)257-5671.

Finney, Vernon L., USDA-Natural Resource Conservation Service, 2121C, 2nd Street, Davis, CA 95616, (916)757-8283, FAX (916)757-8382.

Franzmeier, D.P., Purdue University, Department of Agronomy, 3-419 Lilly Hall, West Lafayette, IN 47907, (317)494-9767, FAX (317)496-1368.

Fulcher, Chris, University of Missouri, Department of Agricultural Economics, 232 Mumford Hall, Columbia, MO 65211, (314)882-0147, FAX (314)882-3958.

Gale, Paula, University of Tennessee, Department of Agriculture and Natural Resources, 131 Brehm Hall, Martin, TN 38238, (901)587-7326, FAX (901)587-7968.

Gassman, Phil, Iowa State University/CARD, 573 Heady Hall, Ames, IA 50011, (515)294-1183,FAX (515)294-6336.

Gburek, W.J., Pasture Systems and Watershed Management Research, USDA-Agricultural Research Service, US Regional Pasture Research Buidling, Curtin Road, University Park, PA 16802-3702, (814)863-8759, FAX (814)863-0935.

Gilliam, J.W., North Carolina State University, Department of Soil Science, 2234 Williams Hall, Raleigh, NC 27695, (919)515-2040, FAX (919)515-2167.

Goldstein, Alan, South Florida Water Management District, 305 East North Park Street, Suite A, Okeechobee, FL 34973-2033, (813)763-2128, FAX (813)763-3872.

Gordillo, Rosa Marie, University of Georgia, Department of Crop and Soil Science, 3111 Plant Science Building, Athens, GA 30602-7272, (706)542-1242, FAX (706)542-0914.

Gould, Charles, University of Georgia, Cooperative Extension Service, 302 West Marion Street, Eatonton, GA 31024, (706)485-8733, FAX (706)485-3820.

Grover, Ray, Harmony Products, 2121 Old Greenbrier Road, Chesapeake, VA 23320, (804)523-2849, FAX (804)523-9567.

Haan, C.T., Oklahoma State University, Department of Biosystems and Agricultural Engineering, 110 Agricultural Hall, Stillwater, OK 74078, (405)744-8397, FAX (405)744-6059.

Haby, Vince A., Texas Agricultural Research and Extension Center, P. O. Box E, Overton, TX 75684, (903)834-6191, FAX (903)845-0456.

Halbach, Thomas, University of Minnesota, Water Quality and Waste Management, 216 Soils Building, St. Paul, MN 55108, (612)625-3135, FAX (612)625-2208.

Hattey, Jeff A., Oklahoma State University, Department of Agronomy, 159 Agricultural Hall, Stillwater, OK 74078-0507, (405)744-6414, FAX (405)744-5269.

Hauck, Roland, (Tennessee Valley Authority-retired), 139 Talisman Drive, Florence, AL 35630, (205)764-7513, FAX (205)386-2191.

Hauck, Larry, Tarleton State University, Texas Institute for Applied Environmental Research, TIAER-Mail Stop T0410, Stephenville, TX 76402, (817)968-9581 or (817)968-9567, FAX (817)968-9568.

Havens, Karl, South Florida Water Management District, Department of Research, P. O. Box 24680, 3301 Gun Club Road, West Palm Beach, FL 33416-4680, (407)686-8800 ext. 6601, FAX (407)687-6442.

Herbert, Stephen, University of Massachusetts, Department of Plant and Soil Sciences, 210 Bowditch Hall, Box 30910, Amherst, MA 01003-0910, (413)545-1843, FAX (413)545-0260.

Hession, W. Cully, Oklahoma State University, Biosystems and Agricultural Engineering, 110 Agricultural Hall, Stillwater, OK 74078, (405)744-8397, FAX (405)744-6059.

Hines, Gregory, Water Quality Demonstration Project-East River, 1221 Bellevue Street, Suite 113, Green Bay, WI 54303, (414)391-3923, FAX (414)465-3767.

Hoag, Dana, Colorado State University, Department of Agricultural and Resource Economics, Clark C306, Fort Collins, CO 80523, (303)491-5549, FAX (303)491-2067.

Hoban, Thomas, North Carolina State University, Department of Sociology and Anthropology, Box 8107, 1911 Building Room 301, Raleigh, NC 27695-8107, (919)515-1676, FAX (919)515-2610.

Holmes, Brian, University of Wisconsin-Madison, Department of Agricultural Engineering, 406 Henry Mall, Madison, WI 54707, (608)262-0096, FAX (608)262-1228.

Holt, John, University of Florida, Department of Food and Resource Economics, 1105 McCarty Hall, Gainesville, FL 32611, (904)392-7844, FAX (904)392-3646.

Hubbard, R.K., SE Watershed, USDA-Agricultural Research Service, P. O. Box 946, Tifton, GA 31794, (912)386-3893, FAX (912)386-7215.

Humenik, Frank J., North Carolina State University, Department of Biological and Agricultural Engineering, Raleigh, NC 27611, (919)515-6767, FAX (919)515-6772.

Huner, Jay, University of Southwestern Louisiana, Department of Renewable Resources, Lafayette, LA 70504, (318)482-6640, FAX (318)482-5395.

Hunt, Pat G., USDA-Agricultural Research Service, P. O. Box 3039, Florence, SC 29502, (803)669-5203, FAX (803)669-6970.

Joern, Brad C., Purdue University, Department of Agronomy, 3-419 Lilly Hall, West Lafayette, IN 47907, (317)494-9767, FAX (317)496-1368.

Johnson, Andy F., Texas A&M University, Department of Soil and Crop Sciences, College Station, TX 77843-2474, (409)845-8795, FAX (409)845-0456.

Johnson, William F., Jr., University of Arkansas, Department of Agronomy, PTSC 115, Fayetteville, AR 72701, (501)575-5739, FAX (501)575-7465.

Johnson, J.C., University of Georgia, Coastal Plains Experiment Station, P. O. Box 748, Tifton, GA 31793-0748, (912)386-7093, FAX (912)386-3219.

Jones, Ordie, USDA-Agricultural Research Service, P. O. Drawer 10, Bushland, TX 79012, (806)356-5745, FAX (806)356-5750.

Jordan, J.D., Virginia Polytechnic Institute and State University, Department of Biological Systems Engineering, 310 Seitz Hall, Blacksburg, VA 24061-0303, (703)231-7600, FAX (703)231-3199.

Jordan, Thomas, Smithsonian Environmental Research Center, P. O. Box 28, Edgewater, MD 21037-2238, (410)798-4424, FAX (301)261-7546.

Kaap, James, Natural Resources Conservation Service, 6515 Watts Road, Suite 200, Madison, WI 53719-2726, (608)264-5585, FAX (608)264-5483.

Kadlec, R.H., University of Michigan, Department of Chemical Engineering, 2300 Hayward, Ann Arbor, MI 48109-2136, (313)764-3362, FAX (313)475-3516.

Kerans, Billi, Indiana University of Pennsylvania, Department of Biology, 114 Weyandt Hall, Indiana, PA 15705-1090, (412)357-2582, FAX (412)357-5700.

Klausner, Stu, Cornell University, Department of Soil, Crop and Atmospheric Science, 149 Emerson Hall, Ithaca, NY 14853, (607)255-1757, FAX (607)255-6143.

Klug, Jay, USDA-Natural Resource Conservation Service, 2121C, 2nd Street, Davis, CA 95616, (916)757-8283, FAX (916)757-8382.

Kolpak, V.Z., Ministry of Environmental Protection of Ukraine, Ukrainian Scientific Center for Proteciton of Waters, 6 Bakulin Str, Kharkov, Ukraine 310888, (0572)45-21-63, FAX (0572)45-50-47.

Kostinec, Rob, University of Maine, Department of Civil and Environmental Engineering, 5711 Boardman Hall, Orona, MA 04469-5711, (207)581-2170, FAX (207)581-2202.

Kraft, George, University of Wisconsin -Stevens Point, Extension Service, Central Wisconsin Groundwater Center, 109 Nelson Hall, Stevens Point, WI 54481-3897, (715)346-4270, FAX (715)346-2965.

Lakshminarayanan, R., Oklahoma State University, Department of Biosystems and Agricultural Engineering, 110 Agricultural Hall, Stillwater, OK 74078, (405)744-8397, FAX (405)744-6059.

Langlinais, Stephen, University of Southwestern Louisiana, Department of Renewable Resources, Lafayette, LA 70504, (318)482-6640, FAX (318)482-5395.

Lindsey, Janice K., Tennessee Valley Authority, P. O. Box 1010, CEB 3A, Muscle Shoals, AL 35660-1010, (205)386-3487, FAX (205)386-2963.

Lowrance, R.R., SE Watershed, USDA-Agricultural Research Service, P. O. Box 946, Tifton, GA 31793, (912)386-3893, FAX (912)386-7215.

Malone, R.F., Louisiana State University, Department of Civil and Environmental Engineering, BatonRouge, LA 70803-6405, (504)388-8549, FAX (504)388-8662.

Massie, Leonard, University of Wisconsin-Madison, Department of Agricultural Engineering, 406 Henry Mall, Madison, WI 54707, (608)262-0096, FAX (608)262-1228.

McCaskey, T.A., Auburn University, Department of Animal and Dairy Sciences, 209 Animal Science Building, Auburn, AL 36849-5415, (334)844-1549, FAX (334)844-1519.

McDonough, Thomas, Water Management, Aquatic Biology Laboratory, Tennessee Valley Authority, Norris, TN 37828, (615)632-1424, FAX (615)632-1693.

McFarland, Anne, Tarleton State University, Texas Institute for Applied Environmental Research, TIAER-Mail Stop T0410, Stephenville, TX 76402, (817)968-9581 or (817)968-9567, FAX (817)968-9568.

Mikkelsen, Rob, North Carolina State University, Department of Soil Science, 2234 Williams Hall, Box 7619, Raleigh, NC 27695-7619, (919)515-2040, FAX (919)515-2167.

Minshall, G. Wayne, Idaho State University, Department of Biological Sciences, Campus Box 8007, Pocatello, ID 83209-8007, (208)236-3765, FAX (208)236-4570.

Moffitt, David, USDA-Natural Resources Conservation Service, Regional Technology Staff, South National Technical Center, P. O. Box 6567, 501 Felix, FWFC Building 23, Forth Worth, TX 76115, (817)334-5242 ext. 3304, FAX (817)334-5584.

Moore, P.A., University of Arkansas, Department of Agronomy, PTSC 115, Fayetteville, AR 72701, (501)575-2848, FAX (501)575-2846.

Moorhead, Stan T., USDA-Natural Resource Conservation Service, 2121C, 2nd Street, Davis, CA 95616, (916)757-8283, FAX (916)757-8382.

Narrod, Clare, University of Pennsylvania, Center for Energy and the Environment, G-29 Meyerson Hall, 210 South 34th Street, Philadelphia, PA 19104, (215)471-1038, FAX (215)573-2034.

Newton, G.Larry, University of Georgia, Coastal Plains Experiment Station, Animal Dairy Science, Moore Highway, Tifton, GA 31793, (912)386-3214, FAX (912)386-3219.

Nichols, D.J., University of Arkansas, Department of Agronomy, PTSC 115, Fayetteville, AR 72701, (501)575-5720, FAX (501)575-7465.

Outlaw, Joe, Texas A&M University, Department of Agricultural Economics, 308D Anthropolgy Building, College Station, TX 77843-2124, (409)845-3062, FAX (409)845-3140.

Pionke, H.B., Pasture Systems and Watershed Management Research, USDA-Agricultural Research Service, US Regional Pasture Research Building, Curtin Road, University Park, PA 16802-3702,(814)863-8759, FAX (814)863-0935.

Powell, Royden III, Office of Resource Conservation, State of Maryland Department of Agriculture, 50 Harry S. Truman Parkway, Annapolis, MD 21401, (410)841-5865, FAX (410)841-5914.

Prato, Tony, University of Missouri, Department of Agricultural Economics, 129 Mumford, Columbia, MO 65211, (314)882-0147, FAX (314)882-3958.

Pray, Carl, Rutgers University, Cook College, Department of Agricultural Economics, P. O. Box 231, Building 6259, Cook Office Building, New Brunswick, NJ 08903, (908)932-9159 ext. 20, FAX (908)932-8887.

Provin, Tony L., Purdue Univeristy, Department of Agronomy, 3-419 Lilly Hall, West Lafayette, IN 47907, (317)494-9767, FAX (317)496-1368.

Purvis, Amy, Texas A&M University, Department of Agricultural Economics, College Station, TX 77843-2124, (409)845-3805, FAX (409)845-4261.

Reddy, K.R., University of Florida, Department of Soil Science, 106 Newell Hall, Soil Science Hall, Box 110510, Gainesville, FL 32611-0510, (904)392-1804, FAX (904)392-3399.

Reijerink, J.G.A., Grontmij, P. O. Box 203, 3730 AE, AFD.Milieu, Utrechtseweg 2, 3732HB, De Bitt, The Netherlands, (31)3404-65619, NO FAX NUMBER AVAIABLE.

Rice, J. Mark, North Carolina State University, Department of Biological and Agricultural Engineering, Raleigh, NC 27611, (919)515-6767, FAX (919)515-6772.

Richard, Tom, Cornell University, Department of Agricultural and Biological Engineering, 207 Riley Robb Hall, Ithaca, NY 14853, (607)255-2488, FAX (607)255-4080.

Ritter, Gary, South Florida Water Management District, 305 East North Park Street, Suite A, Okeechobee, FL 34973-2033, (813)763-2128, FAX (813)763-3872.

Robinson, Chris, Idaho State University, Department of Biological Sciences, Campus Box 8007, Pocatello, ID 83209-8007, (208)236-3765, FAX (308)236-4570.

Robinson, S. J., University of Florida, Department of Soil and Water Science, 106 Newell Hall, Gainesville, FL 32611, (904)392-1804, FAX (904)392-3399.

Rock, Chet, University of Maine, Department of Civil and Environmental Engineering, 5711 Boardman Hall, Orona, MA 04469-5711, (207)581-2170, FAX (207)581-2202.

Roka, Fritz, North Carolina State University, Department of Agriculture and Resource Economics, 210 Patterson Hall, Box 8109, Raleigh, NC 27695-8109, (919)515-4540, FAX (919)515-6268.

Ross, Blake, Virginia Polytechnic Institute and State University, Department of Biological Systems Engineering, Blacksburg, VA 24061-0303, (703)231-4385, FAX (703)231-3199.

Rouquette, F. Monte, Jr., Texas Agricultural Research and Extension Center, P. O. Box E, Overton, TX 75684, (903)834-6191, FAX (903)834-7140.

Sagona, Frank, Water Management, Tennessee Valley Authority, 1101 Market Street, Room CST17D, Chattanooga, TN 37422, (615)751-7334, FAX (615)751-7648.

Sanderson, Matt, Texas Agricultural Extension Station, P. O. Box 292, Stephenville, TX 76401, (817)968-4144, FAX (817)968-3759.

Saylor, Charle, Water Management, Aquatic Biology Laboratory, Tennessee Valley Authority, Norris, TN 37828, (615)632-1424, FAX (615)632-1693.

Schoumans, O.F., DLO Winand Staring Centre for Integrated Land, Soil and Water Research, Marijkeweg 11/22, P. O. Box 125, 6700 AC Wageningen, The Netherlands, (31)8370-74200, FAX (31)8370-24812.

Schwart, Robert B., Jr., Texas A&M University, Department of Agricultural Economics, Room 5 Agricultural Building, College Station, TX 77843-2124, (409)845-5284, FAX (409)845-4261.

Shank, Randall, Virginia Division of Soil and Water Conservation, 203 Governor Street, Suite 206, Richmond, VA 23219-2094, (804)371-8884, FAX (804)371-7490.

Sharpley, Andrew, USDA-Agriclutural Research Service, National Agricultural Water Quality Laboratory, P. O. Box 1430, Durant, OK 74702-1430, (405)5066, FAX (405)294-5307.

Shore, Laurence, Kimron Veterinary Institute, Bet Dagan, Israel, P. O. Box 12, FAX 972-3-9681-753.

Shuyler, Lynn, Chesapeake Bay Program Office, US Environmental Protection Agency, 410 Severn Avenue, Suite 109, Annapolis, MD 21403, (410)267-5700, FAX (410)267-5777.

Sims, J.T., University of Delaware, Department of Plant and Soil Sciences, 149 Townsend Hall, Newark, NJ 19717-1303, (302)831-2531, FAX (302)831-3651.

Skryl'nik, E.V., Ministry of Environmental Protection of Ukraine, Ukrainian Scientific Center for Protection of Waters, 6 Bakulin Str, Kharkov, Ukraine 310888, (0572)45-21-63, FAX (0572)45-50-47.

Smith, Bo, CDR Environmental/Thone Brothers, P. O. Box 2023, Russellville, AR 72811, (501)968-5837, FAX (501)968-2157.

Smith, Samuel J., USDA- Agricultural Research Service, National Water Quality Laboratory, P. O. Box 1430, Durant, OK 74702-1430, (405)924-5066, FAX (405)294-5307.

Smolen, M.D., Oklahoma State University, Department of Biosystems and Agricultural Engineering, 110 Agricultural Hall, Stillwater, OK 74078, (405)744-8397, FAX (405)744-6059.

Steinman, Alan, South Florida Water Management District, Department of Research, P. O. Box 24680, 3301 Gun Club Road, West Palm Beach, FL 33416-4680, (407)686-8800 ext. 6601, FAX (407)687-6442.

Stevenson, Gordon, Wisconsin Department of Natural Resources, Water Resources Management, 2nd Floor, 101 South Webster, Madison, WI 53707, (608)267-9306, FAX (608)267-2800.

Stewart, Bobby A., West Texas A&M University, Dryland Agriculture Institute, WTAMU Box 278, Canyon, TX 79016, (806)656-2299, FAX (806)656-2938.

Stone, Ken C., USDA-Agricultural Research Service, P. O. Box 3039, Florence, SC 29502, (803)669-5203, FAX (803)669-6970.

Storm, Daniel E., Oklahoma State University, Department of Biosystems and Agricultural Engineering, 110 Agricultural Hall, Stillwater, OK 74078, (405)744-8397, FAX (405)744-6059.

Strickland, Richard, Tennessee Valley Authority, CEB 5C-M, Muscle Shoals, AL 35661, (205)386-2542, FAX (205)386-2963.

Sutton, A.L., Purdue University, Department of Agronomy, 3-419 Lilly Hall, West Lafayette, IN 47907, (317)494-9767, FAX (317)496-1368.

Sweeten, John, Department of Agricultural Engineering, Texas Agriculture Extension Service, 303 Agricultrual Building, College Station, TX 77843-8828, (409)845-7451, FAX (409)847-8828.

Swisher, Jerry, Virginia Polytechnic Institute and State University, Virginia Cooperative Extension Service, P. O. Box 590, Verona, VA 24482-0590, (703)245-5750, FAX (703)245-5752.

Taylor, A.G., Illinois Environmental Protection Agency, 2200 Churchill Road, P. O. Box 19276, Springfield, IL 62794-9276, (217)785-0830, FAX (217)782-1431.

Veitch, John D., Tennessee Valley Authority, P. O. Box 1010, CEB 3A, Muscle Shoals, AL 35660-1010, (205)386-2866, FAX (205)386-2963.

Vellidis, George, University of Georgia, Coastal Plains Experiment Station, Biological and Agricultural Engineering, P. O. Box 748, Tifton, GA 31793, (912)386-3893, FAX (912)386-3958.

Veneman, Peter, University of Massachusetts, Department of Plant and Soil Sciences, 210 Bowditch Hall, Box 30910, Amherst, MA 01003-0910, (413)545-1843, FAX (413)545-0260.

Vietor, D.M., Texas A&M University, Department of Soil and Crop Sciences, 430 Heep Center, College Station, TX 77843-2474, (409)845-8795, FAX (409)845-0456.

Weller, Donald, Smithsonian Environmental Research Center, P. O. Box 28, Edgewater, MD 21037-2238, (410)798-4424, FAX (301)261-7546.

Wetzel, Robert, University of Alabama, Department of Biological Sciences, Aquatic Biology Program, 319 Biology, Box 870334, Tuscalossa, AL 35487-0344, (205)348-5960, FAX (205)348-1786.

Whigham, Dennis, Smithsonian Environmental Research Center, P. O. Box 28, Edgewater, MD 21037-0028, (410)269-1412, FAX (301)261-7954.

Willis, William, USDA-Agricultural Research Service, P. O. Drawer 10, Bushland, TX 79012, (806)356-5745, FAX (806)356-5750.

Wolf, Duane C., University of Arkansas, Department of Agronomy, PTSC 115, Fayetteville, AR 72701, (501)575-5739, FAX (501)575-7465.

Wolfe, Mary Leigh, Virginia Polytechnic Institute and State University, Department of Agricultural Engineering, Seitz Hall, Agri Engr Building, Blacksburg, VA 34061-0303, (703)231-6092,FAX (703)231-3199.

Wood, Wes, Auburn University, Department of Agronomy and Soils, 202 Funchess Hall, Auburn University, AL 36849-5412, (205)844-4100, FAX (205)844-3945.

Wooden, Kenneth G., Virginia Cooperative Extension Office, 128 East Court Street, Rocky Mount, VA 24151, (703)483-5161, FAX (703)483-0807.

Wright, Judith, Cornell Cooperative Extension-Cayuga County, 248 Grant Avenue, Auburn, NY 13021, (315)255-1183, FAX (315)255-1187.

Xu, Feng, Iowa State University, Center for Agriculture and Rural Development, 581 Heady Hall, Ames, IA 50011-1070, (515)294-6258, FAX (515)294-6336.

Young, Robert, USDA-Agricultural Research Service, North Central Soil Conservation Research Laboratory, North Iowa Avenue, Morris, MN 56267, (612)589-3411, FAX (612)589-3787.

Younos, Tamim, Virginia Polytechnic Institute and State University, Department of Biological Systems Engineering, Blacksburg, VA 24061-0303, (703)231-4385, FAX (703)231-3199.

Zering, Kelly, North Carolina State University, Department of Agriculture and Resource Economics, 211-D Hillsborough Building, Box 7509, Raleigh, NC 27695-8109, (919)515-6089, FAX (919)515-1824.

Zublena, Joseph, North Carolina State University, Associate State Program Leader for Natural Resources and Rural Development, P. O. Box 7602, 214 Ricks Hall, Raleigh, NC 27695, (919)515-3252, FAX (919)515-5950.

INDEX